Titles in This Series

102 **Mark A. Pinsky,** Introduction to Fourier analysis and wavelets, 2009
101 **Ward Cheney and Will Light,** A course in approximation theory, 2009
100 **I. Martin Isaacs,** Algebra: A graduate course, 2009
99 **Gerald Teschl,** Mathematical methods in quantum mechanics: With applications to Schrödinger operators, 2009
98 **Alexander I. Bobenko and Yuri B. Suris,** Discrete differential geometry: Integrable structure, 2008
97 **David C. Ullrich,** Complex made simple, 2008
96 **N. V. Krylov,** Lectures on elliptic and parabolic equations in Sobolev spaces, 2008
95 **Leon A. Takhtajan,** Quantum mechanics for mathematicians, 2008
94 **James E. Humphreys,** Representations of semisimple Lie algebras in the BGG category $\mathcal{O}$, 2008
93 **Peter W. Michor,** Topics in differential geometry, 2008
92 **I. Martin Isaacs,** Finite group theory, 2008
91 **Louis Halle Rowen,** Graduate algebra: Noncommutative view, 2008
90 **Larry J. Gerstein,** Basic quadratic forms, 2008
89 **Anthony Bonato,** A course on the web graph, 2008
88 **Nathanial P. Brown and Narutaka Ozawa,** C^*-algebras and finite-dimensional approximations, 2008
87 **Srikanth B. Iyengar, Graham J. Leuschke, Anton Leykin, Claudia Miller, Ezra Miller, Anurag K. Singh, and Uli Walther,** Twenty-four hours of local cohomology, 2007
86 **Yulij Ilyashenko and Sergei Yakovenko,** Lectures on analytic differential equations, 2007
85 **John M. Alongi and Gail S. Nelson,** Recurrence and topology, 2007
84 **Charalambos D. Aliprantis and Rabee Tourky,** Cones and duality, 2007
83 **Wolfgang Ebeling,** Functions of several complex variables and their singularities (translated by Philip G. Spain), 2007
82 **Serge Alinhac and Patrick Gérard,** Pseudo-differential operators and the Nash–Moser theorem (translated by Stephen S. Wilson), 2007
81 **V. V. Prasolov,** Elements of homology theory, 2007
80 **Davar Khoshnevisan,** Probability, 2007
79 **William Stein,** Modular forms, a computational approach (with an appendix by Paul E. Gunnells), 2007
78 **Harry Dym,** Linear algebra in action, 2007
77 **Bennett Chow, Peng Lu, and Lei Ni,** Hamilton's Ricci flow, 2006
76 **Michael E. Taylor,** Measure theory and integration, 2006
75 **Peter D. Miller,** Applied asymptotic analysis, 2006
74 **V. V. Prasolov,** Elements of combinatorial and differential topology, 2006
73 **Louis Halle Rowen,** Graduate algebra: Commutative view, 2006
72 **R. J. Williams,** Introduction the the mathematics of finance, 2006
71 **S. P. Novikov and I. A. Taimanov,** Modern geometric structures and fields, 2006
70 **Seán Dineen,** Probability theory in finance, 2005
69 **Sebastián Montiel and Antonio Ros,** Curves and surfaces, 2005

For a complete list of titles in this series, visit the
AMS Bookstore at **www.ams.org/bookstore/**.

Introduction to Fourier Analysis and Wavelets

Introduction to Fourier Analysis and Wavelets

Mark A. Pinsky

Graduate Studies
in Mathematics
Volume 102

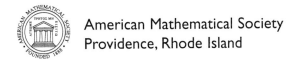

American Mathematical Society
Providence, Rhode Island

EDITORIAL COMMITTEE
David Cox (Chair)
Steven G. Krantz
Rafe Mazzeo
Martin Scharlemann

2000 *Mathematics Subject Classification*. Primary 42–02; Secondary 42C40.

For additional information and updates on this book, visit
www.ams.org/bookpages/gsm-102

Library of Congress Cataloging-in-Publication Data
Pinsky, Mark A., 1940–
 Introduction to fourier analysis and wavelets / Mark A. Pinsky.
  p. cm. — (Graduate studies in mathematics ; v. 102)
 Originally published: Pacific Grove, CA : Brooks/Cole, c2002.
 Includes bibliographical references and index.
 ISBN 978-0-8218-4797-8 (alk. paper)
 1. Fourier analysis. 2. Wavelets (Mathematics) I. Title.

QA403.5.P56 2009
515′.2433—dc22   2008047419

Copying and reprinting. Individual readers of this publication, and nonprofit libraries acting for them, are permitted to make fair use of the material, such as to copy a chapter for use in teaching or research. Permission is granted to quote brief passages from this publication in reviews, provided the customary acknowledgment of the source is given.

Republication, systematic copying, or multiple reproduction of any material in this publication is permitted only under license from the American Mathematical Society. Requests for such permission should be addressed to the Acquisitions Department, American Mathematical Society, 201 Charles Street, Providence, Rhode Island 02904-2294, USA. Requests can also be made by e-mail to reprint-permission@ams.org.

© 2002 held by the American Mathematical Society, All rights reserved.
The American Mathematical Society retains all rights
except those granted to the United States Government.
Printed in the United States of America.

∞ The paper used in this book is acid-free and falls within the guidelines
established to ensure permanence and durability.
Visit the AMS home page at http://www.ams.org/

10 9 8 7 6 5 4 3 2 1   14 13 12 11 10 09

To my parents,
Harry A. Pinsky and Helen M. Pinsky,
who led me to the path of learning

CONTENTS

1	**FOURIER SERIES ON THE CIRCLE**			1
	1.1	Motivation and Heuristics		1
		1.1.1	Motivation from Physics	1
			1.1.1.1 The Vibrating String	1
			1.1.1.2 Heat Flow in Solids	2
		1.1.2	Absolutely Convergent Trigonometric Series	3
		1.1.3	*Examples of Factorial and Bessel Functions	6
		1.1.4	Poisson Kernel Example	7
		1.1.5	*Proof of Laplace's Method	9
		1.1.6	*Nonabsolutely Convergent Trigonometric Series	11
	1.2	Formulation of Fourier Series		13
		1.2.1	Fourier Coefficients and Their Basic Properties	13
		1.2.2	Fourier Series of Finite Measures	19
		1.2.3	*Rates of Decay of Fourier Coefficients	20
			1.2.3.1 Piecewise Smooth Functions	21
			1.2.3.2 Fourier Characterization of Analytic Functions	22
		1.2.4	Sine Integral	24
			1.2.4.1 Other Proofs That $Si(\infty) = 1$	24
		1.2.5	Pointwise Convergence Criteria	25
		1.2.6	*Integration of Fourier Series	29
			1.2.6.1 Convergence of Fourier Series of Measures	30
		1.2.7	Riemann Localization Principle	31
		1.2.8	Gibbs-Wilbraham Phenomenon	31
			1.2.8.1 The General Case	34
	1.3	Fourier Series in L^2		35
		1.3.1	Mean Square Approximation—Parseval's Theorem	35
		1.3.2	*Application to the Isoperimetric Inequality	38

	1.3.3	*Rates of Convergence in L^2	39
		1.3.3.1 Application to Absolutely-Convergent Fourier Series	43
1.4	Norm Convergence and Summability		45
	1.4.1	Approximate Identities	45
		1.4.1.1 Almost-Everywhere Convergence of the Abel Means	49
	1.4.2	Summability Matrices	51
	1.4.3	Fejér Means of a Fourier Series	54
		1.4.3.1 Wiener's Closure Theorem on the Circle	57
	1.4.4	*Equidistribution Modulo One	57
	1.4.5	*Hardy's Tauberian Theorem	59
1.5	Improved Trigonometric Approximation		61
	1.5.1	Rates of Convergence in $C(\mathbb{T})$	61
	1.5.2	Approximation with Fejér Means	62
	1.5.3	*Jackson's Theorem	65
	1.5.4	*Higher-Order Approximation	66
	1.5.5	*Converse Theorems of Bernstein	70
1.6	Divergence of Fourier Series		73
	1.6.1	The Example of du Bois-Reymond	74
	1.6.2	Analysis via Lebesgue Constants	75
	1.6.3	Divergence in the Space L^1	78
1.7	*Appendix: Complements on Laplace's Method		80
		1.7.0.1 First Variation on the Theme-Gaussian Approximation	80
		1.7.0.2 Second Variation on the Theme-Improved Error Estimate	80
	1.7.1	*Application to Bessel Functions	81
	1.7.2	*The Local Limit Theorem of DeMoivre-Laplace	82
1.8	Appendix: Proof of the Uniform Boundedness Theorem		84
1.9	*Appendix: Higher-Order Bessel functions		85
1.10	Appendix: Cantor's Uniqueness Theorem		86

2 FOURIER TRANSFORMS ON THE LINE AND SPACE — 89

2.1	Motivation and Heuristics		89
2.2	Basic Properties of the Fourier Transform		91
	2.2.1	Riemann-Lebesgue Lemma	94
	2.2.2	Approximate Identities and Gaussian Summability	97
		2.2.2.1 Improved Approximate Identities for Pointwise Convergence	100
		2.2.2.2 Application to the Fourier Transform	102
		2.2.2.3 The n-Dimensional Poisson Kernel	106

	2.2.3	Fourier Transforms of Tempered Distributions	108
	2.2.4	*Characterization of the Gaussian Density	109
	2.2.5	*Wiener's Density Theorem	110
2.3	Fourier Inversion in One Dimension		112
	2.3.1	Dirichlet Kernel and Symmetric Partial Sums	112
	2.3.2	Example of the Indicator Function	114
	2.3.3	Gibbs–Wilbraham Phenomenon	115
	2.3.4	Dini Convergence Theorem	115
		2.3.4.1 Extension to Fourier's Single Integral	117
	2.3.5	Smoothing Operations in $\mathbb{R}^1$–Averaging and Summability	117
	2.3.6	Averaging and Weak Convergence	118
	2.3.7	Cesàro Summability	119
		2.3.7.1 Approximation Properties of the Fejér Kernel	121
	2.3.8	Bernstein's Inequality	122
	2.3.9	*One-Sided Fourier Integral Representation	124
		2.3.9.1 Fourier Cosine Transform	124
		2.3.9.2 Fourier Sine Transform	125
		2.3.9.3 Generalized h-Transform	125
2.4	L^2 Theory in $\mathbb{R}^n$		128
	2.4.1	Plancherel's Theorem	128
	2.4.2	*Bernstein's Theorem for Fourier Transforms	129
	2.4.3	The Uncertainty Principle	131
		2.4.3.1 Uncertainty Principle on the Circle	133
	2.4.4	Spectral Analysis of the Fourier Transform	134
		2.4.4.1 Hermite Polynomials	134
		2.4.4.2 Eigenfunction of the Fourier Transform	136
		2.4.4.3 Orthogonality Properties	137
		2.4.4.4 Completeness	138
2.5	Spherical Fourier Inversion in $\mathbb{R}^n$		139
	2.5.1	Bochner's Approach	139
	2.5.2	Piecewise Smooth Viewpoint	145
	2.5.3	Relations with the Wave Equation	146
		2.5.3.1 The Method of Brandolini and Colzani	149
	2.5.4	Bochner-Riesz Summability	152
		2.5.4.1 A General Theorem on Almost-Everywhere Summability	153
2.6	Bessel Functions		154
	2.6.1	Fourier Transforms of Radial Functions	157
	2.6.2	L^2-Restriction Theorems for the Fourier Transform	158
		2.6.2.1 An Improved Result	159
		2.6.2.2 Limitations on the Range of p	161
2.7	The Method of Stationary Phase		162
	2.7.1	Statement of the Result	163
	2.7.2	Application to Bessel Functions	164
	2.7.3	Proof of the Method of Stationary Phase	165
	2.7.4	Abel's Lemma	167

3 FOURIER ANALYSIS IN L^p SPACES — 169

- 3.1 Motivation and Heuristics — 169
- 3.2 The M. Riesz-Thorin Interpolation Theorem — 169
  - 3.2.0.1 Generalized Young's Inequality — 174
  - 3.2.0.2 The Hausdorff-Young Inequality — 174
  - 3.2.1 Stein's Complex Interpolation Theorem — 175
- 3.3 The Conjugate Function or Discrete Hilbert Transform — 176
  - 3.3.1 L^p Theory of the Conjugate Function — 177
  - 3.3.2 L^1 Theory of the Conjugate Function — 179
    - 3.3.2.1 Identification as a Singular Integral — 183
- 3.4 The Hilbert Transform on $\mathbb{R}$ — 184
  - 3.4.1 L^2 Theory of the Hilbert Transform — 185
  - 3.4.2 L^p Theory of the Hilbert Transform, $1 < p < \infty$ — 186
    - 3.4.2.1 Applications to Convergence of Fourier Integrals — 187
  - 3.4.3 L^1 Theory of the Hilbert Transform and Extensions — 188
    - 3.4.3.1 Kolmogorov's Inequality for the Hilbert Transform — 192
  - 3.4.4 Application to Singular Integrals with Odd Kernels — 194
- 3.5 Hardy-Littlewood Maximal Function — 197
  - 3.5.1 Application to the Lebesgue Differentiation Theorem — 200
  - 3.5.2 Application to Radial Convolution Operators — 202
  - 3.5.3 Maximal Inequalities for Spherical Averages — 203
- 3.6 The Marcinkiewicz Interpolation Theorem — 206
- 3.7 Calderón-Zygmund Decomposition — 209
- 3.8 A Class of Singular Integrals — 210
- 3.9 Properties of Harmonic Functions — 212
  - 3.9.1 General Properties — 212
  - 3.9.2 Representation Theorems in the Disk — 214
  - 3.9.3 Representation Theorems in the Upper Half-Plane — 216
  - 3.9.4 Herglotz/Bochner Theorems and Positive Definite Functions — 219

4 POISSON SUMMATION FORMULA AND MULTIPLE FOURIER SERIES — 222

- 4.1 Motivation and Heuristics — 222
- 4.2 The Poisson Summation Formula in $\mathbb{R}^1$ — 223
  - 4.2.1 Periodization of a Function — 223
  - 4.2.2 Statement and Proof — 225
  - 4.2.3 Shannon Sampling — 228
- 4.3 Multiple Fourier Series — 230
  - 4.3.1 Basic L^1 Theory — 231
    - 4.3.1.1 Pointwise Convergence for Smooth Functions — 233
    - 4.3.1.2 Representation of Spherical Partial Sums — 233

	4.3.2 Basic L^2 Theory	235
	4.3.3 Restriction Theorems for Fourier Coefficients	236
4.4	Poisson Summation Formula in $\mathbb{R}^d$	238
	4.4.1 *Simultaneous Nonlocalization	239
4.5	Application to Lattice Points	241
	4.5.1 Kendall's Mean Square Error	241
	4.5.2 Landau's Asymptotic Formula	243
	4.5.3 Application to Multiple Fourier Series	244
	4.5.3.1 Three-Dimensional Case	245
	4.5.3.2 Higher-Dimensional Case	247
4.6	Schrödinger Equation and Gauss Sums	247
	4.6.1 Distributions on the Circle	248
	4.6.2 The Schrödinger Equation on the Circle	250
4.7	Recurrence of Random Walk	252

5 APPLICATIONS TO PROBABILITY THEORY — 256

5.1	Motivation and Heuristics	256
5.2	Basic Definitions	256
	5.2.1 The Central Limit Theorem	260
	5.2.1.1 Restatement in Terms of Independent Random Variables	261
5.3	Extension to Gap Series	262
	5.3.1 Extension to Abel Sums	266
5.4	Weak Convergence of Measures	268
	5.4.1 An Improved Continuity Theorem	269
	5.4.1.1 Another Proof of Bochner's Theorem	270
5.5	Convolution Semigroups	272
5.6	The Berry-Esséen Theorem	276
	5.6.1 Extension to Different Distributions	279
5.7	The Law of the Iterated Logarithm	280

6 INTRODUCTION TO WAVELETS — 284

6.1	Motivation and Heuristics	284
	6.1.1 Heuristic Treatment of the Wavelet Transform	285
6.2	Wavelet Transform	286
	6.2.0.1 Wavelet Characterization of Smoothness	290
6.3	Haar Wavelet Expansion	291
	6.3.1 Haar Functions and Haar Series	291
	6.3.2 Haar Sums and Dyadic Projections	292
	6.3.3 Completeness of the Haar Functions	295
	6.3.3.1 Haar Series in C_0 and L_p Spaces	296
	6.3.3.2 Pointwise Convergence of Haar Series	298

	6.3.4	*Construction of Standard Brownian Motion	299
	6.3.5	*Haar Function Representation of Brownian Motion	301
	6.3.6	*Proof of Continuity	301
	6.3.7	*Lévy's Modulus of Continuity	302
6.4	Multiresolution Analysis	303
	6.4.1	Orthonormal Systems and Riesz Systems	304
	6.4.2	Scaling Equations and Structure Constants	310
	6.4.3	From Scaling Function to MRA	313
		6.4.3.1 Additional Remarks	315
	6.4.4	Meyer Wavelets	318
	6.4.5	From Scaling Function to Orthonormal Wavelet	319
		6.4.5.1 Direct Proof that $V_1 \ominus V_0$ Is Spanned by $\{\Psi(t-k)\}_{k\in\mathbb{Z}}$	324
		6.4.5.2 Null Integrability of Wavelets Without Scaling Functions	325
6.5	Wavelets with Compact Support	326
	6.5.1	From Scaling Filter to Scaling Function	327
	6.5.2	Explicit Construction of Compact Wavelets	330
		6.5.2.1 Daubechies Recipe	331
		6.5.2.2 Hernandez-Weiss Recipe	333
	6.5.3	Smoothness of Wavelets	334
		6.5.3.1 A Negative Result	336
	6.5.4	Cohen's Extension of Theorem 6.5.1	338
6.6	Convergence Properties of Wavelet Expansions	341
	6.6.1	Wavelet Series in L^p Spaces	341
		6.6.1.1 Large Scale Analysis	345
		6.6.1.2 Almost-Everywhere Convergence	346
		6.6.1.3 Convergence at a Preassigned Point	347
	6.6.2	Jackson and Bernstein Approximation Theorems	347
6.7	Wavelets in Several Variables	352
	6.7.1	Two Important Examples	352
		6.7.1.1 Tensor Product of Wavelets	354
	6.7.2	General Formulation of MRA and Wavelets in $\mathbb{R}^d$	354
		6.7.2.1 Notations for Subgroups and Cosets	355
		6.7.2.2 Riesz Systems and Orthonormal Systems in $\mathbb{R}^d$	356
		6.7.2.3 Scaling Equation and Structure Constants	357
		6.7.2.4 Existence of the Wavelet Set	358
		6.7.2.5 Proof That the Wavelet Set Spans $V_1 \ominus V_0$	361
		6.7.2.6 Cohen's Theorem in $\mathbb{R}^d$	362
	6.7.3	Examples of Wavelets in $\mathbb{R}^d$	362

References 365

Notations 369

Index 373

LIST OF FIGURES

Figure 1.1.1 Poisson kernel $P_r(\theta)$ with $r = 0.8$ [Page 8]
Figure 1.1.2 Conjugate Poisson kernel $Q_r(\theta)$ with $r = 0.8$ [Page 9]
Figure 1.2.1 The Dirichlet kernel $D_N(u)$ with $N = 5$ [Page 16]
Figure 1.2.2 Graphs of the partial sums $S_N f(x)$ for $N = 1, 2, 3$ of the Fourier series of $f(x) = x$, $-\pi < x < \pi$ [Page 16]
Figure 1.2.3 The Gibbs-Wilbraham phenomenon for the function $f(x) = \text{sgn}(x)$ [Page 35]
Figure 1.4.1 Relations between Abel and Cesàro summability. Stronger methods are to the right of weaker methods [Page 53]
Figure 1.4.2 The Fejér kernel with $N = 8$ [Page 56]
Figure 1.5.1 The de la Vallée Poussin kernel with $N = 8$ [Page 69]
Figure 2.5.1 The spherical partial sum of the indicator function of the unit ball in $\mathbb{R}^3$, with $M = 99/2\pi$ [Page 144]
Figure 2.5.2 Illustrating Huygens' principle [Page 149]
Figure 6.1.1 Gaussian wavelet [Page 289]
Figure 6.1.2 Mexican hat wavelet [Page 289]

PREFACE

This book provides a self-contained treatment of classical Fourier analysis at the upper undergraduate or begining graduate level. I assume that the reader is familiar with the rudiments of Lebesgue measure and integral on the real line. My viewpoint is mostly classical and concrete, preferring explicit calculations to existential arguments. In some cases, several different proofs are offered for a given proposition to compare different methods.

The book contains more than 175 exercises that are an integral part of the text. It can be expected that a careful reader will be able to complete all of these exercises. Starred sections contain material that may be considered supplementary to the main themes of Fourier analysis. In this connection, it is fitting to comment on the role of Fourier analysis, which plays the dual role of queen and servant of mathematics. Fourier-analytic ideas have an inner harmony and beauty quite apart from any applications to number theory, approximation theory, partial differential equations, or probability theory. In writing this book it has been difficult to resist the temptation to develop some of these applications as a testimonial of the power and flexibility of the subject. The following list of "extra topics" are included in the starred sections: Stirling's formula, Laplace asymptotic method, the isoperimetric inequality, equidistribution modulo one, Jackson/Bernstein theorems, Wiener's density theorem, one-sided heat equation with Robin boundary condition, the uncertainty principle, Landau's asymptotic lattice point formula, Gaussian sums and the Schrödinger equation, the central limit theorem, the Berry-Esséen theorem and the law of the iterated logarithm. While none of these topics is "mainstream Fourier anaysis," each of them has a definite relation to some part of the subject.

A word about the organization of the first two chapters, which are essentially independent of one another. Readers with some sophistication but little previous knowledge of Fourier series can begin with Chapter 2 and anticipate a self-contained treatment of the n-dimensional Fourier transform and many of its applications. By contrast, readers who wish an introductory treatment of Fourier series should begin with Chapter 1, which provides a reasonably complete introduction to Fourier analysis on the circle. In both cases I emphasize the Riesz-Fischer and Plancherel theorems, which demonstrate the natural harmony of Fourier analysis with the Hilbert spaces $L^2(\mathbb{T})$ and $L^2(\mathbb{R}^n)$. However much of modern harmonic analysis is carried out in the L^p spaces for $p \neq 2$, which is the subject of Chapter 3. Here we find the interpolation theorems of Riesz-Thorin and Marcinkiewicz, which are applied to discuss the boundedness of the Hilbert transform and its application to the L^p convergence of Fourier series and integrals. In Chapter 4 I merge the subjects of Fourier series and Fourier transforms by means of the Poisson summation formula in one and several dimensions. This also has applications to number theory and multiple Fourier series, as noted above.

Chapter 5 explores the application of Fourier methods to probability theory. Limit theorems for sums of independent random variables are equivalent to the study of iterated convolutions of a probability measure on the line, leading to the central limit theorem for convergence and the Berry-Esséen theorems for error estimates. These are then applied to prove the law of the iterated logarithm.

The final Chapter 6 deals with wavelets, which form a class of orthogonal expansions that can be studied by means of Fourier analysis—specifically the Plancherel theorem from Chapter 2. In contrast to Fourier series and integral expansions, which require one parameter (the frequency), wavelet expansions involve two indices—the scale and the location parameter. This allows additional freedom and leads to improved convergence properties of wavelet expansions in contrast with Fourier expansions. I include a brief application to Brownian motion, where the wavelet approach furnishes an easy access to the precise modulus of continuity of the standard Brownian motion.

Many of the topics in this book have been "class-tested" to a group of graduate students and faculty members at Northwestern University during the academic years 1998–2000. I am grateful to this audience for the opportunity to develop and improve my original efforts.

I owe a debt of gratitude to Paul Sally, Jr., who encouraged this project from the beginning. Gary Ostedt gave me full editorial support at the initial stages followed by Bob Pirtle and his efficient staff. Further thanks are due to Robert Fefferman, whose lectures provided much of the inspiration for the basic parts of the book. Further assistance and feedback was provided by Marshall Ash, William Beckner, Miron Bekker, Leonardo Colzani, Galia Dafni, George Gasper, Umberto Neri, Cora Sadosky, Aurel Stan, and Michael Taylor. Needless to say, the writing of Chapter 1 was strongly influenced by the classical treatise of Zygmund and the elegant text of Katznelson. The latter chapters were influenced in many ways by the books of Stein and Stein/Weiss. The final chapter on wavelets owes much to the texts of Hernandez/Weiss and Wojtaszczyk.

Mark A. Pinsky

INTRODUCTION TO FOURIER ANALYSIS AND WAVELETS

CHAPTER 1

FOURIER SERIES ON THE CIRCLE

1.1 MOTIVATION AND HEURISTICS

1.1.1 Motivation from Physics

Two major sources of Fourier series are the mathematical models for (i) the vibrating string and (ii) heat flow in solids.

1.1.1.1 The vibrating string

The first systematic use of trigonometric series can be found in the work of Daniel Bernoulli (1753) on the vibrating string. A simple harmonic motion of a string of length π is defined by the formula

$$(1.1.1) \qquad f(x, t) = A \sin nx \cos(nt - \alpha)$$

for suitable constants A, α, and $n = 1, 2, \ldots$. A is the amplitude, n is the angular frequency, and α is the phase shift.

The simple harmonic motion is a solution of the differential equation $f_{tt} = f_{xx}$, which is supposed to describe the small transverse displacement $f(x, t)$ of a tightly stretched string whose ends are fixed at $x = 0$ and $x = \pi$.

More complex, multiple harmonic motions are obtained by linear superposition

$$(1.1.2) \qquad f(x, t) = \sum_{n=1}^{N} A_n \sin nx \cos(nt - \alpha_n).$$

Functions of this form can be used to satisfy a variety of initial conditions, if we are given the values of f and the partial derivative $\partial f / \partial t$ when $t = 0$. This is possible whenever $f(x, 0)$ and $\partial f/\partial t(x, 0)$ are expressed as finite linear combinations $\sum_{n=1}^{N} a_n \sin nx$. This may be less obvious in other cases; for example

$f(x,0) = \sin^3 x = (3\sin x - \sin 3x)/4$, whereas $\sin^2 x$ cannot be so expressed. In order to work with these trigonometric sums, we note the property of orthogonality, expressed as

(1.1.3)
$$\int_0^\pi \sin mx \sin nx \, dx = 0, \qquad m \neq n.$$

If a function has the form $f(x) = \sum_{k=1}^N a_k \sin kx$, then we must have $\int_0^\pi f(x) \sin nx \, dx = 0$ for $n > N$.

Exercise 1.1.1. Show that if N is odd, $\sin^N x$ can be written as a finite sum of the form $\sum_{k=1}^N a_k \sin kx$.

Exercise 1.1.2. Suppose that we have a convergent series expansion of $\sin^2 x = \sum_{k=1}^\infty a_k \sin kx$ on the interval $0 \leq x \leq \pi$. Prove that a_k is nonzero for infinitely many values of k.

Hint: Assume a finite expansion and use the orthogonality relation (1.1.3) to obtain a contradiction.

Exercise 1.1.3. Generalize Exercise 1.1.2 to any even power of $\sin x$, showing the impossibility of an expansion $\sin^n x = \sum_{k=1}^N a_k \sin kx$ for $0 \leq x \leq \pi$ where $n = 4, 6, \ldots$.

Any multiple harmonic motion (1.1.2) is a 2π-periodic function of time: $f(x, t + 2\pi) = f(x, t)$ for all $-\infty < x, \infty$, $-\infty < t < \infty$. It also is a 2π-periodic function of x and is odd with respect to $x = 0$ and $x = \pi$, meaning that $f(-x) = -f(x)$ and $f(\pi + x) = -f(\pi - x)$ for all x.

Exercise 1.1.4. Suppose that $f(x)$, $-\infty < x < \infty$ is given. Show that any two of the following properties imply the third: (i) $f(x + 2\pi) = f(x)$, $\forall x$; (ii) $f(-x) = -f(x)$, $\forall x$; (iii) $f(\pi - x) = -f(\pi + x)$, $\forall x$.

1.1.1.2 Heat flow in solids

The vibrating string suggests the use of sine series, since the ends of the string are fixed. More general trigonometric series are suggested by the study of heat flow in a circular ring, assumed to have circumference 2π. In this model it is natural to assume that the temperature $u(x, t)$ is a 2π-periodic function of x (but not periodic in time). Fourier (1822) formulated the heat equation $u_t = u_{xx}$ to describe the time evolution of the temperature. It is satisfied by any function of the form $(A_n \cos nx + B_n \sin nx)e^{-n^2 t}$ where $n = 0, 1, 2, \ldots$, $t \geq 0$ and $-\pi < x \leq \pi$. Taking linear combinations of these, we arrive at a "general solution"

(1.1.4)
$$u(x, t) = \sum_{n=0}^N (A_n \cos nx + B_n \sin nx) e^{-n^2 t}.$$

This will fit an initial temperature profile $f(x)$ if and only if f is expressed as a finite trigonometric sum

(1.1.5) $$f(x) = \sum_{n=0}^{N} (A_n \cos nx + B_n \sin nx).$$

The coefficients A_n, B_n can be found by using the orthogonality relations

(1.1.6) $$\boxed{\int_{-\pi}^{\pi} \sin mx \sin nx \, dx = 0 \quad m \neq n,}$$

(1.1.7) $$\boxed{\int_{-\pi}^{\pi} \cos mx \cos nx \, dx = 0 \quad m \neq n,}$$

(1.1.8) $$\boxed{\int_{-\pi}^{\pi} \sin mx \cos nx \, dx = 0 \quad \text{all } m, n,}$$

together with the norms: $\int_{-\pi}^{\pi} \sin^2 nx \, dx = \pi = \int_{-\pi}^{\pi} \cos^2 nx \, dx$, $n = 1, 2, \ldots$ Thus

(1.1.9) $$A_0 = \frac{1}{2\pi} \int_{-\pi}^{\pi} f(x) \, dx,$$

(1.1.10) $$A_n = \frac{1}{\pi} \int_{-\pi}^{\pi} f(x) \cos nx \, dx \quad n = 1, 2, \ldots,$$

(1.1.11) $$B_n = \frac{1}{\pi} \int_{-\pi}^{\pi} f(x) \sin mx \, dx \quad n = 1, 2, \ldots.$$

Fourier's thesis is that (1.1.5) will also remain true for $N = \infty$, if the coefficients A_n, B_n are defined by these formulas. This is most easily done in case the series $\sum_{n=0}^{\infty} (|A_n| + |B_n|)$ converges.

1.1.2 Absolutely Convergent Trigonometric Series

We begin the mathematics by considering functions defined by

(1.1.12) $$\boxed{f(\theta) = \sum_{n=0}^{\infty} (A_n \cos n\theta + B_n \sin n\theta)}$$

where $\sum_{n=0}^{\infty} (|A_n| + |B_n|) < \infty$. The values of f are determined on any interval of length 2π. A standard choice is the interval $\mathbb{T} = (-\pi, \pi]$, where we identify 2π-periodic functions on $\mathbb{R}$ with functions on $\mathbb{T}$. A function on $\mathbb{T}$ is considered continuous (resp. differentiable) if the corresponding periodic function on $\mathbb{R}$ is continuous (resp. differentiable). In concrete terms this means that f is continuous (resp. differentiable) on $(-\pi, \pi]$ with $f(\pi - 0) = f(-\pi + 0)$ (resp. $f'(\pi - 0) = f'(-\pi + 0)$).

In order to simplify the notations throughout, we recall the Euler formula for the complex exponential function $e^{i\theta} = \cos\theta + i\sin\theta$ and its consequences

(1.1.13) $$\cos\theta = \frac{1}{2}(e^{i\theta} + e^{-i\theta}), \qquad \sin\theta = \frac{1}{2i}(e^{i\theta} - e^{-i\theta}).$$

This allows us to rewrite the trigonometric series (1.1.12) in the complex form

(1.1.14) $$f(\theta) = \sum_{n=-\infty}^{\infty} C_n e^{in\theta}$$

Complex notation is especially efficient when we multiply two such functions, the formula $e^{i\theta}e^{i\phi} = e^{i(\theta+\phi)}$ being a streamlined expression of the addition formulas for both the sine and cosine functions. Similar efficiency is realized in the integration formulas $\int e^{ax}\,dx = a^{-1}e^{ax}$ when a is a nonzero complex number. The passage from real functions to complex functions also suggests the natural definition of convergence of the series (1.1.14), namely as the limit of the symmetric partial sums $\sum_{-N}^{N}$. We use throughout the notation $\sum_{n\in\mathbb{Z}}$ for this limiting process. If we try to consider more general definitions of convergence, difficulties will arise.

Throughout the text we will systematically use the Lebesgue integral and its many properties. In some cases a more elementary definition of integration will suffice, but we prefer to systematically employ the Lebesgue theory—both for its increased generality and its ease with respect to passage to the limit.

Theorem 1.1.5. *Suppose that $\sum_{n\in\mathbb{Z}} |C_n| < \infty$. Then f defined by (1.1.14) is a continuous function on $\mathbb{T}$. The coefficients are obtained as*

(1.1.15) $$C_n = \frac{1}{2\pi}\int_{-\pi}^{\pi} f(\theta)e^{-in\theta}\,d\theta, \qquad n \in \mathbb{Z}.$$

If g is any other L^1 function on $\mathbb{T}$, we have the Fourier reciprocity formula

(1.1.16) $$\frac{1}{2\pi}\int_{-\pi}^{\pi} f(\theta)\,g(\theta)\,d\theta = \sum_{n\in\mathbb{Z}} C_n D_{-n}$$

where D_n is the Fourier coefficient of g, defined by (1.1.15) with f replaced by g. In particular we have Parseval's identity

(1.1.17) $$\frac{1}{2\pi}\int_{-\pi}^{\pi} |f(\theta)|^2\,d\theta = \sum_{n\in\mathbb{Z}} |C_n|^2.$$

Proof. The uniform convergence of (1.1.14) follows from the Weierstrass M test, since $|C_n e^{in\theta}| = |C_n|$, the general term of a convergent numerical series, so that the limit function is continuous. This uniform convergence also holds for the series defined by $e^{-iN\theta}f(\theta)$, which we may therefore integrate term-by-term. In the process we encounter the integral

$$\int_{-\pi}^{\pi} e^{-iN\theta} e^{in\theta} \, d\theta = 0 \quad (n \neq N, n \in \mathbb{Z}, N \in \mathbb{Z}) \tag{1.1.18}$$

while the integral is 2π for $n = N$. Equation (1.1.18) is known as the *complex orthogonality relation*. This shows that

$$\frac{1}{2\pi} \int_{-\pi}^{\pi} e^{-iN\theta} f(\theta) \, d\theta = C_N, \quad (N \in \mathbb{Z}) \tag{1.1.19}$$

which was to be proved. To prove (1.1.16), we multiply the series (1.1.14) by $g(\theta)$. The partial sums are bounded by a multiple of the integrable function $|g(\theta)|$, hence we can apply Lebesgue's dominated convergence and integrate term-by-term to obtain

$$\frac{1}{2\pi} \int_{-\pi}^{\pi} f(\theta) g(\theta) \, d\theta = \sum_{n \in \mathbb{Z}} \frac{C_n}{2\pi} \int_{-\pi}^{\pi} e^{in\theta} g(\theta) \, d\theta = \frac{1}{2\pi} \sum_{n \in \mathbb{Z}} C_n D_{-n} \tag{1.1.20}$$

which was to be proved. Taking $g = \bar{f}$ gives the Parseval identity (1.1.17). ∎

Exercise 1.1.6. *Suppose that $\sum_{n \in \mathbb{Z}} |nC_n| < \infty$. Prove that the series (1.1.14) defines a differentiable function with $f'(\theta) = \sum_{n \in \mathbb{Z}} inC_n e^{in\theta}$ a continuous function.*

Hint: Use the inequality $|e^{ih} - 1| \leq |h|$ to justify passage to the limit.

Exercise 1.1.7. *Suppose that $\sum_{n \in \mathbb{Z}} |n^k C_n| < \infty$ for some $k = 2, 3, \ldots$. Prove that the series (1.1.14) defines a k-times differentiable function with $f^{(k)}(\theta) = \sum_{n \in \mathbb{Z}} (in)^k C_n e^{in\theta}$ a continuous function.*

The Fourier reciprocity formula (1.1.16) can be rewritten to obtain a useful representation of the *convolution* of an absolutely convergent trigonometric series with an arbitrary integrable function. Taking $g_\phi(\theta) = g(\phi - \theta)$, we compute the Fourier coefficient by writing

$$\int_{\mathbb{T}} e^{-in\theta} g_\phi(\theta) \, d\theta = \int_{\mathbb{T}} e^{in(\psi-\phi)} g(\psi) \, d\psi = e^{-in\phi} \int_{\mathbb{T}} g(\psi) e^{in\psi} \, d\psi$$

which, when substituted into (1.1.16), yields the following.

Corollary 1.1.8. *The convolution of an absolutely convergent trigonometric series f with an arbitrary L^1 function g has the representation*

$$\boxed{\frac{1}{2\pi} \int_{\mathbb{T}} f(\theta) g(\phi - \theta) \, d\theta = \sum_{n \in \mathbb{Z}} C_n D_n e^{in\phi}.} \tag{1.1.21}$$

A large source of examples of absolutely convergent trigonometric series is obtained from power series. Consider a general Laurent series

$$f(z) = \sum_{n=0}^{\infty} a_n z^n + \sum_{n=1}^{\infty} b_n z^{-n}, \tag{1.1.22}$$

assumed to be absolutely convergent in an annular region $r_1 < |z| < r_2$. Then

$$(1.1.23) \qquad f(re^{i\theta}) = \sum_{n=0}^{\infty} a_n r^n e^{in\theta} + \sum_{n=1}^{\infty} b_n r^{-n} e^{-in\theta}, \qquad r_1 < r < r_2$$

is an absolutely convergent trigonometric series. In particular we will apply this to $e^z = \sum_{n=0}^{\infty} z^n/n!$ in the next Section 1.1.3.

1.1.3 *Examples of Factorial and Bessel Functions

We can generate many useful examples beginning with the power series of the exponential function.

Example 1.1.9. Let $C_n = 0$ for $n < 0$ and $C_n = r^n/n!$ where $r \geq 0$ and $n = 1, 2, \ldots$. Then $F(\theta) = e^{re^{i\theta}}$ and the Fourier coefficient formula (1.1.15) and Parseval identity (1.1.17) specialize to

$$(1.1.24) \qquad \frac{r^n}{n!} = \frac{1}{2\pi} \int_{-\pi}^{\pi} e^{re^{i\theta}} e^{-in\theta} d\theta, \qquad r \geq 0, n = 0, 1, 2, \ldots$$

$$(1.1.25) \qquad I_0(2r) := \sum_{n=0}^{\infty} \left(\frac{r^n}{n!}\right)^2 = \frac{1}{2\pi} \int_{-\pi}^{\pi} e^{2r\cos\theta} d\theta, \qquad r \geq 0.$$

Equation (1.1.25) gives an integral formula for the modified Bessel function $I_0(2r)$ defined by the power series on the left side. In particular we will determine the asymptotic behavior of $I_0(2r)$ when $r \to \infty$ by analysis of the integral on the right side of (1.1.25). Meanwhile (1.1.24) gives a useful representation of the factorial function. We will use this to present a self-contained treatment of Stirling's formula in the form

$$(1.1.26) \qquad \boxed{\lim_{n \to \infty} n!/n^{n+\frac{1}{2}} e^{-n} = \sqrt{2\pi}.}$$

To obtain this result we take $r = n$ in (1.1.24), to obtain

$$(1.1.27) \qquad \frac{n^n e^{-n}}{n!} = \frac{1}{2\pi} \int_{-\pi}^{\pi} e^{nF(\theta)} d\theta,$$

where $F(\theta) = e^{i\theta} - 1 - i\theta$. This is set up to apply the Laplace asymptotic method, whose proof will be given at the end of the section, and whose statement follows.

Proposition 1.1.10. Suppose that $F(\theta)$, $-\pi \leq \theta \leq \pi$ is a continuous complex-valued function so that $\operatorname{Re} F(\theta)$ has a unique maximum at $\theta = 0$ with $F(0) = 0$, $\lim_{\theta \to 0} F(\theta)/\theta^2 = -k$ with $\operatorname{Re} k > 0$. Then

$$(1.1.28) \qquad \int_{-\pi}^{\pi} e^{nF(\theta)} d\theta = \sqrt{\frac{\pi}{nk}} + o\left(\frac{1}{\sqrt{n}}\right), \qquad n \to \infty.$$

In the present case $F(\theta) = e^{i\theta} - 1 - i\theta$, so the conditions are satisfied with $\operatorname{Re} F(\theta) = \cos\theta - 1 < 0$ for $\pi > |\theta| > 0$ and $k = -F''(0)/2 = 1/2$.

Exercise 1.1.11. *Prove that the factorial function satisfies the two-sided system of inequalities $C_1 n^n e^{-n} \leq n! \leq C_2(n+1)^{n+1} e^{-(n+1)}$ for positive constants C_1, C_2 and $n = 1, 2, \ldots$.*

Hint: Compare $\sum_{k=1}^{n} \log k$ with the integral of $\log x$ and then exponentiate.

As a second application of Laplace's method, we take $F(\theta) = \cos\theta - 1$ to deduce the asymptotic formula for the modified Bessel function defined by (1.1.25):

$$(1.1.29) \qquad I_0(2r) = e^{2r}\sqrt{\frac{1}{4\pi r}}(1 + o(1)), \qquad r \to \infty.$$

Returning to the theory, one can discuss further general properties of the class of absolutely convergent trigonometric series. They form an *algebra* of functions, meaning that the sum and product of two is again in the same class. In detail

$$(1.1.30) \qquad \sum_{n \in \mathbb{Z}} A_n e^{in\theta} + \sum_{n \in \mathbb{Z}} B_n e^{in\theta} = \sum_{n \in \mathbb{Z}} (A_n + B_n) e^{in\theta},$$

$$(1.1.31) \qquad \sum_{n \in \mathbb{Z}} A_n e^{in\theta} \times \sum_{n \in \mathbb{Z}} B_n e^{in\theta} = \sum_{n \in \mathbb{Z}} \left(\sum_{k \in \mathbb{Z}} A_k B_{n-k} \right) e^{in\theta}.$$

We will prove later that the class of absolutely convergent trigonometric series contains the class of Hölder continuous functions of exponents greater than $\frac{1}{2}$.

1.1.4 Poisson Kernel Example

A second useful example of an absolutely convergent trigonometric series is generated by the function $f(z) = 1/(1-z)$, defined in the unit disc $|z| < 1$. Thus we obtain the absolutely convergent series

$$(1.1.32) \qquad \frac{1}{1 - re^{i\theta}} = \sum_{n=0}^{\infty} r^n e^{in\theta}, \qquad 0 \leq r < 1.$$

When we take the real and imaginary parts, we obtain the real series

$$\frac{1 - r\cos\theta}{1 + r^2 - 2r\cos\theta} = 1 + \sum_{n=1}^{\infty} r^n \cos n\theta = \frac{1}{2} + \frac{1}{2} \sum_{n \in \mathbb{Z}} r^{|n|} e^{in\theta},$$

$$\frac{r\sin\theta}{1 + r^2 - 2r\cos\theta} = \sum_{n=1}^{\infty} r^n \sin n\theta = -\frac{i}{2} \sum_{n \in \mathbb{Z}} r^{|n|} \operatorname{sgn}(n) e^{in\theta},$$

where the signum function $\operatorname{sgn}(n)$ is defined by setting $\operatorname{sgn}(0) = 0$, $\operatorname{sgn}(n) = 1$ if $n \geq 1$ and $\operatorname{sgn}(n) = -1$ if $n \leq -1$. These can be rewritten as

$$(1.1.33) \qquad \boxed{P_r(\theta) := \frac{1 - r^2}{1 + r^2 - 2r\cos\theta} = \sum_{n \in \mathbb{Z}} r^{|n|} e^{in\theta},}$$

$$(1.1.34) \qquad \boxed{Q_r(\theta) := \frac{2r\sin\theta}{1 + r^2 - 2r\cos\theta} = -i \sum_{n \in \mathbb{Z}} \operatorname{sgn}(n) r^{|n|} e^{in\theta}.}$$

P_r is the *Poisson kernel* and Q_r is the *conjugate Poisson kernel*. See Figures 1.1.1 and 1.1.2. It is easily checked that both P_r and Q_r are solutions of Laplace's equation. Equations (1.1.15) and (1.1.17) yield

(1.1.35) $$\frac{1}{2\pi} \int_{-\pi}^{\pi} e^{-in\theta} P_r(\theta) \, d\theta = r^{|n|}, \qquad n \in \mathbb{Z}, \quad 0 \le r < 1,$$

(1.1.36) $$\frac{1}{2\pi} \int_{-\pi}^{\pi} e^{-in\theta} Q_r(\theta) \, d\theta = -i \operatorname{sgn}(n) r^{|n|}, \qquad n \in \mathbb{Z}, \quad 0 \le r < 1,$$

(1.1.37) $$\frac{1}{2\pi} \int_{-\pi}^{\pi} P_r(\theta)^2 \, d\theta = 1 + 2 \sum_{n=1}^{\infty} r^{2n} = \frac{1+r^2}{1-r^2}, \qquad 0 \le r < 1,$$

(1.1.38) $$\frac{1}{2\pi} \int_{-\pi}^{\pi} Q_r(\theta)^2 \, d\theta = 2 \sum_{n=1}^{\infty} r^{2n} = \frac{2r^2}{1-r^2}, \qquad 0 \le r < 1.$$

Setting $n = 0$ in (1.1.35) yields

(1.1.39) $$1 = \frac{1}{2\pi} \int_{-\pi}^{\pi} \frac{1 - r^2}{1 + r^2 - 2r \cos \theta} \, d\theta, \qquad 0 \le r < 1.$$

These integration formulas will be used frequently in dealing with the summability of Fourier series.

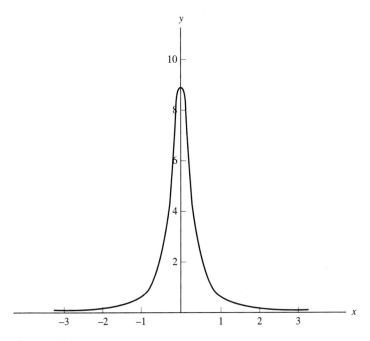

FIGURE 1.1.1
Poisson kernel $P_r(\theta)$ with $r = 0.8$

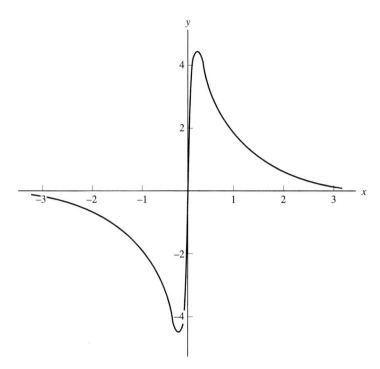

FIGURE 1.1.2
Conjugate Poisson kernel $Q_r(\theta)$ with $r = 0.8$

Exercise 1.1.12. *Prove that P_r and Q_r are both solutions of Laplace's equation $u_{xx} + u_{yy} = 0$ where $x + iy = re^{i\theta}$.*

Hint: Using $(x + iy)^n = r^n e^{in\theta}$ shows that each term in the series is a solution of Laplace's equation. Another approach is to recall the form of Laplace's equation in polar coordinates, namely $u_{rr} + (1/r)u_r + (1/r^2)u_{\theta\theta} = 0$.

1.1.5 *Proof of Laplace's Method

We formulate this in somewhat greater generality as follows: Let $A(x)$, $B(x)$ be continuous functions defined on an interval $a \leq x \leq b$ so that $\operatorname{Re} B(x) < \operatorname{Re} B(x_0)$ for $x \neq x_0 \in (a, b)$. B satisfies the asymptotic relation $(B(x) - B(x_0))/(x - x_0)^2 \to -k$ when $x \to x_0 \in (a, b)$ with $\operatorname{Re}(k) > 0$. We will prove that when $t \to \infty$

$$(1.1.40) \quad C(t) := \int_a^b A(x) e^{tB(x)} \, dx = e^{tB(x_0)} \left(A(x_0) \sqrt{\frac{\pi}{tk}} + o\left(\frac{1}{\sqrt{t}}\right) \right).$$

Proof. Without loss of generality we can assume that $x_0 = 0$ and $B(0) = 0$.

Step 1. Localizing the range of integration: For any $\delta > 0$ the maximum of Re $B(x)$ when $|x| \geq \delta$ is negative so that we can write

$$C(t) = \int_{|x|<\delta} A(x) e^{tB(x)}\, dx + O(e^{-Dt})$$

for some $D > 0$.

Step 2. Replacing B by its quadratic approximation: Now we write the expansion of $B(x)$ around $x = 0$:

$$B(x) = -kx^2 + o(x^2), \qquad k = k_R + ik_I.$$

For any two complex numbers $z_1 = x_1 + iy_1$, $z_2 = x_2 + iy_2$ we have the inequality

(1.1.41) $\qquad |e^{z_1} - e^{z_2}| \leq |z_1 - z_2| e^{z_3}, \qquad z_3 = \max(x_1, x_2).$

Applying this with $z_1 = B(x)$, $z_2 = -kx^2$, we can take $z_3 \leq -k_R x^2/2$ by taking δ small enough. With this choice of δ we can write

$$|e^{tB(x)} - e^{-tkx^2}| \leq tC_3 \epsilon x^2 e^{-tk_R x^2/2}, \qquad |x| \leq \delta.$$

The error in replacing the exponential by the purely quadratic exponential term is bounded by

$$tC_3 \epsilon \int_{|x|<\delta} |x|^2 e^{-tk_R x^2/2}\, dx < tC_3 \epsilon \int_{-\infty}^{\infty} |x|^2 e^{-tk_R x^2/2}\, dx = \epsilon \frac{\text{const}}{\sqrt{t}}$$

by making the substitution $v = x\sqrt{t}$.

Step 3. Replacing $A(x)$ by $A(0)$: Since A is continuous, we can rechoose $\delta > 0$ so that $|A(x) - A(0)| < \epsilon$ when $|x| < \delta$. The error made in replacing $A(x)$ by $A(0)$ is bounded by

$$\epsilon \int_{|x|<\delta} e^{-tk_R x^2/2}\, dx < \epsilon \int_{-\infty}^{\infty} e^{-tk_R x^2/2}\, dx = \epsilon \frac{\text{const}}{\sqrt{t}}$$

also by making the substitution $v = x\sqrt{t}$. From Steps 1, 2 and 3 we have

$$C(t) = A(0) \int_{-\delta}^{\delta} e^{-tkx^2}\, dx + o\left(\frac{1}{\sqrt{t}}\right).$$

Step 4. Integral over the real line: Finally we replace the integral on $-\delta < x < \delta$ by the integral over $-\infty < x < \infty$ with an exponential error, since

$$\left| \int_{|x|>\delta} e^{-tkx^2}\, dx \right| < \frac{1}{\delta} \int_{|x|>\delta} |x| e^{-tk_R x^2}\, dx$$

$$= \frac{1}{\delta k_R t} e^{-tk_R \delta^2}.$$

But the integral over $-\infty < x < \infty$ is

(1.1.42) $\qquad \displaystyle\int_{-\infty}^{\infty} e^{-tkx^2}\, dx = \sqrt{\frac{\pi}{kt}}.$

If k is complex, we must take the square root with positive real part, since the integral clearly has this property. Combining these steps completes the proof. ∎

Exercise 1.1.13. *If* Re $k > 0$*, show that the real part of* $\int_{-\infty}^{\infty} e^{-kx^2}\, dx$ *is positive.*

Hint: First make the change of variable $x^2 = y$ to reduce consideration to $\int_0^\infty (e^{-ay}/y^{1/2}) \cos(by) \, dy$, which is an infinite sum of integrals over $(k\pi/b, (k+1)\pi/b)$. Show that the first of these is positive and the remaining terms alternate in sign and decrease to zero. Alternatively, one may prove this by using complex analysis, considering the analytic function $k \to \phi(k) = \int_0^\infty e^{-kx^2} \, dx$, defined for Re $k > 0$. On the real axis $\phi(k)$ agrees with the branch of the function $k \to (\pi/k)^{1/2}$, which is positive and real on the positive real axis. Hence these two analytic functions agree on the entire halfplane Re $k > 0$.

Exercise 1.1.14. *Suppose, in addition, that $A(x)$ is Lipschitz continuous: $|A(x) - A(y)| \le K|x - y|$ and that the second derivative $B''(x)$ exists and is also Lipschitz continuous. By going through the steps of the above proof, show that in this case the error term $o(1/\sqrt{t})$ in Laplace's method can be replaced by $O(1/t)$.*

Computation of (1.1.42). This is obtained by considering the square of $I := \int_\mathbb{R} e^{-kx^2} \, dx$ in polar coordinates. Thus

$$I^2 = \left(\int_\mathbb{R} e^{-kx^2} \, dx \right) \left(\int_\mathbb{R} e^{-ky^2} \, dy \right)$$
$$= \int_\mathbb{R} \int_\mathbb{R} e^{-k(x^2+y^2)} \, dx \, dy$$
$$= \int_{-\pi}^{\pi} \int_0^\infty e^{-kr^2} r \, dr \, d\theta$$
$$= 2\pi \frac{1}{2k}.$$

1.1.6 *Nonabsolutely Convergent Trigonometric Series

It is possible to deal with trigonometric series with monotonically decreasing coefficients by the method of *summation by parts* to produce convergent series. Given any sequence of complex numbers a_n, $n = 0, 1, 2, \ldots$, define $(\Delta a)_n = a_n - a_{n-1}$ for $n \ge 1$. The basic identity is that for any two sequences a_n, b_n

$$(1.1.43) \quad a_N b_N - a_M b_M = \sum_{k=M+1}^{N} a_k (\Delta b)_k + \sum_{k=M+1}^{N} b_{k-1} (\Delta a)_k, \quad M < N.$$

The proof is left as an exercise.

Exercise 1.1.15. *Prove (1.1.43).*

Hint: Write the left side as a telescoping sum and show that $a_k b_k - a_{k-1} b_{k-1} = a_k (\Delta b)_k + b_{k-1} (\Delta a)_k$.

As a first application of summation-by-parts, we can deduce the convergence of certain trigonometric series.

Proposition 1.1.16. *Suppose that $A_n > 0$ and $A_n \ge A_{n+1}$ with $\lim A_n = 0$. Then the trigonometric series $\sum_{n=0}^{\infty} A_n e^{inx}$ is convergent for $x \ne 0$.*

Proof. We apply (1.1.43) with $a_n = A_n$, $b_n = B_n(x) = \sum_{k=0}^{n} e^{ikx}$. This is a finite geometric sum with

$$(1.1.44) \qquad |B_n(x)| = \left| \frac{1 - e^{i(n+1)x}}{1 - e^{ix}} \right| \leq \frac{1}{\sin(x/2)}.$$

Applying (1.1.43) shows that

$$(1.1.45) \qquad \sum_{k=M+1}^{N} a_k e^{ikx} = a_N B_N(x) - a_M B_M(x) - \sum_{k=M+1}^{N} B_{k-1}(x)(\Delta a)_k.$$

The first two terms tend to zero when $M, N \to \infty$. The sum is estimated by

$$(1.1.46) \qquad \left| \sum_{k=M+1}^{N} B_{k-1}(x)(\Delta a)_k \right| \leq \frac{1}{\sin(x/2)} \sum_{k=M+1}^{N} |\Delta a_k| = (a_M - a_N) \frac{1}{\sin(x/2)}$$

which tends to zero when $M, N \to \infty$. ∎

Example 1.1.17. *The trigonometric series $\sum_{n \geq 2} e^{inx}/\log n$ is convergent for $x \neq 0$. By taking the real and imaginary parts, we see that the series $\sum_{n \geq 2} \cos nx / \log n$ is convergent for $x \neq 0$ and the series $\sum_{n \geq 2} \sin nx / \log n$ is convergent for all x.*

Exercise 1.1.18. *Prove that we have uniform convergence of $\sum_{n \geq 2} e^{inx}/\log n$ on any closed interval not containing $x = 0$.*

In an appendix to this chapter we prove the basic Cantor uniqueness theorem, which allows one to identify the coefficients A_n from the sum of the (conditionally convergent) trigonometric series.

Summation by parts can also be used to estimate the *modulus of continuity* of an absolutely convergent trigonometric series in terms of the *tail sum*, defined as

$$E_n := \sum_{k=n+1}^{\infty} (|A_k| + |A_{-k}|).$$

To do this, let $h > 0$ and write

$$f(x+h) - f(x) = \sum_{n \in \mathbb{Z}} A_n e^{inx} (e^{inh} - 1),$$

$$|f(x+h) - f(x)| \leq \sum_{|n| \leq N} |nhA_n| + 2 \sum_{|n| > N} |A_n|.$$

The second sum is $2E_N$. The first sum is rewritten as

$$\sum_{|n| \leq N} |nhA_n| = -h \sum_{n=0}^{N} n(\Delta E)_n = -hNE_N + h \sum_{n=0}^{N-1} E_n,$$

so that

$$|f(x+h) - f(x)| \leq hN \left(E_N + \frac{1}{N} \sum_{n=0}^{N-1} E_n \right) + 2E_N.$$

The first and third terms are balanced by taking $hN = 1$. The middle term is an average of the last term and can be estimated therefrom. Specific forms of the tail sum will lead to various concrete estimates.

Exercise 1.1.19. *Suppose that the tail sum satisfies $E_n \leq Cn^{-\alpha}$ for some $0 < \alpha < 1$. Prove that f satisfies a Hölder condition: $|f(x+h) - f(x)| \leq Ch^{\alpha}$ for some constant C.*

Exercise 1.1.20. *Suppose that the tail sum satisfies $E_n \leq Cn^{-1}$. Prove that f satisfies $|f(x+h) - f(x)| \leq Ch \log(1/h)$ for some constant C.*

Exercise 1.1.21. *Let $f(x) = \sum_{n=0}^{\infty} a^n \cos(b^n x)$ where $0 < a < 1$, $b \in \mathbb{Z}^+$. Find a modulus of continuity of f. Consider separately the cases $ab < 1$, $ab = 1$ and $ab > 1$.*

Exercise 1.1.22. *Illustrate the results of the previous exercise in the following three cases:*

$$\sum_{n=0}^{\infty} 2^{-n} \cos(3^n x), \quad \sum_{n=0}^{\infty} 3^{-n} \cos(2^n x), \quad \sum_{n=0}^{\infty} 2^{-n} \cos(2^n x).$$

1.2 FORMULATION OF FOURIER SERIES

Armed with some motivation, we now begin the formal study of Fourier series.

1.2.1 Fourier Coefficients and Their Basic Properties

We begin with an integrable function f on $\mathbb{T} = (-\pi, \pi]$. Any such function can naturally be identified with a 2π-periodic function on the entire real line. This extension is helpful in many ways, especially in computing integrals since we may write $\int_{\mathbb{T}} f(x+\theta) d\theta = \int_{\mathbb{T}} f(\theta) d\theta$.

The Fourier coefficients, or discrete Fourier transform, of $f \in L^1(\mathbb{T})$ are defined by the formula

(1.2.1) $$\hat{f}(n) = \frac{1}{2\pi} \int_{\mathbb{T}} f(\theta) e^{-in\theta} d\theta.$$

This is a linear transformation from the space $L^1(\mathbb{T})$ to the space $l^{\infty}(\mathbb{Z})$ of bounded bilateral sequences. We formalize this as a proposition.

Proposition 1.2.1. *The space of discrete Fourier transforms is an algebra, as expressed by*

(1.2.2) $$\hat{f_1}(n) + \hat{f_2}(n) = \widehat{(f_1 + f_2)}(n)$$

(1.2.3) $$\hat{f_1}(n) \times \hat{f_2}(n) = \widehat{(f_1 * f_2)}(n)$$

where the convolution of two L^1 functions is defined by

$$(f_1 * f_2)(\theta) = \frac{1}{2\pi} \int_\mathbb{T} f_1(\phi) f_2(\theta - \phi) \, d\phi$$

and where the integral is convergent for almost every θ. The convolution product is commutative and associative: $f_1 * f_2 = f_2 * f_1$, $(f_1 * f_2) * f_3 = f_1 * (f_2 * f_3)$. Furthermore the mapping $f \to \hat{f}$ is a contraction, meaning that

(1.2.4) $$|\hat{f}(n)| \leq \frac{1}{2\pi} \int_\mathbb{T} |f(\theta)| \, d\theta.$$

Proof. The first formula is immediate from the linearity of the integral. To prove the second formula, we note that the convolution is well defined by the Fubini theorem, since $2\pi |f_1 * f_2|(\theta) \leq \int_\mathbb{T} |f_1(\phi)||f_2(\theta - \phi)| \, d\phi$. The latter integral has a finite integral over $\mathbb{T}$, since by Fubini and a change of variable we find that

$$\int_\mathbb{T} d\theta \int_\mathbb{T} |f_1(\phi)||f_2(\theta - \phi)| \, d\phi = \int_\mathbb{T} \int_\mathbb{T} |f_1(\theta)||f_2(\phi)| \, d\phi \, d\theta = \int_\mathbb{T} |f_1(\theta)| \, d\theta \int_\mathbb{T} |f_2(\theta)| \, d\theta.$$

Therefore the integral defining the convolution of $|f_1| * |f_2|$ is finite almost everywhere, and dominates the convolution $f_1 * f_2$. Having made these preparations, we can multiply the Fourier coefficients and transform by Fubini:

$$4\pi^2 \hat{f}_1(n) \times \hat{f}_2(n) = \int_\mathbb{T} f_1(\theta) e^{-in\theta} \int_\mathbb{T} f_2(\phi) e^{-in\phi} = \int_\mathbb{T} e^{-in\psi} \left(\int_\mathbb{T} f_1(\theta) f_2(\psi - \theta) \, d\theta \right) d\psi$$

which was to be proved. The commutative and associative properties are most easily deduced from these properties of ordinary multiplication, once we have used Fubini to identify the convolution as the unique L^1 function F with the property that for any bounded function h, $\int_\mathbb{T} \int_\mathbb{T} h(x+y) f_1(x) f_2(y) \, dx \, dy = 2\pi \int_\mathbb{T} h(z) F(z) \, dz$. The contraction property is immediate from (1.2.1). ∎

Exercise 1.2.2. Let $f_1, f_2 \in L^1(\mathbb{T})$. Show that for any bounded measurable function h we have

$$\int_\mathbb{T} \int_\mathbb{T} h(x+y) f_1(x) f_2(y) \, dx \, dy = 2\pi \int_\mathbb{T} h(z) (f_1 * f_2)(z) \, dz.$$

One may be tempted to conclude that the convolution $f_1 * f_2$ is represented by the Fourier series $\sum_{n \in \mathbb{Z}} \hat{f}_1(n) \hat{f}_2(n) e^{in\theta}$. This cannot be literally true in general, because the latter series does not converge pointwise or in L^1, for a general L^1 function. Nevertheless the following special case of Fourier reciprocity is quite useful.

Proposition 1.2.3. Suppose that $\sum_{n \in \mathbb{Z}} |A_n| < \infty$ and we have an absolutely convergent Fourier series $K(\theta) = \sum_{n \in \mathbb{Z}} A_n e^{in\theta}$. Then for any $f \in L^1(\mathbb{T})$ we have

(1.2.5) $$(f * K)(\theta) = \sum_{n \in \mathbb{Z}} A_n \hat{f}(n) e^{in\theta}.$$

Proof. Begin with the identity

$$f(\theta - \phi) K(\phi) = \sum_{n \in \mathbb{Z}} A_n e^{in\phi} f(\theta - \phi).$$

By hypothesis, this is a series of L^1 functions that is convergent and bounded pointwise by the integrable function $|f| \times \sum_{n \in \mathbb{Z}} |A_n|$, so that by Lebesgue's dominated convergence theorem we can integrate term-by-term to conclude

$$\int_{\mathbb{T}} f(\theta - \phi) K(\phi)\, d\phi = \sum_{n \in \mathbb{Z}} A_n \int_{\mathbb{T}} e^{in\phi} f(\theta - \phi)\, d\phi$$

$$= \sum_{n \in \mathbb{Z}} A_n \int_{\mathbb{T}} e^{in(\theta - \psi)} f(\psi)\, d\psi$$

$$= 2\pi \sum_{n \in \mathbb{Z}} A_n \hat{f}(n) e^{in\theta}$$

which was to be proved. ∎

This can be immediately applied to the Poisson kernel $K(\theta) = P_r(\theta) = \sum_{n \in \mathbb{Z}} r^{|n|} e^{in\theta}$. If $\sum_{n \in \mathbb{Z}} \hat{f}(n) e^{in\theta}$ is the Fourier series of some L^1 function f, we may now assert that

$$(f * P_r)(\theta) = \sum_{n \in \mathbb{Z}} \hat{f}(n) r^{|n|} e^{in\theta}.$$

Another example is obtained from the *Dirichlet kernel,* defined by the finite trigonometric sum

(1.2.6) $$D_N(\theta) = \sum_{n=-N}^{N} e^{in\theta}.$$

The Fourier partial sum operator

$$S_N f = \sum_{-N}^{N} \hat{f}(n) e^{in\theta} = \sum_{n \in \mathbb{Z}} 1_{[-N,N]}(n) \hat{f}(n) e^{in\theta}$$

is equivalently obtained as the convolution

(1.2.7) $$\boxed{S_N f(\theta) = (D_N * f)(\theta).}$$

Here 1_A is the indicator function, defined by $1_A(n) = 1$ if $n \in A$ and equals zero otherwise. From (1.2.6), it is clear that D_N is an even function and that $\int_{\mathbb{T}} D_N = 2\pi$. See Figure 1.2.1 for the Dirichlet kernel and Figure 1.2.2 for the Fourier partial sums of $f(x) = x$, $-\pi < x < \pi$.

Exercise 1.2.4. Show that the Dirichlet kernel can be equivalently expressed as

(1.2.8) $$\boxed{D_N(\theta) = \frac{\sin(N + \tfrac{1}{2})\theta}{\sin \theta/2} \quad \theta \neq 0, \pm 2\pi, \ldots}$$

by summing a finite geometric series.

Exercise 1.2.5. Suppose that $K \in L^1(\mathbb{T})$ is an even function: $K(-\theta) = -K(\theta)$. Prove that the convolution operator $f \to K * f$ is self-adjoint, meaning that for any $f \in L^1(\mathbb{T})$ and any bounded function g we have $\int_{\mathbb{T}} (K * f) g = \int_{\mathbb{T}} f(K * g)$.

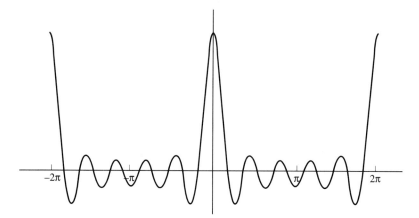

FIGURE 1.2.1
The Dirichlet kernel $D_N(\theta)$ for $N = 5$.
From M. Pinsky, *Partial Differential Equations and Boundary-Value Problems with Applications.*
Reprinted by permission of The McGraw-Hill Companies.

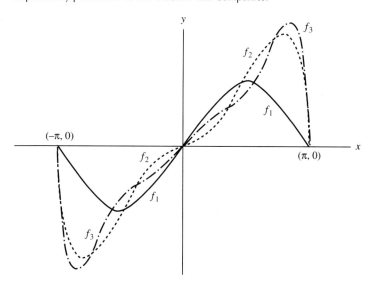

FIGURE 1.2.2
The graphs of the partial sums $f_N(\theta)$ for $N = 1, 2, 3$ of the Fourier series of $f(\theta) = \theta$, $-\pi < \theta < \pi$.
From M. Pinsky, *Partial Differential Equations and Boundary-Value Problems with Applications.*
Reprinted by permission of The McGraw-Hill Companies.

Exercise 1.2.6. *Suppose that $K \in L^1(\mathbb{T})$ is an odd function: $K(-\theta) = -K(\theta)$. Prove that the convolution operator is skew-adjoint, meaning that for any $f \in L^1(\mathbb{T})$ and any bounded function g we have $\int_{\mathbb{T}} (K * f)g = -\int_{\mathbb{T}} f(K * g)$.*

At this point we can formulate the *uniqueness* of the Fourier coefficients, following an elementary argument of Lebesgue. The result will also be deduced as a corollary of the summability of one-dimensional Fourier series in Section 1.3.

Proposition 1.2.7. *Suppose that $f, g \in L^1(\mathbb{T})$ have the property that $\hat{f}(n) = \hat{g}(n)$ for all $n \in \mathbb{Z}$. Then $f = g$ a.e.*

Proof. By the linearity of the map $f \to \hat{f}$ it suffices to prove the result in case $g = 0$. We first prove the result in case f is a continuous function. Writing $f = u + iv$, we have for any $n \in \mathbb{Z}$,

$$0 = 2\pi \hat{f}(n) = \int_{-\pi}^{\pi} (u(x) + iv(x))e^{-inx}\, dx$$

$$= \int_{-\pi}^{\pi} (u(x)\cos nx + v(x)\sin nx)\, dx + i \int_{-\pi}^{\pi} (v(x)\cos nx - u(x)\sin nx)\, dx.$$

For any $m = 0, 1, 2, \ldots$ we apply this to $n = m$ and $n = -m$ to conclude that

$$\int_{-\pi}^{\pi} u(x)\cos mx\, dx = 0 = \int_{-\pi}^{\pi} u(x)\sin mx\, dx,$$

$$\int_{-\pi}^{\pi} v(x)\cos mx\, dx = 0 = \int_{-\pi}^{\pi} v(x)\sin mx\, dx.$$

Therefore we are reduced to the case of a real-valued continuous function f for which

(1.2.9) $$\int_{-\pi}^{\pi} f(x)\cos mx\, dx = 0 = \int_{-\pi}^{\pi} f(x)\sin mx\, dx, \qquad m = 0, 1, 2, \ldots.$$

If f is not identically zero, there exists a point x_0 where $f(x_0) \neq 0$. Replacing f by $f(x + x_0)/f(x_0)$, we may assume that $f(x) \geq \frac{1}{2}$ in a closed interval $I = [-\delta, \delta]$, where $0 < \delta < \pi$. Let $t(x) = 1 + \cos x - \cos \delta$, $T_n(x) = t(x)^n$. Clearly $t(x) \geq 1$ on I while $|t(x)| < 1$ on I^c, so that $T_n(x) \geq 1$ on I while $T_n(x) \to 0$ on I^c. Now T_n is a trigonometric polynomial of degree n, so that from the hypothesis (1.2.9)

$$0 = \int_{-\pi}^{\pi} f(x)T_n(x)\, dx = 0.$$

On the other hand, the dominated convergence theorem shows that $\lim_n \int_{I^c} f(x)T_n(x)\, dx = 0$. Subtracting these, we conclude that $\lim_n \int_I f(x)T_n(x)\, dx = 0$, which contradicts the fact that $f(x)T_n(x) \geq \frac{1}{2}$ on I. Hence $f(x) \equiv 0$. Applying this argument separately to $f = u$ and $f = v$ proves the result for any complex-valued $f \in C(\mathbb{T})$.

Now if $f \in L^1(\mathbb{T})$ satisfies $\hat{f}(n) \equiv 0$, let $F(x) = \int_{-\pi}^{x} f(t)\, dt$, a continuous function. Interchanging the orders of integration shows further that for $n = \pm 1, \pm 2, \ldots$

$$\int_{-\pi}^{\pi} F(x)e^{-inx}\, dx = \int_{-\pi}^{\pi} e^{-inx} \left(\int_{-\pi}^{x} f(t)\, dt \right) dx$$

$$= \int_{-\pi}^{\pi} f(t) \left(\int_{t}^{\pi} e^{-inx}\, dx \right) dt$$

$$= \int_{-\pi}^{\pi} f(t) \left(\frac{e^{-in\pi} - e^{-int}}{-in} \right) dt$$

$$= 0.$$

Hence $F(n) = 0$ for $0 \neq n \in \mathbb{Z}$. Letting $A_0 = 1/2\pi \int_{-\pi}^{\pi} F(x)\, dx$, we can apply the proof in the previous paragraph to the continuous function $x \to F(x) - A_0$, to conclude that $F(x) - A_0 \equiv 0$. But from Lebesgue's theorem on the differentiation of the integral, we have almost everywhere that $f(x) = (d/dx)F(x) = 0$, which completes the proof. ∎

The above properties of the mapping $f \to \hat{f}$ are algebraic in nature. The following fundamental property is analytic. It is valid for an arbitrary approach to infinity, not restricting n to be integer-valued. Here we note that the Fourier coefficients (1.2.1) are well defined for any $n \in \mathbb{R}$.

Theorem 1.2.8. *Riemann-Lebesgue lemma: For any $f \in L^1(\mathbb{T})$, $\lim_{|n| \to \infty} \hat{f}(n) = 0$ and the convergence is uniform on compact subsets of $L^1(\mathbb{T})$.*

We will give two separate proofs of the first statement, then deal with the compactness.

First Proof. By making the change of variable $\phi = \theta + \pi/n$, the integral defining the Fourier coefficient is transformed into

$$2\pi \hat{f}(n) = \int_\mathbb{T} f(\phi + \pi/n) e^{-in(\phi + \pi/n)} \, d\phi = -\int_\mathbb{T} f(\phi + \pi/n) e^{-in\phi} \, d\phi.$$

Adding this to the definition of $\hat{f}(n)$, we obtain

$$4\pi \hat{f}(n) = \int_\mathbb{T} [f(\phi + \pi/n) - f(\phi)] e^{-in\phi} \, d\phi.$$

If f is a continuous function, the integrand tends to zero uniformly when $|n| \to \infty$, hence $\hat{f}(n) \to 0$. In case f is an arbitrary L^1 function, we can find a continuous function g so that $\|f - g\|_1 < \epsilon$. Then

$$\hat{f}(n) = \hat{g}(n) + \widehat{(f - g)}(n).$$

The first term tends to zero when $|n| \to \infty$ whereas the second term is less than ϵ, by virtue of (1.2.4). Thus $\limsup_n |\hat{f}(n)| \leq \epsilon$, which was arbitrary, completing the proof. ∎

Second proof. The result is clearly true for the indicator function of an interval (a, b), since for $n \neq 0$

$$(1.2.10) \qquad \int_a^b e^{-in\theta} = \frac{e^{-inb} - e^{-ina}}{-in} \to 0, \qquad |n| \to \infty.$$

Hence it is also true for the indicator function of a finite union of intervals. Now if E is any measurable set, by the definition of outer measure, there exists a finite union of intervals $\tilde{E}$ so that the symmetric difference $\tilde{E} \Delta E$ has measure less than ϵ. In terms of L^1 norms $\int_\mathbb{T} |1_E - 1_{\tilde{E}}| < \epsilon$. By linearity, we have

$$|\widehat{1_E}(n)| \leq |\widehat{1_{\tilde{E}}}(n)| + |\widehat{1_{\tilde{E} \Delta E}}(n)|.$$

The first term tends to zero while the second term is less than ϵ. Now we can extend to simple functions $f = \sum a_i 1_{E_i}$ by linearity and finally to all L^1 functions by appealing to the density of simple functions. For example $f_N = \sum_{|k| \leq N2^N} k2^{-N} 1_{k2^{-N} \leq f \leq (k+1)2^{-N}}$ is a simple function so that $\int_\mathbb{T} |f - f_N| \to 0$ when $N \to \infty$.

To prove the uniform convergence, let K be a compact set in $L^1(\mathbb{T})$. We first cover K by a union of balls: $K \subset \cup B(f, \epsilon)$. By compactness, we can extract a finite set $f_1, \ldots, f_N$ so that $K \subset \cup_{i=1}^N B(f_i, \epsilon)$. For each $1 \leq i \leq N$ we can apply the Riemann-Lebesgue lemma to conclude $|\hat{f_i}(n)| < \epsilon$ for $n \geq M$ and all i, $1 \leq i \leq N$. Now any other member of K is included in one of the balls B_i, so that by the contraction property, we must have $\limsup_n |\hat{f}(n)| < \epsilon$. This proves the uniform convergence. ∎

This property of uniform convergence is easily applied to show that if $f \in L^1(\mathbb{T})$ and g is a bounded function, then

(1.2.11) $$\int_{\mathbb{T}} g(\theta) f(\phi + \theta) e^{-in\theta} \, d\theta \to 0$$

uniformly for $\phi_1 \leq \phi \leq \phi_2$ when $|n| \to \infty$. To see this, note that for any $f \in L^1(\mathbb{T})$ the mapping $x \to f_x$ from $\mathbb{T}$ to $L^1(\mathbb{T})$ is continuous. In detail

$$\lim_{x \to x_0} \int_{\mathbb{T}} |f(x + \theta) - f(x_0 + \theta)| \, d\theta = 0.$$

Multiplication by the bounded function g preserves this continuity. The continuous image of a compact set is compact. Hence we can apply the uniform convergence on compacts to deduce that (1.2.11) holds uniformly on compact ϕ intervals.

1.2.2 Fourier Series of Finite Measures

The concept of a Fourier series can be easily extended from the class of integrable functions to the class of finite signed measures. Recall that a signed measure on $\mathbb{T}$ is defined by a function of bounded variation, which can be represented as the difference of two monotone functions. The sum of two signed measures is the setwise sum: $(\mu_1 + \mu_2)(A) = \mu_1(A) + \mu_2(A)$. The convolution of two signed measures μ_1, μ_2 is, by definition, the signed measure μ with the property that for every continuous function $h \in C(\mathbb{T})$

$$\int_{\mathbb{T} \times \mathbb{T}} h(x + y) \, d\mu_1(x) \, d\mu_2(y) = 2\pi \int_{\mathbb{T}} h(z) \, d\mu(z).$$

The Fourier coefficients of a signed measure are defined by

$$\hat{\mu}(n) = \frac{1}{2\pi} \int_{\mathbb{T}} e^{-in\theta} \, d\mu(\theta).$$

Proposition 1.2.1 carries over with no essential change.

Proposition 1.2.9. *The space of discrete Fourier transforms is an algebra, as expressed by*

(1.2.12) $$\hat{\mu}_1(n) + \hat{\mu}_2(n) = \widehat{(\mu_1 + \mu_2)}(n),$$

(1.2.13) $$\hat{\mu}_1(n) \times \hat{\mu}_2(n) = \widehat{(\mu_1 * \mu_2)}(n).$$

The convolution product is commutative and associative: $\mu_1 * \mu_2 = \mu_2 * \mu_1$, $(\mu_1 * \mu_2) * \mu_3 = \mu_1 * (\mu_2 * \mu_3)$. *Furthermore the mapping* $\mu \to \hat{\mu}$ *is a contraction, meaning that*

(1.2.14) $$|\hat{\mu}(n)| \leq \frac{1}{2\pi} \text{Var}(\mu).$$

Exercise 1.2.10. *Prove these statements.*

The Riemann-Lebesgue lemma is not true for signed measures. For example, the Dirac measure δ_0, for which $\delta_0(A) = 1$ iff $0 \in A$, has $\hat{\delta}_0(n) = 1$ for all n. One might ask if the Riemann-Lebesgue lemma holds for measures that are continuous, but this is not true either. The Fourier coefficients of the Cantor measure do not tend to zero.

Exercise 1.2.11. *Prove this statement*

Hint: For the Cantor measure, $\hat{\mu}(n) = \Pi_{k=1}^{\infty} \cos(2\pi n/3^k)$, with $\hat{\mu}(3^m) = \hat{\mu}(1)$ for $m = 1, 2, \ldots$. For details, consult Zygmund, 1959, p. 196.

1.2.3 *Rates of Decay of Fourier Coefficients

The Riemann-Lebesgue lemma provides no further quantitative information about the speed of convergence to zero for an arbitrary L^1 function. We can obtain a convenient upper bound from the representation

(1.2.15) $$\hat{f}(n) = \frac{1}{4\pi} \int_{\mathbb{T}} \left[f\left(\theta + \frac{\pi}{n}\right) - f(\theta) \right] e^{-in\theta} d\theta.$$

Hence we have immediately Proposition 1.2.12.

Proposition 1.2.12. *Suppose that $f \in C(\mathbb{T})$ has a modulus of continuity $\omega(\delta) := \sup_{|x-y| \leq \delta} |f(x) - f(y)|$. Then $|\hat{f}(n)| \leq \frac{1}{2}\omega(\pi/n)$. More generally, if $1 \leq p < \infty$ and $\Omega_p(\delta) := \sup_{|h| \leq \delta} \|f_h - f\|_p$ is the L^p modulus of continuity of $f \in L^p(\mathbb{T})$, then $|\hat{f}(n)| \leq \frac{1}{2}\Omega_p(\pi/n)$.*

For example, if f satisfies a Hölder condition with exponent $\alpha \in (0, 1)$, we see that $\hat{f}(n) = O(n^{-\alpha})$, $|n| \to \infty$.

Exercise 1.2.13. *Suppose that the L^p modulus of continuity satisfies $\Omega_p(h) \leq Ch^\alpha$ for $C > 0$ and $\alpha > 1$. Prove that f is a constant, a.e.*

Hint: First show that $\Omega_p(h+k) \leq \Omega_p(h) + \Omega_p(k)$; then iterate this to obtain a contradiction.

If we want to obtain a more precise estimation, we can assume that f is absolutely continuous and integrate by parts, as follows:

Proposition 1.2.14. *Suppose that $f \in C(\mathbb{T})$ is absolutely continuous. Then $\hat{f}(n) = (1/in)\hat{f}'(n)$; in particular $\hat{f}(n) = o(1/|n|)$, $|n| \to \infty$. If in addition $f', f'', \ldots, f^{(k-1)}$ are absolutely continuous, then $\hat{f}(n) = (1/in)^k \hat{f}^{(k)}(n)$; in particular $\hat{f}(n) = o(1/|n|^k)$. If $f^{(k)}$ satisfies a Hölder condition with exponent α, then $\hat{f}(n) = O(1/|n|^{k+\alpha})$, $|n| \to \infty$.*

Exercise 1.2.15. *Prove the above properties by integration by parts.*

It is difficult to characterize differentiability of a fixed degree in terms of the behavior of the Fourier coefficients. In order to obtain some simple characterizations of smoothness, we consider functions that are *infinitely differentiable*. This means that for each $m \in \mathbb{Z}^+$, the derivative $f^{(m)}$ exists and is a continuous function. Then we can integrate-by-parts for $n \neq 0$ and write $2\pi \hat{f}(n) = (1/in)^m \int_{\mathbb{T}} e^{-int} f^{(m)}(t)\, dt$, to conclude that the Fourier coefficients satisfy a system of estimates of the form

$$|\hat{f}(n)| \leq \frac{C_m}{|n|^m} \qquad 0 \neq n \in \mathbb{Z}, \quad m = 0, 1, \ldots. \tag{1.2.16}$$

In other words $\hat{f}(n)$ tends to zero faster than any negative power when $|n| \to \infty$. Conversely, if the Fourier coefficients of $f \in L^1(\mathbb{T})$ satisfy (1.2.16), then we can repeatedly differentiate the absolutely convergent Fourier series to conclude that f is a.e. equal to an infinitely differentiable function. This is summarized as follows.

Proposition 1.2.16. *$f \in L^1(\mathbb{T})$ is a.e. equal to an infinitely differentiable function if and only if its Fourier coefficients are rapidly decreasing, according to (1.2.16).*

1.2.3.1 Piecewise smooth functions

The correspondence between smoothness of a fixed degree and decay of the Fourier coefficients is not sharp in general. More precisely, the converse of Proposition 1.2.14 is false: there exists a nonabsolutely continuous $f \in L^1(\mathbb{T})$ for which $n\hat{f}(n) \to 0$ when $|n| \to \infty$. In order to obtain sharp results, we consider functions that are *piecewise smooth*, described as follows. If there exists a subdivision $-\pi \leq \theta_0 < \theta_1 < \cdots < \theta_K < \pi$ so that f is absolutely continuous on each subinterval with a simple jump at the endpoint, denoted $\delta f(\theta_j) := f(\theta_j + 0) - f(\theta_j - 0)$, then we say that f is *piecewise smooth of degree* 0. In general we say that f is *piecewise smooth of degree* k if there exists such a subdivision so that f is absolutely continuous on $\mathbb{T}$, together with its first $k-1$ derivatives and that $f^{(k)}$ is piecewise absolutely continuous as above, with jumps denoted $\delta f^{(k)}(\theta_j) := f^{(k)}(\theta_j + 0) - f^{(k)}(\theta_j - 0)$.

Proposition 1.2.17. *Suppose that f is piecewise smooth of degree k. Then the Fourier coefficients satisfy the identity*

$$\hat{f}(n) = \frac{1}{2\pi} \sum_{l=0}^{k} \left(\frac{1}{in}\right)^{l+1} \left(\sum_{j=0}^{K} \delta f^{(l)}(\theta_j) e^{-in\theta_j}\right) + \frac{\hat{f}^{(k+1)}(n)}{(in)^{k+1}}. \tag{1.2.17}$$

Furthermore the coefficient of $1/n^{l+1}$ tends to zero if and only if all of the jumps are zero, i.e. $f^{(l)}$ is a continuous function.

Note that, in case $\theta_0 = -\pi$, we interpret $\delta f(\theta_0) = f(-\pi) - f(\pi)$.

Proof. In case $k = 0$ we do an integration by parts on each interval of continuity:

$$\int_{\theta_j}^{\theta_{j+1}} f(\theta) e^{-in\theta}\, d\theta = \int_{\theta_j}^{\theta_{j+1}} f(\theta)\, d\left(\frac{e^{-in\theta}}{-in}\right) = f(\theta) \frac{e^{-in\theta}}{-in}\bigg|_{\theta_j}^{\theta_{j+1}} - \int_{\theta_j}^{\theta_{j+1}} f'(\theta) \frac{e^{-in\theta}}{-in}\, d\theta.$$

When we sum the boundary terms and simplify, the sum is written in terms of the jumps and the Fourier coefficients of f', which proves the result for $k = 0$. If f has $k-1$ absolutely continuous derivatives, we can iterate this to obtain the k terms displayed, together with the Fourier coefficient of $f^{(k)}$. In order to prove the sharpness, we prove a separate lemma.

Lemma 1.2.18. *Suppose that* $-\pi \leq \theta_0 < \theta_1 < \cdots < \theta_K < \pi$ *and that* C_j *are complex numbers so that*

$$\lim_{n\to\infty} \sum_{j=0}^{K} C_j e^{in\theta_j} = 0.$$

Then $C_j = 0$ *for all* j.

Proof. We use the identity that for $\mathbb{T} \ni \theta \neq 0$

$$\sum_{n=0}^{N-1} e^{in\theta} = \frac{1 - e^{iN\theta}}{1 - e^{i\theta}}.$$

Let $\Gamma_n = \sum_{j=0}^{K} C_j e^{in\theta_j}$, so that $\Gamma_n \to 0$, by hypothesis. Then

$$\Gamma_n e^{-in\theta_m} = C_m + \sum_{j \neq m} C_j e^{in(\theta_j - \theta_m)}$$

$$\frac{1}{N} \sum_{n=0}^{N-1} \Gamma_n e^{-in\theta_m} = C_m + \frac{1}{N} \sum_{j \neq m} C_j \frac{1 - e^{iN(\theta_j - \theta_m)}}{1 - e^{i(\theta_j - \theta_m)}}.$$

Taking $N \to \infty$, both the left side and the sum on the right tend to zero, hence $C_m = 0$, as required proving the lemma and Proposition 1.2.17. ∎

Corollary 1.2.19. *Let* f *be piecewise smooth of degree* k *and let* $0 \leq r \leq k$. *Then we have the asymptotic estimate* $\hat{f}(n) = o(|n|^{-r-1})$, $|n| \to \infty$ *if and only if* $f \in C^r(\mathbb{T})$, *i.e.,* f *has* r *continuous derivatives.*

Proof. The Riemann-Lebesgue lemma implies that the last term in (1.2.17) is $o(|n|^{-k-1})$. Therefore the asymptotic behavior of $\hat{f}(n)$ is equivalent to that of the finite sum. If $f \in C^r(\mathbb{T})$, then all of the jump terms in (1.2.17) are zero for $l \leq r$, in particular this sum $= o(|n|^{-r-1})$. Conversely, if $\hat{f}(n) = o(|n|^{-r-1})$, then the same is true of the finite sum. Applying Lemma 1.2.18 repeatedly shows that $\delta f^{(l)}(\theta_j) \equiv 0$ for $l \leq r$, which proves that $f \in C^r(\mathbb{T})$. ∎

This corollary takes a particularly simple form in case $k = \infty$, i.e., f is piecewise C^∞. Within this class of functions we can simply state that $f \in C^r(\mathbb{T})$ if and only if $\hat{f}(n) = o(|n|^{-r-1})$, $|n| \to \infty$.

1.2.3.2 Fourier characterization of analytic functions

The Fourier coefficients of an analytic function can be characterized in terms of the exponential decay of the Fourier coefficients. Recall that a function is said to be *analytic* if it possesses a power series expansion about each point: $f(t) = \sum_{k=0}^{\infty} a_k(t - t_0)^k$, convergent in some interval $|t - t_0| < \delta$. From this, it is immediately concluded that $f(t)$ is infinitely differentiable and that the successive derivatives are obtained as $f^{(m)}(t_0) = m! a_m$. Since the series converges at $t = t_0$, the terms of the series must tend to zero, and in particular are bounded, from which we conclude that

(1.2.18) $\qquad |f^{(m)}(t)| \leq M m! R^m, \qquad t = t_0,\ n = 0, 1, \ldots, R > 1/\delta.$

Equation (1.2.18) is also valid in an open interval $|t - t_0| < \delta/3$, by noting the following system of estimates:

$$f^{(m)}(t) = \sum_{k=0}^{\infty}(k+1)\cdots(k+m)a_k(t-t_0)^k$$

$$\frac{f^{(m)}(t)}{m!} = \sum_{k=0}^{\infty}\binom{k+m}{m}a_{k+m}(t-t_0)^k$$

$$\left|\frac{f^{(m)}(t)}{m!}\right| \leq C\sum_{k=0}^{\infty}2^{k+m}R^{k+m}|t-t_0|^k$$

$$= \frac{C 2^m R^m}{1 - 2R|t - t_0|}$$

valid for $|t - t_0| < 1/3R$, which provides the desired uniform estimate. By covering $\mathbb{T}$ by a finite number of these intervals, we may assume that R, M are independent of t_0. Conversely, if an infinitely differentiable function $f(t)$, $t \in \mathbb{T}$ satisfies (1.2.18), then the series $\sum_{m=0}^{\infty} f^{(m)}(t_0)(t-t_0)^m/m!$ converges in the interval $|t-t_0| < 1/R$ and is therefore an analytic function. Summarizing, we see that f is analytic if and only if (1.2.18) holds.

In order to characterize analyticity in terms of the Fourier coefficients, first suppose that f is analytic on $\mathbb{T}$. In particular f is infinitely differentiable and we can integrate-by-parts to write for $n \neq 0$,

$$2\pi\hat{f}(n) = \int_{\mathbb{T}} e^{-int} f(t)\, dt$$

$$= \left(\frac{1}{in}\right)^m \int_{\mathbb{T}} e^{-int} f^{(m)}(t)\, dt$$

$$|\hat{f}(n)| \leq \frac{1}{|n|^m} M m! R^m.$$

Since this is valid for all m, we choose the optimal value $m = [n/R]$ and apply Stirling's formula to conclude that $|\hat{f}(n)| \leq M e^{-c|n|}$ for any $c < 1/R$.

Conversely, suppose that f is an infinitely differentiable function on $\mathbb{T}$ whose Fourier coefficients satisfy a system of inequalities of the form $|\hat{f}(n)| \leq A e^{-c|n|}$ for some positive constants A, c. In particular, by modifying f on a null set, we have the convergent Fourier series $f(t) = \sum_{n \in \mathbb{Z}} \hat{f}(n) e^{int}$, which can be differentiated m times to obtain $f^{(m)}(t) = \sum_{n \in \mathbb{Z}} (in)^m \hat{f}(n) e^{int}$. Applying the hypothesis, we have $|f^{(m)}(t)| \leq A \sum_{n \in \mathbb{Z}} |n|^m e^{-c|n|}$. Comparing this sum with the integral $\int_0^\infty x^m e^{-cx}\, dx$ shows that the successive derivatives satisfy the estimates $|f^{(m)}(t)| \leq A m!/c^m$.

Hence we obtain the following Fourier-analytic characterization of analytic functions on the circle.

Proposition 1.2.20. *$f \in L^1(\mathbb{T})$ is a.e. equal to an analytic function if and only if its Fourier coefficients satisfy the system of inequalities*

(1.2.19) $$|\hat{f}(n)| \leq A e^{-c|n|} \qquad n \in \mathbb{Z}$$

for positive constants c, A.

Exercise 1.2.21. *Carry out the details of the replacement of the sum by the integral.*

1.2.4 Sine Integral

In order to treat pointwise convergence of Fourier series and integrals in one dimension, a fundamental role is played by the function

(1.2.20) $$\mathrm{Si}(x) = \frac{2}{\pi} \int_0^x \frac{\sin t}{t}\, dt, \qquad 0 \le x < \infty.$$

Its basic properties are listed as follows:

(i) $\mathrm{Si}(0) = 0$, $\lim_{x \to \infty} \mathrm{Si}(x) = 1$.
(ii) $\mathrm{Si}(x) \le \mathrm{Si}(\pi) = 1.18\ldots$ for all $x \ge 0$.
(iii) $x \to \mathrm{Si}(x)$ has relative maxima at the points $\pi, 3\pi, 5\pi, \ldots$ and relative minima at the points $2\pi, 4\pi, 6\pi, \ldots$

To prove these properties, we first note that $\mathrm{Si}((n+1)\pi) - \mathrm{Si}(n\pi) = 2/\pi \int_{n\pi}^{(n+1)\pi} (\sin t/t)\, dt$ and that these numbers alternate in sign and decrease to zero in absolute value. Hence the improper integral defining $\lim_x \mathrm{Si}(x)$ exists. To compute its value, we note that by the Riemann-Lebesgue lemma,

$$\lim_n \int_{\mathbb{T}} \left[\frac{1}{\sin(\phi/2)} - \frac{2}{\phi} \right] \sin\left(n + \frac{1}{2}\right)\phi\, d\phi = 0.$$

But $\int_{\mathbb{T}} D_n(\phi)\, d\phi = 2\pi$, which shows that

$$\lim_n \int_0^\pi \frac{\sin\left(n+\frac{1}{2}\right)\phi}{\sin(\phi/2)}\, d\phi = \pi$$

or equivalently $\mathrm{Si}\left(n\pi + \frac{\pi}{2}\right) \to 1$ when $n \to \infty$.

1.2.4.1 Other proofs that $\mathrm{Si}(\infty) = 1$

We offer two other proofs of this important improper integral.

Proof using complex analysis. Consider the integral of the analytic function e^{iz}/z on the (counterclockwise) contour defined by the two semicircular arcs $z = \epsilon e^{i\theta}$, $0 \le \theta \le \pi$; $z = Re^{i\theta}$, $0 \le \theta \le \pi$; and the two segments of the real axis defined by $\epsilon \le |x| \le R$. By Cauchy's theorem the total line integral is zero. When $R \to \infty$, $\epsilon \to 0$, the integral on the large semicircle tends to zero, while the integral on the small semicircle is

$$-i \int_0^\pi e^{i\epsilon e^{i\theta}}\, d\theta \to -i\pi, \qquad \epsilon \to 0.$$

Hence we have

$$\lim_{\epsilon \to 0, R \to \infty} \int_{\epsilon < |x| < R} \frac{e^{ix}}{x}\, dx = i\pi,$$

from which we see that the improper integral $\int_{-\infty}^{\infty}(\sin x/x)\,dx = \pi$, from which it follows that $\mathrm{Si}(\infty) = 1$. ∎

Proof using real analysis. Define $S(t) = \int_0^t (\sin s)/s \, ds$, assumed to converge to a limit $S(\infty)$. For $x > 0$, define

$$G(x) = \int_0^\infty e^{-xt} \frac{\sin t}{t} dt = \int_0^\infty e^{-xt} dS(t), \qquad x > 0.$$

On the one hand we may differentiate under the integral for $x > 0$ to obtain

$$G'(x) = -\int_0^\infty e^{-xt} \sin t \, dt = -\frac{1}{1+x^2}, \qquad x > 0.$$

Hence $G(x) = C - \arctan x$ for some constant C. But the estimation $|G(x)| \leq 1/x$ shows that $0 = G(\infty) = C - \pi/2$, which shows that $C = G(0) = \pi/2$. On the other hand we can transform the integral defining $G(x)$ by a partial integration:

$$\int_0^M e^{-xt} dS(t) = e^{-xt} S(t)|_{t=0}^{t=M} + \int_0^M x e^{-xt} S(t) \, dt$$

$$\frac{\pi}{2} - \arctan x = G(x) = \int_0^\infty x e^{-xt} S(t) \, dt.$$

The latter integral tends to $S(\infty)$ when $x \to 0$. Hence $\mathrm{Si}(\infty) = 2/\pi \, S(\infty) = 2/\pi \, G(0) = 1$. ∎

1.2.5 Pointwise Convergence Criteria

We are now in a position to prove some criteria for the convergence of the partial sums of a Fourier series at a given point. All of the results described below will be in the form of *sufficient* conditions. It is not possible to formulate any effective necessary and sufficient conditions for the convergence, as we will discuss below.

The first step is to recall that the partial sum is expressed as a convolution with the Dirichlet kernel:

$$(1.2.21) \qquad S_N f(\theta) = (f * D_N)(\theta) = \frac{1}{2\pi} \int_\mathbb{T} \frac{\sin\left(N + \frac{1}{2}\right)(\phi)}{\sin(\phi/2)} f(\theta - \phi) \, d\phi.$$

This formula will be simplified in two ways. First, we will show that the factor $\sin(\phi/2)$ in the denominator can be replaced by the simpler function $\phi \to \phi/2$. Secondly we will show that the integral over the circle $\mathbb{T}$ can be replaced by the integral over a small interval about $\phi = 0$. The details follow.

The Dirichlet kernel is an even function and satisfies the normalization

$$\frac{1}{2\pi} \int_\mathbb{T} D_N(\theta) \, d\theta = 1,$$

so that we can write

$$(1.2.22) \qquad S_N f(\theta) = \frac{1}{2\pi} \int_0^\pi [f(\theta + \phi) + f(\theta - \phi)] \frac{\sin\left(N + \frac{1}{2}\right)(\phi)}{\sin(\phi/2)} d\phi,$$

and hence for any constant S

$$(1.2.23) \quad S_N f(\theta) - S = \frac{1}{2\pi} \int_0^\pi [f(\theta + \phi) + f(\theta - \phi) - 2S] \frac{\sin\left(N + \frac{1}{2}\right)(\phi)}{\sin(\phi/2)} d\phi.$$

As a first reduction, we can replace the function $1/\sin(\phi/2)$ by the function $2/\phi$ with an error that tends to zero, uniformly in θ. This comes from the fact that

$$\phi \to \frac{1}{\sin(\phi/2)} - \frac{2}{\phi}$$

is bounded and continuous on the interval $[-\pi, \pi]$. The only possible difficulty is at $\phi = 0$, where we can apply l'Hospital's rule to show that the difference tends to zero.

Exercise 1.2.22. *Show that*

$$\left| \frac{1}{\sin x} - \frac{1}{x} \right| \leq \frac{\pi^2}{24}, \quad -\frac{\pi}{2} \leq x \leq \frac{\pi}{2}.$$

Hint: First show that $|x - \sin x| \leq |x|^3/6$ and then simplify the fractions.

For each θ, the function

$$\phi \to F_\theta(\phi) = \left[\frac{1}{\sin(\phi/2)} - \frac{2}{\phi} \right] (f(\theta + \phi) + f(\theta - \phi))$$

is an L^1 function, and the map $\theta \to F_\theta$ is continuous from $(-\pi, \pi)$ to $L^1(\mathbb{T})$. Hence by the Riemann-Lebesgue lemma the integral of $F_\theta(\phi) \sin(N + 1/2)\phi$ tends to zero uniformly in θ when $n \to \infty$.

Thus we have reduced the problem of pointwise convergence to proving that

$$(1.2.24) \quad \lim_N \int_0^\pi [f(\theta + \phi) + f(\theta - \phi) - 2S] \frac{\sin\left(N + \frac{1}{2}\right)\phi}{\phi} d\phi = 0.$$

For the second simplification, we consider the contribution to (1.2.24) from the interval $\delta \leq \phi \leq \pi$. Noting that the function $\phi \to [f(\theta + \phi) + f(\theta - \phi)]/\phi$ is integrable on the interval $[\delta, \pi]$, therefore by the Riemann-Lebesgue lemma this contribution to (1.2.24) tends to zero when $N \to \infty$. We summarize this as the following necessary and sufficient condition for the convergence of a Fourier series.

Theorem 1.2.23. *Let $f \in L^1(\mathbb{T})$. A necessary and sufficient condition that the partial sums $S_N f(\theta)$ converge to a limit S when $N \to \infty$ is that for some $\delta \in (0, \pi)$,*

$$(1.2.25) \quad \lim_N \int_0^\delta [f(\theta + \phi) + f(\theta - \phi) - 2S] \frac{\sin\left(N + \frac{1}{2}\right)\phi}{\phi} d\phi = 0.$$

One may note that if (1.2.25) holds for some $\delta \in (0, \pi)$, then it holds for *every* $\delta \in (0, \pi)$, since the difference of two such integrals is an integral over an interval $\delta_1 \leq \phi \leq \delta_2$, which tends to zero by the Riemann-Lebesgue lemma.

We can now state and prove Dini's theorem:

Theorem 1.2.24. *Suppose that f satisfies a Dini condition at θ, meaning that for some $\delta > 0$ and some real number S*

$$\int_0^\delta \frac{|f(\theta + \phi) + f(\theta - \phi) - 2S|}{\phi} \, d\phi < \infty.$$

Then $\lim_N S_N f(\theta) = S$.

Proof. This is an immediate application of Theorem 1.2.23 and the Riemann-Lebesgue lemma. ∎

Specific conditions that imply the Dini condition can be obtained from a symmetric form of the Hölder condition, as follows:

Corollary 1.2.25. *Suppose that f satisfies a symmetric Hölder condition at θ, in the form*

$$|f(\theta + \phi) + f(\theta - \phi) - 2f(\theta)| \leq C|\phi|^\alpha$$

for $0 < \phi < \delta$, where $0 < \alpha < 1$. Then $\lim_N S_N f(\theta) = f(\theta)$.

If f satisfies a Dini condition at θ with $S = f(\theta)$, we say that f is *normalized*. The following exercise applies to functions defined on the entire real line, not necessarily periodic.

Exercise 1.2.26. *Let $f \in L^1_{\text{loc}}(\mathbb{R})$ satisfy a Dini condition at $x = 0, \pm 2\pi, \ldots$, where it is normalized. Then if $M \notin \mathbb{Z}^+$,*

$$\lim_{N \to \infty} \frac{1}{2\pi} \int_{-M}^{M} D_N(x) f(x) \, dx = \sum_{m \in \mathbb{Z}, |m| \leq M} f(2\pi m)$$

whereas for $M \in \mathbb{Z}^+$ we must add $\frac{1}{2} f(2\pi M) + \frac{1}{2} f(-2\pi M)$ to the right side.

The next result is historically the first theorem on convergence of Fourier series, attributed to Dirichlet (1829) for monotone functions and to Jordan (1881) for the general case. We will use the properties of Si (x).

Theorem 1.2.27. *Suppose that f is of bounded variation on $[\theta - \delta, \theta + \delta]$ for some $\delta > 0$. Then $\lim_N S_N f(\theta) = \frac{1}{2}[f(\theta + 0) + f(\theta - 0)]$.*

Proof. This is done by integration by parts: We write $m = N + 1/2$, $F(\phi) = \frac{1}{2}f(\theta + \phi) + \frac{1}{2}f(\theta - \phi)$, $S = F(0+0)$; from (1.2.24) and the remarks that follow, we have

$$S_N f(\theta) - S = \frac{2}{\pi} \int_0^\delta \left(\frac{f(\theta + \phi) + f(\theta - \phi)}{2} - S \right) \frac{\sin m\phi}{\phi} d\phi + o(1)$$

(1.2.26)
$$= \int_0^\delta (F(\phi) - S) \, d\text{Si}(m\phi) + o(1)$$

$$= [F(\delta - 0) - S]\text{Si}(m\delta) - \int_0^\delta \text{Si}(m\phi) \, dF(\phi) + o(1),$$

where the final integration is with respect to the finite measure defined by this function of bounded variation. Now we apply the dominated convergence theorem and the properties of Si (x) to conclude that

$$\lim_N [S_N f(\theta) - S] = [F(\delta - 0) - S] - [F(\delta - 0) - S] = 0,$$

which was to be proved. ■

Since every absolutely continuous function is of bounded variation, we obtain the following corollary.

Corollary 1.2.28. *If f is absolutely continuous, then the Fourier series converges to f everywhere.*

It should be noted that the Dini condition and the condition of bounded variation are not comparable with one another. For example, the function $\phi \to 1/\log(1/\phi)$ is of bounded variation but does not satisfy a Dini condition. On the other hand, the function $\phi \to \phi \sin(1/\phi)$ satisfies a Dini condition but is not of bounded variation. Finally we remark on the difficulty of finding necessary conditions. The formula (1.2.22) shows that the Fourier partial sum at θ depends on the symmetrized function $\phi \to f(\theta + \phi) + f(\theta - \phi)$. If this function is identically zero, then we have $\lim_N S_N f(\theta) = 0$ regardless of whatever other behavior is present. In particular any odd function ($f(-x) = -f(x)$) has a convergent Fourier series at $\theta = 0$.

The technique used in the proof of Theorem 1.2.27 can also be used to prove the uniform boundedness of the partial sums of the Fourier series of a function of bounded variation. To do this, write

$$2\pi S_N f(\theta) = \int_0^\pi \left[\frac{1}{\sin(\phi/2)} - \frac{2}{\phi} \right] \sin m\phi \, [f(\theta + \phi) + f(\theta - \phi)] \, d\phi$$

$$+ \int_0^\pi \frac{\sin m\phi}{\phi/2} [f(\theta + \phi) + f(\theta - \phi)] \, d\phi.$$

The first term is less than $(\pi^3/12) \max |f|$, while the second term can be integrated by parts in terms of the Sine Integral:

$$\frac{2}{\pi} \int_0^\pi \frac{\sin m\phi}{\phi} f(\theta \pm \phi) \, d\phi = f(\theta \pm \pi)\text{Si}(m\pi) - \int_0^\pi \text{Si}(m\phi) \, df(\theta \pm \phi) \, d\phi.$$

The first term is bounded by Si $(\pi) \max |f|$ and the second term is bounded by Si (π) Var f. We summarize as follows:

Proposition 1.2.29. *If f is of bounded variation on $\mathbb{T}$, then the partial sums of the Fourier series are uniformly bounded, in the form*

$$|S_n f(\theta)| \leq C_1 \max_{\mathbb{T}} |f| + C_2 \operatorname{Var} f$$

where C_1, C_2 are universal constants and $\operatorname{Var} f$ denotes the total variation of the signed measure defined by f.

Exercise 1.2.30. Find values for the constants C_1, C_2.

Exercise 1.2.31. Suppose that f is of bounded variation and continuous in a neighborhood of θ. Prove that the partial sums converge uniformly in a closed interval containing θ.

Hint: Without loss of generality assume f monotone increasing. Reconsider the representation (1.2.26) of the partial sums and break up the integral into $\int_0^\delta$ and $\int_\delta^\pi$. Apply the mean value theorem for integrals to each term and recall that $\operatorname{Si}(x) \leq \operatorname{Si}(\pi)$ for all x.

1.2.6 *Integration of Fourier Series

In many applications of Fourier series, f represents the density of mass, charge, or probability. In such situations, the quantity of interest is the integral of f over an interval. It is reassuring that, at this level, there are no obstructions to convergence.

Theorem 1.2.32. *Suppose $f \in L^1(\mathbb{T})$. Then for any interval (a, b) we have*

$$(1.2.27) \qquad \lim_n \int_a^b (S_n f)(\theta)\, d\theta = \int_a^b f(\theta)\, d\theta.$$

Proof. The partial sum on the left is given by the convolution with the Dirichlet kernel, an even kernel that defines a self-adjoint operator. Thus

$$\int_a^b (S_n f)(\theta)\, d\theta = \int_{\mathbb{T}} (S_n f)(\theta) 1_{(a,b)}(\theta)\, d\theta = \int_{\mathbb{T}} f(\theta)(S_n 1_{(a,b)})(\theta)\, d\theta.$$

Since $1_{(a,b)}$ is of bounded variation, $S_n 1_{(a,b)}$ converges boundedly to $1_{(a,b)}$ except at the endpoints $x = a, b$, which have Lebesgue measure zero. Applying the dominated convergence theorem completes the proof. ∎

Theorem 1.2.32 can also be proved by considering the integral

$$F(\theta) = \int_{-\pi}^{\theta} (f(\phi) - \hat{f}(0))\, d\phi.$$

This function is absolutely continuous with $F(-\pi) = 0 = F(\pi)$, $F'(\theta) = f(\theta) - \hat{f}(0)$, so that $in\hat{F}(n) = \hat{f}(n) - \delta_{n0}\hat{f}(0)$. Hence the Fourier coefficients of F are given by $\hat{F}(n) = \hat{f}(n)/in$ for $n \neq 0$ and we have the everywhere convergent Fourier series

(1.2.28) $$F(\theta) = \hat{F}(0) + \sum_{0 \neq n \in \mathbb{Z}} \frac{\hat{f}(n)}{in} e^{in\theta}.$$

In particular, for any $(a, b) \subset [-\pi, \pi]$, we have

$$F(b) - F(a) = \sum_{0 \neq n \in \mathbb{Z}} \frac{\hat{f}(n)}{in} (e^{inb} - e^{ina})$$

$$= \sum_{0 \neq n \in \mathbb{Z}} \hat{f}(n) \int_a^b e^{in\theta}\, d\theta$$

or equivalently

$$\int_a^b f(\theta)\, d\theta - (b-a)\hat{f}(0) = \lim_N \int_a^b [S_N f(\theta) - \hat{f}(0)]\, d\theta$$

which reduces to (1.2.27).

Another by-product of the integration identity (1.2.28) is the following necessary condition on the Fourier coefficients of an L^1 function.

Corollary 1.2.33. *For any $f \in L^1(\mathbb{T})$, the series $\sum_{0 \neq n \in \mathbb{Z}} \hat{f}(n)/n$ converges.*

This corollary can be used to manufacture trigonometric series that are not Fourier series.

Exercise 1.2.34. *Prove that $\sum_{n \geq 2} (\sin n\theta / \log n)$ is not the Fourier series of an L^1 function.*

Hint: Identify $\hat{f}(n)$ as a suitable odd function.

Later we will prove that the series $\sum_{n \geq 2} (\cos n\theta / \log n)$ is the Fourier series of an integrable function.

Exercise 1.2.35. *Suppose that g is a function of bounded variation on $\mathbb{T}$. Prove that for any $f \in L^1(\mathbb{T})$,*

$$\lim_{N \to \infty} \int_\mathbb{T} g(\theta) S_N f(\theta)\, d\theta = \int_\mathbb{T} g(\theta) f(\theta)\, d\theta.$$

1.2.6.1 Convergence of Fourier series of measures

The above ideas can also be used to give a quick treatment of the convergence of Fourier series of any finite signed measure on $\mathbb{T}$. The partial sum of its Fourier series is written

$$S_n \mu(\theta) = \sum_{-n}^{n} \hat{\mu}(k) e^{ik\theta}.$$

Proposition 1.2.36. *Suppose that μ is a finite signed measure on $\mathbb{T}$. Then*

$$\lim_N \int_a^b S_N \mu(\theta)\, d\theta = \mu((a,b)) + \frac{1}{2}\mu(\{a\}) + \frac{1}{2}\mu(\{b\}).$$

Proof. The partial sum on the left is written in terms of the convolution with the Dirichlet kernel. Thus

$$\int_a^b (S_N \mu)(\theta)\, d\theta = \int_{\mathbb{T}} (S_N \mu)(\theta) 1_{(a,b)}(\theta)\, d\theta = \int_T (S_N 1_{(a,b)})(\theta) \mu(d\theta).$$

Since $1_{(a,b)}$ is of bounded variation, $S_N 1_{(a,b)}$ converges boundedly to $1_{(a,b)} + \frac{1}{2} 1_{\{a\}} + \frac{1}{2} 1_{\{b\}}$. Applying the dominated convergence theorem completes the proof. ∎

1.2.7 Riemann Localization Principle

Fourier series in one dimension have the property that the limiting behavior of the partial sums at a point depends only on the values of the function in a neighborhood of the point, no matter how small. This is expressed as follows.

Proposition 1.2.37. *Suppose that $f \in L^1(\mathbb{T})$ is identically zero in an open interval (a,b). Then for any compact subinterval the Fourier partial sums tend uniformly to zero when $n \to \infty$.*

Proof. From formula (1.2.24) we have

(1.2.29) $$S_N f(\theta) = o(1) + \frac{1}{\pi} \int_0^\pi [f(\theta + \phi) + f(\theta - \phi)] \frac{\sin\left(N + \frac{1}{2}\right)\phi}{\phi}\, d\phi.$$

If $[a_1, b_1]$ is a subinterval of (a,b), let $2\delta = \min(a_1 - a, b - b_1)$. By hypothesis $f(\theta + \phi) + f(\theta - \phi) = 0$ if $\theta \in [a_1, b_1], \phi < \delta$. Hence the integrand is identically zero when $\theta \in [a_1, b_1], \phi \le \delta$. On the other hand, the integral on $[\delta, \pi]$ tends to zero uniformly when $\theta \in [a_1, b_1]$, by the Riemann-Lebesgue lemma. ∎

The Riemann localization principle allows us to infer that if two functions agree on an interval, then the Fourier series are *equiconvergent* meaning that $\lim_n (S_n f_1 - S_n f_2) = 0$ on that interval. This phenomenon is no longer present in higher dimensional Fourier analysis, as we shall see.

1.2.8 Gibbs-Wilbraham Phenomenon

In the neighborhood of a discontinuity one cannot expect uniform convergence of the Fourier partial sums. The specific form of nonuniform convergence is best illustrated by the example

(1.2.30) $$f(x) = (\pi - x)/2\pi, \quad 0 < x < \pi,$$
$$f(x) = -(\pi + x)/2\pi, \quad -\pi < x < 0.$$

This is the simplest function of bounded variation that has a single jump of unit size. Now f is an odd function whose Fourier coefficients are

$$\hat{f}(n) = \frac{1}{2\pi} \int_{\mathbb{T}} f(x) e^{-inx} \, dx$$

$$= \frac{-i}{\pi} \int_0^\pi \frac{\pi - x}{2\pi} \sin nx \, dx$$

$$= \frac{-i}{2n\pi}$$

so that the partial sum of the Fourier series is

$$S_N f(x) = \frac{1}{\pi} \left[\sin x + \cdots + \frac{\sin Nx}{N} \right].$$

This may be written as a definite integral by first computing the derivative

$$(S_N f)'(x) = \frac{1}{\pi} [\cos x + \cdots + \cos Nx]$$

$$= \frac{1}{2\pi} (D_N(x) - 1).$$

Thus

(1.2.31) $\quad S_N f(x) = \frac{1}{2\pi} \int_0^x [D_N(t) - 1] \, dt = \frac{-x}{2\pi} + \frac{1}{2\pi} \int_0^x \frac{\sin(N + 1/2)t}{\sin(t/2)} \, dt.$

Defining $g(t) = 1/\sin(t/2) - 2/t$, we have by a single integration-by-parts,

$$\int_0^x g(t) \sin\left(N + \tfrac{1}{2}\right) t \, dt = O\left(\frac{1}{N}\right), \qquad N \to \infty$$

uniformly for $0 \le x \le \pi$, since g is a C^1 function on the interval $[0, \pi]$. Thus

(1.2.32) $\quad S_N f(x) = \frac{-x}{2\pi} + \frac{1}{\pi} \int_0^x \frac{\sin(N + \tfrac{1}{2})t}{t} \, dt + O\left(\frac{1}{N}\right).$

If $0 < x < \pi$, we see clearly that

$$\lim_N S_N f(x) = \frac{-x}{2\pi} + \frac{1}{\pi} \int_0^\infty \frac{\sin t}{t} \, dt = \frac{-x}{2\pi} + \frac{1}{2} = f(x).$$

To study the behavior when $x \to 0$, we note that on the one hand

$$\sup_{0 \le x \le \pi} \left[S_N f(x) + \frac{x}{2\pi} \right] \le \frac{1}{2} \sup_{0 \le x < \infty} \operatorname{Si}(x) + O\left(\frac{1}{N}\right) = \frac{1}{2} \operatorname{Si}(\pi) + O\left(\frac{1}{N}\right),$$

which gives an upper bound to the fluctuations of the partial sums. On the other hand, if $x_N \to 0$ so that $N x_N \to \pi$, then

$$\lim_N S_N f(x_N) = \lim_N \frac{1}{\pi} \int_0^{(N+\tfrac{1}{2})x_N} \frac{\sin t}{t} \, dt = \frac{1}{2} \operatorname{Si}(\pi).$$

These computations are summarized as follows.

Proposition 1.2.38. *The set of accumulation points of the partial sums $S_N f(x_N)$ when $N \to \infty$, $x_N \to 0$ with $0 \leq x_N \leq \pi$ is described as follows:*

$$\limsup_N S_N f(x_N) \leq \tfrac{1}{2}\operatorname{Si}(\pi) = 0.59\ldots. \tag{1.2.33}$$

If $x_N \to 0$ so that $N x_N \to \pi$, then

$$\lim_N S_N f(x_N) = \tfrac{1}{2}\operatorname{Si}(\pi). \tag{1.2.34}$$

In particular, any point in the interval $[0, \tfrac{1}{2}\operatorname{Si}(\pi)]$ is an accumulation point of partial sums $S_N f(x_N)$ when $N \to \infty$, $x_N \to 0$.

Proof. We have shown the first two statements above. For the final statement, apply the intermediate value theorem to the continuous function $S_N f$. ∎

The number $\tfrac{1}{2}\operatorname{Si}(\pi) - \tfrac{1}{2} = .09\ldots$ is called the *overshoot* of the partial sums in the right neighborhood of $x = 0$. We will show below that for any function of bounded variation with $f(x+0) > f(x-0)$ we can discuss the overshoot in a right neighborhood of x, defined as $\limsup_{N \to \infty, x_N \to x} S_N f(x_N) - f(x+0)$, which will be proportional to the jump $f(x+0) - f(x-0)$.

The numerical value of $\operatorname{Si}(\pi) = 1.18\ldots$ can be computed by expanding $\sin t/t$ in a Taylor series and integrating term-by-term. This is carried out in Section 2.3.3.

Exercise 1.2.39. *Compute $\lim_N S_N f(k\pi/N + \tfrac{1}{2})$ for $k = 2, 3, \ldots$*

Strichartz (2000) discovered a corresponding behavior with respect to the arc length of the curve $\{(x, S_N f(x)), -\pi \leq x \leq \pi\}$. The proof anticipates the behavior of the *Lebesgue constants*, to be studied in Section 1.5.

Proposition 1.2.40. *Let f be defined by (1.2.30). Then when $N \to \infty$ we have*

$$\int_{-\pi}^{\pi} \sqrt{1 + [(S_N f)'(x)]^2}\, dx = C_1 \log N + O(1),$$

where C_1 is a positive constant.

Proof. From the computations following (1.2.30) we have $(S_N f)'(x) = (1/2\pi)(D_N(x) - 1)$, so that

$$\sqrt{1 + [(S_N f)'(x)]^2} - |(S_N f)'(x)| = \frac{1}{\sqrt{1 + [(S_N f)'(x)]^2} + |(S_N f)'(x)|}$$

$$\leq 1 \tag{1.2.35}$$

so that the error term in (1.2.35) is $O(1)$, $N \to \infty$. But Proposition 1.5.1 also shows that $\int_{-\pi}^{\pi} |\cos x + \cdots + \cos Nx|\, dx = C_1 \log N + O(1)$ for a positive constant C_1. ∎

Exercise 1.2.41. *Suppose that $f(x)$, $-\pi \leq x \leq \pi$ is a piecewise C^1 function and that f is continuous on $\mathbb{T}$. Prove that when $N \to \infty$ we have*

$$\int_{-\pi}^{\pi} \sqrt{1 + [(S_N f)'(x)]^2}\, dx \to \int_{-\pi}^{\pi} \sqrt{1 + [f'(x)]^2}\, dx.$$

1.2.8.1 The general case

For a general $f \in L^1(\mathbb{T})$, the Gibbs-Wilbraham phenomenon at a point $\phi \in \mathbb{T}$ describes the set of accumulation points of $S_N f(\phi_N)$ when $N \to \infty$ and $\phi_N \to \phi$. In the above example it is seen that the set of accumulation points consists of the interval $(-0.59\ldots, 0.59\ldots)$, whereas if $\phi \neq 0$, then $S_N f \to f$ uniformly in a neighborhood of ϕ and the set of accumulation points consists of the single point $f(\phi)$.

To study the Gibbs-Wilbraham phenomenon more generally, let $f \in L^1(\mathbb{T})$ be a function of bounded variation with discontinuities labeled $\theta_1, \theta_2, \ldots$. Let the jumps be denoted $\delta f(\theta_i) := f(\theta_i + 0) - f(\theta_i - 0)$. Let $J(\theta) = (\pi - \theta)/2\pi$ for $0 < \theta < \pi$ and extended periodically to the entire line. Then J has a unit jump at $\theta = 0$, which will be used to study the general case. Form the function

$$(1.2.36) \qquad f_{\text{jump}}(\theta) := \sum_{i=1}^{\infty} (\delta f)(\theta_i) J_{\theta_i}(\theta) := \sum_{i=1}^{\infty} (\delta f)(\theta_i) J(\theta - \theta_i).$$

The series (1.2.36) is uniformly convergent to a function of bounded variation whose discontinuitites are precisely the points $\{\theta_i\}$. The function $\tilde{f} := f - f_{\text{jump}}$ is a continuous function of bounded variation, since all of its jumps are zero. From Exercise 1.2.31, the Fourier series of $\tilde{f}$ is uniformly convergent. Therefore the analysis of the Gibbs-Wilbraham phenomenon for f is reduced to that of f_{jump}. Given $\epsilon > 0$, let M be so large that $\sum_{i=M}^{\infty} |(\delta f)(\theta_i)| \|S_N J_{\theta_i}\|_{\infty} < \epsilon$, which is possible since the sum $\sum_{i=1}^{\infty} |\delta f(\theta_i)| < \infty$ and the partial sums $S_N J_{\theta_i}$ are uniformly bounded by a constant for all i, N. To study the Gibbs-Wilbraham phenomenon near θ_1, let $\delta f(\theta_1) > \phi$ and write

$$S_N f(\theta) \leq (\delta f)(\theta_1) S_N J_{\theta_1}(\theta) + \sum_{i=2}^{M} (\delta f)(\theta_i) S_N J_{\theta_i}(\theta) + S_N \tilde{f}(\theta) + \epsilon.$$

From Exercise 1.2.31, the finite sum is uniformly convergent in a closed interval about θ_1. The Fourier series of $\tilde{f}$ is uniformly convergent. Therefore when $\phi_N \to \theta_1$, we have

$$S_N f(\phi_N) \leq (\delta f)(\theta_1) S_N J_{\theta_1}(\phi_N) + [f(\theta_1 + 0) - \tfrac{1}{2}(\delta f)(\theta_1)] + \epsilon.$$

Letting $\phi_N \to \theta_1$, we have

$$\limsup_N S_N f(\phi_N) \leq \tfrac{1}{2}(\delta f)(\theta_1) \text{Si}(\pi) + [f(\theta_1 + 0) - \tfrac{1}{2}(\delta f)(\theta_1)] + \epsilon,$$

$$\liminf_N S_N f(\phi_N) \geq [f(\theta_1 + 0) - \tfrac{1}{2}(\delta f)(\theta_1)] - \epsilon.$$

But ϵ was arbitrary. The same computation applies to any of the discontinuity points θ_i. We summarize as follows:

Proposition 1.2.42. *Let $f \in L^1(\mathbb{T})$ be a function of bounded variation with a simple discontinuity at $\phi \in (-\pi, \pi)$, where $(\delta f)(\phi) > 0$. The set of accumulation points of the partial sums $S_N f(x_N)$ when $N \to \infty$ and $x_N \to \phi$ with $\phi < x_N < \pi$ is described as follows:*

$$(1.2.37) \qquad \limsup_N S_N f(x_N) \leq f(\phi + 0) + \frac{\text{Si}(\pi) - 1}{2} (\delta f)(\phi).$$

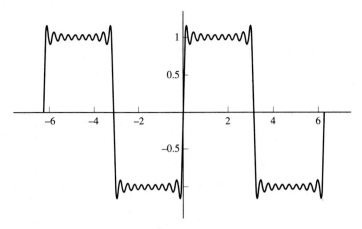

FIGURE 1.2.3
The Gibbs-Wilbraham phenomenon for the function $f(\theta) = \text{sgn}(\theta)$.
From M. Pinsky, *Partial Differential Equations and Boundary-Value Problems with Applications*. Used by the permission of The McGraw-Hill Companies.

If $x_N \to \phi$ so that $N(x_N - \phi) \to \pi$, then

$$(1.2.38) \qquad \lim_N S_N f(x_N) = f(\phi + 0) + \frac{\text{Si}(\pi) - 1}{2} (\delta f)(\phi).$$

In particular, any point in the interval $[\frac{1}{2}(f(\phi + 0) + f(\phi - 0)), f(\phi + 0) + (\text{Si}(\pi) - 1)/2)(\delta f)(\phi)]$ is an accumulation point of partial sums $S_N f(x_N)$ for some sequence x_N when $N \to \infty$, $x_N \to \phi$ with $x_N > \phi$.

1.3 FOURIER SERIES IN L^2

1.3.1 Mean Square Approximation—Parseval's Theorem

Fourier series are well adapted to deal with the Euclidean geometry of the space $L^2(\mathbb{T})$ where the inner product is defined by

$$(1.3.1) \qquad \langle f, g \rangle = \frac{1}{2\pi} \int_{\mathbb{T}} f(\theta) \bar{g}(\theta) \, d\theta.$$

We can measure the degree of approximation by the mean square distance

$$\|f - g\|_2^2 = \langle f - g, f - g \rangle.$$

In particular, if $g(\theta) = \sum_{n=-N}^{N} b_n e^{in\theta}$ is a trigonometric polynomial then

$$\|f - g\|_2^2 = \|f\|_2^2 - \langle f, g \rangle - \langle g, f \rangle + \|g\|_2^2$$

$$= \|f\|_2^2 - \sum_{n=-N}^{N} \left(\hat{f}(n) \bar{b}_n + \bar{\hat{f}}(n) b_n - |b_n|^2 \right)$$

$$= \sum_{n=-N}^{N} |b_n - \hat{f}(n)|^2 + \|f\|_2^2 - \sum_{n=-N}^{N} |\hat{f}(n)|^2.$$

This has the following consequences.

Proposition 1.3.1. *Suppose that $f \in L^2(\mathbb{T})$. Then the minimum mean square error is attained when b_n is the Fourier coefficient $b_n = \hat{f}(n)$. The mean square distance is given by*

$$(1.3.2) \qquad \|f - g\|_2^2 = \|f\|_2^2 - \sum_{n=-N}^{N} |\hat{f}(n)|^2.$$

In particular we have for each $N \in \mathbb{Z}^+$ the inequality

$$\sum_{n=-N}^{N} |\hat{f}(n)|^2 \leq \|f\|_2^2$$

and in particular Bessel's inequality:

$$\sum_{n \in \mathbb{Z}} |\hat{f}(n)|^2 \leq \|f\|_2^2.$$

The space $L^2(\mathbb{T})$ is distinguished in the theory of Fourier series because of this characterization of the Fourier partial sums. If we consider optimal trigonometric approximation in the space $C(\mathbb{T})$ of continuous functions or the space $L^p(\mathbb{T}), p \neq 2$, then the Fourier partial sum no longer provides the best approximation in norm.

Fourier series in $L^2(\mathbb{T})$ have the remarkable property that Bessel's inequality is in fact true with the sign of equality. This is Parseval's identity, stated as a theorem.

Theorem 1.3.2. *Parseval's theorem: For any $f \in L^2(\mathbb{T})$, the Fourier series converges in $L^2(\mathbb{T})$ and we have Parseval's identity*

$$(1.3.3) \qquad \frac{1}{2\pi} \int_{-\pi}^{\pi} |f(\theta)|^2 \, d\theta = \sum_{n \in \mathbb{Z}} |\hat{f}(n)|^2.$$

Proof. We first show that the partial sums converge in the norm of $L^2(\mathbb{T})$. We have

$$\|S_N f - S_M f\|^2 = \frac{1}{2\pi} \int_{\mathbb{T}} |S_N f - S_M f|^2 = \sum_{M+1}^{N} |\hat{f}(n)|^2 \to 0 \qquad M, N \to \infty$$

by the Bessel inequality. But the space $L^2(\mathbb{T})$ is complete, hence there exists $F = \lim_N S_N f$ in the metric of $L^2(\mathbb{T})$. It remains to show that $F = f$ a.e. To do this, we compute the Fourier coefficient by writing

$$2\pi \hat{F}(n) = \int_{\mathbb{T}} [F(\theta) - S_N f(\theta)] e^{-in\theta} \, d\theta + \int_{\mathbb{T}} S_N f(\theta) e^{-in\theta} \, d\theta.$$

If $N > |n|$, then the last integral $= 2\pi \hat{f}(n)$. Applying Cauchy-Schwarz, we have

$$|\hat{F}(n) - \hat{f}(n)| \leq \|F - S_N f\|_2 \qquad (N > n).$$

Letting $N \to \infty$ shows that $\hat{F}(n) = \hat{f}(n)$ for all $n \in \mathbb{Z}$. Hence $F = f$ a.e. and we have proved that $S_n f \to f$ in the norm of $L^2(\mathbb{T})$. In particular the L^2 norms converge and we have

$$\|f\|_2^2 = \lim_N \sum_{n=-N}^{N} |\hat{f}(n)|^2,$$

which is the statement of Parseval's equality. ∎

Exercise 1.3.3. Suppose that f is absolutely continuous with $f' \in L^2(\mathbb{T})$. Use Parseval's identity to prove that $\|f'\|_2^2 = \sum_{n \in \mathbb{Z}} n^2 |\hat{f}(n)|^2$.

Exercise 1.3.4. Suppose that f is absolutely continuous with $f' \in L^2(\mathbb{T})$. Use Parseval's identity to prove that the Fourier series converges absolutely: $\sum_{n \in \mathbb{Z}} |\hat{f}(n)| < \infty$.

Hint: Apply the Cauchy-Schwarz inequality.

Theorem 1.3.2 shows that the mapping $f \to \hat{f}$, from $L^2(\mathbb{T})$ to $L^2(\mathbb{Z})$ preserves the respective L^2 norms. The next proposition shows that the mapping $f \to \hat{f}$ is onto the set of square-summable sequences. Put otherwise, we have an isomorphism between $L^2(\mathbb{T})$ and $L^2(\mathbb{Z})$, the set of square-summable sequences.

Theorem 1.3.5. *Riesz-Fischer theorem:* Suppose that $\{c_n\}, n \in \mathbb{Z}$ is a bilateral sequence of complex numbers with $\sum_{n \in \mathbb{Z}} |c_n|^2 < \infty$. Then there is a unique $f \in L^2(\mathbb{T})$ with $\hat{f}(n) = c_n$ for all $n \in \mathbb{Z}$.

Proof. Let $T_n(\theta) = \sum_{m=-n}^{n} c_m e^{im\theta}$. Then T_n is a Cauchy sequence in $L^2(\mathbb{T})$ since for $M < N$, $\|T_N - T_M\|_2^2 = \sum_{M < |m| \leq N} |c_m|^2 \to 0$ when $M, N \to \infty$. By the completeness of $L^2(\mathbb{T})$, there exists an L^2 limit f, where $\|f - T_n\|_2 \to 0$ when $n \to \infty$. The Fourier coefficients of f are obtained by writing

$$2\pi \hat{f}(n) = \int_{\mathbb{T}} f(\theta) e^{-in\theta} d\theta$$
$$= \int_{\mathbb{T}} [f(\theta) - T_N(\theta)] e^{-in\theta} d\theta + \int_{\mathbb{T}} T_N(\theta) e^{-in\theta} d\theta.$$

The absolute value of the first term is less than or equal to $2\pi \|f - T_N\|_2$. The second term is $2\pi \times$ the Fourier coefficient of the trigonometric polynomial T_N; if $N > |n|$ this is equal to c_n. Thus we conclude

$$|\hat{f}(n) - c_n| \leq \|f - T_N\|_2 \quad \text{for } N > |n|.$$

Taking $N \to \infty$ completes the proof of existence. The uniqueness of f follows from the uniqueness of Fourier coefficients in the space $L^1(\mathbb{T})$. ∎

Theorem 1.3.5 can be used to identify the smoothness of a function from the Fourier coefficients, as follows.

Proposition 1.3.6. Suppose that $f \in L^2(\mathbb{T})$ has Fourier coefficients that satisfy $\sum_{n \in \mathbb{Z}} n^2 |\hat{f}(n)|^2 < \infty$. Then f is a.e. equal to an absolutely continuous function, with $f' \in L^2(\mathbb{T})$ with Fourier series $f'(\theta) \sim \sum_{n \in \mathbb{Z}} in\hat{f}(n) e^{in\theta}$.

Proof. By Theorem 1.3.5, $\sum_{n \in \mathbb{Z}} in\hat{f}(n)e^{in\theta}$ is the Fourier series of an L^2 function g. Furthermore $\int_{\mathbb{T}} g(\theta) d\theta = 0$. Let $F(\theta) = \int_{-\pi}^{\theta} g(\phi) d\phi$. This is an absolutely continuous function on T with $F' = g$ almost everywhere. Therefore the Fourier coefficients of F are obtained from $\hat{g}(n) = in\hat{F}(n)$. But $\hat{g}(n) = in\hat{f}(n)$, from which we conclude that $\hat{F}(n) = \hat{f}(n)$ for $n \neq 0$. Hence $F - f$ is a constant a.e. In particular f is a.e. equal to an absolutely continuous with $f' = g \in L^2(\mathbb{T})$. ∎

Exercise 1.3.7. *Suppose that $f \in L^2(\mathbb{T})$ has Fourier coefficients that satisfy $\sum_{n \in \mathbb{Z}} n^{2k} |\hat{f}(n)|^2 < \infty$ for some integer $k \geq 1$. Then $f, f', \ldots, f^{(k-1)}$ are absolutely continuous with $f^{(k)} \in L^2(\mathbb{T})$.*

1.3.2 *Application to the Isoperimetric Inequality

We now give an application of Parseval's theorem to geometry. Suppose that we have a closed curve in the xy plane that encloses an area A and has perimeter P. We will prove that

$$P^2 \geq 4\pi A,$$

with equality if and only if the curve is a circle.

To do this, suppose that the curve is described by parametric equations $x = x(t)$, $y = y(t)$ where $-\pi \leq t \leq \pi$. The functions $x(t), y(t)$ are supposed to be absolutely continuous with derivatives in the space $L^2(\mathbb{T})$ and to satisfy the normalization conditions $x(-\pi) = x(\pi), y(-\pi) = y(\pi)$, since the curve is closed. From calculus, the perimeter and area are given by the formulas

$$P = \int_{-\pi}^{\pi} \sqrt{x'(t)^2 + y'(t)^2}\, dt, \qquad A = \int_{-\pi}^{\pi} x(t) y'(t)\, dt,$$

where $x' = dx/dt$, $y' = dy/dt$. By reparametrizing the curve, we may suppose that $x'(t)^2 + y'(t)^2$ is constant; in fact it must be

$$x'(t)^2 + y'(t)^2 = \frac{P^2}{4\pi^2}.$$

Since the functions are real-valued, we work with the original trigonometric form of Fourier series. Since the functions $x(t), y(t)$ are supposed absolutely continuous, they are also of bounded variation, and we have the uniformly convergent Fourier series

(1.3.4) $$x(t) = a_0 + \sum_{n=1}^{\infty} (a_n \cos nt + b_n \sin nt), \qquad -\pi \leq t \leq \pi,$$

(1.3.5) $$y(t) = c_0 + \sum_{n=1}^{\infty} (c_n \cos nt + d_n \sin nt), \qquad -\pi \leq t \leq \pi.$$

The Fourier series of the derivatives are not necessarily pointwise convergent, but they do converge in L^2 and we can apply Parseval's theorem. We have

(1.3.6) $\quad x'(t) \sim \sum_{n=1}^{\infty} n(-a_n \sin nt + b_n \cos nt), \quad -\pi \leq t \leq \pi,$

(1.3.7) $\quad y'(t) \sim \sum_{n=1}^{\infty} n(-c_n \sin nt + d_n \cos nt), \quad -\pi \leq t \leq \pi.$

Applying Parseval's theorem, we have

$$\frac{P^2}{2\pi} = \int_{-\pi}^{\pi} [x'(t)^2 + y'(t)^2] dt = \pi \sum_{n=1}^{\infty} n^2 (a_n^2 + b_n^2 + c_n^2 + d_n^2)$$

$$A = \int_{-\pi}^{\pi} x(t) y'(t) dt$$

$$= \frac{1}{4} \int_{-\pi}^{\pi} \{[x(t) + y'(t)]^2 - [x(t) - y'(t)]^2\} dt$$

$$= \pi \sum_{n=1}^{\infty} n(a_n d_n - b_n c_n).$$

Performing the necessary algebraic steps, we have

$$\frac{P^2}{2\pi} - 2A = \pi \sum_{n=1}^{\infty} \left[n(a_n - d_n)^2 + n(b_n + c_n)^2 + n(n-1)(a_n^2 + b_n^2 + c_n^2 + d_n^2) \right].$$

The right side is a sum of squares with nonnegative coefficients; thus $P^2/2\pi - 2A \geq 0$. If the sum is zero, then all of the terms are zero, in particular $a_n^2 + b_n^2 + c_n^2 + d_n^2 = 0$ for $n > 1$ and $a_1 - d_1 = 0$, $b_1 + c_1 = 0$. This means that

(1.3.8) $\quad x(t) = a_0 + a_1 \cos t - c_1 \sin t, \quad -\pi \leq t \leq \pi,$

(1.3.9) $\quad y(t) = c_0 + c_1 \cos t + a_1 \sin t, \quad -\pi \leq t \leq \pi,$

which is the equation of a circle of radius $\sqrt{a_1^2 + c_1^2}$ with center at (a_0, c_0). The proof is complete.

Exercise 1.3.8. Show that if $x = x(t), y = y(t)$ describes a plane curve of finite length, then we can re-parametrize with $t = t(s)$ so that $(dx/ds)^2 + (dy/ds)^2$ is a constant.

1.3.3 *Rates of Convergence in L^2

Parseval's theorem allows one to reduce the study of rates of convergence to the estimation of series. The nth Fourier coefficient of $S_N f - f$ is zero for $|n| \leq N$, therefore from (1.3.3), the mean square error is

(1.3.10) $$\|S_N f - f\|_2^2 = \sum_{|n|>N} |\hat{f}(n)|^2.$$

This can be used to estimate the mean square error in terms of the smoothness of f. If, for example, $f \in C^j(\mathbb{T})$, then $\hat{f}(n) = O(|n|^{-j})$ and

$$\|S_N f - f\|^2 \leq C \sum_{k>N} k^{-2j} = O(N^{1-2j}), \qquad N \to \infty$$

which gives an upper bound for the mean square error when $N \to \infty$.

In order to obtain more precise estimates, we introduce the L^2-translation error, defined by

(1.3.11) $$\|f_h - f\|_2^2 = \frac{1}{2\pi} \int_{\mathbb{T}} |f(x+h) - f(x)|^2 \, dx, \qquad h > 0, \quad f \in L^2(\mathbb{T}).$$

This can be expressed directly in terms of the Fourier coefficients by using Parseval's identity to write

(1.3.12) $$\|f_h - f\|_2^2 = \sum_{n \in \mathbb{Z}} |e^{inh} - 1|^2 |\hat{f}(n)|^2.$$

The next theorem describes equivalent norms to measure the smoothness of f.

Theorem 1.3.9. *Suppose that $f \in L^2(\mathbb{T})$, $0 < \alpha < 1$. Then $\|f - S_n f\|_2 \leq C n^{-\alpha}$ if and only if f satisfies the L^2 Hölder condition $\|f_h - f\|_2 \leq K h^\alpha$ for suitable constants $C, K > 0$.*

Proof. From (1.3.12) we have for any M

$$\|f_h - f\|_2^2 = \left(\sum_{|n| \leq M} + \sum_{|n| > M} \right) |e^{inh} - 1|^2 |\hat{f}(n)|^2$$

$$\leq \sum_{|n| \leq M} n^2 h^2 |\hat{f}(n)|^2 + 4 \sum_{|n| > M} |\hat{f}(n)|^2.$$

If $\|S_n f - f\| \leq C n^{-\alpha}$, then the second sum is $O(M^{-2\alpha})$. To estimate the first sum, we sum by parts, writing $E_n := \sum_{|k| \leq n} |\hat{f}(k)|^2$:

$$\sum_{|n| \leq M} n^2 |\hat{f}(n)|^2 = M^2(E_M - E_\infty) + \sum_{n=1}^{M} (2n-1)(E_\infty - E_{n-1}).$$

By hypothesis $E_\infty - E_n = O(n^{-2\alpha})$, so that both of these terms are $O(M^{2-2\alpha})$, therefore

$$\|f_h - f\|_2^2 \leq C^2 [h^2 M^{2-2\alpha} + 4M^{-2\alpha}].$$

Choosing $M = 1/h$ completes the proof.

Conversely, if f satisfies an L^2 Hölder condition, we can write

$$2 \sum_{n \in \mathbb{Z}} [1 - \cos(nh)] |\hat{f}(n)|^2 = \|f_h - f\|_2^2 \leq K^2 h^{2\alpha}.$$

Integrating this inequality over the interval $[0, k]$ and dividing by k, we have

$$\sum_{n \in \mathbb{Z}} \left[1 - \frac{\sin(nk)}{nk} \right] |\hat{f}(n)|^2 \leq C k^{2\alpha}, \qquad k > 0.$$

The terms on the left side are nonnegative. Restricting the sum to indices n for which $|n|k \geq 2$, we have

$$Ck^{2\alpha} \geq \sum_{n\in\mathbb{Z}:|n|k\geq 2}\left[1 - \frac{\sin(nk)}{nk}\right]|\hat{f}(n)|^2 \geq \frac{1}{2}\sum_{n\in\mathbb{Z}:|n|k\geq 2}|\hat{f}(n)|^2.$$

The proof is complete. ∎

We can paraphrase Theorem 1.3.9 in terms of equivalent norms. On the one hand we have the L^2 Hölder norm

$$N_\alpha(f) := \sup_{h\neq 0} \|f_h - f\|_2/|h|^\alpha.$$

On the other hand we have the normalized mean square error, defined by

$$R_\alpha(f) := \sup_{n\geq 0} n^\alpha \|S_n f - f\|_2.$$

Theorem 1.3.9 asserts that for $0 < \alpha < 1$, there exists a constant $C = C_\alpha$ so that

$$C^{-1}N_\alpha(f) \leq R_\alpha(f) \leq CN_\alpha(f).$$

In case $\alpha = 1$ one cannot expect the above equivalence to hold, as shown in the next example.

Example 1.3.10. *Let f be defined by the absolutely convergent trigonometric series*

$$f(x) = \sum_{k=1}^{\infty} \frac{e^{ikx}}{k^{3/2}}.$$

Then

$$\|f - S_n f\|_2^2 = \sum_{k=n+1}^{\infty} \frac{1}{k^3} \leq \frac{1}{2n^2}.$$

On the other hand, from (1.3.12)

$$\|f_h - f\|_2^2 = 4\sum_{k=1}^{\infty} \frac{1}{k^3} \sin^2(kh/2)$$

$$\geq 4\sum_{k<\pi/h} \frac{1}{k^3} \sin^2(kh/2)$$

$$\geq 4\sum_{k<\pi/h} \frac{1}{k^3} (kh/\pi)^2$$

$$\geq \frac{4h^2}{\pi^2} \log(\pi/h),$$

which shows that f cannot satisfy an L^2 Hölder condition with $\alpha = 1$.

Exercise 1.3.11. *Suppose that f satsifies an L^2 Hölder condition with $\alpha = 1$. Prove that $\|f - S_n f\|_2 = O(1/n)$.*

Hint: Go back through the steps of the second part of the proof of Theorem (1.3.9) with $\alpha = 1$.

Exercise 1.3.12. *Suppose that f satisfies an L^2 Hölder condition with $\alpha = 1$. Prove that $\sum_{n \in \mathbb{Z}} |n|^2 |\hat{f}(n)|^2 < \infty$.*

Hint: Apply Fatou's lemma to formula (1.3.12).

The relation between the L^2 Hölder condition and the pointwise Hölder condition is not symmetrical. Indeed, if $|f(x+h) - f(x)| \leq Ch^\alpha$ for some $0 < \alpha < 1$, then clearly $\|f_h - f\|_2 \leq Ch^\alpha$. The converse is false, as shown by the following example.

Example 1.3.13. *Let $f(x) = |x|^\alpha$ for $-\pi \leq x \leq \pi$, where $0 < \alpha < 1$.*

We will show that the L^2 Hölder condition is a strict improvement of the pointwise Hölder condition. To see this, we first compute the Fourier coefficients:

$$2\pi \hat{f}(n) = \int_{-\pi}^{\pi} |x|^\alpha \cos(nx)\, dx$$

$$= 2 \int_0^\pi x^\alpha \cos(|n|x)\, dx$$

$$= -\frac{2\alpha}{|n|} \int_0^\pi x^{\alpha-1} \sin(|n|x)\, dx$$

$$= -\frac{2\alpha}{|n|^{1+\alpha}} \int_0^{|n|\pi} y^{\alpha-1} \sin y\, dy.$$

But the final integral converges to a constant when $n \to \infty$ as shown by another partial integration. Therefore we have the asymptotic formula $\hat{f}(n) = C/|n|^{1+\alpha}(1 + o(1))$, $|n| \to \infty$. Applying the formula (1.3.12), we have

$$\|f_h - f\|_2^2 = \sum_{n \in \mathbb{Z}} |e^{inh} - 1|^2 |\hat{f}(n)|^2$$

(1.3.13)
$$\leq \sum_{|n| \leq M} (nh)^2 |\hat{f}(n)|^2 + 4 \sum_{|n| > M} |\hat{f}(n)|^2.$$

We consider three cases.

Case I: $\alpha > \frac{1}{2}$. The first sum in (1.3.13) is convergent when $M \to \infty$, while the second sum is $O(M^{-1-2\alpha})$. Taking $M = 1/h$, we see that $\|f_h - f\|^2 \leq Ch^2$, hence the L^2 modulus satisfies $\|f_h - f\|_2 \leq Ch$, irrespective of α. Hence we have an improvement of the Hölder exponent in the amount $1 - \alpha$.

Case II: $\alpha = \frac{1}{2}$. In this case the first sum in (1.3.13) $\sim h^2 \log M$, whereas the second sum $= O(1/M^2)$. Again taking $M = 1/h$, we have $\|f_h - f\|_2 \leq Ch\sqrt{\log(1/h)}$, thus an improvement of the Hölder exponent by nearly $\frac{1}{2}$.

Case III: $0 < \alpha < \frac{1}{2}$. In this case the first sum diverges, asymptotic to a multiple of $M^{1-2\alpha}$. Again taking $M = 1/h$, the two terms are now balanced and both are asymptotic to a constant multiple of $h^{2\alpha+1}$. Thus in this case we have $\|f_h - f\|_2 \leq Ch^{\alpha+\frac{1}{2}}$, an improvement of the Hölder exponent by $\frac{1}{2}$.

These examples give a concrete indication of the disparity between the L^2 Hölder classes and the pointwise Hölder classes. A more systematic approach is contained in the proof of Corollary 1.3.19 below.

The next two exercises give a relation between the mean square error and the fractional Sobolev classes on the circle.

Exercise 1.3.14. *Suppose that $f \in L^2(\mathbb{T})$ and that for some $\alpha > 0$, $\sum_{n \in \mathbb{Z}} |n|^{2\alpha} |\hat{f}(n)|^2 < \infty$. Prove that $\|S_N f - f\|_2 = o(N^{-\alpha}), N \to \infty$.*

Hint: Begin with (1.3.10).

Exercise 1.3.15. *Suppose that $f \in L^2(\mathbb{T})$ and that for some $\alpha > 0$, $\|S_N f - f\|_2 = O(N^{-\alpha}), N \to \infty$. Prove that $\sum_{n \in \mathbb{Z}} |n|^{2\beta} |\hat{f}(n)|^2 < \infty$ for any $\beta < \alpha$.*

Hint: Apply summation-by-parts to the sum $\sum_{|n| \leq M} |n|^{2\beta} |\hat{f}(n)|^2$.

At the beginning of this section it was noted that if a function has additional smoothness, then we may expect that the mean square error decays more rapidly when $N \to \infty$. The following exercise gives an extension of Theorem 1.3.9 to higher derivatives.

Exercise 1.3.16. *Let $k \in \mathbb{Z}^+$ and $0 < \alpha < 1$. In order that the mean square error satisfy the estimate $\|S_N f - f\|_2 = O(N^{-(k+\alpha)}), N \to \infty$, it is necessary and sufficient that $f, f', \ldots, f^{(k-1)}$ be absolutely continuous and that $f^{(k)}$ satisfy the L^2 Hölder condition $\|f_h^{(k)} - f^{(k)}\|_2 \leq Kh^\alpha$.*

1.3.3.1 Application to absolutely convergent Fourier series

We can also use the L^2 Hölder condition to give a sufficient condition for f to be represented as an absolutely convergent Fourier series.

Theorem 1.3.17. *Bernstein:* *Suppose that f satisfies an L^2 Hölder condition with exponent $\alpha > \frac{1}{2}$. Then the Fourier series is absolutely convergent: $\sum_{n \in \mathbb{Z}} |\hat{f}(n)| < \infty$.*

Proof. We estimate dyadic blocks by the Cauchy-Schwarz inequality. Thus

$$\left(\sum_{2^m \leq |n| < 2^{m+1}} |\hat{f}(n)| \right)^2 \leq 2^{m+1} \sum_{2^m \leq |n| < 2^{m+1}} |\hat{f}(n)|^2.$$

By Parseval's identity, we have for any m

$$\|f_h - f\|_2^2 = \sum_{n \in \mathbb{Z}} |e^{inh} - 1|^2 |\hat{f}(n)|^2 \geq \sum_{2^m \leq |n| < 2^{m+1}} |e^{inh} - 1|^2 |\hat{f}(n)|^2.$$

Writing $|e^{inh} - 1|^2 = 4\sin^2(nh/2)$, we see that if $h = \frac{1}{3}\pi 2^{-m}$, then $|e^{inh} - 1| \geq 1$ for $2^m \leq |n| < 2^{m+1}$, so that

$$\|f_h - f\|_2^2 \geq \sum_{2^m \leq |n| < 2^{m+1}} |\hat{f}(n)|^2, \quad h = \frac{\pi}{3} 2^{-m}.$$

Now we apply Cauchy-Schwarz to each dyadic block and sum to obtain

$$\sum_{0 \neq n \in \mathbb{Z}} |\hat{f}(n)| = \sum_{m=0}^{\infty} \sum_{2^m \leq |n| < 2^{m+1}} |\hat{f}(n)|$$

$$\leq \sum_{m=0}^{\infty} 2^{(m+1)/2} \Omega_2(f; \frac{1}{3}\pi 2^{-m})$$

$$\leq C \left(\frac{\pi}{3}\right)^\alpha \sum_{m=0}^{\infty} 2^{m/2} 2^{-m\alpha}$$

$$= C \left(\frac{\pi}{3}\right)^\alpha \sum_{m=0}^{\infty} 2^{\frac{m}{2}(1-2\alpha)} < \infty,$$

since $\alpha > \frac{1}{2}$. ∎

Exercise 1.3.18. *Suppose that f satisfies an L^2 Hölder condition with exponent $\alpha > \frac{1}{2}$. Prove that $\sum_{n \in \mathbb{Z}} |n|^\beta |\hat{f}(n)| < \infty$ for any $\beta < \alpha - \frac{1}{2}$.*

If f is of bounded variation, then the absolute convergence of the Fourier series holds under any pointwise Hölder condition, according to the next corollary.

Corollary 1.3.19. *Zygmund: Suppose that $f \in BV(\mathbb{T})$ and that f satisfies a pointwise Hölder condition: $|f(x + y) - f(x)| \leq C|y|^\alpha$ for $0 < \alpha < 1$. Then $\sum_{n \in \mathbb{Z}} |\hat{f}(n)| < \infty$.*

Proof. Letting V_f denote the total variation of f, we can estimate the L^2 modulus of continuity by writing

$$\|f_{\pi/3N} - f\|_2^2 = \frac{1}{2N\pi} \sum_{k=1}^{N} \int_{\mathbb{T}} \left| f\left(x + \frac{k\pi}{3N}\right) - f\left(x + \frac{(k-1)\pi}{3N}\right) \right|^2 dx$$

$$\leq \frac{V_f}{N} C \left(\frac{\pi}{3N}\right)^\alpha$$

$$= \frac{C}{N^{1+\alpha}}.$$

Therefore f satisfies the L^2 Hölder condition with exponent $(1 + \alpha)/2 > \frac{1}{2}$, at least along the values $h = \pi/3N$. But this is sufficient to apply the proof of Bernstein's theorem and thus conclude that $\sum_{n \in \mathbb{Z}} |\hat{f}(n)| < \infty$. ∎

If f satisfies only a Hölder condition with $\alpha < 1/2$, the Fourier series is not absolutely convergent in general. There exist many examples in the literature. An alternative treatment is to look at *random* Fourier series and to prove that almost every realization is not absolutely convergent, but satisfies a Hölder condition with $\alpha < 1/2$ (see Kahane (1968)).

1.4 NORM CONVERGENCE AND SUMMABILITY

The tools introduced thus far do not permit us to deal with the norm convergence within a Banach space of functions. Indeed, the oscillatory properties of the Dirichlet kernel will allow us to show that there exists a continuous function whose Fourier series diverges at a point. Hence one cannot prove uniform convergence, for example, within the class of continuous functions. Furthermore, we would like to deal with convergence in the norm of L^p, where $p \geq 1$. This turns out to be impossible in the space L^1, but can be dealt with nicely if $p > 1$. In order to launch a systematic theory, we consider the Cesàro averages of the Fourier partial sums. These are called the *Fejér means* and defined as the arithmetic means

(1.4.1) $$\sigma_N f(\theta) = \frac{1}{N+1}(S_0 f(\theta) + \cdots + S_N f(\theta)), \qquad N = 0, 1, 2, \ldots$$

We also consider the Abel means

(1.4.2) $$A_r f(\theta) = (1-r)\sum_{n=0}^{\infty} r^n S_n f(\theta), \qquad 0 \leq r < 1.$$

The Abel means can be written as a Fourier series by proving the identity

(1.4.3) $$(1-r)\sum_{n=0}^{\infty} r^n S_n f(\theta) = \sum_{n \in \mathbb{Z}} \hat{f}(n) r^{|n|} e^{in\theta}, \qquad 0 \leq r < 1.$$

Exercise 1.4.1. *Prove (1.4.3).*

Exercise 1.4.2. *Prove that if the Fourier partial sums converge pointwise, then the Fejér means converge pointwise. Prove that if the Fourier partial sums converge uniformly, then the Fejér means converge uniformly. Prove that if the Fourier partial sums converge in L^p, $p \geq 1$, then the Fejér means converge in L^p.*

Hint: Use the triangle inequality and the fact that if a numerical sequence s_n satisfies $\lim_n s_n = s$, then $\lim_n (s_0 + \cdots + s_n)/n + 1 = s$.

Exercise 1.4.3. *Prove that if the Fourier partial sums converge pointwise, then the Abel means converge pointwise. Prove that if the Fourier partial sums converge uniformly, then the Abel means converge uniformly. Prove that if the Fourier partial sums converge in L^p, $p \geq 1$, then the Abel means converge in L^p.*

Hint: First prove that if a numerical sequence s_n satisfies $\lim_n s_n = s$, then $\lim_{r \to 1} (1-r)\sum_{n=0}^{\infty} r^n s_n = s$.

1.4.1 Approximate Identities

In order to deal systematically with these procedures, we define the general notion of *approximate identity*.

Definition 1.4.4. *An approximate identity on the circle* $\mathbb{T}$ *is a function* $k(r, \theta)$ *defined for* $\theta \in \mathbb{T}$ *and* r *in some directed index set* I, *so that*

$$\lim_r \frac{1}{2\pi} \int_\mathbb{T} k(r, \theta) \, d\theta = 1 \tag{1.4.4}$$

$$\int_\mathbb{T} |k(r, \theta)| \, d\theta \leq C, \qquad \forall r \in I \tag{1.4.5}$$

$$\lim_r \int_{|\theta| > \delta} |k(r, \theta)| \, d\theta = 0, \qquad \forall \delta > 0. \tag{1.4.6}$$

Here C *is a constant independent of* r. *In case* $k(r, \theta) \geq 0$, *then* (1.4.5) *is superfluous and we can take* $C = 2\pi$.

By definition, a *directed set* is a set I together with a collection of subsets $\{A_i\}$ with the property that for each (i, j) there exists k with $A_k \subset A_i \cap A_j$. In case $I = [0, 1)$, the subsets can be taken in the form $A_k = (1 - \frac{1}{k}, 1)$. A complex-valued function f on a directed set has a limit L, by definition, if for each $\epsilon > 0$, there exists a subset A_j so that $|f(x) - L| < \epsilon$ for all $x \in A_j$. With this definition it is immediate that limits obey the usual laws for sums, products, and composition of functions.

Remark. We choose the formulation with a general directed set in order to have maximum flexibility in the applications. For example, for the Fejér means we have the index set $\{1, 2, \ldots\}$ with $n \to \infty$, whereas for the Poisson kernel associated with the Abel means we have the index set $[0, 1)$ with $r \to 1$. In the first case we may take $A_k = (k, k+1, \ldots)$ whereas in the second case we take A_k as the open interval $(1 - \frac{1}{k}, 1)$.

Example 1.4.5. *The Poisson kernel* $P_r(\theta)$ *is an approximate identity.*

Indeed, we showed in (1.1.39) that $\int_\mathbb{T} P_r(\theta) \, d\theta = 2\pi$. Since $P_r(\theta) \geq 0$, the second property is automatically satisfied. To prove the third property, note that for $|\theta| > \delta$ the denominator $1 + r^2 - 2r \cos \theta = (1-r)^2 + 2r(1 - \cos \theta) > 2r(1 - \cos \delta)$. Therefore in this interval we have $P_r(\theta) \leq (1 - r^2)/2r(1 - \cos \delta)$, which tends to zero when $r \to 1$. The fundamental use of approximate identities is described as follows:

Proposition 1.4.6. *Suppose that* $k(r, \theta)$ *is an approximate identity.*

- *If* $\Phi \in L^\infty(\mathbb{T})$ *with* $\lim_{\theta \to 0} \Phi(\theta) = L$, *then*

$$\lim_r \frac{1}{2\pi} \int_\mathbb{T} k(r, \theta) \Phi(\theta) \, d\theta = L. \tag{1.4.7}$$

- *If, in addition, for each* $\delta > 0$ $\sup_{|\theta| \geq \delta} |k(r, \theta)| \to 0$, *then* (1.4.7) *holds for all* $\Phi \in L^1(\mathbb{T})$ *with* $\lim_{\theta \to 0} \Phi(\theta) = L$.

Proof. For any $\delta > 0$, we have

$$\frac{1}{2\pi} \int_\mathbb{T} k(r, \theta) \Phi(\theta) \, d\theta - L = \frac{1}{2\pi} \left(\int_{|\theta| > \delta} + \int_{|\theta| \leq \delta} \right) k(r, \theta)(\Phi(\theta) - L) \, d\theta + o(1).$$

The first integral tends to zero, for any $\delta > 0$. Given $\epsilon > 0$, the second integral can be made less than ϵ by taking δ sufficiently small, which proves the first statement. To prove the second statement, note that the first integral is bounded by $\sup_{|\theta|>\delta} |k(r,\theta)| \times (L + \|\Phi\|_1)$, which tends to zero by hypothesis. The second integral is bounded by $\epsilon \times \int_{\mathbb{T}} |k(r,\theta)|\,d\theta$, which completes the proof. ∎

Exercise 1.4.7. *Suppose that the approximate identity $k(r,\theta)$ has the additional property that k is even: $k(r,\theta) = k(r,-\theta)$ for all $\theta \in \mathbb{T}$. Suppose that $\Phi \in L^\infty(\mathbb{T})$ with $\lim_{\theta \to 0}[\Phi(\theta) + \Phi(-\theta)] = 2L$, for some complex number L. Prove that formula (1.4.7) holds.*

Exercise 1.4.8. *Suppose that the approximate identity $k(r,\theta)$ is even and has the property that for each $\delta > 0$, $\sup_{|\theta| \geq \delta} |k(r,\theta)| \to 0$. Suppose that $\Phi \in L^1(\mathbb{T})$ with $\lim_{\theta \to 0}[\Phi(\theta) + \Phi(-\theta)] = 2L$, for some complex number L. Prove that formula (1.4.7) holds.*

In order to apply approximate identites to norm convergence, we recall the notation f_ϕ for the translate of $f \in L^1(\mathbb{T})$, defined by $f_\phi(\theta) = f(\theta - \phi)$. The following definition is essential.

Definition 1.4.9. *A subspace $B \subset L^1(\mathbb{T})$ with norm $\|\cdot\|_B$ is called a homogeneous Banach subspace if we have $\|f\|_1 \leq \|f\|_B$, the map $f \to f_\theta$ is B-norm preserving, and the map $\theta \to f_\theta$ is continuous in the B norm. In detail we require that $\|f_\theta\|_B = \|f\|_B$ for all $f \in B$ and all $\theta \in \mathbb{T}$, and that $\lim_{\theta \to 0} \|f_\theta - f\|_B \to 0$ for all $f \in B$.*

Example 1.4.10. *The space $C(\mathbb{T})$ with the supremum norm is a homogeneous Banach subspace. The space $L^p(\mathbb{T})$ for $1 \leq p < \infty$ is also a homogeneous Banach subspace.*

Exercise 1.4.11. *Prove these properties. Then prove that $L^\infty(\mathbb{T})$ with the supremum norm is not a homogeneous Banach space.*

This notion is very effective for dealing with norm convergence, when we represent the convolution of two functions as a vector-valued integral. If $K \in L^1(\mathbb{T})$, we can write

$$(K * f)(\theta) = \frac{1}{2\pi} \int_{\mathbb{T}} K(\phi) f(\theta - \phi)\, d\phi = \frac{1}{2\pi} \int_{\mathbb{T}} K(\phi) f_\phi(\theta)\, d\phi.$$

The final integral is a vector-valued integral, defined as a limit in norm of Riemann sums. In particular, if $f \in B$, then $K * f$ is an element of B and we can estimate the B-norm of the vector-valued integral by the inequality

$$\left\| \int_{\mathbb{T}} K(\phi) f_\phi\, d\phi \right\|_B \leq \int_{\mathbb{T}} |K(\phi)|\, \|f_\phi\|_B\, d\phi \leq C\|f\|_B,$$

which follows from the triangle inequality for finite sums. Similarly

$$\|K * f - f\|_B \leq \frac{1}{2\pi} \int_{\mathbb{T}} |K(\phi)| \times \|f_\phi - f\|_B\, d\phi,$$

which can be analyzed by the more elementary techniques of Proposition 1.4.6. We formalize this as follows.

Theorem 1.4.12. *If B is a homogeneous Banach subspace and $k(r, \theta)$ is an approximate identity, then*

$$\lim_r \left\| \frac{1}{2\pi} \int_{\mathbb{T}} k(r, \phi) f_\phi \, d\phi - f \right\|_B = 0.$$

Proof. The required norm is less than or equal to

$$\frac{1}{2\pi} \int_{\mathbb{T}} |k(r, \phi)| \, \|f_\phi - f\|_B \, d\phi,$$

which tends to zero by Proposition 1.4.6. ∎

The first application of approximate identities is to the sequence of Abel means of a Fourier series. To make the connection between Abel means and the Poisson kernel, we recall the basic identity of Fourier reciprocity, Proposition 1.2.3, which states in this case that for any $f \in L^1(\mathbb{T})$

$$\frac{1}{2\pi} \int_{\mathbb{T}} f(\theta - \phi) P_r(\phi) \, d\phi = \sum_{n \in \mathbb{Z}} r^{|n|} \hat{f}(n) e^{in\theta}.$$

Theorem 1.4.13. *If $f \in C(\mathbb{T})$, then the Abel means converge uniformly to f. If $f \in L^p(\mathbb{T})$, $1 \leq p < \infty$ then the Abel means converge to f in the norm of L^p. If $f \in L^1(\mathbb{T})$ has right and left limits at $\theta \in \mathbb{T}$, then the Abel means converge to $\frac{1}{2}[f(\theta + 0) + f(\theta - 0)]$.*

Proof. We have shown that the Abel means are defined by the Poisson kernel, which satisfies the conditions of an approximate identity. Hence the first two statements follow immediately from Proposition 1.4.12. The third statement is a direct application of Proposition 1.4.6. ∎

We will prove in the next section that the Fejér means are also represented by an approximate identity. This will allow us to prove the norm convergence properties for Fejér means also.

Theorem 1.4.12 admits a sort of converse, expressed as follows:

Proposition 1.4.14. *Suppose that B is a homogeneous Banach subspace of $L^1(\mathbb{T})$ and $k(r, \theta)$ is an approximate identity with the property that for some $f \in L^1(\mathbb{T})$, $k * f \in B$ for all $r \in I$ and $k * f$ converges in the B norm. Then $f \in B$.*

Proof. Letting $g = \lim_r k * f$ (in the B norm), we must also have $g = \lim_r k * f$ in the L^1 norm. But from Theorem 1.4.12, $f = \lim_r k * f$ in the L^1 norm. Hence $f = g$ a.e. But the space B is a closed subspace of $L^1(\mathbb{T})$, therefore $f \in B$ as required. ∎

Applying this to the Poisson kernel, we have the following useful converse statements:

Corollary 1.4.15. *(i) Suppose that $f \in L^1(\mathbb{T})$ is such that $P_r f$ converges in the norm of $L^p(\mathbb{T})$ for some $1 < p < \infty$. Then $f \in L^p(\mathbb{T})$.*

(ii) Suppose that $f \in L^1(\mathbb{T})$ is such that $P_r f$ converges uniformly. Then $f \in C(\mathbb{T})$.

Proof. It suffices to remark that for any $f \in L^1(\mathbb{T})$, $P_r f$ is a continuous function, in particular a member of the space $L^p(\mathbb{T})$. ∎

We will see later that condition (i) can be weakened to *boundedness* in the L^p norm. However condition (ii) cannot be weakened to uniform boundedness; consider the example of the Poisson integral of a bounded but discontinuous function.

1.4.1.1 Almost everywhere convergence of the Abel means

We have shown that the Abel means of an L^1 function converge at every point where the right and left limits exist. This condition can be weakened to the existence of the limits of the averages:

$$(1.4.8) \qquad L_\theta = \lim_{\epsilon \to 0} \frac{1}{2\epsilon} \int_{\theta-\epsilon}^{\theta+\epsilon} f(y)\,dy.$$

From the fundamental theorem of calculus (Lebesgue's differentiation theorem), it is known that these limits exists with $L_\theta = f(\theta)$ except for a θ-set of measure zero.

The almost everywhere convergence of the Abel means will be deduced as a corollary of the following general theorem on a class of approximate identities on $[0, \pi]$. The definition of the latter simply amounts to replacing $\mathbb{T}$ by $[0, \pi]$ in the original definition of approximate identity.

Theorem 1.4.16. *Suppose that $k(r, \theta)$ is an approximate identity on $[0, \pi]$, with the property that $k(r, \theta) \geq 0$ and $\theta \to k(r, \theta)$ is absolutely continuous with $k'(r, \theta) := (dk/d\theta) \leq 0$ for $0 \leq \theta \leq \pi$ and all $r \in I$. Suppose that $\bar{f} \in L^1[0, \pi]$ satisfies $\lim_{\theta \to 0} \theta^{-1} \int_0^\theta \bar{f}(\phi)\,d\phi = L$. Then $\lim_r (1/\pi) \int_0^\pi k(r, \theta) \bar{f}(\theta)\,d\theta = L$.*

Proof. We will show that $\theta \to -\theta k'(r, \theta)$ is an approximate identity on $[0, \pi]$. To see this we first note that, since $k' \leq 0$, we have for $0 < \delta_1 < \delta_2 \leq \pi$,

$$(\delta_1 - \delta_2) k(r, \delta_2) \leq \int_{\delta_1}^{\delta_2} k(r, \theta)\,d\theta \to 0$$

which shows that $\lim_r k(r, \theta) = 0$ for $0 \leq \theta \leq \pi$. From the normalization of k, we have

$$1 \leftarrow \frac{1}{\pi} \int_0^\pi k(r, \theta)\,d\theta = k(r, \pi) - \frac{1}{\pi} \int_0^\pi \theta k'(r, \theta)\,d\theta$$

which shows that $\lim_r \int_0^\pi -\theta k'(r, \theta)\,d\theta = 1$. Furthermore, for any $\delta \in (0, \pi)$,

$$\int_\delta^\pi |\theta k'|\,d\theta = -\int_\delta^\pi \theta k'(r, \theta)\,d\theta = \delta k(r, \delta) - \pi k(r, \pi) + \int_\delta^\pi k(r, \theta)\,d\theta \to 0$$

which shows that $\theta \to -\theta k'(r, \theta)$ is an approximate identity.

Now define $F(\theta) = \int_0^\theta [\bar{f}(\phi) - L] d\phi$, a bounded function on $[0, \pi]$ with $\lim_{\theta \to 0} F(\theta)/\theta = 0$. We integrate by parts, replacing k by $c_r k$ with $\lim_r c_r = 1$, thus

$$\int_0^\pi k(r, \theta)[\bar{f}(\theta) - L] d\theta = k(r, \pi) F(\pi) - \int_0^\pi F(\theta) k'(r, \theta) d\theta$$

$$= o(1) + \int_0^\pi \frac{F(\theta)}{\theta} [-\theta k'(r, \theta)] d\theta.$$

The final integral tends to zero by Proposition 1.4.6. ∎

Corollary 1.4.17. *Fatou: The Abel means $P_r f(\theta)$ converge to $f(\theta)$ when $r \to 1$ whenever the limit (1.4.8) exists.*

Proof. On the interval $0 \leq \theta \leq \pi$, the functions $2 P_r(\theta)$ form an approximate identity and satisfy the conditions of Theorem 1.4.16. Now define $\bar{f}(\phi) = \frac{1}{2}[f(\theta + \phi) + f(\theta - \phi)]$. If $\theta \in \mathbb{T}$ satisfies (1.4.8), then $\lim_{\phi \to 0} F(\phi)/\phi = 0$, so that we may apply Theorem 1.4.16 with $L = f(\theta)$ to conclude that $\lim_r P_r f(\theta) = f(\theta)$, as required. ∎

Alternative (explicit) proof. One can avoid Theorem 1.4.16 and work directly as follows. Define $F(\phi) = \int_{-\phi}^\phi (f(u + \theta) - f(\theta)) du$. Then the Poisson integral of f can be integrated by parts as follows:

$$P_r f(\theta) - f(\theta) = \frac{1}{2\pi} \int_0^\pi \frac{1 - r^2}{1 + r^2 - 2r \cos \phi} (f(\theta + \phi) + f(\theta - \phi) - 2 f(\phi)) d\phi$$

$$= \frac{1}{2\pi} \int_0^\pi \frac{1 - r^2}{1 + r^2 - 2r \cos \phi} dF(\phi)$$

$$= \frac{1}{2\pi} \frac{1 - r}{1 + r} F(\pi) + \frac{1}{2\pi} \int_0^\pi K_r(\phi) \frac{F(\phi)}{\sin \phi} d\phi$$

where

$$K_r(\phi) = \frac{(1 - r^2) \sin^2 \phi}{(1 + r^2 - 2r \cos \phi)^2}$$

∎

is an approximate identity on $[0, \pi]$, since $K_r(\phi) \geq 0$ and we have from (1.1.38)

$$\frac{1}{\pi} \int_0^\pi \frac{(1 - r^2) \sin^2 \theta}{(1 + r^2 - 2r \cos \theta)^2} d\theta = 1, \quad \lim_{r \to 1} \sup_{\theta \geq \delta} \frac{(1 - r^2) \sin^2 \theta}{(1 + r^2 - 2r \cos \theta)^2} = 0.$$

Exercise 1.4.18. *Show that Theorem 1.4.16 can be generalized as follows. Instead of assuming absolute continuity, simply assume that $\theta \to k(r, \theta)$ is monotone, decreasing for each r. By suitably applying integration-by-parts and suitably modifying the definition of approximate identity to include a sequence of measures, show that the conclusion holds exactly as stated.*

1.4.2 Summability Matrices

Closely related to the notion of approximate identity on $\mathbb{T}$ is the notion of a *summability matrix*, which is the discrete analogue for sequences of real or complex numbers. In this section we give the basic notions of summability, which includes the Abel and Fejér means of the Fourier series as special instances of this general notion.

Definition 1.4.19. *A summability matrix is a doubly infinite array of real numbers a_{mn} defined for $m, n \geq 0$ with the following properties:*

(i) $\lim_{m \to \infty} a_{mn} = 0$ *for each* $n = 0, 1, 2, \ldots$
(ii) $\sum_{n=0}^{\infty} a_{mn} = 1$ *for each* $m = 0, 1, 2, \ldots$
(iii) $\sum_{n=0}^{\infty} |a_{mn}| \leq C$ *for some constant C and all* $m = 0, 1, 2, \ldots$.

A summability matrix defines a linear transformation on the space of bounded sequences

$$s \to A(s), \qquad A_m(s) = \sum_{m=0}^{\infty} a_{mn} s_n.$$

The basic property of consistency is expressed as follows:

Proposition 1.4.20. *If $\lim_n s_n = s$, then $\lim_m A_m(s) = s$.*

Proof. We use (ii) to write

$$A_m(s) - s = \sum_{n=0}^{\infty} a_{mn}(s_n - s).$$

Given $\epsilon > 0$, choose $N = n(\epsilon)$ so that $|s_n - s| < \epsilon$ for $n > n(\epsilon)$. Then

$$|A_m(s) - s| \leq \sum_{n=0}^{\infty} |a_{mn}| \, |s_n - s| \leq \left(\sum_{n=0}^{N} + \sum_{n>N}\right) |a_{mn}| \, |s_n - s|.$$

The last sum is less than $C\epsilon$. Now we can let $m \to \infty$ and use (i) to conclude that

$$\limsup_m |A_m(s) - s| \leq C\epsilon.$$

But ϵ was arbitrary, which completes the proof. ∎

Exercise 1.4.21. *Prove that the conclusion of Proposition 1.4.20 still holds if condition (ii) is weakened to the relation that $\lim_{m \to \infty} \sum_{n=0}^{\infty} a_{mn} = 1$.*

Basic examples of summability matrices are provided by the Cesàro means and Abel means:

$$C_m(s) = \frac{1}{m+1} \sum_{n=0}^{m} s_n, \qquad m = 0, 1, \ldots$$

$$A_r(s) = (1-r) \sum_{n=0}^{\infty} r^n s_n, \qquad 0 \leq r < 1.$$

In the first case we have $a_{mn} = 1/(m+1)$ for $n \leq m$ and zero otherwise. In the second case we pick a sequence of $r_m \to 1$, for example $r_m = 1 - 1/m$ thus defining $a_{mn} = r_m^n(1 - r_m)$.

In order to work with Abel means, it is useful to note the following transformation formula, proved by summation-by-parts for $0 \leq r < 1$:

$$(1.4.9) \qquad (1-r)\sum_{n=0}^{\infty} r^n s_n = \sum_{n=0}^{\infty} r^n a_n$$

where $s_n = a_0 + \cdots + a_n$ for $n = 0, 1, \ldots$. This identity allows one to go back and forth between a sequence a_n and its partial sums. The sequence $\{s_n\}$ is Abel-summable to L if and only if we have $\lim_{r \to 1} \sum_{n=0}^{\infty} a_n r^n = L$.

Exercise 1.4.22. *Prove (1.4.9).*

The following concrete examples are useful for reference.

Example 1.4.23. *The negative binomial series is $1/(1+r)^k = \sum_{n=0}^{\infty} \binom{-k}{n} r^n$ for any $0 \leq r < 1$ and k is any real number. Hence the numerical series $\sum_{n=0}^{\infty} \binom{-k}{n}$ is Abel-summable to the value 2^{-k} for any real number k.*

For example, with $k = 0, 1, 2$, we have

$$\frac{1}{1+r} = 1 - r + r^2 - r^3 + \cdots \implies \text{Abel}\sum_{n=0}^{\infty}(-1)^n = \frac{1}{2}$$

$$\frac{1}{(1+r)^2} = 1 - 2r + 3r^2 - 4r^3 + \cdots \implies \text{Abel}\sum_{n=0}^{\infty}(n+1)(-1)^n = \frac{1}{4}$$

$$\frac{1}{(1+r)^3} = 1 - 3r + 6r^2 - 10r^3 + \cdots \implies \text{Abel}\sum_{n=0}^{\infty}\frac{(n+1)(n+2)}{2}(-1)^n = \frac{1}{8}.$$

On the other hand, the series $\sum_{n=0}^{\infty} \binom{-k}{n}$ is Cesàro-summable if $k \leq 1$ but not for $k > 1$.

To see that $k > 2$ does not yield a Cesàro-summable sequence, consider the following exercise.

Exercise 1.4.24. *Suppose that the sequence s_n is Cesàro-summable to s. Prove that $s_n = O(n), n \to \infty$*

Hint: Letting $\sigma_n = C_n(s)$, check that $\sigma_n \to s$ implies that $s_n = (n+1)\sigma_n - n\sigma_{n-1} = O(n)$. This is violated for the example in case $k > 2$.

It is natural to compare different methods of summability. We say that method A is *stronger* than method C if the matrix A can be factored in the form $A = BC$ where B is another summability matrix. This ensures that any sequence that is C-summable is also A-summable.

The previous examples suggest that Abel-summability is stronger than Cesàro-summability. Even more is true:

Proposition 1.4.25. *For any k there exists a summability matrix B_k so that the Abel matrix can be factored as $A = B_k C^k$. Otherwise put, Abel is stronger than any power of Cesàro.*

Proof. We factor the Abel matrix as follows:

$$(1-r)\sum_{n=0}^{\infty} r^n s_n = (1-r)\left[\sigma_0 + r(2\sigma_1 - \sigma_0) + \cdots + r^n((n+1)\sigma_n - n\sigma_{n-1}) + \cdots\right]$$

$$= (1-r)\left[(1-r)\sigma_0 + \cdots + (n+1)(r^n - r^{n+1})\sigma_n + \cdots\right]$$

$$= (1-r)^2 \sum_{n=0}^{\infty} (n+1)\sigma_n r^n,$$

so that the matrix B_1 is defined by the coefficients $(1-r)^2(n+1)r^n$, which satisfy the conditions of a summability matrix. Continuing inductively, we write the second Cesàro means $\sigma_n^{(2)} = (\sigma_0 + \cdots + \sigma_n)/(n+1)$ and its "inverse" $\sigma_n = (n+1)\sigma_n^{(2)} - n\sigma_{n-1}^{(2)}$ to obtain

$$(1-r)\sum_{n=0}^{\infty} r^n s_n = (1-r)^3 \sum_{n=0}^{\infty} \sigma_n^{(2)} (n+1)(n+2) r^n / 2.$$

In general we show by induction that

$$(1-r)\sum_{n=0}^{\infty} r^n s_n = (1-r)^k \sum_{n=0}^{\infty} \binom{-k}{n} (-1)^n \sigma_n^{(k-1)} r^n,$$

which exhibits the matrix B_k explicitly. ∎

Remark. The matrix C^k defines the kth order *Hölder means*. This summabillity method is distinct from the kth order Cesàro means, usually denoted (C, k). For details see Hardy (1949), p. 94 ff.

Figure 1.4.1 shows the relation between Abel and Cesàro summability.

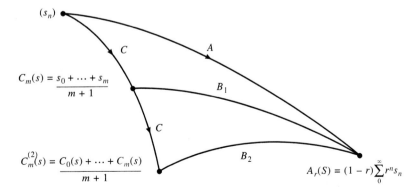

FIGURE 1.4.1
Relations between Abel- and Cesàro-summability. Stronger methods are to the right of weaker methods.

Any theorem that affirms a method of summability is called an *Abelian theorem*. Theorems in the converse direction are called *Tauberian theorems*. The following is a simple Tauberian theorem.

Proposition 1.4.26. *Suppose that $\sum_{n=0}^{\infty} a_n$ is Cesàro-summable to s and that the terms satisfy $\lim_n na_n = 0$. Then the series converges to the same sum s.*

Proof. In terms of the original sequence we can write

$$\sigma_n = \frac{s_0 + \cdots + s_n}{n+1} = \sum_{k=0}^{n}\left(1 - \frac{k}{n+1}\right)a_k.$$

Hence we have

$$s_n - \sigma_n = \frac{1}{n+1}\sum_{k=0}^{n} ka_k.$$

But this is the average of a sequence which tends to zero, hence also tends to zero, which completes the proof. ∎

Exercise 1.4.27. *Suppose that the terms satisfy $k|a_k| \leq M$ for some constant M and $k = 0, 1, 2, \ldots$. Show that $|\sigma_n - s_n| \leq M$ for all $n = 1, 2, \ldots$.*

Exercise 1.4.28. *The Riemann means of a series $\sum_{n=0}^{\infty} a_n$ are defined by*

$$R_h(a) = \sum_{m=0}^{\infty}\left(\frac{\sin mh}{mh}\right)^2 a_m,$$

where $0 < h \to 0$. Show that this is defined by a summability matrix that is not positive.

Hint: Write $R_h(a) = \sum_{m=0}^{\infty}[K(mh) - K((m+1)h)]s_m$ where $K(x) = (\sin x/x)^2$; check that $K(0) = 1$, $K(\infty) = 0$ and $\sum_{n=0}^{\infty}|K(nh) - K((n+1)h)| < \infty$.

1.4.3 Fejér Means of a Fourier Series

The Fejér means are defined by

$$\sigma_N(f) := \frac{S_0 f + \cdots + S_N f}{N+1}.$$

Writing this out in detail, we see

$$(N+1)\sigma_N f = \sum_{n=0}^{N}\sum_{k=-n}^{n} \hat{f}(k)e^{ik\theta}$$

$$= \sum_{k=-N}^{N}\left[\sum_{n=|k|}^{N} \hat{f}(k)e^{ik\theta}\right]$$

$$= \sum_{k=-N}^{N} (N+1-|k|)\hat{f}(k)e^{ik\theta},$$

which gives the useful representation

(1.4.10) $$\sigma_N(f) = \sum_{k=-N}^{N}\left(1 - \frac{|k|}{N+1}\right)\hat{f}(k)e^{ik\theta}$$

The Fejér kernel is

$$K_N(\theta) := \sum_{k=-N}^{N}\left(1 - \frac{|k|}{N+1}\right)e^{ik\theta}.$$

This is a finite Fourier series of a discrete convolution. To see this, define $I_M(j) = 1$ if $|j| \leq M$ and zero otherwise. Then by counting points along the 45 degree lines in the square of side M, we see that $(I_M * I_M)(j) = (2M + 1 - |j|)I_{2M+1}(j)$ if $|j| \leq 2M$ and zero otherwise. This will allow us to factor K_N in case N is even. To deal with the case of N odd, define $I_M^{\text{odd}}(j) = 1$ if $j = \pm\frac{1}{2}, \ldots, \pm(M - \frac{1}{2})$ and zero otherwise. Again we count the points in the square to see that $I_M^{\text{odd}} * I_M^{\text{odd}} = (2M - |j|)I_{2M}$. In either case K_N has been written as the Fourier series of a self-convolved sequence, hence it must be the square of the trigonometric sum obtained from the original sequence. In detail, we have

$$(2M+1)K_{2M}(\theta) = \sum_{k=-2M}^{2M}(2M+1-|k|)I_{2M+1}(k)e^{ik\theta} = \left(\sum_{k=-M}^{M} I_M(k)e^{ik\theta}\right)^2,$$

$$2M\, K_{2M-1}(\theta) = \sum_{k=-2M}^{2M}(2M-|k|)I_{2M+1}(k)e^{ik\theta} = \left(\sum_{k=-M}^{M} I_M^{\text{odd}}(k)e^{ik\theta}\right)^2.$$

The trigonometric sum $\sum_{-M}^{M} I_M(k)e^{ik\theta}$ was evaluated as the Dirichlet kernel $\sin[(M+\frac{1}{2})\theta]/\sin(\theta/2)$, which gives a closed form for K_{2M}. But K_{2M-1} can also be expressed in this form by doing a finite geometric sum:

(1.4.11) $$e^{-it(M-\frac{1}{2})} + \cdots + e^{it(M-1/2)} = \frac{\sin Mt}{\sin(t/2)}.$$

We conclude that for all N we have the formula

(1.4.12) $$K_N(\theta) = \frac{1}{N+1}\left(\frac{\sin[(N+1)\theta/2]}{\sin(\theta/2)}\right)^2.$$

Exercise 1.4.29. *Prove the trigonometric identity (1.4.11).*

Figure 1.4.2 shows the graph of the Fejér kernel with $N = 8$.
The above computations permit us to conclude

Proposition 1.4.30. $K_N(\theta)$ *is an approximate identity.*

Proof. From the definition we see that $\frac{1}{2\pi}\int_{\mathbb{T}} K_N = 1$. From the formula (1.4.12) we see that $K_N \geq 0$, so that the L^1 norms are bounded by 1. Finally, we see from (1.4.12) that for

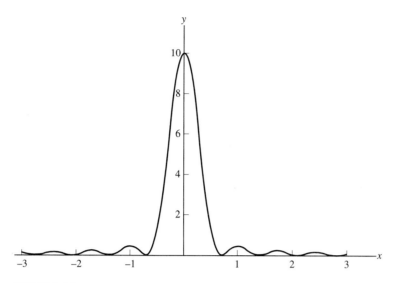

FIGURE 1.4.2
The Fejér kernel with $N = 8$

any $\delta > 0$, $K_N(\theta) \leq 1/(N+1)\sin^2(\delta/2)$ whenever $|\theta| > \delta$. Hence $\int_{|\theta|>\delta} K_N(\theta)\,d\theta \to 0$ when $N \to \infty$. ∎

This leads us to a general statement of Fejér's theorem.

Theorem 1.4.31. *If $f \in L^p(\mathbb{T})$, $1 \leq p < \infty$, then the Fejér means converge in $L^p(\mathbb{T})$. If $f \in C(\mathbb{T})$, then the Fejér means converge uniformly. If $f \in L^1(\mathbb{T})$ has right and left limits $f(\theta \pm 0)$ at a point θ_0, then $(\sigma_n f)(\theta_0) \to \frac{1}{2}[f(\theta_0 + 0) + f(\theta_0 - 0)]$ when $N \to \infty$. If, in addition, $f \in L^\infty(\mathbb{T})$, then $|\sigma_n(f)| \leq \|f\|_\infty$.*

Proof. The first two statements follow immediately from the fact that K_N is an approximate identity and that $L^p(\mathbb{T})$ and $C(\mathbb{T})$ are homogeneous Banach spaces. For the third statement, note that the kernel K_N is even, so that we can write

$$E_n(\theta_0) := \sigma_N(f)(\theta_0) - \tfrac{1}{2}[f(\theta_0 + 0) + f(\theta_0 - 0)]$$
$$= \frac{1}{2\pi}\int_0^\pi (f(\theta_0 + \phi) + f(\theta_0 - \phi) - f(\theta_0 + 0) - f(\theta_0 - 0))\,K_N(\phi)\,d\phi.$$

The convergence now follows by Proposition 1.4.6. Finally, if $f \in L^\infty(\mathbb{T})$, then

$$\sigma_n(f) \leq \frac{\|f\|_\infty}{2\pi}\int_\mathbb{T} K_n(\theta)\,d\theta = \|f\|_\infty,$$

which completes the proof. ∎

Remark. The proof of pointwise convergence of the Fejér means only requires that the symmetrized function $\phi \to f(\theta + \phi) + f(\theta - \phi)$ have a limit when $\phi \to 0$. This principle applies more generally to operations that are defined by an even kernel, thus expressed in terms of the symmetrized function. Such extensions of the basic results are helpful and can be deduced when necessary.

Corollary 1.4.32. *The space of trigonometric polynomials is dense in the spaces $C(\mathbb{T})$ and $L^p(\mathbb{T})$ for $1 \leq p < \infty$.*

Proof. The nth Fejér mean $\sigma_n(f)$ is a trigonometric polynomial of degree n and converges to f in the norm of $C(\mathbb{T})$ and in the norm of $L^p(\mathbb{T})$. ∎

This leads to a new proof of the uniqueness of the Fourier coefficients, as follows:

Corollary 1.4.33. *If $f \in L^1(\mathbb{T})$ has Fourier coefficients identically zero, then $f = 0$ a.e.*

Proof. From formula (1.4.10) we see that $\sigma_n(f) \equiv 0$, hence $f = 0$ a.e. from the previous corollary. ∎

Exercise 1.4.34. *Suppose that f is bounded above and below: $m \leq f(\theta) \leq M$ for all $\theta \in \mathbb{T}$. Prove that $m \leq \sigma_n(f)(\theta) \leq M$ for all $n = 1, 2, \ldots, \theta \in \mathbb{T}$.*

Exercise 1.4.35. *Suppose that f is a continuous function with nonnegative Fourier coefficients: $\hat{f}(n) \geq 0, n \in \mathbb{Z}$. Prove that $\sum_{n \in \mathbb{Z}} \hat{f}(n) < \infty$.*

Hint: Apply Fatou's lemma to the Fejér means.

Exercise 1.4.36. *Suppose that f is continuous on a closed subinterval I. Prove that the Fejér means converge uniformly on I.*

1.4.3.1 Wiener's closure theorem on the circle

Fejér's theorem can be used to discuss the L^1 closure of the set of translates of a given $f \in L^1(\mathbb{T})$. This is identical to the closure of

$$\mathcal{M}_f := \{f * g : g \in L^1(\mathbb{T})\}.$$

Proposition 1.4.37. $\mathcal{M}_f$ *is dense on $L^1(\mathbb{T})$ if and only if $\hat{f}(n) \neq 0, \forall n \in \mathbb{Z}$.*

Proof. If $\hat{f}(n_0) = 0$ then for any $g \in L^1(\mathbb{T})$ we have $(\hat{f} * g)(n_0) = 0$. Therefore $\mathcal{M}_f$ lies in the proper closed subspace consisting of $\{F \in L^1(\mathbb{T}) : \hat{F}(n_0) = 0\}$. Conversely, if $\hat{f}(n) \neq 0, \forall n \in \mathbb{Z}$, let $F \in L^1(\mathbb{T})$ be given. By Fejér's theorem, $\sigma_{N-1}(F) \to F$ in $L^1(\mathbb{T})$ when $N \to \infty$. Let $g_N(\theta) = \sum_{-N}^{N} \hat{F}(k)/\hat{f}(k)(1 - (|k|/N))e^{ik\theta} \in L^1(\mathbb{T})$. Computing the Fourier coefficients, we see that $f * g_N = \sigma_{N-1}(F)$. Taking $N \to \infty$ completes the proof. ∎

1.4.4 *Equidistribution Modulo One

The density of the trigonometric polynomials can be used to give a direct treatment of a simple model in ergodic theory. The integer part and fractional part of a real number x are denoted $[x]$ and (x) respectively; thus $x = [x] + (x)$ with $[x] \in \mathbb{Z}$ and $0 \leq (x) < 1$.

If $\alpha \in (0, 2\pi)$ is a real number, we consider
$$N_n(a, b) = \text{card}\{0 \leq k \leq n-1 : (k\alpha/2\pi) \in (a, b)/2\pi\}$$
where $(a, b) \subset [0, 2\pi]$. We propose to show the following.

Proposition 1.4.38. *If $\alpha/2\pi$ is an irrational number and $0 \leq a < b \leq 2\pi$, then $N_n(a, b)/n \to (b-a)/2\pi$ when $n \to \infty$.*

Proof. To do this, we first write the counting function as
$$N_n(a, b) = \sum_{k=0}^{n-1} f(k\alpha)$$
where f is the 2π-periodic extension of the indicator function $1_{(a,b)}$. This leads us to study the class of measurable functions f for which

(1.4.13)
$$\lim_{n \to \infty} \frac{1}{n} \sum_{k=0}^{n-1} f(k\alpha) = \frac{1}{2\pi} \int_{\mathbb{T}} f(\theta)\, d\theta.$$

We will first prove (1.4.13) for trigonometric polynomials. If $f(\theta) = e^{im\theta}$, with $m \neq 0$, then
$$\frac{1}{n} \sum_{k=0}^{n-1} f(k\alpha) = \frac{1}{n} \sum_{k=0}^{n-1} e^{imk\alpha} = \frac{1 - e^{imn\alpha}}{n(1 - e^{im\alpha})} \to 0, \qquad n \to \infty$$
since the numerator is less than 2 and the denominator is a nonzero multiple of n. Now if $f(\theta) = \sum_{-N}^{N} a_k e^{ik\theta}$ is a trigonometric polynomial, we see immediately that
$$\frac{1}{n} \sum_{k=0}^{n-1} f(k\alpha) \to a_0 = \frac{1}{2\pi} \int_{\mathbb{T}} f(\theta)\, d\theta = \hat{f}(0), \qquad n \to \infty.$$

Now let f be any continuous function on $\mathbb{T}$, extended 2π-periodically to $\mathbb{R}$. By the corollary to Fejér's theorem there exists, for any $\epsilon > 0$, a trigonometric polynomial g so that $|f - g| < \epsilon$ on the real line. Therefore
$$\frac{1}{n} \sum_{k=0}^{n-1} g(k\alpha) - \epsilon \leq \frac{1}{n} \sum_{k=0}^{n-1} f(k\alpha) \leq \frac{1}{n} \sum_{k=0}^{n-1} g(k\alpha) + \epsilon.$$

Taking $n \to \infty$, the extreme members tend to $\hat{g}(0)$. Since this holds for any ϵ, we conclude that $\lim_{n \to \infty} \frac{1}{n} \sum_{k=0}^{n-1} f(k\alpha) = \hat{f}(0)$ whenever f is continuous and periodic. Finally, if $f = 1_{(a,b)}$, there exists a sequence of continuous functions $F_j^{\pm}$ with trapezoidal profiles so that $F_j^- \leq 1_{(a,b)} \leq F_j^+ \leq 1$, $\lim_j F_j^- = 1_{(a,b)}$ and $\lim_j F_j^+ = 1_{[a,b]}$ when $j \to \infty$. For each j we have
$$\frac{1}{n} \sum_{k=0}^{n-1} F_j^-(k\alpha) \leq \frac{N_n(a,b)}{n} \leq \frac{1}{n} \sum_{k=0}^{n-1} F_j^+(k\alpha)$$
$$\hat{F}_j^-(0) \leq \liminf_n \frac{N_n(a,b)}{n} \leq \limsup_n \frac{N_n(a,b)}{n} \leq \hat{F}_j^+(0).$$

Taking $j \to \infty$ we have
$$\frac{b-a}{2\pi} \leq \liminf_n \frac{N_n(a,b)}{n} \leq \limsup_n \frac{N_n(a,b)}{n} \leq \frac{b-a}{2\pi},$$
which completes the proof. ∎

Corollary 1.4.39. *If $\alpha/2\pi$ is irrational, then the set $\{(k\alpha/2\pi) : k \in \mathbb{Z}\}$ is dense in $[0, 1]$.*

Remark. It's clear that this result is true only if $\alpha/2\pi$ is irrational. If $\alpha/2\pi$ is rational, the set of $\{(k\alpha/2\pi) : k \in \mathbb{Z}\}$ is a finite subset of $(0, 1)$. In this case $N_n(a, b) = 0$ if (a, b) is in the complement of this finite set.

1.4.5 *Hardy's Tauberian Theorem

Having proved the convergence of the Fejér means, we can obtain results on the convergence of $S_n f$ by means of Tauberian techniques. We now prove the Tauberian theorem of Hardy, which applies to any sequence of complex numbers a_n with partial sums and Cesàro means denoted

$$s_n := \sum_{j=0}^n a_j, \quad \sigma_n := \frac{1}{n+1} \sum_{j=0}^n s_j = \frac{1}{n+1} \sum_{j=0}^n (n+1-j) a_j.$$

Theorem 1.4.40. *Suppose that $\{a_k\}_{k \geq 0}$ is a sequence of complex numbers such that $k|a_k| \leq M$ for some constant M. Then the convergence of the Cesàro means implies the convergence of the original sums: $\lim_n (s_n - \sigma_n) = 0$. If a_k depends on a parameter x so that the convergence of σ_k is uniform in x and $|ka_k| \leq M$ for all x, then s_n converges uniformly.*

Proof. For $n < m$ we study the expression

$$(m+1)\sigma_m - (n+1)\sigma_n - \sum_{j=n+1}^m (m+1-j) a_j$$

$$= \sum_{j=0}^n (m+1-j) a_j - \sum_{j=0}^n (n+1-j) a_j$$

$$= (m-n) \sum_{j=0}^n a_j$$

$$= (m-n) s_n.$$

Subtracting $(m-n)\sigma_n$ from both sides and dividing by $m-n$, we have the useful identity

(1.4.14) $$s_n - \sigma_n = \frac{m+1}{m-n} (\sigma_m - \sigma_n) - \frac{m+1}{m-n} \sum_{n+1}^m \left(1 - \frac{j}{m+1}\right) a_j.$$

The first term tends to zero whenever $m, n \to \infty$ with $(m+1)/(m-n)$ bounded, for example if $m/n \to K > 1$. To examine the second term, write

(1.4.15) $$(m+1) \sum_{n+1}^m \left(1 - \frac{j}{m+1}\right) |a_j| \leq M \left[(m+1) \log(m/n) - (m-n)\right].$$

Now we let $m, n \to \infty$ so that $m/n \to (1+\delta)$ and use the Taylor expansion of the logarithm; we see that the final term in (1.4.15) is less than $(m+1)\delta^2$, so that when we

divide by $(m-n)$ the sum is eventually less than 2δ. Thus we have
$$\limsup_n |s_n - \sigma_n| \leq 2\delta.$$
But δ was arbitrary, so that we conclude $s_n - \sigma_n \to 0$, as desired. ∎

How did Hardy think of the representation (1.4.14) for s_n in terms of σ_n? Although we cannot say with certainty, we can surely provide a natural motivation in terms of the continuous-parameter analogue. Suppose that three functions $a(t), s(t), \sigma(t)$ are related by the formulas
$$s(t) = \int_0^t a(x)\,dx, \qquad \sigma(t) = \frac{1}{t}\int_0^t s(x)\,dx.$$
To study $s(t)$ in terms of $a(t), \sigma(t)$, we can use Taylor's formula with remainder:
$$t_2\sigma(t_2) - t_1\sigma(t_1) = (t_2 - t_1)s(t_1) + \int_{t_1}^{t_2}(t_2 - x)a(x)\,dx, \qquad t_2 > t_1.$$
Solving for $s(t_1)$, we obtain
$$s(t_1) - \sigma(t_1) = t_2\frac{\sigma(t_2) - \sigma(t_1)}{t_2 - t_1} - \frac{\int_{t_1}^{t_2}(t_2 - x)a(x)\,dx}{t_2 - t_1},$$
which is the exact analogue of (1.4.14) in the continuous-parameter context.

Exercise 1.4.41. *Suppose that $f(t), t > 0$ is absolutely continuous with an absolutely continuous first derivative and that $f''(t) = O(1/t)$, $f(t)/t \to s$ when $t \to \infty$. Prove that $f'(t) \to s$ when $t \to \infty$.*

Hint: Take $f(t) = t\sigma(t)$ above.

We can now reap some consequences of Hardy's theorem.

Corollary 1.4.42. *Suppose that f is continuous on $\mathbb{T}$ and that its Fourier coefficients satisfy $\hat{f}(j) = O(1/|j|)$, $|j| \to \infty$. Then the Fourier series of f converges uniformly on $\mathbb{T}$ to f.*

Proof. We let $a_0 = \hat{f}(0)$ and $a_k = \hat{f}(k)e^{ik\theta} + \hat{f}(-k)e^{-ik\theta}$ for $k = 1, 2, \ldots$. The Fejér means converge uniformly, so that σ_n satisfies the hypotheses of Hardy's theorem, together with $a_j = O(1/|j|)$ when $j \to \infty$, uniformly in $\theta \in \mathbb{T}$. ∎

In particular, Hardy's theorem gives a new proof of the uniform convergence of the Fourier series of a continuous function of bounded variation.

Corollary 1.4.43. *Suppose that f is continuous and of bounded variation on $\mathbb{T}$. Then the Fourier series of f converges uniformly on $\mathbb{T}$ to f.*

Proof. We need to check the Fourier coefficients. By partial integration, we have
$$\hat{f}(n) = \frac{1}{2\pi}\int_{\mathbb{T}} f(\theta)e^{-in\theta}\,d\theta = \frac{1}{2in\pi}\int_{\mathbb{T}} e^{-in\theta}\,df(\theta) = O\left(\frac{1}{n}\right).$$
Here we used the fact that the Fourier coefficients of a finite measure are bounded. ∎

1.5 IMPROVED TRIGONOMETRIC APPROXIMATION

1.5.1 Rates of Convergence in $C(\mathbb{T})$

We now consider the Fourier approximation in the space $C(\mathbb{T})$. In contrast with the space $L^2(\mathbb{T})$, the Fourier partial sum is not the closest trigonometric polynomial in this norm. In order to study the rate of convergence of the Fourier partial sum, we recall the representation of the Fourier partial sum in terms of the Dirichlet kernel:

$$(1.5.1) \qquad S_N f = D_N * f \implies |S_N f(\theta)| \leq \frac{\max_\mathbb{T} |f|}{2\pi} \times \int_\mathbb{T} |D_N(\phi)|\, d\phi.$$

Suppose that g_N is another trigonometric polynomial of degree N. Then clearly $g_N = S_N g_N$ so that we have $S_N f - f = (S_N f - S_N g_N) + (g_N - f)$, from which we obtain

$$(1.5.2) \qquad |S_N f(\theta) - f(\theta)| \leq \frac{\max_\mathbb{T} |f - g_N|}{2\pi} \times \int_\mathbb{T} |D_N(\phi)|\, d\phi + \max_\mathbb{T}|f - g_N|.$$

Therefore the discrepancy $|S_N f - f|$ is measured in terms of the best trigonometric approximation and the L^1 norm of the Dirichlet kernel, which we now estimate. The *Lebesgue constants* are defined by the integrals

$$(1.5.3) \qquad L_n = \frac{1}{2\pi} \int_\mathbb{T} |D_n(\phi)|\, d\phi = \frac{1}{2\pi} \int_{-\pi}^{\pi} \frac{|\sin(n+\frac{1}{2})t|}{\sin(t/2)}\, dt.$$

The asymptotic behavior of the Lebesgue constants is provided as follows.

Proposition 1.5.1. *When $n \to \infty$, $L_n = (4 \log n / \pi^2) + O(1)$. Furthermore $L_n \leq 4 + \log n$ for all $n \geq 1$.*

Proof. Recall that on the interval $(0, \pi)$ the function

$$t \to \frac{1}{\sin(t/2)} - \frac{2}{t}$$

is bounded. Therefore we can write

$$L_n = \frac{1}{\pi} \int_0^\pi \frac{|\sin(n+\frac{1}{2})t|}{\sin(t/2)}\, dt$$

$$= \frac{2}{\pi} \int_0^\pi \frac{|\sin(n+\frac{1}{2})t|}{t}\, dt + O(1)$$

$$= \frac{2}{\pi} \int_0^{(n+\frac{1}{2})\pi} \frac{|\sin v|}{v}\, dv + O(1).$$

We are reduced to examining the integral of $|\sin v|/v$ on the interval $0 \leq v \leq (n+\frac{1}{2})\pi$. This is the sum of the integrals on intervals over $(k\pi, (k+1)\pi)$ for $0 \leq k \leq n$ plus a term over the last half interval, which tends to zero. The separate terms, apart from sign, are of the form

$$-\int_{k\pi}^{(k+1)\pi} \frac{\sin v}{v}\,dv = \int_{k\pi}^{(k+1)\pi} \frac{1}{v}\,d\cos v$$

$$= \frac{(-1)^{k+1}}{\pi k} + \frac{(-1)^{k+1}}{\pi(k+1)} + \int_{k\pi}^{(k+1)\pi} \frac{\cos v}{v^2}\,dv$$

$$= \frac{2(-1)^{k+1}}{\pi k} + O\!\left(\frac{1}{k^2}\right),$$

so that

$$L_n = \frac{2}{\pi}\sum_{k=1}^n \left(\frac{2}{\pi k} + O(k^{-2})\right) = \frac{4\log n}{\pi^2} + O(1).$$

To obtain the upper bound, we underestimate the denominator of the integrand in (1.5.3), replacing $\sin(t/2)$ by t/π and considering separately the integrals on $0 \le v \le 1$ and $1 \le v \le m\pi$, $m := n + \frac{1}{2}$. ∎

Referring to formula (1.5.2), we obtain the following useful fact.

Proposition 1.5.2. *Let $f \in C(\mathbb{T})$. Then the maximum discrepancy between f and its Nth Fourier partial sum is bounded by $\log N$ times the best trigonometric approximation of f by any trigonometric polynomial of degree N:*

$$|f(x) - S_N f(x)| \le [5 + \log N] \times \inf_{g \in P_N} \max_{\mathbb{T}} |f - g|$$

1.5.2 Approximation with Fejér Means

In parallel with the L^2 theory developed in Section 1.3.3, we wish to develop corresponding results on the speed of trigonometric approximation in the supremum norm on the space $C(\mathbb{T})$. The Fejér means are an efficient device for obtaining the first results of this type. In this section we will develop results for Lipschitz and Hölder continuity. Higher order smoothness will be treated in the subsequent sections.

To begin, we write the Fejér approximation of order $m - 1$ (to simplify the formulas):

$$(\sigma_{m-1}f)(x) - f(x) = \frac{1}{2\pi m}\int_{-\pi}^{\pi}\left(\frac{\sin mt/2}{\sin t/2}\right)^2 [f(x+t) - f(x)]\,dt$$

$$= \frac{1}{\pi m}\int_{-\pi/2}^{\pi/2}\left(\frac{\sin mu}{\sin u}\right)^2 [f(x+2u) - f(x)]\,du$$

(1.5.4)
$$= \frac{1}{\pi m}\int_0^{\pi/2}\left(\frac{\sin mu}{\sin u}\right)^2 [f(x+2u) + f(x-2u) - 2f(x)]\,du.$$

Using this form of the approximation, we can now state and prove some properties of the Fejér approximation.

Theorem 1.5.3. *Suppose that f is Lipschitz continuous with constant K. Then*

$$\|\sigma_n(f) - f\|_\infty \le C_1 K \frac{\log n}{n}$$

where C is an absolute constant. More generally, if f satisfies a Hölder condition with exponent $\alpha < 1$ and Holder constant K_α, then

$$\|\sigma_n(f) - f\|_\infty \leq C_\alpha K_\alpha n^{-\alpha}$$

where C_α depends only on α.

Proof. Beginning with formula (1.5.4), we integrate by parts, writing $\Phi(y) = \int_0^y |f(x+u) + f(x-u) - 2f(x)|\, du$. From the hypothesis of Lipschitz continuity, we have $\Phi(y) \leq K_1 y^2$. Now the Fejér kernel is bounded everywhere by m and by $\pi^2/4mu^2$ on the interval $u > 1/m$. Therefore we can write

$$\pi|\sigma_{m-1}(f) - f| \leq \frac{1}{m}\left(\int_0^{1/m} + \int_{1/m}^{\pi/2}\right)\left(\frac{\sin mu}{\sin u}\right)^2 d\Phi(u).$$

The first integral is estimated as

(1.5.5) $$\frac{1}{m}\left|\int_0^{1/m}\left(\frac{\sin mu}{\sin u}\right)^2 d\Phi(u)\right| \leq m\Phi\left(\frac{1}{m}\right) \leq \frac{K_1}{m},$$

while the second integral is estimated by

$$\frac{4}{m\pi^2}\left|\int_{1/m}^{\pi/2}\left(\frac{\sin mu}{\sin u}\right)^2 d\Phi(u)\right| \leq \int_{1/m}^{\pi/2} \frac{1}{mu^2} d\Phi(u)$$

(1.5.6) $$= (4/m\pi^2)\Phi(\pi/2) - m\Phi(1/m) + \int_{1/m}^{\pi/2} \frac{2\Phi(u)}{mu^3}\, du.$$

The first two of these terms is $O(1/m)$, $m \to \infty$, and the last integral is $O(\log m/m)$, completing the first part. In case f is only Hölder continuous, we have $\Phi(u) \leq K_\alpha u^{1+\alpha}$. Appealing again to (1.5.5) and (1.5.6), we see that the desired conclusion holds. ∎

Remark. One may inquire on the choice of the cutoff level $1/m$ in the above proof. It can be checked that this is optimal in terms of balancing the size of the two error terms.

Exercise 1.5.4. *Suppose that f satisfies the symmetric Hölder condition $|f(x+u) + f(x-u) - 2f(x)| \leq K_\alpha |u|^\alpha$ for some K, where $0 < \alpha < 1$. Show that we still have $\|\sigma_n(f) - f\|_\infty \leq C_\alpha K_\alpha n^{-\alpha}$. Formulate the corresponding result for $\alpha = 1$.*

Exercise 1.5.5. *Suppose that f satisfies the continuity condition $|f(x+h) - f(x)| \leq C_1/\log(1/h)$ for $0 < h < \frac{1}{2}$. Prove that $\|\sigma_n(f) - f\|_\infty \leq C_2/\log n$ for $n \geq 2$, for a suitable constant C_2.*

The above transformations can be used to prove a theorem of Lebesgue.

Theorem 1.5.6. *Suppose that $f \in L^1(\mathbb{T})$ satisfies the condition that*

$$\frac{1}{h}\int_0^h |f(x+u) + f(x-u) - 2S|\, du \to 0, \quad h \to 0,$$

for some $x \in T$, $S \in \mathbb{C}$. Then the Fejér means converge to S: $\lim_n \sigma_n(f)(x) = S$.

Proof. Using the same notations as above, we let

$$\Phi(h) = \int_0^h |f(x+2u) + f(x-2u) - 2S|\, du,$$

so that $\Phi(h) = o(h)$, $h \to 0$. Referring to (1.5.5) and (1.5.6) we have

(1.5.7)
$$\frac{1}{m}\int_0^{1/m} \left(\frac{\sin mu}{\sin u}\right)^2 d\Phi(u) \le m\Phi(1/m) \le \epsilon,$$

$$\frac{4}{m\pi^2}\left|\int_{1/m}^{\pi/2}\left(\frac{\sin mu}{\sin u}\right)^2 d\Phi(u)\right| \le \int_{1/m}^{\pi/2} \frac{1}{mu^2} d\Phi(u)$$

(1.5.8)
$$= \frac{4}{m\pi^2}\Phi(\pi/2) - m\Phi(1/m) + \int_{1/m}^{\pi/2} \frac{2\Phi(u)}{mu^3}\, du.$$

The first term is $O(m^{-1})$ and the second term is $o(1)$. The final integral clearly tends to zero if we change the notation and write $\Phi(u)/u = \epsilon(1/v) \to 0$, so that

$$\int_{1/m}^{\pi/2} \frac{\Phi(u)}{mu^3}\, du = \frac{1}{m}\int_{2/\pi}^{m} \epsilon(1/v)\, dv \to 0.\qquad\blacksquare$$

Corollary 1.5.7. *The Fejér means of an L^1 function converge almost everywhere.*

Proof. The strong form of Lebesgue's differentiation theorem states that for almost all $x \in \mathbb{T}$, $\lim_{h\to 0}(1/h)\int_0^h |f(x+u) - f(x)|\, du \to 0$. Therefore on this set we can take $S = f(x)$ in the previous theorem. $\blacksquare$

Corollary 1.5.8. *The Abel means of an L^1 function converge almost everywhere.*

Proof. If a sequence $\{s_n\}$ is Cesàro-summable, then it is also Abel-summable to the same sum, from Proposition 1.4.25. Since $\sigma_n(f)$ converges almost everywhere, the same is true of the Abel means $P_r f$. $\blacksquare$

We close this section with a negative result, showing that the Fejér means have an inherent limitation in their ability to approximate functions to a higher order of approximation.

Proposition 1.5.9. *Suppose that $f \in C(\mathbb{T})$ satisfies $\|\sigma_n(f) - f\|_\infty = o(1/n)$, $n \to \infty$. Then f is a constant, almost everywhere.*

Proof. Recall that

$$\sigma_n(f) = \sum_{-n}^{n}\left(1 - \frac{|k|}{n+1}\right)\hat{f}(k)e^{ik\theta}.$$

Hence

$$\frac{|k|\hat{f}(k)}{n+1} = \frac{1}{2\pi}\int_{\mathbb{T}}[f(\theta) - \sigma_n(f)]e^{-ik\theta}\, d\theta, \qquad |k| \le n,$$

$$|k|\,|\hat{f}(k)| \le \frac{n+1}{2\pi}\int_{\mathbb{T}} |f(\theta) - \sigma_n(f)|\, d\theta, \qquad |k| \le n.$$

For any fixed k, the right side tends to zero when $|n| \to \infty$, hence $\hat{f}(k) = 0$ for all $k \neq 0$. By the uniqueness of Fourier coefficients, we conclude that a.e. $f \equiv \hat{f}(0)$. ∎

Exercise 1.5.10. *Suppose that $f \in L^1(\mathbb{T})$ satisfies $\|\sigma_n(f) - f\|_1 = o(1/n)$, $n \to \infty$. Prove that f is a constant, almost everywhere.*

Finally, we note that the Fejér approximation holds with the rate $O(n^{-1})$ for well-behaved functions.

Exercise 1.5.11. *Suppose that the Fourier coefficients of $f \in L^1(\mathbb{T})$ satisfy $\sum_{n \in \mathbb{Z}} |n| |\hat{f}(n)| < \infty$. Prove that $\|\sigma_n(f) - f\|_\infty \leq C/n$ for some constant C.*

Exercise 1.5.12. *Suppose that the Fourier coefficients of $f \in L^1(\mathbb{T})$ satisfy $\sum_{n \in \mathbb{Z}} |n| |\hat{f}(n)| < \infty$. Prove that the uniform limit of $n(\sigma_n(f) - f)$ exists and compute its Fourier series.*

In the next two sections we introduce other approximate identities to obtain higher-order trigonometric polynomial approximations.

1.5.3 *Jackson's Theorem

If f has additional smoothness properties, we can obtain quantitive estimates for $f - g_N$ by working with the Jackson means.

The *Jackson means of order four* are defined by

$$J_N f(x) = \frac{1}{2h_N} \int_0^{\pi/2} \left(\frac{\sin Nu}{\sin u}\right)^4 (f(x+2u) + f(x-2u)) \, du$$

where

$$h_N := \int_0^{\pi/2} \left(\frac{\sin Nu}{\sin u}\right)^4 du.$$

By examining the transformations in (1.5.4), it is clear that $J_N f$ is the convolution of f with the square of the Fejér kernel, a trigonometric polynomial of degree $2N - 2$. Therefore $J_N f$ is a trigonometric polynomial of degree $2N - 2$.

Recall that f satisfies a *Lipschitz condition* if there exists a constant K so that

$$|f(x) - f(y)| \leq K|x - y|.$$

Theorem 1.5.13. *If f has Lipschitz constant K_1, then*

$$|J_N f(x) - f(x)| \leq C_1 \frac{K_1}{N},$$

where C_1 is a universal constant. If, in addition, the derivative f' is Lipschitz continuous with Lipschitz constant K_2, then

$$|J_N f(x) - f(x)| \leq C_2 \frac{K_2}{N^2},$$

where C_2 is a universal constant.

Proof. We write

$$J_N f(x) - f(x) = \frac{1}{2h_N} \int_0^{\pi/2} \left(\frac{\sin Nu}{\sin u} \right)^4 [f(x+2u) + f(x-2u) - 2f(x)] \, du$$

$$|J_N f(x) - f(x)| \leq \frac{1}{2h_N} \int_0^{\pi/2} \left(\frac{\sin Nu}{\sin u} \right)^4 4K_1 |u| \, du.$$

The denominator is estimated by

$$h_N \geq \int_0^{\pi/2} \left(\frac{\sin Nu}{u} \right)^4 du = N^3 \int_0^{N\pi/2} \left(\frac{\sin v}{v} \right)^4 dv \sim \text{const } N^3.$$

To estimate the numerator, note that, on the interval $[0, \pi/2]$, $\sin u$ is bounded below by $2u/\pi$. Making the change of variable $v = Nu$, the integral is no more than

$$\frac{\pi^4 K_1}{16} \int_0^{\pi/2} \left(\frac{\sin Nu}{u} \right)^4 u \, du = \frac{\pi^4 K_1}{16} N^2 \int_0^{N\pi/2} \left(\frac{\sin v}{v} \right)^4 v \, dv,$$

which proves the first statement and identifies the constant as

$$C_1 = \frac{\pi^4}{32} \frac{\int_0^\infty (\sin v/v)^4 v \, dv}{\int_0^\infty (\sin v/v)^4 \, dv}.$$

If, in addition, f' is Lipschitz, then the mean-value theorem provides the estimate $|f(x+2u) + f(x-2u) - 2f(x)| \leq 8K_2|u|^2$, which gives the improved estimate

$$|J_N f(x) - f(x)| \leq \frac{1}{h_N} \int_0^{\pi/2} \left(\frac{\sin Nu}{\sin u} \right)^4 8K_2 |u|^2 \, du.$$

Again we make the change of variable $v = Nu$ and find the required estimate, with

$$C_2 = \frac{\pi^4}{2} \frac{\int_0^\infty (\sin v/v)^4 v^2 \, dv}{\int_0^\infty (\sin v/v)^4 \, dv}.$$

∎

Exercise 1.5.14. *Suppose that f is Hölder continuous with exponent α: $|f(x) - f(y)| \leq K|x-y|^\alpha$ for some $0 < \alpha < 1$. Prove that $|J_N f(x) - f(x)| \leq C_\alpha K N^{-\alpha}$ for a universal constant C_α.*

Exercise 1.5.15. *Suppose that f is absolutely continuous and that f' is Hölder continuous with exponent α, $0 < \alpha < 1$. Prove that $|J_N f(x) - f(x)| \leq C'_\alpha K N^{-1-\alpha}$ for a universal constant C'_α.*

1.5.4 *Higher-Order Approximation

If $r \in \mathbb{Z}^+$, the space $C^r(\mathbb{T})$ consists of functions whose rth derivative $f^{(r)}$ is a continuous function. If $r \in \mathbb{Z}^+$ and $0 < \alpha \leq 1$, the space $C^{r,\alpha}(\mathbb{T})$ consists of functions $f \in C^r(\mathbb{T})$ such that $f^{(r)}$ satisfies a Hölder condition of order α. Since any differentiable function is Lipschitz continuous, we have the inclusion $C^{r+1}(\mathbb{T}) \subset C^{r,1}(\mathbb{T})$. If $0 < \alpha < 1$, we often write, by abuse of notation, $C^{r,\alpha}(\mathbb{T}) = C^{r+\alpha}(\mathbb{T})$.

It is natural to expect that if $f \in C(\mathbb{T})$ has derivatives of higher order, then we will obtain an improved rate of approximation by suitable trigonometric polynomials.

To make this concrete, consider for any even integer $2k$, the difference operator

$$\Delta_{2k}f(x;u) = \sum_{j=0}^{2k}\binom{2k}{j}(-1)^j f(x+u(k-j))$$

(1.5.9)
$$= f(x+ku) - 2kf(x+(k-1)u) + \cdots + (-1)^k\binom{2k}{k}f(x) + \cdots$$
$$\quad - 2kf(x-(k-1)u) + f(x-ku).$$

The coefficients are those that occur in the binomial expansion of $(e^{i\theta}-1)^{2k}$, which vanishes to order $2k$ at $\theta = 0$. Hence if $f \in C^{2k}(\mathbb{T})$, the derivatives are

$$\left(\frac{d}{du}\right)^j \Delta_{2k}f(x;u)|_{u=0} = 0, \qquad 0 \le j \le 2k-1,$$

$$\left(\frac{d}{du}\right)^{2k} \Delta_{2k}f(x;u)|_{u=0} = (2k)! f^{(2k)}(x),$$

so that we have the bound

$$|\Delta_{2k}f(x;u)| \le C u^{2k}.$$

More generally, we can apply Taylor's theorem with remainder to prove that when $u \to 0$

$$f \in C^{2k}(\mathbb{T}) \implies \Delta_{2k}f(x;u) = O(u^{2k}), \qquad u \to 0$$
$$f \in C^{2k-1,\alpha}(\mathbb{T}) \implies \Delta_{2k}f(x;u) = O(u^{2k-1+\alpha}), \qquad u \to 0, \quad 0 < \alpha \le 1$$
$$f \in C^{2k-1}(\mathbb{T}) \implies \Delta_{2k}f(x;u) = O(u^{2k-1}), \qquad u \to 0$$
$$f \in C^{2k-2,\alpha}(\mathbb{T}) \implies \Delta_{2k}f(x;u) = O(u^{2k-2+\alpha}), \qquad u \to 0, \quad 0 < \alpha \le 1.$$

In order to construct improved trigonometric approximations, we consider

$$E_{2k,m}f(x) := \int_0^{\pi/2} \Delta_{2k}f(x;u)\left(\frac{\sin mu}{\sin u}\right)^{2k+2} du.$$

Noting that $(\sin(mu)/\sin u)^{2k+2}$ is a trigonometric polynomial of degree $(m-1)(k+1)$, we see that the same is true for each term that figures in the definition of $E_{2k,m}f$, save for the middle term with $j = k$. If $f \in C^{2k}(\mathbb{T})$, we have

(1.5.10)
$$|E_{2k,m}f(x)| \le C \int_0^{\pi/2} u^{2k}\left(\frac{\sin mu}{\sin u}\right)^{2k+2} du$$

(1.5.11)
$$\le C \int_0^{\pi/2} u^{2k}\left(\frac{\sin mu}{2u/\pi}\right)^{2k+2} du$$

(1.5.12)
$$\le C \int_0^{\infty} u^{2k}\left(\frac{\sin mu}{2u/\pi}\right)^{2k+2} du$$

(1.5.13)
$$= mC \int_0^{\infty} v^{2k}\left(\frac{\sin u}{2u/\pi}\right)^{2k+2} dv,$$

while

$$C_{km} := \int_0^{\pi/2} \left(\frac{\sin mu}{\sin u}\right)^{2k+2} du$$

$$\geq \int_0^{\pi/2} \left(\frac{\sin mu}{u}\right)^{2k+2} du$$

$$= m^{2k+1} \int_0^{m\pi/2} \left(\frac{\sin v}{v}\right)^{2k+2} dv.$$

These two estimates are combined to prove the following.

Proposition 1.5.16. *Suppose that* $f \in C^{2k}(\mathbb{T})$. *Then there exists a sequence of trigonometric polynomials f_m of degree $(m-1)(k+1)$, so that*

$$\|f - f_m\|_\infty \leq C_k m^{-2k} \|f^{(2k)}\|_\infty.$$

Proof. It suffices to set

(1.5.14) $$f_m(x) = f(x) - \frac{E_{2k,m} f(x)}{D_{km}},$$

(1.5.15) $$D_{km} = (-1)^k \binom{2k}{k} \int_0^{\pi/2} \left(\frac{\sin mu}{\sin u}\right)^{2k+2} du = (-1)^k \binom{2k}{k} C_{km}.$$

Note that f_m is a trigonometric polynomial of degree $(m-1)(k+1)$. Then $D_{km}(f(x) - f_m(x)) = (E_{2k,m} f)(x)$. Applying the above estimates gives the result. ∎

In the general case of functions with a Hölder continuous derivative, we have the following general result.

Theorem 1.5.17. *Suppose that* $f \in C^{r,\alpha}(\mathbb{T})$. *Then there exists a sequence of trigonometric polynomials f_m of degree $\leq m(r+2)$ so that*

$$\|f - f_m\|_\infty \leq C m^{-(r+\alpha)} \|f\|_{r,\alpha}$$

where the norm is defined as

$$\|f\|_{r,\alpha} = \sup_{x \in \mathbb{T}} |f(x)| + \cdots + |f^{(r)}(x)| + \sup_{x \neq y \in \mathbb{T}} \frac{|f^{(r)}(x) - f^{(r)}(y)|}{|x - y|^\alpha}$$

and the constant C depends only upon r and α.

Exercise 1.5.18. *Prove the estimates (1.5.10).*

Exercise 1.5.19. *Complete the steps of the proof of Theorem 1.5.17.*

It is often convenient to have a *universal* sequence of trigonometric polynomials, which can be used at every level of differentiablity. These are provided by the de la Vallée

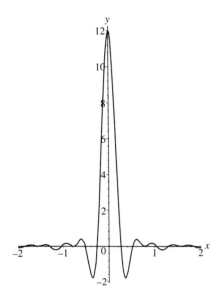

FIGURE 1.5.1
The de la Vallée Poussin kernel with $N = 8$

Poussin means (Figure 1.5.1), which are defined in terms of the Fejér means by

$$\tau_n(x) = 2\sigma_{2n-1}(x) - \sigma_{n-1}(x).$$

Let $E_n(f) = \inf_{T \in \mathcal{P}_n} \|f - T\|_\infty$ be the sup-norm distance between f and the trigonometric polynomials of degree n.

Theorem 1.5.20. *The de la Vallée Poussin means satisfy the estimate*

$$\|\tau_n - f\|_\infty \leq 4E_n(f).$$

Proof. The above infimum is attained by some (possibly nonunique) $T_n^* \in \mathcal{P}_n$. Indeed, any minimizing sequence p_k must be uniformly bounded, in particular have uniformly bounded Fourier coefficients. But these reside in the finite dimensional space $\mathbb{C}^{2n+1}$, where one can apply the Bolzano-Weierstrass theorem to obtain a convergent subsequence, for which the corresponding trigonometric polynomials are uniformly convergent to some $T_n^* \in \mathcal{P}_n$. Now we write

$$f(x) = T_n^*(x) + R(x) \qquad \text{where} \qquad |R(x)| \leq E_n(f).$$

The Fourier partial sum operators S_k and Fejér means σ_{k-1} are defined by

$$S_k f(x) = \sum_{j=-k}^{k} \hat{f}(j) e^{ijx}, \qquad S_k R(x) = \sum_{j=-k}^{k} \hat{R}(j) e^{ijx}$$

$$\sigma_{k-1} f(x) = \frac{1}{k} \sum_{j=0}^{k-1} S_j f(x), \qquad \sigma_{k-1} R(x) = \frac{1}{k} \sum_{j=0}^{k-1} S_k R(x).$$

If $k \geq n$ we have $S_k T_n^* = T_n^*$, so that $S_k f = T_n^* + S_k R$ and hence

$$\frac{1}{n}\sum_{k=n}^{2n-1} S_k f = T_n^* + \frac{1}{n}\sum_{k=n}^{2n-1} S_k R,$$

which can be written in terms of the delayed Fejér means:

$$2\sigma_{2n-1}f - \sigma_{n-1}f = T_n^* + 2\sigma_{2n-1}R - \sigma_{n-1}R.$$

But the Fejér kernel is a contraction in $L^\infty(\mathbb{T})$, thus $|\sigma_k(R)| \leq \|R\|_\infty \leq E_n(f)$. In particular

$$|\tau_n(x) - T_n^*(x)| = |2\sigma_{2n-1}f(x) - \sigma_{n-1}f(x) - T_n^*(x)| \leq 2E_n(f) + E_n(f) = 3E_n(f)$$

so that

$$\|\tau_n - f\|_\infty \leq \|\tau_n - T_n^*\|_\infty + \|T_n^* - f\|_\infty$$
$$\leq 3E_n(f) + E_n(f)$$
$$= 4E_n(f),$$

and the proof is complete. ∎

The de la Vallée Poussin approximation can be written in terms of the Fourier coefficients as follows:

$$\tau_n(x) = 2\sum_{|j|\leq 2n-1}\left(1 - \frac{|j|}{2n}\right)\hat{f}(j)e^{ijx} - \sum_{|j|\leq n}\left(1 - \frac{|j|}{n}\right)\hat{f}(j)e^{ijx}$$

$$= \sum_{|j|\leq n-1}\hat{f}(j)e^{ijx} + 2\sum_{n\leq |j|<2n}\left(1 - \frac{|j|}{2n}\right)\hat{f}(j)e^{ijx}.$$

In terms of the Fejér kernel, we have

$$2K_{2n-1}(t) - K_{n-1}(t) = \frac{2}{2n}\left(\frac{\sin nt}{2\sin t/2}\right)^2 - \frac{2}{n}\left(\frac{\sin nt/2}{2\sin t/2}\right)^2$$

$$= \frac{2\sin^2 nt - 2\sin^2(nt/2)}{4n\sin^2(t/2)}$$

$$= \frac{\cos nt - \cos 2nt}{4n\sin^2(t/2)}.$$

Unlike the Fejér kernel, the kernel of de la Vallée Poussin is not positive. However it does satisfy the general properties of approximate identities.

Exercise 1.5.21. *Prove directly that the de la Vallée Poussin kernel satisfies the three properties for an approximate identity.*

1.5.5 *Converse Theorems of Bernstein

Bernstein showed that the rates of convergence of trigonometric approximation can be used to characterize the degree of smoothness of a function. The key to proving these converse theorems is the following inequality for trigonometric polynomials.

Lemma 1.5.22. **Bernstein:** If $f(\theta) = \sum_{k=-M}^{M} a_k e^{ik\theta}$, then

(1.5.16) $$\sup_{\theta \in \mathbb{T}} |f'(\theta)| \leq 2M \times \sup_{\theta \in \mathbb{T}} |f(\theta)|.$$

Proof. Recall the Fejér kernel

$$K_{M-1}(\theta) = \frac{1}{M}\left(\frac{\sin M\theta/2}{\sin \theta/2}\right)^2 = \sum_{-M}^{M}\left(1 - \frac{|k|}{M}\right)e^{ik\theta}.$$

This will allow us to represent $f'(\theta)$ as an integral transform of f. We begin with the Fourier coefficients

$$\Delta(k) := \frac{1}{2\pi}\int_{-\pi}^{\pi} K_{M-1}(\phi)e^{-ik\phi}\,d\phi = \begin{cases} 1 - |k|/M & |k| \leq M \\ 0 & |k| > M \end{cases}.$$

Hence $\Delta(k - M) - \Delta(k + M) = k/M$ for $|k| \leq M$. Thus

(1.5.17) $$\frac{k}{M} = \frac{1}{2\pi}\int_{-\pi}^{\pi} K_{M-1}(\phi)e^{-ik\phi}2i\sin M\phi\,d\phi, \qquad |k| \leq M.$$

Multiply (1.5.17) by $iMa_k e^{ik\theta}$ and sum for $-M \leq k \leq M$:

$$f'(\theta) = \sum_{k=-M}^{M} ika_k e^{ik\theta}$$

$$= \frac{iM}{2\pi}\int_{-\pi}^{\pi} K_{M-1}(\phi)2i\sin M\phi\left(\sum_{k=-M}^{M} a_k e^{ik(\theta-\phi)}\right)d\phi$$

$$= \frac{M}{2\pi}\int_{-\pi}^{\pi} K_{M-1}(\phi)(-2\sin M\phi)f(\theta - \phi)\,d\phi$$

which is the desired representation. Hence

$$|f'(\theta)| \leq \frac{M}{2\pi}\left(\int_{-\pi}^{\pi} 2K_{M-1}(\phi)\,d\phi\right) \times \sup_{\phi \in \mathbb{T}}|f(\phi)|,$$

$$= 2M\sup_{\phi \in \mathbb{T}}|f(\phi)|$$

which completes the proof. ∎

To prove converse theorems in the supremum norm, it is convenient to work with trigonometric sums of order 2^k, known as *dyadic sums*. This subsequence is useful because of the following simple estimates that pertain to tail sums and finite sums:

$$\alpha > 0 \implies \sum_{k=N}^{\infty} 2^{-k\alpha} = \frac{2^{-N\alpha}}{1 - 2^{-\alpha}} = C_\alpha 2^{-N\alpha},$$

$$\beta > 0 \implies \sum_{k=0}^{N-1} 2^{k\beta} = \frac{2^{N\beta} - 1}{2^\beta - 1} \leq C_\beta 2^{N\beta}.$$

To study the approximation in detail, let $f \in C(\mathbb{T})$ and $T_n(x)$ be the trigonometric polynomial of degree n, which achieves the best approximation: $E_n(f) = \|f - T_n\|_\infty$.

Theorem 1.5.23. *If $E_n(f) \leq Cn^{-\alpha}$ for $C > 0$, $0 < \alpha < 1$, then f satisfies a Hölder condition: $|f(x+h) - f(x)| \leq Kh^\alpha$.*

Proof. We examine the limit in terms of dyadic sums. For any $n_0 \in Z^+$,

$$f(x) = \lim_{n \to \infty} T_{2^n}(x) = T_{2^{n_0}}(x) + \sum_{i=n_0+1}^{\infty} \Phi_i(x)$$

where $\Phi_i(x) = T_{2^i}(x) - T_{2^{i-1}}(x)$. From the triangle inequality we have $|\Phi_i(x)| \leq 3C2^{-\alpha i}$. Now define $m = m(h) \in Z^+$ by the inequalities $2^{m-1} < 1/h \leq 2^m$. Then

$$\left| \sum_{i=m(h)}^{\infty} \Phi_i(x) \right| \leq \sum_{i=m(h)}^{\infty} 3C2^{-i\alpha}$$

$$\leq C_\alpha 2^{-m\alpha}$$

$$\leq C_\alpha h^\alpha.$$

For the remaining sum we use the mean value theorem and Bernstein's lemma (1.5.16) to write

$$\sum_{i=n_0+1}^{m(h)-1} |\Phi_i(x+h) - \Phi_i(x)| \leq h \sum_{i=n_0+1}^{m(h)-1} \sup_{x \in \mathbb{T}} |\Phi_i'(x)|$$

$$\leq h \sum_{i=n_0+1}^{m(h)-1} C2^i \, 2^{-\alpha i}$$

$$\leq hC_\alpha 2^{m(h)(1-\alpha)}$$

$$\leq C_\alpha h^\alpha.$$

Hence $|f(x+h) - f(x)| \leq |T_{2^{n_0}}(x+h) - T_{2^{n_0}}(x)| + C_\alpha h^\alpha$ as required. ■

Exercise 1.5.24. *Use Theorem 1.5.23 and the proof of Theorem 1.5.3 to show that if f satisfies a symmetric Hölder condition: $|f(x+h) + f(x-h) - 2f(x)| \leq ch^\alpha$ for some $0 < \alpha < 1$, then f satisfies the usual one-sided Hölder condition: $|f(x+h) - f(x)| \leq Ch^\alpha$. (For $\alpha = 1$ this is false; see Zygmund (1959).)*

If $\alpha = 1$, the above estimates break down. Indeed, it is not generally true that $E_n(f) \leq C/n$ implies that $f \in \text{Lip}(\mathbb{T})$. The difficulty is in the estimate of

$$\sum_{i=n_0+1}^{m(h)-1} |\Phi_i(x+h) - \Phi_i(x)| \leq h \sum_{i=n_0+1}^{m(h)-1} \sup_{x \in \mathbb{T}} |\Phi_i'(x)|$$

$$\leq h \sum_{i=n_0+1}^{m(h)-1} C2^i \, 2^{-i}$$

$$\leq hCm$$

$$\leq Ch \log(1/h).$$

Thus we have the general implication

$$E_n(f) \leq \frac{C}{n} \implies |f(x+h) - f(x)| \leq Kh \log(1/h).$$

Bernstein's inequality can also be used to characterize the differentiability of f in terms of rates of convergence of $E_n(f)$. Indeed, suppose that $E_n(f) \leq C/n^\beta$ for some $\beta > 1$. Then the triangle inequality gives $|\Phi_i(x)| \leq 3C2^{-\beta i}$ and Bernstein's inequality (1.5.16) shows that for $r \in \mathbb{Z}^+, r < \beta$,

$$|\Phi_i^{(r)}(x)| \leq 2^{(i+1)r} 3C 2^{-\beta i} = C_r 2^{i(r-\beta)}.$$

Hence the series $\sum_{i=n_0+1}^{\infty} \Phi_i(x)$ can be differentiated term-by-term, since the numerical series $\sum_{i=1}^{\infty} 2^{i(r-\beta)}$ converges. We formalize this discussion as a theorem.

Theorem 1.5.25. *Suppose that $E_n(f) \leq C/n^{r+\alpha}$ where $r \in \mathbb{Z}^+$ and $0 < \alpha < 1$. Then $f \in C^r(\mathbb{T})$ and $f^{(r)}$ satisfies a Hölder condition of order α.*

Proof. From the above discussion, we have the uniformly convergent series

$$f^{(r)}(x) = \left(\frac{d}{dx}\right)^r T_{n_0}(x) + \sum_{i=n_0+1}^{\infty} \Phi_i^{(r)}(x).$$

The first term is infinitely differentiable, hence Hölder continuous. The second term is handled exactly as in the proof of Theorem 1.5.23, replacing Φ_i by $\Phi_i^{(r)}$, to which the same estimates apply. ∎

Exercise 1.5.26. Suppose that $E_n(f) \leq Cn^{-r}$ where $r \in \mathbb{Z}^+$. Show that $f \in C^{(r-1)}(\mathbb{T})$ and that $|f^{(r-1)}(x+h) - f^{(r-1)}(x)| \leq Kh\log(1/h)$.

Exercise 1.5.27. Suppose that $E_n(f) \leq C_1/\log n$ for all $n \geq 2$ for some constant C_1. Prove that f satisfies the continuity condition $|f(x+h) - f(x)| \leq C_2/\log(1/h)$ for $0 < h < \frac{1}{2}$. Compare with the result of Exercise 1.5.5.

1.6 DIVERGENCE OF FOURIER SERIES

In this section we turn to some negative results, which have been instrumental in the development of harmonic analysis. In 1873 du Bois-Reymond showed that there exists a continuous function whose Fourier series diverges at a point. This was further developed to show that any preassigned set of Lebesgue measure zero can be the set of divergence of the Fourier series of a continuous function (Kahane and Katznelson, 1966). Meanwhile, in 1915 Lusin had posed the problem of proving the almost-everywhere convergence of the Fourier series of an arbitrary $f \in L^2(\mathbb{T})$. This was proved by Carleson (1966) and extended by Hunt (1968) to the class $L^p(\mathbb{T})$ for $p > 1$. Another proof of Carleson's theorem by C. Fefferman (1973) has been useful in more recent developments of harmonic analysis. Many years earlier Kolmogorov (1926) had proved the existence of an L^1 function whose Fourier series diverges *at every point* of $\mathbb{T}$. These results and counterexamples are beyond the scope of this book. We will prove, by two different methods, the existence of continuous functions with Fourier partial sums unbounded at a point. We will also construct L^1 functions whose Fourier series do not converge in the L^1 norm. In Chapter 3 we will prove the theorem of M. Riesz (1927) that for any function in $L^p(\mathbb{T}), 1 < p < \infty$ the Fourier series converges in the L^p norm.

The upshot of these results and counterexamples is that $L^p, p > 1$ is a good space for one-dimensional Fourier series, both in the a.e. sense and in the sense of norm

1.6.1 The Example of du Bois-Reymond

Proof. By suitable grouping of terms, we will construct a continuous function whose Fourier series diverges at a preassigned point. Without loss of generality, we will do this at $\theta = 0$. The desired function will be sought in the form

$$(1.6.1) \qquad f(\theta) = \sum_{n=1}^{\infty} e^{iN_k\theta} \frac{B_k(\theta)}{k^2} \quad \text{where} \quad B_k(\theta) = \sum_{j=-m_k}^{m_k} a_j e^{ij\theta},$$

and where the integers m_k, N_k will be chosen. $B_k(\theta)$ is the partial sum of a Fourier series of a function of bounded variation with a jump discontinuity at the point $\theta = 0$ and otherwise smooth. For example, one may take $f(\theta) = \pi - \theta$ for $0 < \theta < \pi$, extended as an odd function. In this case $a_k = -i/k$. Since f is of bounded variation, the partial sums are uniformly bounded: $|B_k(\theta)| \le M$ for a constant M, simultaneously for all $\theta \in T$, $k = 1, 2, \ldots$. Hence the series (1.6.1) is uniformly convergent to a continuous function, by the Weierstrass M test. In order that the different blocks B_k involve different frequencies, we will choose the integers m_k, N_k so that

$$N_k + m_k < N_{k+1} - m_{k+1}.$$

Although (1.6.1) is not written as a Fourier series, we claim that the nth Fourier coefficient of f is given by the coefficient of $e^{in\theta}$ in the series. Indeed, since the series (1.6.1) is uniformly convergent, when we multiply by $e^{-in\theta}$ we still obtain a uniformly convergent series that we can integrate term-by-term:

$$\int_T f(\theta) e^{-in\theta} \, d\theta = \sum_{k=1}^{\infty} \frac{1}{k^2} \int_T B_k(\theta) e^{iN_k\theta} e^{-in\theta} \, d\theta.$$

Since the terms of B_k contain different frequencies, all of the integrals will be zero, save for the value of k satisfying $|n - N_k| \le m_k$, if there is one with $a_{n-N_k} \ne 0$. In that case the integral is $2\pi a_{n-N_k}$ and zero otherwise. Hence the nth Fourier coefficient is given by a_{n-N_k}/k^2 or zero, which completes the required identification.

Now we examine the partial sum at level N_k:

$$S_{N_k} f(0) = \sum_{j=1}^{N_k-1} \frac{B_j(0)}{j^2} + \frac{1}{k^2} \sum_{j=1}^{m_k} a_j.$$

The first sum converges by the Weierstrass M test. The second sum can be evaluated exactly in the case $a_k = -i/k$:

$$\sum_{j=1}^{m_k} \frac{1}{j} = \log m_k + O(1).$$

We now choose m_k so that $(\log m_k)/k^2 \to \infty$, for example $m_k = 2^{k^3}$ will do. Having chosen m_k, we choose the sequence N_k so that $N_{k+1} - N_k > m_k + m_{k+1}$ which is possible, for example by taking $N_1 = 1$ and $N_{k+1} = \sum_{j=0}^{k-1}(m_j + m_{j+1} + 1)$ for $k = 1, 2, \ldots$. The proof is complete. ∎

The example of du Bois-Reymond depends critically on the fact that the one-sided sums defined by $\sum_{j=0}^{k} a_j e^{ij\theta}$ are divergent when $k \to \infty$, whereas the corresponding two sided sums defined by $B_k(\theta)$ are convergent when $k \to \infty$.

Exercise 1.6.1. *Suppose that f is a function of bounded variation that has a jump discontinuity at $\theta = 0$ and is otherwise of class C^2 on the circle. Prove the asymptotic formula $\hat{f}(n) = C/n + O(1/n^2)$ and identify C in terms of the jump.*

1.6.2 Analysis via Lebesgue Constants

In this section we re-examine the questions of convergence and divergence in a more general setting. A *Banach space* is a complete normed linear space. For example $C(\mathbb{T})$, $L^p(\mathbb{T})$ for $p \geq 1$ are familiar Banach spaces. A mapping $T: B_1 \to B_2$ is a *bounded linear operator* if it satisfies the conditions that

(1.6.2) $\quad T(f+g) = Tf + Tg, \quad T(cf) = cT(f), \quad \|Tf\|_{B_2} \leq K\|f\|_{B_1}$

Here c is any complex number and K is a positive real number. For example, if $B_1 = B_2 = L^p(\mathbb{T})$ and $g \in L^1(\mathbb{T})$, the convolution $Tf = f * g$ defines a bounded linear operator. This is immediate from the B-valued norm computation

$$\|Tf\|_{B_2} = \left\|\int_\mathbb{T} f(x-y)g(y)\,dy\right\|_{B_2} = \left\|\int_\mathbb{T} g(y) f_y\,dy\right\|_{B_1}$$

$$\leq \int_\mathbb{T} |g(y)|\|f_y\|_{B_1}\,dy = \|f\|_{B_1}\int_\mathbb{T} |g(y)|\,dy = \|f\|_{B_1}\|g\|_{L^1(\mathbb{T})}.$$

The *operator norm* of a bounded linear operator is the smallest number K so that (1.6.2) holds. Equivalently,

$$\|T\| = \sup_{\|f\|\leq 1} \|Tf\|_{B_2}.$$

If T_n is a sequence of bounded linear operators, the pointwise convergence to a limiting operator T is defined by the requirement that

$$\lim_n \|T_n f - Tf\|_{B_2} = 0$$

for all $f \in B_1$. The following condition is clearly sufficient.

Proposition 1.6.2. *Suppose that T_n is a sequence of bounded linear operators and T is a bounded linear operator so that*

(i) *For some dense set $\mathcal{D}$, $\lim T_n f = Tf$ for all $f \in \mathcal{D}$.*
(ii) *There exists a constant K so that $\|T_n\| \leq K$ for all n and $\|T\| \leq K$.*

Then the sequence converges pointwise on all of B.

Proof. If $f \in B$, there exists $g \in \mathcal{D}$ so that $\|f - g\|_{B_1} < \epsilon$. Now

$$T_n f - Tf = (T_n f - T_n g) + (T_n g - Tg) + (Tg - Tf),$$

$$\|T_n f - Tf\|_{B_2} \leq \|T_n f - T_n g\|_{B_2} + \|T_n g - Tg\|_{B_2} + \|Tg - Tf\|_{B_2}$$

$$\leq K\|f - g\|_{B_1} + \|T_n g - Tg\|_{B_2} + K\|f - g\|_{B_1}.$$

Hence $\limsup_n \|T_n f - Tf\|_{B_2} \leq 2K\|f - g\|_{B_1} \leq 2K\epsilon$. But ϵ was arbitrary, completing the proof. ∎

This may be applied to give a new proof that the Abel and Cesàro means converge in $L^p(\mathbb{T})$. In each case we can take $K = 1$ and note that we have convergence on the space of trigonometric polynomials, since in each case we have $T_n f = T_{n_0} f$ for $n \geq n_0$, the degree of f.

It is surprising that the converse is true: If a sequence of linear operators converges pointwise, then the operator norms remain bounded. This is embodied in the *uniform boundedness principle*, stated as follows. The proof is given in an appendix to this chapter.

Theorem 1.6.3. *Suppose that T_n is a sequence of bounded linear operators so that for each $f \in B_1$, $\sup_n \|T_n f\|_{B_2} < \infty$. Then the sequence of operator norms remains bounded: $\sup_n \|T_n\| < \infty$.*

This is applied to study the Fourier partial sum operators on $C(\mathbb{T})$:

$$f \mapsto S_N f = \sum_{n=-N}^{N} \hat{f}(n) e^{in\theta}$$

$$= \frac{1}{2\pi} \int_{\mathbb{T}} D_N(\theta - \phi) f(\phi) \, d\phi.$$

The supremum norm is estimated by

$$\|S_N f\|_\infty \leq \|f\|_\infty \times \frac{1}{2\pi} \int_{\mathbb{T}} |D_n(\phi)| \, d\phi.$$

In fact this bound is sharp, since we can take the bounded function $f_0(\phi) = \operatorname{sgn} |D_n(\phi)|$ and achieve equality. This choice of f_0 is not continuous, but can be approximated boundedly be a sequence of continuous functions, for example by using the Fejér means to define $f_n = K_n f_0$. Then $2\pi S_N f_n(0) = \int_{\mathbb{T}} f_n D_N \to \int_{\mathbb{T}} f_0 D_N = \int_{\mathbb{T}} |D_N|$ when $n \to \infty$.

We now recall the estimation of Proposition 1.5.1, which states that $\|S_N\| \sim (4 \log N)/\pi^2$. This allows us to infer, without any computations, that there exists a continuous function whose partial sums do not remain uniformly bounded, especially not uniformly convergent.

Proposition 1.6.4. *There exists a continuous function whose Fourier partial sums do not remain uniformly bounded. For any $\theta \in \mathbb{T}$, there exists a continuous function whose Fourier partial sums are unbounded at θ.*

Proof. The first statement is evident from the previous discussion where we take $B_1 = B_2 = C(\mathbb{T})$. For the second statement simply take $B_1 = C(\mathbb{T})$ and $B_2 = \mathbb{C}$, the complex numbers. The operators $f \to S_n f(\theta)$ are each bounded and linear from B_1 to B_2. The operator norms are computed exactly as above: $\|S_n\| = L_n = (4 \log n)/\pi^2 + O(1)$, $n \to \infty$. Therefore, by the uniform boundedness principle, there exists a continuous function whose Fourier series is unbounded at θ. ∎

The following exercise shows that the Lebesgue constants can be used to estimate the rate of divergence of the Fourier partial sums for a function that has a local average.

Exercise 1.6.5. *Supppose that $f \in L^1(\mathbb{T})$ and for some $\theta \in \mathbb{T}$, $s := \lim_{\phi \to 0}[f(\theta + \phi) + f(\theta - \phi)]/2$ exists. Prove that $S_N f(\theta) = o(\log N)$ when $N \to \infty$.*

Hint: Given $\epsilon > 0$, choose $\delta > 0$ so that $|f(\theta + \phi) + f(\theta - \phi) - 2s| < 2\epsilon$ for $0 < \phi < \epsilon$. From the representation of partial sums we have

$$|S_N f(\theta) - s| = \frac{1}{\pi} \left| \int_0^\pi D_N(\phi)(\bar{f}_\theta(\phi) - s) \, d\phi \right|$$

$$\leq \epsilon \int_0^\delta |D_N(\phi)| \, d\phi + \left| \int_\delta^\pi D_N(\phi)(\bar{f}_\theta(\phi) - s) \, d\phi \right|$$

$$\leq \epsilon \log N + o(1).$$

For a general $f \in L^1(\mathbb{T})$, one can establish the above rate of divergence at almost every $\theta \in \mathbb{T}$. The proof is very similar to the proof of almost-everywhere Cesàro-summability in the space $L^1(\mathbb{T})$.

Proposition 1.6.6. *Let $f \in L^1(\mathbb{T})$. Then for almost every $\theta \in \mathbb{T}$, $S_N f(\theta) = o(\log N)$, $N \to \infty$.*

Proof. Let $F_\theta(\phi) = \int_0^\phi |f(\theta + u) + f(\theta - u) - 2f(\theta)| \, du$. From Lebesgue's differentiation theorem, we have for almost every $\theta \in \mathbb{T}$, $F_\theta(u)/u \to 0$ when $u \to 0$. Now we write $s = 2f(\theta)$ and, as above,

$$S_N f(\theta) - f(\theta) = \frac{1}{\pi} \left(\int_0^{\pi/N} + \int_{\pi/N}^\pi \right) D_N(\phi)(\bar{f}_\theta(\phi) - s) \, d\phi = I_1 + I_2.$$

On the first integral we use the fact that $|D_N(\phi)| \leq 2N + 1$, $0 < \phi < \pi$; thus

$$|I_1| \leq \frac{2N+1}{\pi} F_\theta(\pi/N) = o(1), \qquad N \to \infty.$$

On the interval $\pi/N \leq \phi \leq \pi$ we can write $D_N(\phi) \leq \pi^2/24 + 2/\phi$, so that

$$|I_2| \leq \frac{\pi}{24} + \frac{2}{\pi} \int_{\pi/N}^\pi \frac{F'(\phi)}{\phi} \, d\phi.$$

A partial integration shows further that

$$\int_{\pi/N}^\pi \frac{F'(\phi)}{\phi} \, d\phi = \frac{F(\phi)}{\phi} \bigg|_{\pi/N}^\pi + \int_{\pi/N}^\pi \frac{F(\phi)}{\phi^2} \, d\phi.$$

The first term is bounded when $N \to \infty$. On the other hand we write $\epsilon(\phi) := F(\phi)/\phi$, which tends to zero when $\phi \to 0$, so that

$$\int_{\pi/N}^{\pi} \frac{F'(\phi)}{\phi} d\phi = \int_{\pi/N}^{\pi} \frac{\epsilon(\phi)}{\phi} d\phi = o(\log N), \qquad N \to \infty.$$

The proof is complete. ∎

1.6.3 Divergence in the Space L^1

The analysis of divergence in the space $L^1(\mathbb{T})$ is entirely parallel. First we outline a qualitative argument to prove the existence of an L^1 divergent Fourier series, then we proceed to construct a class of examples.

In order to exploit the principle of uniform boundedness, we need to compute the norm of the operator S_N, which maps $L^1(\mathbb{T})$ to $L^1(\mathbb{T})$. On the one hand, for any $f \in L^1(\mathbb{T})$, we have

$$\|S_N f\|_1 = \|D_N * f\|_1 \leq \|D_N\|_1 \|f\|_1,$$

so that the operator norm $\|S_N\|_{1,1}$ is bounded by the Lebesgue constant L_N. On the other hand, if we take the Fejér kernel $f = K_n$ with $n \geq N$, then by the properties of the Fejér kernel

$$\|S_N f\|_1 = \|D_N * K_n\|_1 = \|\sigma_n(D_N)\|_1 \to \|D_N\|_1, \qquad n \to \infty,$$

since for any fixed N, $\sigma_n(D_N)$ converges boundedly to D_N when $n \to \infty$. Hence we conclude that the operator norm is given by

$$\|S_N\|_{1,1} = \|D_N\|_1 = L_N = \frac{4}{\pi^2} \log N + O(1).$$

By the uniform boundedness theorem, we conclude that there exists $f \in L^1(\mathbb{T})$ such that $\|S_N f\|_1$ is unbounded when $N \to \infty$. This completes the qualitative argument and proves the following proposition.

Proposition 1.6.7. *There exists $f \in L^1(\mathbb{T})$ whose Fourier series diverges in the L^1 norm:* $\sup_N \|S_N f\|_1 = +\infty$.

We now use Lebesgue constants to construct explicit examples of functions in the space $L^1(\mathbb{T})$ whose Fourier series do not converge in the L^1 norm. We begin with a sequence $\{a_n\}$ of positive real numbers tending to zero so that the second differences are nonnegative; in detail $(\Delta^2 a)_n := a_{n+1} + a_{n-1} - 2a_n \geq 0$. Such a sequence is termed *convex*.

Lemma 1.6.8. *For any convex sequence, we have $\lim_n n(a_n - a_{n+1}) = 0$ and the series $\sum_{n=0}^{\infty} n(\Delta^2 a)_n = a_0 < \infty$.*

Proof. Let $b_n = a_n - a_{n+1}$. From the convexity condition we see that $b_n \geq b_{n+1}$ for all n. Also $b_n \to 0$, since $a_n \to 0$.

To prove that $b_n \geq 0$ for all n, we assume that $b_{n_0} < 0$ for some n_0. By convexity we have $b_n \leq b_{n_0}$ for $n \geq n_0$. Then $a_n - a_{n_0} \geq (n_0 - n) b_{n_0}$, which implies $a_n \to \infty$, a contradiction.

Given $\epsilon > 0$, there exists $N(\epsilon)$ so that $a_n < \epsilon$, $b_n < \epsilon$ for $n > N$. Thus $\epsilon > a_N - a_n = b_{N+1} + \cdots + b_n \geq (n - N)b_n$, which proves that $\limsup_n nb_n \leq \epsilon$, which was to be proved. To prove the convergence of the series of second differences, use summation by parts to write

$$\sum_{n=1}^{N} n(\Delta^2 a)_n = a_0 - a_N - Nb_N.$$

When $N \to \infty$, the right side tends to a_0, as required. ∎

We can use these techniques to construct trigonometric series in $L^1(\mathbb{T})$. Let $\{a_n\}$ be a convex sequence and consider the sequence of functions

$$s_N(\theta) = a_0 + 2\sum_{n=1}^{N} a_n \cos(n\theta).$$

Recall the Dirichlet kernel and Fejér kernels:

$$D_n(\theta) = 1 + 2\cos\theta + \cdots + 2\cos(n\theta), \qquad K_n(\theta) = \frac{D_0(\theta) + \cdots + D_n(\theta)}{n+1}.$$

Inversely,

$$2\cos(n\theta) = (\Delta D)_n(\theta), \qquad D_n(\theta) = \Delta((n+1)K)_n(\theta).$$

Using summation by parts we can write

(1.6.3) $$s_N(\theta) = \sum_{n=1}^{N-1} n(\Delta^2 a)_n K_{n-1}(\theta) - NK_{N-1}(\theta)(\Delta a)_N + a_N D_N(\theta).$$

For any $\theta \neq 0$, the last two terms tend to zero, while the first sum remains bounded in L^1. We define $f \in L^1(\mathbb{T})$ by the L^1 convergent sum

(1.6.4) $$f(\theta) = \sum_{n=1}^{\infty} n(\Delta^2 a)_n K_{n-1}(\theta).$$

From the above lemma and the normalization of K_n we see that this series of nonnegative terms converges in L^1, hence the sum is finite almost everywhere and defines an L^1 function. It remains to compute the Fourier coefficients. To do this we multiply $f(\theta)$ by $e^{-im\theta}$ and integrate term-by-term, by dominated convergence. Thus we have

$$\hat{f}(m) = \sum_{n=0}^{\infty} n(\Delta^2 a)_n \hat{K}_{n-1}(m)$$

$$= \sum_{n=0}^{\infty} n(\Delta^2 a)_n \left(1 - \frac{|m|}{n}\right) = a_{|m|},$$

the last step being the result of summing by parts twice. Having established that $a_{|m|} = \hat{f}(m)$, we can investigate the L^1 norms.

Proposition 1.6.9. *Suppose that $\{a_n\}$ is a convex sequence and let f be defined by (1.6.4). Then the partial sums s_N remain bounded in L^1 if and only if $\{a_n \log n\}$ is a* bounded sequence. The partial sums are convergent in L^1 if and only if $a_n \log n \to 0$ when $n \to \infty$.

Proof. Returning to (1.6.3) we see that the first two terms are bounded in L^1. Therefore if s_N also remains bounded in L^1 the same must be true of the final term $a_n D_N$. But its L^1 norm is asymptotic to $(4/\pi^2) a_N \log N$, which proves the first statement. To prove the second statement, write

$$s_N(\theta) - f(\theta) = \sum_{n=N}^{\infty} n(\Delta^2 a)_n K_{n-1}(\theta) - N K_{N-1}(\theta)(\Delta a)_N + a_N D_N(\theta).$$

The first sum tends to zero in L^1 when $N \to \infty$, as does the second term. Therefore the convergence to zero of $\|s_N - f\|_1$ is equivalent to the convergence to zero of $\|a_N D_N\|_1$, which is equivalent to $a_N \log N \to 0$, which completes the proof. ∎

1.7 *APPENDIX: COMPLEMENTS ON LAPLACE'S METHOD

1.7.0.1 First variation on the theme-Gaussian approximation

The proof of Laplace's method can be modified at no expense to handle integrals of the form

$$C(t) = \int_a^b A(\mu) e^{tB(\mu)} e^{i\mu c} \, d\mu.$$

Assume that $A(\mu)$ and $B''(\mu)$ are Lipschitz continuous. Without loss of generality, we may assume that the maximum of B is attained at $\mu = 0 \in (a, b)$ and that $B(0) = 0$. Thus we apply Steps 1, 2 and 3 of Section 1.1.5 to reduce to

$$C(t) = A(0) \int_{-\delta}^{\delta} e^{-tk\mu^2} e^{i\mu c} \, d\mu + O\left(\frac{1}{t}\right).$$

Applying Step 4 we replace the limits by $-\infty < \mu < \infty$ and incur an exponential error so that

$$C(t) = A(0) \int_{-\infty}^{\infty} e^{-tk\mu^2} e^{i\mu c} \, d\mu + O\left(\frac{1}{t}\right).$$

But this is the Fourier transform of the Gaussian density, hence we have the asymptotic formula

$$C(t) = A(0) \sqrt{\frac{\pi}{kt}} e^{-c^2/4kt} + O\left(\frac{1}{t}\right).$$

All of the error estimates are independent of c, so that this can be used in cases when c depends on t. Of course, to provide useful information, it is only interesting when c is restrained, for example $c = O(\sqrt{t})$; if c is too large the exponential term will be smaller than the error term when $t \to \infty$.

1.7.0.2 Second variation on the theme-improved error estimate

If we have additional information on the function $B(\mu)$, we can refine the error estimate to $O(1/n^{3/2})$, which comes up in many problems. Specifically, assume that $A(\mu) \equiv 1$, and that $B(\mu)$ is four times differentiable with $B'(0) = 0$, $B''(0) = -k < 0$, $B'''(0) = ib_3$ for some b_3 real. We begin with

$$C(t) = \int_a^b e^{tB(\mu)} d\mu.$$

Assuming that the maximum is attained at $\mu = 0$ and that $B(0) = 0$ we immediately reduce to,

$$C(t) = \int_{-\delta}^{\delta} e^{tB(\mu)} d\mu + O(e^{-Ct}), \quad C > 0.$$

Now we use the inequality (1.1.41) with $z_1 = B(\mu)$, $z_2 = -k\mu^2 + i\mu^3 b_3/3$

$$|e^{-tB(\mu)} - e^{-t(k\mu^2/2 - i\mu^3 b_3/6)}| \leq tC_4 \mu^4 e^{-tk\mu^2/3}, \quad -\delta < \mu < \delta.$$

The integral of the remainder term is

$$\int_{-\delta}^{\delta} tC_4 \mu^4 e^{-tk\mu^2/3} d\mu < \int_{-\infty}^{\infty} tC_4 \mu^4 e^{-tk\mu^2/3} d\mu = \frac{C_5}{t^{3/2}},$$

as required. Now we use the inequality $|e^{ia} - 1 - ia| \leq a^2/2$ for a real to write

$$|e^{-t(k\mu^2/2 - i\mu^3 b_3)} - e^{-tk\mu^2/2} - i\mu^3 b_3 e^{-tk\mu^2/2}| \leq t^2 \mu^6 b_3^2 e^{-tk\mu^2/2}.$$

The integral of the cubic term is zero, since this is an odd function over a symmetric interval. The integral of the remainder term is

$$t^2 b_3^2 \int_{-\delta}^{\delta} \mu^6 e^{-tk\mu^2/2} d\mu < t^2 b_3^2 \int_{-\infty}^{\infty} \mu^6 e^{-tk\mu^2/2} d\mu = \frac{\text{const}}{t^{3/2}}.$$

Finally we replace the integral over $-\delta < \mu < \delta$ by an integral over $-\infty < \mu < \infty$ with an exponential error. Thus we have shown that

(1.7.1) $$C(t) = \sqrt{\frac{2\pi}{-tB''(\mu_0)}} + O\left(\frac{1}{t^{3/2}}\right), \quad t \to \infty.$$

1.7.1 *Application to Bessel Functions

The modified Bessel functions $I_m(t)$ are obtained from the absolutely convergent trigonometric series

$$e^{t \cos \theta} = \sum_{m=-\infty}^{\infty} e^{im\theta} I_{|m|}(t),$$

or explicitly as the Fourier coefficients

$$I_m(t) = \frac{1}{2\pi} \int_{-\pi}^{\pi} e^{-im\theta} e^{t \cos \theta} d\theta \quad m = 0, 1, 2, \ldots.$$

Laplace's method can be applied here with the choice $A(\theta) = e^{-im\theta}$, which is Lipschitz continuous and $B(\theta) = \cos \theta$, which is twice differentiable with B'' Lipschitz continuous.

B has a unique maximum at $\theta = 0$ with $B(0) = 1$, $B'(0) = 0$, $B''(0) = -1$, $B'''(0) = 0$. Applying (1.7.1) gives the asymptotic result

$$I_m(t) = \frac{e^t}{\sqrt{2\pi t}}\left[1 + O\left(\frac{1}{t}\right)\right], \qquad t \to \infty.$$

1.7.2 *The Local Limit Theorem of DeMoivre-Laplace

Laplace's method is naturally adopted to problems in the theory of probability. In the simplest probability model, one considers independent trials of an event whose probability p of success is assumed known with $0 < p < 1$. For example, if we have a fair coin it is natural to take $p = 1/2$. If we are rolling dice then it is natural to take $p = 1/6$, if success corresponds to a given face showing.

More general systems of probability distributions will be studied in Chapter 5, where we prove a more general form of the central limit theorem, generalizing the theorem of DeMoivre-Laplace.

Assuming n trials, this random experiment has 2^n possible outcomes, consisting of strings of zeros and ones, where 1 corresponds to success and 0 corresponds to failure. The probability of a given string with k successes is defined to be $p^k(1-p)^{n-k}$ and the number of such strings is the binomial coefficient $\binom{n}{k} = n!/k!(n-k)!$. This is summarized as the statement

The probability of k successes in n trials is $P_{k,n} := \binom{n}{k}p^k(1-p)^{n-k}$.

When we try to compute $P_{k,n}$ we find that the maximum value occurs at the integer closest to np, and that this maximum tends to zero when $n \to \infty$.

The local limit theorem of DeMoivre-Laplace is the following quantitative asymptotic statement:

Theorem 1.7.1. *Let $x = (k - np)/\sqrt{np(1-p)}$. Then for $n \to \infty$, uniformly in $0 \le k \le n$*

$$\binom{n}{k}p^k(1-p)^{n-k} = \frac{e^{-x^2/2}}{\sqrt{2\pi np(1-p)}} + O\left(\frac{1}{n^{3/2}}\right).$$

This theorem shows that the individual probabilities tend to zero, while approximating a bell-shaped curve. In particular if $p = 1/2$ and $x = 0$, we see that the probability of an equal number of successes and failures is asymptotic to $\sqrt{2/n\pi}$ when $n \to \infty$.

Proof. First we represent the binomial probability function as the Fourier coefficient of an elementary function. Writing $q = 1 - p$, we let $f(\theta) = pe^{i\theta} + q$. Then from the binomial theorem

(1.7.2) $$f(\theta)^n = \left(pe^{i\theta} + q\right)^n$$

(1.7.3) $$= \sum_{k=0}^{n} \binom{n}{k} p^k q^{n-k} e^{ik\theta},$$

from which we conclude

(1.7.4) $$P_{k,n} = \binom{n}{k} p^k q^{n-k} = \frac{1}{2\pi} \int_{-\pi}^{\pi} e^{-ik\theta} f(\theta)^n \, d\theta.$$

We cannot apply Laplace's method directly, since the integrand is not presented as an exponential. To find an equivalent exponential form, we note first that for $\theta \neq 0$, $f(\theta)$ lies on the segment joining two distinct points of the unit circle, hence $|f(\theta)| < 1$ for $0 < |\theta| \leq \pi$. Therefore, at the expense of an error which is exponentially small, we can restrict attention to the integral over $(-\delta, \delta)$. The Taylor expansion of $f(\theta)$ is

$$f(\theta) = 1 + ip\theta - p\theta^2/2 - ip\theta^3/6 + O(\theta^4), \qquad \theta \to 0.$$

Choosing δ so that $|1 - f(\theta)| < 1/2$ for $|\theta| < \delta$, we can define the logarithm

$$\log f(\theta) = ip\theta - p\theta^2/2 - ip\theta^3/6 - (1/2)[ip\theta - p\theta^2/2]^2 + (1/3)[ip\theta - p\theta^2/2]^3 + O(\theta^4)$$
$$= ip\theta - pq\theta^2/2 - i(\theta^3/6)pq(q-p) + O(\theta^4).$$

The Fourier representation (1.7.4) gives, up to an exponentially small error

$$P_{k,n} = \frac{1}{2\pi} \int_{-\delta}^{\delta} \exp\big(n[ip\theta - pq\theta^2/2 - i(\theta^3/6)pq(q-p) + O(\theta^4)] - i\theta(np + x\sqrt{npq})\big) \, d\theta$$
$$= \frac{1}{2\pi} \int_{-\delta}^{\delta} \exp\big[-npq\theta^2/2 - ix\theta\sqrt{npq} - i(n\theta^3/6)pq(q-p) + O(n\theta^4)\big] \, d\theta.$$

We can apply the first variation of Laplace's method to obtain the Gaussian approximation with $c = x\sqrt{npq}$ and obtain the required result with an error of $O(1/n)$. To obtain the sharper error, we argue directly as follows:

If we ignore the error term in the exponent, we incur an error of at most

$$\int_{-\delta}^{\delta} n\theta^4 \exp[-npq\theta^2/2] \, d\theta = O\left(\frac{1}{n^{3/2}}\right).$$

To dispose of the cubic term we again employ the inequality $|e^{ia} - 1 - ia| \leq a^2/2$ with $a = \theta^3/3$. The integral involving θ^3 is zero and the new error term is at most

$$\int_{-\delta}^{\delta} n^2 \theta^6 \exp[-npq\theta^2/2] \, d\theta = O\left(\frac{1}{n^{3/2}}\right).$$

Having done all this, we have

$$P_{k,n} = \int_{-\delta}^{\delta} \exp[-npq\theta^2/2 - ix\theta\sqrt{npq}] \, d\theta + O\left(\frac{1}{n^{3/2}}\right).$$

Finally, we can replace the integral over $(-\delta, \delta)$ by an integral over $(-\infty, \infty)$ at the expense of an error

$$\int_{|\theta| \geq \delta} \exp[-npq\theta^2/2 - ix\theta\sqrt{npq}] \, d\theta = O\left(e^{-npq\delta^2}\right).$$

From Chapter 2, Example 2.2.7, we borrow the Fourier transform of the Gaussian density:

$$\int_{-\infty}^{\infty} \exp[-npq\theta^2/2 - ix\theta\sqrt{npq}] \, d\theta = \sqrt{\frac{2\pi}{npq}} e^{-x^2/2}$$

which completes the proof. ∎

Exercise 1.7.2. *Show by direct calcuation that $|f(\theta)| < 1$ for $0 < |\theta| \leq \pi$.*

Exercise 1.7.3. *Suppose that $x = (k - np)/\sqrt{npq} \to \infty$, so that $x^2/2 - \log n \to -\infty$. Show that*

$$P_{k,n} \sim \frac{e^{-x^2/2}}{\sqrt{2\pi npq}}$$

in the sense that the ratio tends to 1 when $n \to \infty$.

1.8 APPENDIX: PROOF OF THE UNIFORM BOUNDEDNESS THEOREM

The norm of a linear operator L is defined as

(1.8.1) $$\|L\| := \sup\{|Lf| : |f| = 1\} = \sup\left\{\frac{|Lf|}{|f|} : f \neq 0\right\}.$$

An operator is *bounded* if $\|L\| < \infty$. The uniform boundedness theorem is the following statement.

Theorem 1.8.1. *Suppose that $\mathcal{L}$ is a collection of bounded linear operators from a Banach space B to a normed linear space Y with the property that for each $f \in B$*

(1.8.2) $$\sup\{|Lf| : L \in \mathcal{L}\} < \infty.$$

Then $\sup\{\|L\| : L \in \mathcal{L}\} < \infty$.

To simplify the proof, we first prepare a lemma that will allow us to make a proof by contraposition. Specifically, we assume that

(1.8.3) $$\sup\{\|L\| : L \in \mathcal{L}\} = +\infty.$$

Lemma 1.8.2. *Suppose that formulas (1.8.2) and (1.8.3) both hold. Then for each $n \geq 1$ there exist $L_n \in \mathcal{L}$ and $f_n \in B$ so that*

(1.8.4) $$|f_n| = 4^{-n}$$

(1.8.5) $$|L_n f_n| > \tfrac{2}{3}\|L_n\| \, |f_n|$$

(1.8.6) $$|L_n f_n| > 2(M_{n-1} + n)$$

where $M_0 = 1$ and for $k \geq 1$, $M_k = \sup\{|L(f_1 + \cdots + f_k)| : L \in \mathcal{L}\}$.

Proof. From (1.8.3), there exists $L_1 \in \mathcal{L}$ with $\|L_1\| > 24$. From the definition (1.8.1), there exists $\tilde{f}_1 \in B$ with $|\tilde{f}_1| = 1$ and $|L_1 \tilde{f}_1| > \tfrac{2}{3}\|L_1\|$. Setting $f_1 = \tilde{f}_1/4$ shows that (1.8.4), (1.8.5) and (1.8.6) are all satisfied with $n = 1$. Assuming that $f_1, \ldots, f_{n-1}, L_1, \ldots, L_{n-1}$, have been defined, choose $L_n \in \mathcal{L}$ so that $\|L_n\| > 3 \cdot 4^n (M_{n-1} + n)$, which is possible by hypothesis (1.8.3). With this choice of L_n, there exists $\tilde{f}_n \in B$ so that $|\tilde{f}_n| = 1$ and $|L_n \tilde{f}_n| > \tfrac{2}{3}\|L_n\|$.

Setting $f_n = \tilde{f}_n/4^n$, we clearly have $|f_n| = 4^{-n}$, proving (1.8.4). Now

$$|L_n f_n| > \tfrac{2}{3} 4^{-n} \|L_n\| > \tfrac{2}{3} 4^{-n} \cdot 3 \cdot 4^n (M_{n-1} + n) = 2(M_{n-1} + n),$$

which proves (1.8.5) and (1.8.6) for the value n. This completes the proof of the lemma by mathematical induction.

To complete the proof of the theorem, we let $f = \sum_{n=1}^{\infty} f_n$, which is a well-defined element in the Banach space B, by virtue of (1.8.4). We will show that $\sup_n |L_n f| = +\infty$. To do this, first note that

$$\left| L_n \left(\sum_{k=n+1}^{\infty} f_k \right) \right| \leq \|L_n\| \sum_{k=n+1}^{\infty} |f_k|$$

$$= \|L_n\| \sum_{k=n+1}^{\infty} 4^{-k}$$

$$= \frac{1}{3} \|L_n\| \, |f_n|$$

so that by the triangle inequality we have

$$|L_n f| = \left| L_n \left(\sum_{k=1}^{n-1} f_k + f_n + \sum_{k=n+1}^{\infty} f_k \right) \right|$$

$$\geq |L_n f_n| - \left| L_n \left(\sum_{k=1}^{n-1} f_k \right) \right| - \left| L_n \left(\sum_{k=n+1}^{\infty} f_k \right) \right|$$

$$\geq |L_n f_n| - M_{n-1} - \tfrac{1}{3} \|L_n\| \, |f_n|$$

$$\geq |L_n f_n| - M_{n-1} - \tfrac{1}{2} |L_n f_n|$$

$$= \tfrac{1}{2} |L_n f_n| - M_{n-1}$$

$$\geq n,$$

which proves that $\sup_n |L_n f| = +\infty$, the desired contradiction. ∎

1.9 *APPENDIX: HIGHER-ORDER BESSEL FUNCTIONS

Higher-order Bessel functions of integral order are easily constructed beginning with the series representation of $I_0(2r)$. Differentiating term-by-term, we find

(1.9.1) $$\frac{d}{dr} I_0(2r) = \sum_{n=1}^{\infty} \frac{2n r^{2n-1}}{n!^2} = \frac{1}{\pi} \int_{-\pi}^{\pi} \cos\theta \, e^{2r\cos\theta} \, d\theta.$$

The right side can be integrated by parts to obtain the identity

$$\sum_{n=0}^{\infty} \frac{r^{2n}}{n!(n+1)!} = \frac{1}{\pi} \int_{-\pi}^{\pi} \sin^2\theta \, e^{2r\cos\theta} \, d\theta := \frac{I_1(2r)}{r}.$$

Proceeding inductively, we find for each $m \geq 1$ the identity

(1.9.2) $$\sum_{n=0}^{\infty} \frac{r^{2n}}{n!(n+m)!} = \frac{C_m}{2\pi} \int_{-\pi}^{\pi} \sin^{2m}\theta \, e^{2r\cos\theta} \, d\theta := \frac{I_m(2r)}{r^m}$$

for a suitable choice of the constant C_m.

Exercise 1.9.1. *Prove that $C_0 = 1$, $C_{m+1}/C_m = 2/(2m+1)$ and conclude that $C_m = m! 2^{2m}/(2m)!$ for $m = 0, 1, \ldots$.*

Equation (1.9.2) provides a definition of the Bessel function I_m for $m = 0, 1, 2, \ldots$ as a power series convergent in the entire complex r-plane. If we interpret the factorial in terms of the gamma function, the definition can be extended to all complex values of m.

Exercise 1.9.2. *Prove that for $r > 0$ we have the inequality $|I_m(2r)| \le C_m r^m e^{2r}$.*

The higher-order Bessel functions can be recognized from the trigonometric series for the even function $f(\theta) = e^{2r\cos\theta}$. We begin with the power series expansion

$$e^{2r\cos\theta} = \sum_{k=0}^{\infty} \frac{(2r\cos\theta)^k}{k!}.$$

From the binomial theorem we have

$$(2\cos\theta)^k = \sum_{j=0}^{k} \binom{k}{j} (e^{i\theta})^j (e^{-i\theta})^{k-j} = \sum_{j=0}^{k} \binom{k}{j} e^{i(2j-k)\theta}$$

leading to the absolutely convergent double series

$$e^{2r\cos\theta} = \sum_{k=0}^{\infty} \frac{r^k}{k!} \sum_{j=0}^{k} \binom{k}{j} e^{i(2j-k)\theta} = \sum_{k=0}^{\infty} \sum_{j=0}^{k} \frac{r^k}{j!(k-j)!} e^{i(2j-k)\theta}.$$

If $m = 0, 1, 2, \ldots$, the coefficient of $e^{-im\theta}$ is obtained by summing over those indices (j, k) for which $2j - k = -m$. This is a line of slope $+2$ in the (j, k) plane, written as $k = 2j + m$. Thus

$$\sum_{(j,k):2j-k=m} \frac{r^k}{j!(k-j)!} = \sum_{j=0}^{\infty} \frac{r^{2j+m}}{j!(j+m)!} = r^m \sum_{j=0}^{\infty} \frac{r^{2j}}{j!(j+m)!} = I_m(2r) \qquad m = 0, 1, 2, \ldots$$

Noting that f is even, if $m = -1, -2, \ldots$ we obtain the same result with $|m|$ in place of m. Hence we obtain the absolutely convergent trigonometric series

$$e^{2r\cos\theta} = \sum_{m\in\mathbb{Z}} I_{|m|}(2r) e^{im\theta} \qquad r \in \mathbb{C}.$$

From Parseval's identity we obtain the unexpected dividend that for r real

$$\sum_{m\in\mathbb{Z}} I_{|m|}(2r)^2 = \frac{1}{2\pi} \int_{\mathbb{T}} e^{4r\cos\theta} d\theta = I_0(4r).$$

Bessel functions will be useful in Chapter 2 when we consider Fourier transforms in $\mathbb{R}^n$.

1.10 APPENDIX: CANTOR'S UNIQUENESS THEOREM

Trigonometric series can be considered apart from the framework of Fourier series, as we have indicated in the opening section of this chapter. In most of our work we operate in the framework of Fourier series. Nevertheless it is still useful and instructive to explore a larger context. The next result is a basic theorem in this direction.

Theorem 1.10.1. *Suppose that we have two trigonometric series $\sum_{n \in \mathbb{Z}} A_n e^{inx}$ and $\sum_{n \in \mathbb{Z}} B_n e^{inx}$ that converge to the same sum for every $x \in \mathbb{T}$. Then $A_n = B_n$ for all $n \in \mathbb{Z}$.*

By considering the difference, we can immediately reduce to the case $B_n \equiv 0$. Thus we are given

$$(1.10.1) \qquad \sum_{n=-N}^{N} A_n e^{inx} \to 0, \qquad N \to \infty, \, x \in \mathbb{T},$$

and we must prove that $A_n \equiv 0$.

Lemma 1.10.2. **Cantor-Lebesgue theorem:** *If $\sum_{n \in \mathbb{Z}} A_n e^{inx}$ converges for all x in a set E of positive measure, then $\lim_{|n| \to \infty} A_n = 0$.*

Proof. Since the series converges, we have $\lim_{n \to \infty}[A_n e^{inx} + A_{-n} e^{-inx}] = 0$. By taking the real and imaginary parts we obtain two terms that may be written in the form $a_n \cos(nx + \phi_n)$, for which we must prove that $a_n \to 0$. Assume not; then there exists $\epsilon > 0$ and infinitely many indices $n_1 < n_2 < \cdots$ so that $|a_{n_k}| \geq \epsilon$. Dividing by $|a_{n_k}|$, we conclude that $\cos(n_k x + \phi_{n_k}) \to 0$. Squaring this and using the double-angle formula, we have $\frac{1}{2}(1 + \cos(2n_k x)) \to 0$ on E, with $|E| > 0$. This is a sequence of uniformly bounded functions, which we can integrate and take the limit. But if we apply the Riemann-Lebesgue lemma to I_E, we conclude that the second integral tends to zero, thus $|E| = 0$, a contradiction. ∎

Riemann introduces the function

$$(1.10.2) \qquad F(x) = \frac{A_0 x^2}{2} - \sum_{n \neq 0} \frac{A_n}{n^2} e^{inx}.$$

Since $A_n \to 0$, this is an absolutely and uniformly convergent trigonometric series, especially a continuous function whose Fourier coefficients may be retrieved by integration. In order to implement the basic hypothesis of convergence, Riemann considers the second difference quotient

$$(1.10.3) \qquad \frac{F(x+h) + F(x-h) - 2F(x)}{h^2} = A_0 + \sum_{n \neq 0} A_n e^{inx} \left(\frac{\sin nh/2}{nh/2} \right)^2.$$

From exercise (1.4.28), the right side defines a regular method of summability; given (1.10.1) we see that (1.10.3) tends to zero when $h \to 0$. To complete the proof we develop some ideas of convexity to prove the following lemma.

Lemma 1.10.3. *Suppose that F is a continuous function defined on some interval so that $\lim_{h \to 0}(F(x+h) + F(x-h) - 2F(x))/2h^2 = 0$. Then F is a linear function: $F(x) = ax + b$ for suitable constants a, b.*

Proof. Introduce $F_\epsilon(x) = F(x) + \epsilon x^2$, a continuous function that satisfies

(1.10.4) $$\lim_{\epsilon \to 0}[F_\epsilon(x+h) + F_\epsilon(x-h) - 2F_\epsilon(x)]/h^2 = 2\epsilon > 0.$$

We will prove that $F_\epsilon(x)$ is a convex function. Suppose not; then there is an interval (a,b) and a linear function $g(x) = rx + s$ so that $F_\epsilon(a) = g(a)$, $F_\epsilon(b) = g(b)$ and $F_\epsilon(x_0) > g(x_0)$ for some $x_0 \in (a,b)$. The difference $F_\epsilon(x) - g(x)$ has a positive maximum at some point $x_{\max} \in (a,b)$. At this point we must have the inequality $F_\epsilon(x+h) + F_\epsilon(x-h) - 2F_\epsilon(x) \leq 0$ for small h. But this contradicts (1.10.4), so we have shown that F_ϵ is a convex function. But F is the limit of the sequence of convex functions F_ϵ, hence also convex. Applying the same reasoning to $-F$, we conclude that $-F$ is also convex, hence F must be a linear function.

To complete the proof of the uniqueness theorem, from (1.10.3) for any $h \neq 0$, we may retrieve the Fourier coefficients as

$$A_n \left(\frac{\sin nh/2}{nh/2}\right)^2 = \frac{1}{2\pi} \int_{\mathbb{T}} \frac{F(x+h) + F(x-h) - 2F(x)}{h^2} e^{-inx} dx = 0 \quad (n \neq 0)$$

whereas

$$A_0 = \frac{1}{2\pi} \int_{\mathbb{T}} \frac{F(x+h) + F(x-h) - 2F(x)}{h^2} dx = 0.$$

The proof is complete. ■

Exercise 1.10.4. Suppose that $\sum_{n \in \mathbb{Z}} A_n e^{inx}$ converges to zero for all $x \in \mathbb{T} \setminus \{x_1, \ldots, x_k\}$. Modify the above proof to show that $A_n = 0$ for all $n \in \mathbb{Z}$.

Hint: $F(x)$, defined by (1.10.2), will be piecewise linear.

CHAPTER 2

FOURIER TRANSFORMS ON THE LINE AND SPACE

2.1 MOTIVATION AND HEURISTICS

In parallel with Chapter 1, we can motivate the theory of the Fourier transform on the real line by considering an absolutely convergent trigonometric integral

(2.1.1) $$f(x) = \int_{\mathbb{R}} C(\xi) e^{i\xi x} d\xi, \quad \text{where} \quad \int_{\mathbb{R}} |C(\xi)| \, d\xi < \infty.$$

In order to retrieve the coefficient function $C(\xi)$ we multiply (2.1.1) by $e^{-i\eta x}$ and integrate over $x \in [-T, T]$;

$$\int_{-T}^{T} e^{-i\eta x} f(x) \, dx = \int_{\mathbb{R}} C(\xi) \left(\int_{-T}^{T} e^{i(\xi-\eta)x} \, dx \right) d\xi$$

$$= 2 \int_{\mathbb{R}} C(\xi) \frac{\sin T(\xi - \eta)}{\xi - \eta} \, d\xi.$$

One cannot immediately take the limit $T \to \infty$ without further hypotheses. If, for example, $C(\xi)$ satisfies a Dini condition, then one can show that the right side converges to $2\pi C(\xi)$ so that

(2.1.2) $$C(\xi) = \lim_{T \to \infty} \frac{1}{2\pi} \int_{-T}^{T} f(x) e^{-ix\xi} \, d\xi,$$

which can be used to motivate the definition of the Fourier transform. The following exercise shows that $C(\xi)$ can be retrieved under the hypothesis of continuity.

Exercise 2.1.1. *Show that*

$$\frac{1}{T}\int_0^T \left(\int_{-t}^t e^{-i\eta x} f(x)\, dx\right) dt = 2\int_{\mathbb{R}} \frac{1-\cos T(\xi-\eta)}{(\xi-\eta)^2} C(\xi)\, d\xi.$$

Conclude that if C is continuous, then the right side converges to $2\pi C(\eta)$ when $T\to\infty$.

Example 2.1.2. *Let $C(\xi) = e^{-y|\xi|}$ for $y > 0$, $x \in \mathbb{R}$. Then*

$$f(x) = \int_{\mathbb{R}} e^{-y|\xi|} e^{-ix\xi}\, d\xi$$

$$= 2\,\text{Re}\int_0^\infty e^{-y\xi} e^{-ix\xi}$$

$$= 2\,\text{Re}\,\frac{1}{y+ix}$$

$$= \frac{2y}{x^2+y^2}.$$

Apart from a constant, this is the Poisson kernel in the setting of the real line. The normalization is obtained from the elementary calculus integral for $\int_{\mathbb{R}} dx/(1+x^2) = \pi$. Thus we have the normalized Poisson kernel

$$P_y(x) = \frac{1}{\pi}\frac{y}{y^2+x^2} \qquad y > 0,\ x \in \mathbb{R},$$

which has the properties of a positive approximate identity:

(2.1.3) $\qquad P_y(x) \geq 0, \qquad \int_{\mathbb{R}} P_y(x)\, dx = 1, \qquad \int_{|x|>\delta} P_y(x)\, dx \to 0 \quad (y\downarrow 0).$

Exercise 2.1.3. *Use the previous example and formula (2.1.2) to compute*

$$\int_{\mathbb{R}} e^{-ix\xi} \frac{1}{1+x^2}\, dx \qquad \xi \in \mathbb{R}.$$

A more classical motivation, which goes back to Fourier, is to begin with a Fourier series on the interval $[-L, L]$:

(2.1.4) $\qquad f(x) \sim \sum_{n\in\mathbb{Z}} C_n e^{in\pi x/L},$

(2.1.5) $\qquad C_n = \frac{1}{2L}\int_{-L}^L f(x) e^{-in\pi x/L}.$

These formulas will now be rewritten in terms of the functions

$$F_L(\mu) = \int_{-L}^L f(x) e^{-i\mu x}\, dx, \qquad F(\mu) = \int_{\mathbb{R}} f(x) e^{-i\mu x}\, dx.$$

Letting $\mu_n = n\pi/L$, we have $(\Delta\mu)_n = \pi/L$ so that we can write (2.1.4) as a formal Riemann sum

$$f(x) \sim \frac{1}{2\pi} \sum_{n \in \mathbb{Z}} F_L(\mu_n) e^{i\mu_n x} (\Delta\mu)_n. \qquad (2.1.6)$$

Formally taking $L \to \infty$, we find

$$f(x) \sim \frac{1}{2\pi} \int_{\mathbb{R}} F(\mu) e^{i\mu x} \, d\mu, \qquad F(\mu) = \int_{\mathbb{R}} f(x) e^{-i\mu x} \, dx.$$

In order to have a more symmetrical theory, we let $\mu = 2\pi\xi$ in the first integral, to obtain

$$f(x) \sim \int_{\mathbb{R}} F(2\pi\xi) e^{2\pi i \xi x} \, d\xi, \qquad F(2\pi\xi) = \int_{\mathbb{R}} f(x) e^{-2\pi i \xi x} \, dx. \qquad (2.1.7)$$

This symmetrical form will be in force in the systematic approach beginning in Section 2.2. In case f is real-valued, (2.1.7) can also be written in terms of real-valued functions by writing $F(\mu) = \frac{1}{2}[A(\mu) - iB(\mu)]$ to obtain

$$f(x) \sim \int_0^\infty [A(2\pi\xi) \cos(2\pi\xi x) + B(2\pi\xi) \sin(2\pi\xi x)] \, d\xi. \qquad (2.1.8)$$

The above transformations are purely heuristic, with no pretense of rigor. We will show in the following sections that they can be systematically developed to obtain a powerful theory of Fourier analysis on the real line and in Euclidean space.

2.2 BASIC PROPERTIES OF THE FOURIER TRANSFORM

In order to formulate an unambiguous theory, we begin with a complex-valued, Lebesgue integrable function on Euclidean space, denoted by $f(x)$, $x \in \mathbb{R}^n$. The Fourier transform is the complex-valued function $\hat{f}(\xi)$, $\xi \in \mathbb{R}^n$ defined by the integral

$$\hat{f}(\xi) = (\mathcal{F}f)(\xi) := \int_{\mathbb{R}^n} f(x) e^{-2\pi i \xi \cdot x} \, dx. \qquad (2.2.1)$$

The Euclidean dot product is $\xi \cdot x = \sum_{j=1}^n \xi_j x_j$.

The basic elementary properties of the Fourier transform are summarized in the next statement.

Proposition 2.2.1. *Let $f \in L^1(\mathbb{R}^n)$. Then*

- *Continuity:* $\xi \mapsto \hat{f}(\xi)$ *is a uniformly continuous function.*
- *Contraction: The mapping $f \to \hat{f}$ is norm-decreasing from L^1 to L^∞, in the sense that*

$$|\mathcal{F}f(\xi)| \leq \int_{\mathbb{R}^n} |f(x)| \, dx := \|f\|_1. \qquad (2.2.2)$$

- *Linearity: If F_1 is the Fourier transform of f_1 and F_2 is the Fourier transform of f_2, then $a_1 F_1 + a_2 F_2$ is the Fourier transform of $a_1 f_1 + a_2 f_2$, for any choice of the complex constants a_1, a_2.*

- *Translation and Phase Factor:* If F is the Fourier transform of f, then $e^{-2\pi i a \cdot \xi} F(\xi)$ is the Fourier transform of $f(x-a)$, and $F(\xi + b)$ is the Fourier transform of $e^{-2\pi i b \cdot x} f(x)$.
- *Multiplication and Convolution:* If F_1 is the Fourier transform of f_1 and F_2 is the Fourier transform of f_2, then $F_1 F_2$ is the Fourier transform of $(f_1 * f_2)$, where the **convolution** of two functions is defined by the integral

$$(f_1 * f_2)(x) = \int_{\mathbb{R}^n} f_1(y) f_2(x-y)\, dy.$$

- *Differentiation and Multiplication:* If the partial derivative $(\partial f / \partial x_j)$ exists and is in $L^1(\mathbb{R}^n)$, then $2\pi i \xi_j \hat{f}(\xi)$ is the Fourier transform of $(\partial f / \partial x_j)$.
  If $x_j f \in L^1(\mathbb{R}^n)$, then $(\partial \hat{f} / \partial \xi_j)(\xi)$ exists and is the Fourier transform of $-2\pi i x_j f(x)$.
- *Fourier Transform of Radial Functions:* If $f(x) = \varphi(|x|)$ for some $\varphi \in L^1(\mathbb{R}^+; r^{n-1}dr)$, then $\hat{f}(\xi) = \psi(|\xi|)$ for some $\psi \in C(\mathbb{R}^+)$. Restated, the Fourier transform of a radial function is again a radial function.

Proof. To prove the uniform continuity, we write

$$\hat{f}(\xi + h) - \hat{f}(\xi) = \int_{\mathbb{R}^n} [e^{-2\pi i (\xi + h) \cdot x} - e^{-2\pi i \xi \cdot x}] f(x)\, dx$$

$$= \int_{\mathbb{R}^n} e^{-2\pi i \xi \cdot x} [e^{-2\pi i h \cdot x} - 1] f(x)\, dx$$

$$|\hat{f}(\xi + h) - \hat{f}(\xi)| \leq \int_{\mathbb{R}^n} |e^{-2\pi i h \cdot x} - 1|\, |f(x)|\, dx$$

$$= \left(\int_{|x| \leq M} + \int_{|x| > M} \right) |e^{-2\pi i h \cdot x} - 1|\, |f(x)|\, dx.$$

Given $\epsilon > 0$, the second integral can be made less than ϵ by taking M sufficiently large. The first integral is majorized by

$$2\pi |h| \int_{|x| \leq M} |x|\, |f(x)|\, dx.$$

Therefore with the above choice of M, we have

$$\limsup_{h \to 0} \sup_{\xi \in \mathbb{R}^n} |\hat{f}(\xi + h) - \hat{f}(\xi)| \leq \epsilon.$$

But ϵ was arbitrary, which proves the uniform continuity. The linearity and phase factor properties are direct computations. The existence of the convolution of two L^1 functions is guaranteed by the Fubini theorem, when we consider the function $f_1(z) f_2(y)$ in $\mathbb{R}^{2n}$ and make the substitution $z = x - y$:

$$\infty > \int_{\mathbb{R}^{2n}} |f_2(y) f_1(z)|\, dz\, dy = \int_{\mathbb{R}^n} |f_2(y)| \left(\int_{\mathbb{R}^n} |f_1(x-y)|\, dx \right) dy$$

$$= \int_{\mathbb{R}^n} \left(\int_{\mathbb{R}^n} |f_1(x-y) f_2(y)|\, dy \right) dx.$$

The joint measurability of the product $f_1(z)f_2(y)$ is established by writing $f_1(z) = \lim_N \sum_{k \in \mathbb{Z}} k 2^{-N} 1_{k 2^{-N} \leq f < (k+1)2^{-N}}$ and similarly for $f_2(y)$. The product is then written as a pointwise limit of simple functions, especially jointly measurable.

Therefore the convolution is finite almost everywhere and defines an L^1 function with $\|f_1 * f_2\| \leq \|f_1\|_1 \|f_2\|_1$. Now we can compute

$$F_2(\xi)F_1(\xi) = \int_{\mathbb{R}^{2n}} e^{-2\pi i \xi \cdot (z+y)} f_2(y) f_1(z) \, dz \, dy$$

$$= \int_{\mathbb{R}^n} e^{-2\pi i \xi \cdot y} f_2(y) \left(\int_{\mathbb{R}^n} e^{-2\pi i \xi \cdot (x-y)} f_1(x-y) \, dx \right) dy$$

$$= \int_{\mathbb{R}^n} e^{-2\pi i \xi \cdot x} \left(\int_{\mathbb{R}^n} f_1(x-y) f_2(y) \, dy \right) dx.$$

The differentiation and multiplication properties will be proved below in a more amplified context. Finally, if f is a radial function, then we consider a rotation $\mathcal{R}$ in the ξ-space, making the change-of-variable $y = \mathcal{R}^t x$ with $|y| = |x|$, $dy = dx$:

$$\hat{f}(\mathcal{R}\xi) = \int_{\mathbb{R}^n} \varphi(|x|) e^{-2\pi i \mathcal{R}\xi \cdot x} \, dx = \int_{\mathbb{R}^n} \varphi(|x|) e^{-2\pi i \xi \cdot \mathcal{R}^t x} \, dx$$

$$= \int_{\mathbb{R}^n} \varphi(|y|) e^{-2\pi i \xi \cdot y} \, dy = \hat{f}(\xi),$$

which shows that $\hat{f}$ is invariant by rotations, hence a function of $|\xi|$. ∎

We will also need a form of the Fourier reciprocity formula.

Lemma 2.2.2. *Suppose that* $f \in L^1(\mathbb{R}^n)$ *and* $\psi \in L^1(\mathbb{R}^n)$ *are integrable functions with Fourier transforms* $\hat{f}$ *and* $\hat{\psi}$. *Then we have the identity*

$$(2.2.3) \qquad \int_{\mathbb{R}^n} \psi(\xi) \hat{f}(\xi) \, d\xi = \int_{\mathbb{R}^n} f(x) \hat{\psi}(x) \, dx.$$

Proof. We use the Fubini theorem to write

$$\int_{\mathbb{R}^n} \psi(\xi) \hat{f}(\xi) \, d\xi = \int_{\mathbb{R}^n} \psi(\xi) \left(\int_{\mathbb{R}^n} f(x) e^{-2\pi i \xi \cdot x} \, dx \right) d\xi$$

$$= \int_{\mathbb{R}^n} f(x) \left(\int_{\mathbb{R}^n} \psi(\xi) e^{-2\pi i \xi x} \, d\xi \right) dx$$

$$= \int_{\mathbb{R}^n} f(x) \hat{\psi}(x) \, dx,$$

which was to be proved. ∎

This applies in particular to a convolution operator with respect to a kernel which is written as an L^1 Fourier integral. In detail, if $K(x) = \int_{\mathbb{R}^n} k(\xi) e^{2\pi i \xi \cdot x} \, d\xi$ with $k \in L^1(\mathbb{R}^n)$, then

$$(2.2.4) \qquad \boxed{\int_{\mathbb{R}^n} f(x-y) K(y) \, dy = \int_{\mathbb{R}^n} k(\xi) \hat{f}(\xi) e^{2\pi i \xi \cdot x} \, d\xi.}$$

Exercise 2.2.3. *If μ is a finite Borel measure on $\mathbb{R}^n$, its Fourier transform is defined as $\hat{\mu}(\xi) := \int_{\mathbb{R}^n} e^{-2\pi i \xi \cdot x} \mu(dx)$. Prove that if μ, ν are finite Borel measures on $\mathbb{R}^n$, then we have the Fourier reciprocity formula $\int_{\mathbb{R}^n} \hat{\nu}(x) \mu(dx) = \int_{\mathbb{R}^n} \hat{\mu}(\xi) \nu(d\xi)$.*

2.2.1 Riemann-Lebesgue Lemma

The most basic analytic fact about the Fourier transform is the Riemann-Lebesgue lemma, expressed as follows.

Theorem 2.2.4. *For any $f \in L^1(\mathbb{R}^n)$*

$$\lim_{|\xi| \to \infty} \hat{f}(\xi) = 0,$$

and the convergence is uniform on compact subsets of $L^1(\mathbb{R}^n)$.

Proof. From inequality (2.2.2), we need only prove this for a dense set of functions in the L^1 norm. From the dominated convergence theorem,

$$\int_{|x|>M} |f(x)| \, dx \to 0, \qquad \int_{\{x:|f(x)|>M\}} |f(x)| \, dx \to 0$$

when $M \to \infty$, so that we can approximate f in the L^1 norm by a bounded measurable function $\tilde{f}$, which is supported on the cube $[-M, M]^n$. By rescaling we may suppose that $M = 1$. Now $\tilde{f}$ can be uniformly approximated by the simple function

$$\tilde{\tilde{f}} = \sum_{k=-2^N}^{2^N} k 2^{-N} 1_{\{(k-1)2^{-N} < \tilde{f} \leq k 2^{-N}\}}$$

so that $\|\tilde{\tilde{f}} - \tilde{f}\|_1 \leq 2 \times 2^{-N}$. Now we will prove that the theorem holds for any simple function, $f = \sum_{k=1}^{N} c_k 1_{E_k}$ where E_k are measurable subsets of $[-1, 1]^n$ and c_k are arbitrary complex numbers.

In case of an indicator function $f(x) = \prod_{j=1}^{n} 1_{(a_j, b_j)}(x)$, the Fourier transform is computed explicitly as

$$\hat{f}(\xi) = \prod_{j=1}^{n} \int_{a_j}^{b_j} e^{-2\pi i \xi_j x_j} dx_j = \prod_{j=1}^{n} \frac{e^{-2\pi i \xi_j b_j} - e^{-2\pi i \xi_j a_j}}{-2\pi i \xi_j}, \qquad \xi_j \neq 0,$$

which clearly tends to zero if at least one of the coordinates $\xi_j \to \infty$. By Proposition 2.2.1, the same is true if f is the indicator function of a finite union of intervals.

Now if E is any measurable subset of $[-1, 1]^n$, there exists a finite collection of open cubes $\{C_j\}$ with union $U = \cup_j C_j$ so that the Lebesgue measure of the symmetric difference $E \Delta U$ is less than ϵ. In particular $\|1_U - 1_E\|_1 < \epsilon$ so that

$$|\mathcal{F}(1_U)(\xi) - \mathcal{F}(1_E)(\xi)| \leq \|1_U - 1_E\|_1 < \epsilon.$$

Therefore, $\mathcal{F}1_E(\xi) \to 0$ when $|\xi| \to \infty$. This extends immediately to finite sums $f = \sum_{k=1}^{N} c_i 1_{E_i}$ and the proof is complete. To check the uniform convergence on compact subsets, follow the same reasoning as in the proof in Chapter 1. ∎

For a general L^1 function, there is no universal rate of decrease to zero. However, if we asssume some differentiability, we have the following fact.

Proposition 2.2.5. *If the partial derivative $(\partial f / \partial x_j)$ exists and is in $L^1(\mathbb{R}^n)$, then $2\pi i \xi_j \hat{f}(\xi)$ is the Fourier transform of $(\partial f / \partial x_j)(x)$. In particular*

$$\lim_{|\xi| \to \infty} \xi_j \hat{f}(\xi) = 0.$$

More generally if $\alpha = (\alpha_1, \ldots, \alpha_n)$ is a multiindex, and the mixed partial derivative $D^\alpha f \in L^1(\mathbb{R}^n)$, then $(2\pi i \xi)^\alpha \hat{f}(\xi) = \mathcal{F}(D^\alpha f)(\xi)$.

Proof. Without loss of generality suppose that $j = 1$. First, note that by Lebesgue theory, $f(x, x_2, \ldots, x_n) - f(y, x_2, \ldots, x_n) = \int_y^x f_{x_1}(z, x_2, \ldots, x_n) \, dz$, which shows that f must have a limit when $x_1 \to \infty$, which can only be zero, since $f \in L^1(\mathbb{R}^n)$. Now we apply partial integration for $\xi_1 \neq 0$:

$$\hat{f}(\xi) = \int_{\mathbb{R}^n} f(x) e^{-2\pi i \xi \cdot x} \, dx = \lim_{M \to \infty} \int_{|x_1| \leq M} f(x) e^{-2\pi i \xi \cdot x} \, dx$$

$$\int_{-M}^M f(x) d_{x_1} \left(\frac{e^{-2\pi i \xi \cdot x}}{-2\pi i \xi_1} \right) = f(x) \frac{e^{-2\pi i \xi \cdot x}}{-2\pi i \xi_1} \bigg|_{-M}^M - \int_{-M}^M f_{x_1}(x) \frac{e^{-2\pi i \xi \cdot x}}{-2\pi i \xi_1} \, dx_1$$

$$\to \int_{-\infty}^\infty f_{x_1}(x) \frac{e^{-2\pi i \xi \cdot x}}{2\pi i \xi_1} \, dx_1.$$

Integrating this over $x_2, \ldots, x_n$ proves the identity $2\pi i \xi_1 \hat{f}(\xi) = \mathcal{F}(f_{x_1})(\xi)$. Repeating this procedure and using mathematical induction shows that for any multiindex α, $(2\pi i \xi)^\alpha \hat{f}(\xi) = \mathcal{F}(D^\alpha f)(\xi)$ from which the result follows by the Riemann-Lebesgue lemma. ∎

A dual property is afforded by the next statement, where $|\alpha| := \alpha_1 + \cdots + \alpha_n$.

Proposition 2.2.6. *Suppose that $\int_{\mathbb{R}^n} |x|^k |f(x)| \, dx < \infty$. Then $\hat{f}$ is differentiable to order k and*

$$D^\alpha \hat{f}(\xi) = \int_{-\infty}^\infty (-2\pi i x)^\alpha f(x) e^{-2\pi i \xi \cdot x} \, dx \qquad |\alpha| \leq k.$$

Proof. To prove the formula for $k = 1$ we apply the dominated convergence theorem, using the inequality $|e^{i\theta} - 1| \leq |\theta|$ for θ real. The general result follows by mathematical induction. ∎

Example 2.2.7. *A basic example with $n = 1$ is provided by the Gaussian density function*

$$f(x) = e^{-\pi x^2}, \qquad \hat{f}(\xi) = e^{-\pi \xi^2}.$$

To see this, we differentiate the Fourier transform:

$$\hat{f}'(\xi) = \int_{-\infty}^\infty (-2\pi i x) e^{-2\pi i \xi x} f(x) \, dx.$$

But the Gaussian density satisfies the differential equation $f' = -2\pi x f(x)$, so that we may apply Proposition 2.2.5 and integrate by parts to obtain

$$\hat{f}'(\xi) = \int_{\mathbb{R}} (-2\pi i x) e^{-2\pi i \xi x} f(x)\,dx$$
$$= i \int_{\mathbb{R}} e^{-2\pi i \xi x} f'(x)\,dx$$
$$= -2\pi \xi \int_{\mathbb{R}} e^{-2\pi i \xi x} f(x)\,dx$$
$$= -2\pi \xi \hat{f}(\xi).$$

The unique solution of this equation with the condition $\hat{f}(0) = 1$ is $\hat{f}(\xi) = e^{-\pi \xi^2}$, and the example is complete.

Exercise 2.2.8. *Show that $\hat{f}(0) = 1$.*

Hint: Do the double integral of $f(x)f(y)$ in polar coordinates.

This example may be immediately extended to $\mathbb{R}^n$, to obtain the Fourier pair

(2.2.5) $$\boxed{f(x) = e^{-\pi |x|^2}, \qquad \hat{f}(\xi) = e^{-\pi |\xi|^2}.}$$

The Gaussian density provides a concrete example of Fourier inversion.

(2.2.6) $$\int_{\mathbb{R}^n} \hat{f}(\xi) e^{2\pi i \xi \cdot x}\,d\xi = f(x)$$

Exercise 2.2.9. *Check the previous statement from the formulas above.*

The Gaussian density is a simple example of a *rapidly decreasing function*. To define this notion in general, introduce the seminorms

$$\|f\|_{k,m} := \sup_{x \in \mathbb{R}^n} (1 + |x|)^k |D^\alpha f(x)|$$

where $\alpha = (\alpha_1, \ldots, \alpha_n)$ is a multiindex, $D^\alpha = (\partial/\partial x_1)^{\alpha_1} \cdots (\partial/\partial x_n)^{\alpha_n}$, and $m = |\alpha| := \alpha_1 + \cdots + \alpha_n$. The *Schwartz class* $\mathcal{S}$ of rapidly decreasing functions is defined as the set of complex-valued functions that are infinitely differentiable and for which $\|f\|_{k,m} < \infty$ for all m, k. From Proposition 2.2.5 and Proposition 2.2.6, the following proposition is immediate.

Proposition 2.2.10. *The Fourier transform maps the space $\mathcal{S}$ into itself.*

In the next subsection we will show that this mapping is 1:1 *onto* the space $\mathcal{S}$.

The Gaussian density example in $n = 1$ can be transformed into additional examples by successive differentiation of the Fourier transform.

Thus

$$\int_{\mathbb{R}} e^{-\pi x^2} e^{-2\pi i \xi x} \, dx = e^{-\pi \xi^2},$$

$$\int_{\mathbb{R}} e^{-\pi x^2} (-2\pi i x) e^{-2\pi i \xi x} \, dx = (-2\pi \xi) e^{-\pi \xi^2},$$

$$\int_{\mathbb{R}} e^{-\pi x^2} (-2\pi i x)^2 e^{-2\pi i \xi x} \, dx = [(-2\pi \xi)^2 - 2\pi] e^{-\pi \xi^2},$$

and so forth. This is closely related to the Hermite functions, which will be discussed in Section 2.4.5.

2.2.2 Approximate Identities and Gaussian Summability

On the circle we had the Abel and Cesàro means to regularize the convergence of Fourier series. Suitable analogues of these exist in the context of one-dimensional Fourier transforms, and will be treated. However there is a more natural and symmetrical approximate identity, the *Gauss-Weierstrass kernel*, which applies in higher dimensions as well as in one dimension. From the example of equation (2.2.5), we replace x by $x/\sqrt{4\pi t}$ to obtain the identity

(2.2.7) $$H_t(x) := \int_{\mathbb{R}^n} e^{-4\pi^2 t |\xi|^2} e^{2\pi i \xi \cdot x} d\xi = \frac{\exp(-|x|^2/4t)}{(4\pi t)^{n/2}}.$$

This is also called the *heat kernel*, since it is a solution of the equation $\partial H/\partial t = \sum_{j=1}^n \partial^2 H/\partial x_j^2$. It has the following three basic properties:

(2.2.8) $\quad H_t(x) \geq 0, \quad \int_{\mathbb{R}^n} H_t(x) \, dx = 1, \quad \int_{|x| > \delta} H_t(x) \, dx \to 0 \quad (t \to 0).$

Exercise 2.2.11. *Prove that $H_t(x)$, defined by the integral (2.2.7), satisfies the n-dimensional heat equation.*

Exercise 2.2.12. *Check the three properties (2.2.8).*

Notation. We will use the notation $H_t f$ for the convolution $H_t * f$. It will always be clear from context whether we are dealing with the kernel or with the convolution operator.

Applying the Fourier reciprocity formula (2.2.4) to (2.2.7) yields a useful identity for $H_t f$:

(2.2.9) $$H_t f(x) = \int_{\mathbb{R}^n} f(x-y) H_t(y) \, dy = \int_{\mathbb{R}^n} e^{-4\pi^2 t |\xi|^2} \hat{f}(\xi) e^{2\pi i \xi \cdot x} d\xi \quad f \in L^1(\mathbb{R}^n).$$

Definition 2.2.13. *An approximate identity in $\mathbb{R}^n$ is a family of functions $k_t(y)$ defined for t in some directed index set, with the following three properties:*

(2.2.10) $$\sup_t \int_{\mathbb{R}^n} |k_t(y)| \, dy < \infty$$

(2.2.11) $$\lim_t \int_{\mathbb{R}^n} k_t(y) \, dy = 1$$

(2.2.12) $$\lim_t \int_{\{|y|>\delta\}} |k_t(y)| \, dy = 0.$$

These properties are clearly satisfied by H_t, for example.

Exercise 2.2.14. *Show that the Fejér kernel, which is defined on $\mathbb{R}$ by $k_T(x) = (1 - \cos Tx)/\pi T x^2$, is an approximate identity, where $T \in (0, \infty)$ the limits are taken as $T \to \infty$. Assume known that $\int_{\mathbb{R}} (1 - \cos x)/x^2 \, dx = \pi$.*

Exercise 2.2.15. *Show that the Poisson kernel, which is defined on $\mathbb{R}$ by $k_y(x) = y/[\pi(y^2 + x^2)]$, is an approximate identity, where $y \in (0, \infty)$ and the limits are taken as $y \to 0$. Assume known that $\int_{\mathbb{R}} 1/(1 + x^2) \, dx = \pi$.*

Exercise 2.2.16. *Suppose that $k_t(x)$ is an approximate identity. Prove that if f is a bounded function with $\lim_{x \to 0} f(x) = L$, then $\lim_t \int_{\mathbb{R}^n} k_t(x) f(x) \, dx = L$.*

The examples of Gauss, Poisson and Fejér can be abstracted as follows.

Exercise 2.2.17. *Suppose that $K \in L^1(\mathbb{R}^n)$ with $K(x) \geq 0$, $\int_{\mathbb{R}^n} K(x) \, dx = 1$ and set $k_t(x) = t^{-n} K(x/t)$. Prove that k_t is an approximate identity, where the limits are taken as $t \to 0$.*

A *homogeneous Banach space* B is a Banach space of complex-valued functions on $\mathbb{R}^n$ whose norm satisfies the properties

$$\|f_y\|_B = \|f\|_B, \quad \forall y \in \mathbb{R}^n \quad \text{and} \quad \lim_{y \to 0} \|f_y - f\|_B = 0$$

where f_y is the translation of f, defined as $f_y(x) = f(x - y)$.

The following examples of homogeneous Banach spaces occur frequently.

Example 2.2.18. $B = L^p(\mathbb{R}^n)$, $1 \leq p < \infty$ *is a homogeneous Banach space.*

This follows from the translation invariance of the Lebesgue integral and a density argument, beginning with continuous functions with compact support.

Example 2.2.19. $B = B_{uc}(\mathbb{R}^n)$, *the space of bounded and uniformly continuous functions on $\mathbb{R}^n$ with the supremum norm, is a homogeneous Banach space.*

Clearly the norm is translation invariant. The continuity of $y \to f_y$ is equivalent to the definition of uniform continuity of f.

Example 2.2.20. $B = C_0(\mathbb{R}^n)$, the space of continuous functions vanishing at infinity, is a homogeneous Banach space.

As a closed subspace of $B_{uc}(\mathbb{R}^n)$, B is a homogeneous Banach space.

In the following theorem we will assume, without loss of generality, that the limit in the directed index set is taken as $t \to 0$.

Theorem 2.2.21. *Suppose that B is a homogeneous Banach space and k_t is an approximate identity. Then $\int_{\mathbb{R}^n} k_t(y) f_y \, dy \in B$ whenever $f \in B$ and $\int_{\mathbb{R}^n} k_t(y) f_y \, dy \to f$ in the norm of B when $t \to 0$.*

Proof. $\int_{\mathbb{R}^n} k_t(y) f_y \, dy$ can be computed as the limit of Riemann sums. But any finite sum is a linear combination of elements of B, hence in B. The B-norm of any finite Riemann sum is bounded by $\|f\|_B$ times the Riemann sum for $\int_{\mathbb{R}^n} |k_t(y)| \, dy$ which is uniformly bounded. From this it follows that the Riemann sums converge in the B-norm. To study the convergence when $t \to 0$, we write

$$\left\| \int_{\mathbb{R}^n} k_t(y)(f_y - f) \, dy \right\|_B \le \int_{\mathbb{R}^n} k_t(y) \|f_y - f\|_B \, dy$$
$$= \left(\int_{\{|y| \le \delta\}} + \int_{\{|y| > \delta\}} \right) |k_t(y)| \, \|f_y - f\|_B \, dy.$$

The integral over $|y| > \delta$ tends to zero for any $\delta > 0$. Given $\epsilon > 0$, we choose $\delta > 0$ so that $\|f_y - f\|_B < \epsilon$ for $|y| < \delta$. Then the first integral is less than $C\epsilon$. With this choice of δ, we have

$$\limsup_{t \to 0} \left\| \int_{\mathbb{R}^n} k_t(y)(f_y - f) \, dy \right\|_B \le C\epsilon$$

for any $\epsilon > 0$. The proof is complete. ∎

Corollary 2.2.22. *i): If $f \in B_{uc}(\mathbb{R}^n)$, then $H_t f$ converges uniformly to f: $\|H_t f - f\|_\infty \to 0$ when $t \to 0$. ii): If $f \in L^p(\mathbb{R}^n)$, $1 \le p < \infty$, then $H_t f$ converges in L^p: $\|H_t f - f\|_p \to 0$ when $t \to 0$.*

Proof. These follow immediately from Theorem 2.2.21 when applied to the heat kernel on the spaces $B = B_{uc}(\mathbb{R}^n)$ and $B = L^p(\mathbb{R}^n)$. ∎

Corollary 2.2.23. *If $f \in L^1(\mathbb{R}^n)$ and $\hat{f} \equiv 0$, then $f = 0$ a.e.*

Proof. By Fourier reciprocity, the heat kernel convolution operator can be written

(2.2.13) $$\int_{\mathbb{R}^n} H_t(y) f(x - y) \, dy = \int_{\mathbb{R}^n} e^{-4\pi^2 t |\xi|^2} \hat{f}(\xi) e^{2\pi i \xi \cdot x} \, d\xi.$$

If $\hat{f} = 0$, then $H_t f = 0$ and by Corollary 2.2.22 $f = 0$ a.e., which was to be proved. ∎

Exercise 2.2.24. Let B be the space of complex-valued functions for which $\int_{\mathbb{R}} (|f(t)|)/(1 + t^2) \, dt < \infty$. Consider the Poisson kernel operator $P_y: f \to \pi^{-1} \int_{\mathbb{R}} y f(t) \, dt / (y^2 + (x - t)^2)$. Show that P_y maps B to B and that $P_y f \to f$ when $y \to 0$.

2.2.2.1 Improved approximate identities for pointwise convergence

In the case of one dimension, we can obtain pointwise convergence at additional points under additional conditions on the functions k_t, assumed to be of the form $t^{-1}K(x/t)$. From Exercise 2.2.17 any such kernel is an approximate identity, hence we have convergence at points of continuity of a test function f. The following theorem formulates supplementary conditions for convergence at additional points.

Theorem 2.2.25. *Suppose that K is absolutely continuous with $K(x) \geq 0$, $\int_{\mathbb{R}} K(x)\,dx = 1$, $K(-x) = K(x)$ and $xK'(x) \leq 0$. Then*

- *(i): $x \to (-2x/t^2)K'(x/t)$ is an approximate identity on $[0, \infty)$.*
- *(ii): $\lim_{t \to 0} t^{-1} \int_{\mathbb{R}} K(x/t) f(x)\,dx = L$ if $\lim_{x \to 0} (1/2x) \int_{-x}^{x} f(t)\,dt = L$.*

Proof. By partial integration, for any $M > 0$ we have

$$(2.2.14) \qquad MK(M) = \int_0^M K(x)\,dx + \int_0^M xK'(x)\,dx.$$

When $M \to \infty$, the first term on the right tends to $1/2$ and the second term has a limit $\in [-\infty, 0)$. But $K(x) \geq 0$, hence there exists $C = \lim_{M \to \infty} MK(M) \in [0, \infty)$. Since $\int_0^\infty K(x)\,dx = 1/2 < \infty$, we must have $C = 0$. Taking $M \to \infty$ in (2.2.14) shows that $\int_0^\infty xK'(x)\,dx = -1/2$. Now for any $\delta > 0$,

$$-\int_{|x| \geq \delta} \frac{x}{t^2} K'\left(\frac{x}{t}\right) dx = -\int_{|y| \geq \delta/t} yK'(y)\,dy$$

which tends to zero when $t \to 0$. In addition,

$$-\int_{\mathbb{R}} \frac{x}{t^2} K'\left(\frac{x}{t}\right) dx = -\int_{\mathbb{R}} yK'(y)\,dy = 1$$

which proves *(i)*. To prove *(ii)*, we define $F(x) = \int_{-x}^{x} f(u)\,du = \int_0^x [f(u)+f(-u)]\,du$, which satisfies $F(0) = 0$, $|F(x)| \leq 2\|f\|_1$ and $F(x)/x \to 2L$ when $x \to 0$. Partial integration yields

$$\int_{\mathbb{R}} \frac{1}{t} K\left(\frac{x}{t}\right) f(x)\,dx = \int_0^\infty \frac{1}{t} K\left(\frac{x}{t}\right) dF(x)$$

$$= -\int_0^\infty \frac{x}{t^2} K'\left(\frac{x}{t}\right) \frac{F(x)}{x}\,dx$$

where we have used the bounds on $F(x)$ and $K(x)$ to discard the term at the limits. But $F(x)/x \to 2L$ and $(-2x/t^2)K'(x/t)$ is an approximate identity on $[0, \infty)$, so that *(ii)* follows from the basic properties of approximate identities (Exercise 2.2.16). ∎

Replacing $f(\cdot)$ by $f(x + \cdot)$, we obtain a more generally applicable form of the theorem.

Corollary 2.2.26. *Under the above conditions on K, we have for any $f \in L^1(\mathbb{R})$,*

$$\lim_{t \to 0} \frac{1}{t} \int_{\mathbb{R}} K\left(\frac{y}{t}\right) f(x+y)\,dy = \lim_{h \to 0} \frac{1}{2h} \int_{x-h}^{x+h} f(u)\,du$$

wherever the latter limit exists, in particular for almost all $x \in \mathbb{R}$.

Note that the set of admissible x is precisely the set of points where the fundamental theorem of calculus applies, in the sense of the symmetrical limit. In general this is *strictly larger* than the Lebesgue set, defined as the set of points x where $\lim_{h \to 0} (1/2h) \int_{x-h}^{x+h} |f(u) - f(x)| \, du = 0$.

Exercise 2.2.27. Show that the Poisson kernel, with $K(x) = 1/\pi(1+x^2)$ satisfies the conditions of Theorem 2.2.25. In particular

$$\lim_{t \to 0} \frac{1}{\pi} \int_{\mathbb{R}} \frac{t}{t^2 + (x-y)^2} f(y) \, dy = \lim_{h \to 0} \frac{1}{2h} \int_{x-h}^{x+h} f(y) \, dy$$

wherever the latter limit exists, in particular almost everywhere.

Exercise 2.2.28. *Prove that the one-dimensional Gauss kernel satisfies the conditions of Theorem 2.2.25, in particular*

$$\lim_{t \to 0} \frac{1}{\sqrt{4\pi t}} \int_{\mathbb{R}} e^{-(x-y)^2/4t} f(y) \, dy = \lim_{h \to 0} \frac{1}{2h} \int_{x-h}^{x+h} f(y) \, dy$$

wherever the latter limit exists, in particular almost everywhere.

Exercise 2.2.29. *Prove that the Fejér kernel does not satisfy the conditions of Theorem 2.2.25.*

Theorem 2.2.25 does not apply directly to the Fejér kernel, where $K(x) = (1 - \cos x)/\pi x^2$. In order to deal with this and other oscillatory kernels, we introduce the notion of *monotone majorant*. This is a function $\bar{K} \in L^1(\mathbb{R})$, which satisfies

(2.2.15) $\qquad |K(x)| \leq \bar{K}(x), \quad \bar{K}(-x) = \bar{K}(x), \quad x\bar{K}'(x) \leq 0.$

Note that we do not require that $\int_{\mathbb{R}} \bar{K}(x) \, dx = 1$, so that we cannot expect, for example, that $t^{-1}\bar{K}(x/t)$ be an approximate identity. However we have the following useful replacement for Theorem 2.2.25.

Lemma 2.2.30. *Suppose that $\bar{K} \in L^1(\mathbb{R})$ satisfies (2.2.15). If $|\epsilon(x)| \leq M$ for $x \geq 0$ with $\lim_{x \to 0} \epsilon(x) = 0$, then*

$$\lim_{t \to 0} \int_0^\infty \frac{x}{t^2} \bar{K}'\left(\frac{x}{t}\right) \epsilon(x) \, dx = 0.$$

Proof. Following the steps of the proof of Theorem 2.2.25 with K replaced by $\bar{K}$, we see that $\lim_{x \to \infty} x\bar{K}(x) = 0$, $-\int_0^\infty x\bar{K}'(x) \, dx \leq C < \infty$. For any $\delta \geq 0$

(2.2.16) $\qquad -\int_\delta^\infty \frac{x}{t^2} \bar{K}'\left(\frac{x}{t}\right) dx = \frac{\delta}{t} \bar{K}\left(\frac{\delta}{t}\right) + \frac{1}{t} \int_\delta^\infty \bar{K}\left(\frac{x}{t}\right) dx.$

Taking $\delta = 0$ shows that $-\int_0^\infty (x/t^2) \bar{K}'(x/t) \, dx \leq C$. For any $\delta > 0$, both terms on the right side of (2.2.16) tend to zero when $t \to 0$. Therefore

$$\left| \int_0^\infty \frac{x}{t^2} \bar{K}'\left(\frac{x}{t}\right) \epsilon(x) \, dx \right| \leq C \sup_{0 \leq x \leq \delta} |\epsilon(x)| + M \left| \int_\delta^\infty \frac{x}{t^2} \bar{K}'\left(\frac{x}{t}\right) dx \right|.$$

First take $t \to \infty$ and then $\delta \to 0$ to complete the proof. ∎

This allows us to reformulate and prove Theorem 2.2.25 as follows.

Theorem 2.2.31. *Suppose that K defines an approximate identity with a monotone majorant satisfying (2.2.15). Then*

$$\lim_{t \to 0} \frac{1}{t} \int_{\mathbb{R}} K\left(\frac{y}{t}\right) f(x+y)\, dy = L \qquad (2.2.17)$$

at every x for which $\lim_{h \to 0}(1/2h) \int_{-h}^{h} |f(x+y) - L|\, dy = 0$.

Proof. Without loss of generality, we take $x = 0$ in the proof. Since $\int_{\mathbb{R}} K(x)\, dx = 1$, we can write

$$\frac{1}{t} \int_{\mathbb{R}} K\left(\frac{y}{t}\right) f(y)\, dy = L + \frac{1}{t} \int_{\mathbb{R}} K\left(\frac{y}{t}\right) (f(y) - L)\, dy.$$

It remains to show that the last term tends to zero. For this purpose, define $F(x) = \int_{-x}^{x} |f(u) - L|\, du$, so that $F(0) = 0$, $F(x)/|x| \to 0$ when $x \to 0$ and $|F(x)| \le 2L|x| + \|f\|_1$. Then

$$\frac{1}{t} \left| \int_{\mathbb{R}} K\left(\frac{y}{t}\right) (f(y) - L)\, dy \right| \le \frac{1}{t} \int_{\mathbb{R}} \bar{K}\left(\frac{y}{t}\right) |f(y) - L|\, dy$$

$$= \frac{1}{t} \int_0^\infty \bar{K}\left(\frac{y}{t}\right) dF(y)$$

$$= -\frac{1}{t^2} \int_0^\infty \bar{K}'\left(\frac{y}{t}\right) F(y)\, dy$$

where we have used the bounds on $\bar{K}(x)$, $F(x)$ to discard the terms at the upper limit in the integration-by-parts. Now we use Lemma 2.2.30 with $\epsilon(y) = F(y)/y$:

$$\frac{1}{t^2} \int_0^\infty \bar{K}'\left(\frac{y}{t}\right) F(y)\, dy = \frac{1}{t^2} \int_0^\infty y \bar{K}'\left(\frac{y}{t}\right) \frac{F(y)}{y}\, dy \to 0,$$

completing the proof. ∎

Exercise 2.2.32. *Show that for the Fejér kernel, with $K(x) = (1 - \cos x)/\pi x^2$ we may take $\bar{K}(x) = 1/(1 + x^2)$ as a monotone majorant.*

Hint: Check that $1 - \cos x \le \pi x^2/(1 + x^2)$ for all x. Consider separately $|x| \le 1$ and $|x| > 1$.

Corollary 2.2.33. *If the approximate identity defined by $K(x)$ has a monotone majorant, then for any $f \in L^1(\mathbb{R})$, and $x \in \mathrm{Leb}(f)$*

$$\lim_{t \to 0} \frac{1}{t} \int_{\mathbb{R}} K\left(\frac{y}{t}\right) f(x+y)\, dy = f(x).$$

We now return to the development of the Fourier transform.

2.2.2.2 Application to the Fourier transform

Theorem 2.2.34. *The Fourier transform is a 1:1 map from the space $\mathcal{S}$ onto itself. In particular every $f \in \mathcal{S}$ can be recovered from its Fourier transform as*

$$f(x) = \int_{\mathbb{R}^n} e^{2\pi i \xi \cdot x} \hat{f}(\xi)\, d\xi.$$

Proof. Since the Fourier transform is a linear map, it suffices to show that $\hat{f} = 0$ implies $f = 0$ a.e., which was just proved in Corollary 2.2.23. To prove the onto property, note that since $\hat{f} \in S$, in particular in L^1, so that we can take the limit under the integral in (2.2.13) to obtain $f(x) = \int_{\mathbb{R}^n} \hat{f}(\xi) e^{2\pi i \langle \xi, x\rangle} d\xi$, which represents f as the Fourier transform of the *reflected function* $x \to \hat{f}(-x)$. This proves the onto property. ∎

The almost everywhere convergence of the heat kernel transform cannot be proved from the abstract properties of approximate identities. The one-dimensional case is covered by Exercise 2.2.28. For the n-dimensional case the statement and proof follow:

Proposition 2.2.35. *If $f \in L^1(\mathbb{R}^n)$, then $\lim_{t\to 0} H_t f(x) = f(x)$ where $\lim_{r\to 0} r^{-n} \int_{|y-x|\leq r} (f(y) - f(x)) \, dy = 0$, in particular almost everywhere.*

Proof. We let
$$\Phi(r) := \int_{|y-x|\leq r} (f(y) - f(x)) \, dy = \int_0^r \int_{S^{n-1}} (f(x+\rho\omega) - f(x)) \rho^{n-1} \, d\rho \, d\omega,$$
a continuous function of bounded variation. Clearly $|\Phi(r)| \leq \int_{\mathbb{R}^n} |f| + C_n r^n |f(x)|$ and by hypothesis $\Phi(r)/r^n \to 0$ when $r \to 0$. Now we can write
$$H_t f(x) - f(x) = \int_{\mathbb{R}^n} (f(x+y) - f(x)) \frac{e^{-|y|^2/4t}}{(4\pi t)^{n/2}} \, dy$$
$$= \int_0^\infty \frac{e^{-r^2/4t}}{(4\pi t)^{n/2}} \, d\Phi(r)$$
$$= \int_0^\infty \Phi(r) \frac{r e^{-r^2/4t}}{2t(4\pi t)^{n/2}} \, dr.$$

To prove that this tends to zero, let $\epsilon(r) = \Phi(r)/r^n$, a bounded function that tends to zero when $r \to 0$. Then we make the change of variable $r = st^{1/2}$ to obtain
$$|H_t f(x) - f(x)| \leq \int_0^\infty r^n |\epsilon(r)| \frac{r e^{-r^2/4t}}{2t(4\pi t)^{n/2}} \, dr$$
$$= \int_0^\infty |\epsilon(s\sqrt{t})| s^{n+1} \frac{e^{-s^2/4}}{2(4\pi)^{n/2}} \, ds.$$

The integrand tends to zero and is bounded by an integrable function, completing the proof. ∎

The *Lebesgue set* of a function $f \in L^1(\mathbb{R}^n)$ is defined by

(2.2.18) $\quad \text{Leb}(f) = \{x \in \mathbb{R}^n : \lim_{r\to 0} r^{-n} \int_{|y-x|\leq r} |f(y) - f(x)| \, dy = 0\}.$

Clearly
$$x \in \text{Leb}(f) \implies \lim_{r\to 0} r^{-n} \int_{|y-x|\leq r} f(y) \, dy = C_n f(x)$$

where the dimension constant C_n is the volume of the unit ball in $\mathbb{R}^n$. But the converse is not true, in general. [Consider $f(x) = 1_{[-1,1]} \operatorname{sgn}(x)/\sqrt{|x|}$ at $x = 0$ in $\mathbb{R}$.] In this terminology, Proposition 2.2.35 implies that $H_t f(x) \to f(x)$ for all $x \in \mathrm{Leb}(f)$. This example is an instance of the fact that many of the almost-everywhere results in harmonic analysis are proved on the Lebesgue set.

Exercise 2.2.36. *Extend the result of Proposition 2.2.35 to $f \in L^p(\mathbb{R}^n)$ where $1 \leq p \leq \infty$.*

Hint: Use the Hölder inequality to check the bound on $\Phi(r)$.

We can use Gaussian summability to extend Fourier inversion beyond the space $\mathcal{S}$.

Proposition 2.2.37. *(i) Suppose that $f \in L^1(\mathbb{R}^n)$ has an integrable Fourier transform: $\hat{f} \in L^1(\mathbb{R}^n)$. Then f is almost everywhere equal to a continuous function, and we have*

$$f(x) = \int_{\mathbb{R}^n} \hat{f}(\xi) e^{2\pi i \xi \cdot x} \, d\xi \qquad a.e. \tag{2.2.19}$$

(ii) Conversely, suppose that $S(x) = \lim_M \int_{|\xi| \leq M} \hat{f}(\xi) e^{2\pi i \xi \cdot x} \, d\xi$ exists. Then $S(x) = f(x)$ a.e.

Proof. From Fourier reciprocity, we have

$$H_t f(x) = \int_{\mathbb{R}^n} \hat{f}(\xi) e^{-4\pi^2 t |\xi|^2} e^{2\pi i \xi \cdot x} \, d\xi.$$

Appealing to Proposition 2.2.35 shows that the left side tends to $f(x)$ almost everywhere, whereas the right side converges by the dominated convergence theorem.

For the converse, let $S_r = \int_{|\xi| \leq r} \hat{f}(\xi) e^{2\pi i \xi \cdot x} \, d\xi$, the so-called spherical partial sum. Using $S_0 = 0$ and $S_\infty = S(x)$, we can write

$$H_t f(x) = \int_0^\infty e^{-4\pi^2 t r^2} \, dS_r$$

$$= \int_0^\infty 8\pi^2 t r e^{-4\pi^2 t r^2} S_r \, dr$$

$$H_t f(x) - S(x) = \int_0^\infty 8\pi^2 t r e^{-4\pi^2 t r^2} [S_r - S(x)] \, dr.$$

It is easily checked that $r \to k_t(r) := 8\pi^2 t r e^{-4\pi^2 t r^2}$ has total integral 1 and that for any $M > 0$, $\lim_{t \to 0} \int_0^M k_t(r) \, dr = 0$, from which the result follows immediately. ∎

Proposition 2.2.38. *Suppose that $f \in L^1(\mathbb{R}^n)$ has a nonnegative Fourier transform: $\hat{f}(\xi) \geq 0$ and $0 \in \mathrm{Leb}(f)$. Then $\hat{f} \in L^1(\mathbb{R}^n)$, in particular f is a.e. equal to a continuous function.*

Proof. Take $x = 0$ in (2.2.13) and apply Fatou's lemma. ∎

Warning. The alert reader may note the difference between the notion of "almost everywhere equal to a continuous function" and the weaker notion of "continuous almost everywhere." For example, the indicator function of an interval is continuous almost everywhere, but it is not a.e. equal to a continuous function. Continuity a.e. is a *local* notion, whereas the stronger concept is a *global* notion.

Exercise 2.2.39. *Suppose that $f \in L^1(\mathbb{R}^n)$ has integrable partial derivatives of order $n + 1$: $D^\alpha f \in L^1(\mathbb{R}^n)$ for any multiindex with $|\alpha| \leq n + 1$. Prove that $\hat{f} \in L^1(\mathbb{R}^n)$ and hence Proposition 2.2.37 applies to give Fourier inversion.*

We will see in the next section that pointwise Fourier inversion holds with $n/2$ derivatives instead of $n + 1$.

The heat kernel convolution operator, which was initally defined on the space $L^1(\mathbb{R}^n)$, can be naturally extended to functions satisfying either of the growth conditions

(2.2.20) $$|f(x)| \leq A e^{B|x|^2}$$

or

(2.2.21) $$\int_{\mathbb{R}^n} |f(x)| e^{-B|x|^2} dx < \infty.$$

Proposition 2.2.40. *Suppose that f satisfies either (2.2.20) or (2.2.21). Then $u = H_t f$ is defined for $0 < t < B/4$ and is a solution of the heat equation $u_t = u_{xx}$ for which $\lim_{t \to 0} H_t f(x) = f(x)$ at every point of continuity of f.*

Exercise 2.2.41. *Prove these statements, noting that one must work directly with the heat kernel transform, since the Fourier transform is not applicable to this class of functions.*

The heat kernel can be used to prove that $\mathcal{S}$ is dense in each of the L^p spaces, $1 \leq p < \infty$. First note that the set of L^p functions with compact support is dense in L^p. Now if f has compact support, the Fourier transform $\hat{f}$ is infinitely differentiable with bounded derivatives. For such an f and any multiindices α, β we can write

$$H_t f(x) = \int_{\mathbb{R}^n} e^{2\pi i x \cdot \xi} \hat{f}(\xi) e^{-4\pi^2 t |\xi|^2} d\xi$$

$$D_x^\alpha H_t f(x) = \int_{\mathbb{R}^n} (2\pi i \xi)^\alpha e^{2\pi i x \cdot \xi} \hat{f}(\xi) e^{-4\pi^2 t |\xi|^2} d\xi$$

$$(2\pi i x)^\beta D_x^\alpha H_t f(x) = \int_{\mathbb{R}^n} D_\xi^\beta \left(e^{2\pi i x \cdot \xi} \right) \left[(2\pi i \xi)^\alpha \hat{f}(\xi) e^{-4\pi^2 t |\xi|^2} \right] d\xi$$

$$= (-1)^{|\beta|} \int_{\mathbb{R}^n} e^{2\pi i x \cdot \xi} D_\xi^\beta \left((2\pi i \xi)^\alpha \hat{f}(\xi) e^{-4\pi^2 t |\xi|^2} \right) d\xi,$$

which is a bounded function. This proves that $H_t f \in \mathcal{S}$. But we proved that for any L^p function $\|H_t f - f\|_p \to 0$ when $t \to 0$. We summarize as follows.

Proposition 2.2.42. *The space $\mathcal{S}$ is dense in each L^p space, $1 \leq p < \infty$.*

Note that the case $p = \infty$ is excluded, because we cannot assert that $H_t f \to f$ for every $f \in L^\infty$.

Exercise 2.2.43. Let $f = 1_{[-1,1]}$ with $n = 1$. Show that $\|H_t f - f\|_\infty$ does not tend to zero when $t \to 0$

One should not get the impression from the above proof that the heat kernel operator maps L^p into $\mathcal{S}$. We expect that for small t the decay of $H_t f$ at infinity should mimick that of f.

Exercise 2.2.44. Let f be the Poisson kernel in $n = 1$: $f(x) = 1/\pi(1 + x^2)$. Show that $H_t f(x) \sim C/x^2$, $|x| \to \infty$, for a positive constant C.

Hint: Write

$$H_t f(x) = \int_{\mathbb{R}} e^{-4\pi^2 t \xi^2} e^{-2\pi|\xi|} e^{2\pi i x \xi} \, d\xi = 2 \int_0^\infty e^{-4\pi^2 t \xi^2} e^{-2\pi|\xi|} \cos(2\pi x \xi) \, d\xi.$$

Integrate by parts twice and identify the constant.

When we pass out of the space $\mathcal{S}$ the problem of pointwise Fourier inversion becomes nontrivial. In the next section we will treat the one-dimensional case, in parallel with the treatment of Fourier series in Chapter 1. We will return to the higher-dimensional case in a separate section.

2.2.2.3 The n-dimensional Poisson kernel
The n-dimensional Poisson kernel is defined as the absolutely convergent Fourier integral

$$(2.2.22) \qquad P(x, y) = \int_{\mathbb{R}^n} e^{2\pi i \xi \cdot x} e^{-2\pi y |\xi|} \, d\xi \qquad y > 0, \, x \in \mathbb{R}^n.$$

By differentiation under the integral sign, it is immediate that $u = P(x, y)$ is a solution of the Laplace equation $u_{yy} + \sum_{i=1}^n u_{x_i x_i} = 0$ in the half space $\{(x, y) : x \in \mathbb{R}^n, y > 0\}$.

Exercise 2.2.45. Suppose that $F \in L^\infty(\mathbb{R}^n)$. Prove that

$$u(x, y) := \int_{\mathbb{R}^n} e^{2\pi i \xi \cdot x} e^{-2\pi y |\xi|} F(\xi) \, d\xi$$

is a solution of Laplace's equation $u_{yy} + \sum_{i=1}^n u_{x_i x_i} = 0$ in the half space.

We have already shown, in case $n = 1$, that $P(x, y)$ is an approximate identity, by an explicit computation. We now obtain an explicit formula for the n-dimensional case. The general idea is called *Bochner's method of subordination*, which allows us to obtain new kernels as suitable transforms of the heat kernel in the t-variable. In terms of operational calculus, the heat kernel operator H_t is the exponential of t times the Laplace operator, whereas the Poisson kernel operator P_y is the negative exponential of y times the square root of the negative of the Laplace operator. Since the exponential is a simple and basic function, it is natural to expect that other kernels can be obtained by suitably transforming the heat kernel. The method will also be applied in later sections to compute the Newtonian kernel associated with Laplace's equation and the more general *Riesz kernels*.

To compute the n-dimensional Poisson kernel, we begin with the Laplace transform of the one-dimensional heat kernel:

$$(2.2.23) \qquad \int_0^\infty \frac{e^{-|\xi|^2/4t}}{\sqrt{4\pi t}} e^{-\lambda t}\, dt = \frac{e^{-|\xi|\sqrt{\lambda}}}{2\sqrt{\lambda}} \qquad \lambda > 0,\ \xi \in \mathbb{R}.$$

This is proved by taking the one-dimensional Fourier transform of both sides, which is justified by the Fubini theorem. Indeed, for the left side we have

$$\int_{\mathbb{R}} e^{2\pi i \xi x} \int_0^\infty \left(\frac{e^{-|\xi|^2/4t}}{\sqrt{4\pi t}} e^{-\lambda t}\, dt\right) d\xi = \int_0^\infty e^{-\lambda t} e^{-4\pi^2 t x^2}\, dt = \frac{1}{\lambda + 4\pi^2 x^2},$$

whereas the Fourier transform of the right side is

$$\int_{\mathbb{R}} e^{2\pi i \xi x} \frac{e^{-|\xi|\sqrt{\lambda}}}{2\sqrt{\lambda}}\, d\xi = \frac{1}{2\sqrt{\lambda}}\left(\frac{1}{\sqrt{\lambda} - 2\pi i x}\right) + \frac{1}{2\sqrt{\lambda}}\left(\frac{1}{\sqrt{\lambda} + 2\pi i x}\right)$$

$$= \frac{1}{\lambda + 4\pi^2 x^2},$$

which proves (2.2.23). Now we apply (2.2.23) with $\lambda^{1/2} = 2\pi y$ and $\xi \in \mathbb{R}^n$ to obtain

$$(2.2.24) \qquad \frac{e^{-|\xi|2\pi y}}{4\pi y} = \int_0^\infty \frac{e^{-|\xi|^2/4t}}{\sqrt{4\pi t}} e^{-4\pi^2 y^2 t}\, dt, \qquad y > 0,\ \xi \in \mathbb{R}^n.$$

Finally, we compute the n-dimensional Fourier transform of both sides. In detail, we multiply (2.2.24) by $e^{2\pi i \xi \cdot x}$ and integrate over $\xi \in \mathbb{R}^n$. On the right side we recognize the n-dimensional Fourier transform of the heat kernel, corrected by the factor $(4\pi t)^{(n-1)/2}$. Using the definition of the Gamma function, we obtain

$$\frac{1}{4\pi y} \int_{\mathbb{R}^n} e^{-|\xi|2\pi y} e^{2\pi i \xi \cdot x}\, d\xi = \int_0^\infty e^{-4\pi^2 y^2} \left(\int_{\mathbb{R}^n} \frac{e^{-|\xi|^2/4t}}{(4\pi t)^{1/2}} e^{2\pi i \xi \cdot x}\, d\xi\right) dt$$

$$= \int_0^\infty (4\pi t)^{(n-1)/2} e^{-4\pi^2 t |x|^2} e^{-4\pi^2 t y^2}\, dt$$

$$= \frac{(4\pi)^{(n-1)/2} \Gamma((n+1)/2)}{(4\pi^2)^{(n+1)/2}[y^2 + |x|^2]^{(n+1)/2}}.$$

Therefore we have the following explicit formula

$$(2.2.25) \qquad \boxed{P(x,y) := \int_{\mathbb{R}^n} e^{-|\xi|2\pi y} e^{2\pi i \xi \cdot x}\, d\xi = \frac{y\Gamma((n+1)/2)}{\pi^{(n+1)/2}[y^2 + |x|^2]^{(n+1)/2}}.}$$

Formula (2.2.25) shows that the functions $x \to P(x,y)$ define an approximate identity with $y \to 0$, since clearly $P(x,y) \geq 0$, $\int_{\mathbb{R}^n} P(x,y)\, dx = 1$ and $\int_{|x|>\delta} P(x,y) \to 0$ when $y \to 0$.

The method used to obtain (2.2.25) is a special case of *Bochner's method of subordination*. In this case, we see that the Poisson kernel is subordinate to the heat kernel, since we can obtain one as an integral transform of the other. In the next section we will see that the same idea can be used to obtain the *Riesz potential kernel* associated with fractional powers of the Laplacian, where the exponential in (2.2.23) is replaced by a suitable monomial.

2.2.3 Fourier Transforms of Tempered Distributions

The space S of rapidly decreasing functions is a linear metric space, when we define the metric by

$$d(\phi, \psi) = \sum_{m,k=0}^{\infty} 2^{-m-k} \frac{d_{mk}}{1+d_{mk}}, \qquad d_{mk} := \sup_{x \in \mathbb{R}^n, |\alpha|=k} |x|^m |D^\alpha \phi(x) - D^\alpha \psi(x)|.$$

A *tempered distribution* is a continuous linear functional L on the space S. The collection of all tempered distributions is denoted S'. The Fourier transform of a tempered distribution is defined by

$$\hat{L}(\phi) = L(\hat{\phi}). \tag{2.2.26}$$

Clearly $\hat{L}$ is again a tempered distribution and the mapping $L \to \hat{L}$ is injective: if $\hat{L}(\phi) = 0$ for all $\phi \in S$, then the distribution L is identically zero. Convergence of tempered distributions is defined in the pointwise sense: $L_k \to L$ if and only if $L_k(\phi) \to L(\phi)$ for each $\phi \in S$.

Exercise 2.2.46. *Show that S is a complete metric space with the metric d defined above.*

Example 2.2.47. *Any locally integrable function f defines a tempered distribution by setting*

$$L(\phi) = \int_{\mathbb{R}^n} f(x)\phi(x)\, dx.$$

A class of interesting examples is provided by the *Riesz potentials*, obtained as follows.

Example 2.2.48. *Let $f(x) = |x|^{-\alpha}$ where $0 < \alpha < n$.*

Then f is locally integrable and the Fourier transform can be computed by beginning with the heat kernel applied to $\phi \in S$:

$$\int_{\mathbb{R}^n} \phi(x-y) \frac{\exp(-|y|^2/4t)}{(4\pi t)^{n/2}}\, dy = \int_{\mathbb{R}^n} e^{-4\pi^2 t |\xi|^2} e^{2\pi i \xi \cdot x} \hat{\phi}(\xi)\, d\xi. \tag{2.2.27}$$

We multiply each side of (2.2.27) by t^k, integrate on $[0, \infty]$, and apply Fubini. The inner integral on the left converges if and only if $k < (n-2)/2$ while the inner integral on the right converges if and only if $k > -1$. In detail, the inner integral on the left side of (2.2.27) is transformed with $s = 1/4t$ to obtain

$$\int_0^\infty t^k \frac{\exp(-|y|^2/4t)}{(4\pi t)^{n/2}} dt = \frac{1}{\pi^{n/2} 4^{k+1}} \int_0^\infty e^{-s|y|^2} s^{n/2-k-2} ds$$

$$= \frac{1}{\pi^{n/2} 4^{k+1}} \frac{((n/2)-k-2)!}{|y|^{n-2k-2}},$$

while the inner integral on the right side of (2.2.27) is

$$\int_0^\infty t^k e^{-4\pi^2 t |\xi|^2} dt = \frac{k!}{(4\pi^2 |\xi|^2)^{k+1}},$$

which gives the result

(2.2.28) $$\boxed{\frac{((n/2)-k-2)!}{\pi^{n/2}} \int_{\mathbb{R}^n} \frac{\phi(x-y)}{|y|^{n-2k-2}} dy = \frac{k!}{\pi^{2k+2}} \int_{\mathbb{R}^n} \hat{\phi}(\xi) \frac{e^{2\pi i \xi \cdot x}}{|\xi|^{2k+2}} d\xi}$$

valid for all $\phi \in \mathcal{S}$. Taking $x = 0$, the right side of (2.2.28) expresses the definition of the Fourier transform of the tempered distribution defined by the function $\xi \to |\xi|^{-2k-2}$. The left side shows that this is equivalent to integration with a constant multiple of the locally integrable function $y \to |y|^{2k+2-n}$. The identification becomes complete when we set $\alpha = 2k + 2$, and the result is paraphrased in the statement

(2.2.29) $$\boxed{\text{The Fourier transform of } |\xi|^{-\alpha} \text{ is } C_{n\alpha}|x|^{\alpha-n}, \text{ where } 0 < \alpha < n.}$$

Note that the case $\alpha = 2$ corresponds to the Newtonian potential kernel associated to the Laplace operator of $\mathbb{R}^n$ when $n \geq 3$. Then $k = 0$ and (2.2.28) takes the form

(2.2.30) $$\frac{((n/2)-2)!}{\pi^{n/2}} \int_{\mathbb{R}^n} \frac{\phi(x-y)}{|y|^{n-2}} dy = \frac{1}{\pi^2} \int_{\mathbb{R}^n} \hat{\phi}(\xi) \frac{e^{2\pi i \xi \cdot x}}{|\xi|^2} d\xi.$$

Exercise 2.2.49. Show that the previous example can be generalized to complex numbers α satisfying $0 < \mathrm{Re}(\alpha) < n$.

Exercise 2.2.50. Let $\phi \in \mathcal{S}(\mathbb{R}^n)$ where $n \geq 3$. Show that (2.2.30) is a solution of the equation $\Delta u = -4\pi^2 \phi$, where Δ denotes the n-dimensional Laplace operator.

2.2.4 *Characterization of the Gaussian Density

We can use the Fourier transform to prove J. C. Maxwell's (1860) characterization of the Gauss density, as follows.

Proposition 2.2.51. *Suppose that $n \geq 2$ and that $f(x), x \in \mathbb{R}^n$ is an integrable function which has the properties that there exist $f_0, f_1, \ldots, f_n$ so that*

$$f(x_1, \ldots, x_n) = f_1(x_1) \cdots f_n(x_n) = f_0(\sqrt{x_1^2 + \cdots + x_n^2}) \qquad x \in \mathbb{R}^n.$$

Then $f(x) = Ae^{-B|x|^2}$ for some $B > 0$.

Proof. From Fubini's theorem we see that $f_i \in L^1(\mathbb{R}^1)$ for $1 \leq i \leq n$. Taking polar coordinates, we see that $f_0 \in L^1(\mathbb{R}^+; r^{n-1} dr)$ and that the Fourier transform $\hat{f}$ can be expressed as a function g of the square of the Euclidean norm. Taking the Fourier transform, we see that

$$\hat{f}(\xi_1, \ldots, \xi_n) = \hat{f}_1(\xi_1) \cdots \hat{f}_n(\xi_n) = g(\xi_1^2 + \cdots + \xi_n^2).$$

If $\hat{f}_i(0) = 0$ for some i, then we would have $g(x) \equiv 0$ and we could take $A = 0$. Hence we can suppose that $\hat{f}_i(0) \neq 0$. Setting $\xi_j = 0$ for $j \neq i$ shows that $\hat{f}_i(\xi) = c_i g(\xi^2)$ for some constant $c_i \neq 0$. Letting $\xi_j \equiv 0$ identifies the constant $c_1 \cdots c_n = g(0)^{1-n}$. Setting $G(x) = g(x)/g(0)$ we obtain the functional equation

(2.2.31) $\qquad G(x+y) = G(x)G(y), \quad G(0) = 1, \quad x \to G(x)$ continuous.

Let $\delta > 0$ so that $G(x) > 0$ for $0 < x \leq \delta$. From this we can compute G on the rationals multiples of δ: $G(\delta m/n) = G(\delta)^{m/n}$ and by continuity this formula extends to all real numbers in $[0, \delta]$ in the form $G(x) = e^{-Bx}$, $B := -\delta^{-1} \log G(\delta)$. Now we can use the functional equation (2.2.31) to extend this to all $x > 0$. Since G is a bounded function, we must have $B > 0$, which completes the proof. ∎

In Chapter 5 we will see that this characterization of the Gaussian density is true in the wider context of probability measures on $\mathbb{R}^n$, not necessarily absolutely continuous with respect to Lebesgue measure.

Exercise 2.2.52. *Suppose that G is a measurable and locally integrable function on $\mathbb{R}$ and satisfies the functional equation $G(x+y) = G(x)G(y)$ a.e. Prove that either $G(x) \equiv 0$ or $G(x) = e^{ax}$ for some $a \in \mathbb{R}$.*

Hint: First show that G is a continuous function.

2.2.5 *Wiener's Density Theorem

The Fourier transform in one dimension can be effectively used to study the L^1 closure of the set of translates of a given L^1 function

(2.2.32) $$\sum_{k=1}^{N} a_k f(x - x_k)$$

where a_k are complex numbers and x_k are real. A closely related set is formed by functions written as convolutions

(2.2.33) $$\int_{\mathbb{R}} a(y) f(x - y) \, dy$$

where $a \in L^1(\mathbb{R})$. From Lebesgue's differentiation theorem it follows that any finite sum of the form (2.2.32) can be written as an L^1 limit of functions of the form (2.2.33); conversely, any convolution can be written as the limit of Riemann sums. Hence we see that the L^1 closure of (2.2.33) is identical to the L^1 closure of (2.2.32). Wiener's theorem characterizes this in terms of the Fourier transform. The following proof is adapted from Garding (1997).

Theorem 2.2.53. *Let $f \in L^1(\mathbb{R})$. Then the L^1 closure of (2.2.33) is the full space $L^1(\mathbb{R})$ if and only if the Fourier transform is never zero: $\hat{f}(\xi) \neq 0$ for all $\xi \in \mathbb{R}$.*

Proof. The necessity of the condition is immediate, since the Fourier transform of $a * f$ is $\hat{a}(\xi)\hat{f}(\xi)$. If $\hat{f}(\xi_0) = 0$, then the same is true for all convolutions and, by continuity, for all elements in the L^1 closure. Therefore the L^1 closure of $\{a * f : a \in L^1(\mathbb{R})\}$ is a proper subset of $L^1(\mathbb{R})$. To prove the sufficiency, we first note that it's enough to prove that there exists a dense subset of $L^1(\mathbb{R})$, all of whose elements can be written $a * f$ for some $a \in L^1$. We let

$$A_0 = \{h \in L^1(\mathbb{R}) : \hat{h} \text{ has compact support}\},$$

$$B_0 = \{h \in A_0 : \hat{h} \text{ is piecewise } C^2\}.$$

Clearly A_0 is dense in $L^1(\mathbb{R})$ since Fejér's theorem guarantees that we have the L^1 convergence

$$h(x) = \lim_M \int_{-M}^{M} \hat{h}(\xi) \left(1 - \frac{|\xi|}{M}\right) e^{2\pi i x \xi} \, d\xi. \qquad \blacksquare$$

To proceed further, we introduce the notation $\mathcal{A} = \{\hat{f} : f \in L^1(\mathbb{R})\}$, which consists of continuous functions vanishing at infinity, with the norm

$$\|\hat{f}\|_{\mathcal{A}} := \|f\|_1.$$

We state and prove the following basic lemma.

Lemma 2.2.54. *For any $f \in L^1(\mathbb{R})$ and $g \in B_0$, let $G = \hat{g}$, $G_\delta(\xi) = G(\xi/\delta)$. Then we have*

$$\lim_{\delta \to 0} \|(\hat{f} - \hat{f}(0))G_\delta\|_{\mathcal{A}} = 0.$$

Proof. $\hat{f}G_\delta$ is the Fourier transform of

$$t \to \int_{\mathbb{R}} f(s) \delta g(\delta(t-s)) \, ds,$$

whereas $\hat{f}(0)G_\delta$ is the Fourier transform of

$$t \to \delta g(\delta t) \left(\int_{\mathbb{R}} f(s) \, ds\right).$$

Therefore the required $\mathcal{A}$ norm is estimated as

$$\|(\hat{f} - \hat{f}(0))G_\delta\|_{\mathcal{A}} \leq \int_{\mathbb{R}} \left(\int_{\mathbb{R}} |f(s)(g(t - \delta s) - g(t))| \, ds\right) dt$$

where we have made the substitution $t \to \delta t$. The final integral tends to zero by the dominated convergence theorem.

To complete the proof of the theorem, we let P be the piecewise linear function such that

$$P(\xi) = \begin{cases} 1 & \text{if } |\xi| < 1 \\ 0 & \text{if } |\xi| > 2. \end{cases}$$

Notice that $P(\xi/\delta) \neq 0$ implies that $P(\xi/2\delta) = 1$, so that we can use the hypothesis $\hat{f}(\xi) \neq 0$ to write

$$P_\delta(\xi) \doteq \rho(\xi/\delta)\hat{f}(\xi)\frac{P(\xi/\delta)}{\hat{f}(\xi)}$$

$$= \hat{f}(\xi)\frac{P(\xi/\delta)}{\hat{f}(0) + P(\xi/2\delta)(\hat{f}(\xi) - \hat{f}(0))}.$$

From the lemma we see that for sufficiently small $\delta > 0$, the term $P(\xi/2\delta)(\hat{f}(\xi)-\hat{f}(0))/\hat{f}(0)$ has norm less than $1/2$, so that we can make the following convergent Taylor series expansion in the space $\mathcal{A}$:

$$P_\delta = \hat{f}\frac{P_\delta}{\hat{f}(0)}\sum_{k=0}^{\infty}(-1)^k\left(P_{2\delta}(\hat{f}-\hat{f}(0))/\hat{f}(0)\right)^k.$$

We have proved that $P_\delta = \hat{f}Q_\delta$ where $Q_\delta \in \mathcal{A}$. In the same manner we can apply this to any translate of P_δ to obtain

$$P\left(\frac{\xi-b}{\delta}\right) = \hat{f}(\xi)Q_\delta(\xi;b)$$

for some $Q_\delta \in \mathcal{A}$. Note that $\sum_{-N}^{N} P(\xi + 3k) = 1$ for $|\xi| \leq 3N + 1$. Hence for any $h \in A_0$ we can write

$$\hat{h}(\xi) = \hat{h}(\xi)1_{[-M,M]}(\xi) = \hat{f}(\xi)\left(\sum_b \hat{h}(\xi)Q_\delta(\xi;b)\right),$$

which exhibits h as the convolution of f and an L^1 function, which completes the proof of the theorem. ∎

2.3 FOURIER INVERSION IN ONE DIMENSION

In this section we give a self-contained treatment of convergence theorems for the Fourier integral in one dimension. Readers who have followed the treatment of Fourier series in Chapter 1 may wish to omit much of the current section, since many of the theorems are direct analogues of the corresponding theorems for Fourier series. An exception is the discussion of one-sided Fourier representations in Section 2.3.9, but this is not used in the sequel.

2.3.1 Dirichlet Kernel and Symmetric Partial Sums

The partial sum operator applied to $f \in L^1(\mathbb{R})$ is defined by

(2.3.1) $$S_M f(x) := \int_{-M}^{M} \hat{f}(\xi)e^{2\pi i\xi x}\,d\xi.$$

We now rewrite the integral defining the partial sum so that it makes no reference to the Fourier transform. This is called the *explicit representation* via the Dirichlet kernel.

In order to do this, we use Fubini to write

$$S_M f(x) = \int_{-M}^{M} \hat{f}(\xi) e^{2\pi i \xi x} \, d\xi = \int_{-M}^{M} \left(\int_{-\infty}^{\infty} e^{-2\pi i \xi y} f(y) \, dy \right) e^{2\pi i \xi x} \, d\xi$$

$$= \int_{-\infty}^{\infty} \left(\int_{-M}^{M} e^{2\pi i \xi (x-y)} \, d\xi \right) f(y) \, d\xi$$

$$= \int_{-\infty}^{\infty} \frac{\sin 2\pi M(x-y)}{\pi(x-y)} f(y) \, dy$$

$$= \int_{-\infty}^{\infty} \frac{\sin 2\pi Mz}{\pi z} f(x-z) \, dz,$$

the required formula. The previous computation is summarized by writing

(2.3.2) $$\boxed{S_M f(x) = \int_{-\infty}^{\infty} \frac{\sin 2\pi Mz}{\pi z} f(x-z) \, dz}$$

or equivalently, since the kernel is an even function

$$S_M f(x) = \int_{0}^{\infty} [f(x+z) + f(x-z)] \frac{\sin 2\pi Mz}{\pi z} \, dz.$$

The function $z \to (\sin 2\pi Mz)/\pi z$ is called the *Dirichlet kernel* and the integral operator is a *convolution* with the Dirichlet kernel. We recognize the Dirichlet kernel as the Fourier transform of the indicator function of the interval $[-M, M]$. As a first application, we use the Gaussian identity (2.2.9) to compute the (improper) integral of the Dirichlet kernel. Applying (2.2.9) with $x = 0$ and $f = 1_{[-1,1]}$

$$\lim_{t \to 0} \int_{-\infty}^{\infty} \frac{\sin 2\pi \xi}{\pi \xi} e^{-4\pi^2 t \xi^2} \, d\xi = 1,$$

or equivalently by changing variables to $z = \xi \sqrt{t}$, $N = 1/\sqrt{t}$ we have

$$\lim_{N \to \infty} \int_{-\infty}^{\infty} \frac{\sin 2\pi Nz}{\pi z} e^{-4\pi^2 z^2} \, dz = 1.$$

This can be applied first to compute

$$\int_{-1}^{1} \frac{\sin 2\pi Nz}{\pi z} e^{-4\pi^2 z^2} \, dz = \left(\int_{-\infty}^{\infty} - \int_{|z| \geq 1} \right) \frac{\sin 2\pi Nz}{\pi z} e^{-4\pi^2 z^2} \, dz.$$

The first integral tends to 1, while the second integral tends to zero, by the Riemann-Lebesgue lemma. Now

$$\int_{-1}^{1} \frac{\sin 2\pi Nz}{\pi z} \, dz = \int_{-1}^{1} \frac{\sin 2\pi Nz}{\pi z} (1 - e^{-4\pi^2 z^2}) \, dz + \int_{-1}^{1} \frac{\sin 2\pi Nz}{\pi z} e^{-4\pi^2 z^2} \, dz.$$

The first integral tends to zero by the Riemann-Lebesgue lemma and the final integral tends to 1, by the previous step. We have proved that

$$\lim_{N \to \infty} \frac{1}{\pi} \int_{-2N\pi}^{2N\pi} \frac{\sin t}{t} \, dt = \lim_{N \to \infty} \int_{-1}^{1} \frac{\sin 2\pi Nz}{\pi z} \, dz = 1.$$

This is the famous *sine integral*, which is often computed from Cauchy's theorem on complex integration. In Chapter 1 this was computed from the properties of Fourier series. Now we have redone this using the Riemann-Lebesgue lemma and the explicit Gaussian example. It is customary to write

$$\text{Si}(x) = \frac{2}{\pi} \int_0^x \frac{\sin t}{t} \, dt,$$

so that $\lim_{x \to \infty} \text{Si}(x) = 1$, $\lim_{x \to -\infty} \text{Si}(x) = -1$.

Exercise 2.3.1. *Prove the inequality* $|1 - \text{Si}(x)| \leq (4/\pi x)$ *for all* $x > 0$.

Hint: Integrate-by-parts $\int_x^M (\sin t/t) \, dt$ and let $M \to \infty$.

2.3.2 Example of the Indicator Function

We now consider in detail the case of the indicator function $f(x) = 1_{(a,b)}(x)$. The Fourier transform is

$$\hat{f}(\xi) = \int_a^b e^{-2\pi i \xi x} dx = \frac{e^{-2\pi i \xi b} - e^{-2\pi i \xi a}}{-2\pi i \xi}$$

for $\xi \neq 0$ and $\hat{f}(0) = (b-a)$, by definition. Now we consider the nonsymmetric partial sum

$$S_{M,N}f(x) = \int_{-N}^{M} e^{2\pi i \xi x} \hat{f}(\xi) \, d\xi = \left(\int_0^M + \int_{-N}^0 \right) e^{2\pi i \xi x} \hat{f}(\xi) \, d\xi.$$

The first term is written

$$\int_0^M e^{2\pi i \xi x} \hat{f}(\xi) \, d\xi = \int_0^M \frac{e^{2\pi i \xi (x-a)} - e^{2\pi i \xi (x-b)}}{2\pi i \xi} \, d\xi$$

while the second term has an identical structure. When we take the real and imaginary part, we see that the real part may be written in terms of the sine integral $\text{Si}(x) = (2/\pi) \int_0^x (\sin t/t) \, dt$, hence convergent. But the imaginary part is written in terms of integrals involving $\int_0^M [\cos \xi (x-b) - \cos \xi (x-a)]/\xi \, d\xi$, which is convergent if $x \neq a$, $x \neq b$, but otherwise diverges logarithmically. Therefore the nonsymmetric partial sum $S_{M,N}f$ does not converge in general. Put otherwise, the improper Riemann integral will not suffice for the Fourier inversion of this function.

This apparently anomalous behavior may be attributed to the generality of complex notation. Indeed, if we had begun with the basic trigonometric form of the Fourier integral (2.1.8) this would not occur, since the corresponding complex form will necessarily be the symmetric partial sum.

Exercise 2.3.2. *Show that for the above example,* $S_{0,N}f(a) \sim C \log N$ *and identify the constant* C.

We now show explicitly that the symmetric partial sums converge.

$$S_M f(x) = \int_{-M}^{M} \hat{f}(\xi) e^{2\pi i \xi x} d\xi$$

$$= \int_0^M [\sin 2\pi \xi (x-b) - \sin 2\pi \xi (x-a)]/\pi \xi \, d\xi$$

$$= \tfrac{1}{2}[\text{Si}(2M\pi(b-x)) - \text{Si}(2M\pi(a-x))].$$

It is immediate that if $a < x < b$ this converges to 1, while if $x < a$ or $x > b$ it converges to zero. At the endpoints $x = a$, $x = b$ it converges to $\tfrac{1}{2}$. Furthermore these approximating functions are uniformly bounded by 3.

Exercise 2.3.3. *Check these statements.*

2.3.3 Gibbs-Wilbraham Phenomenon

The Fourier inversion of the indicator function provides the simplest occurence of the *Gibbs-Wilbraham phenomenon*. This is the detailed statement of nonuniform convergence that is present in the Fourier analysis of discontinuous functions. Indeed, if we had uniform convergence, then the above sequence of continuous functions would have a continuous limit. But the indicator fails to be continuous at its endpoints. In order to see this in more detail, we take the special case $a = 0, b = 1$. Applying the previous discussion, we see that

$$S_M f(x) = \tfrac{1}{2}[\text{Si}(2M\pi(1-x)) + \text{Si}(2M\pi x)].$$

For any fixed $x \in (0, 1)$ this converges to 1, when $M \to \infty$. But if we take $x = 1/2M \to 0$, then $f_M(1/2M) \to (1/2)[1 + \text{Si}(\pi)]$, which is now shown to be larger than 1. Indeed

$$\text{Si}(\pi) = \frac{2}{\pi} \int_0^\pi \frac{\sin x}{x} dx = \frac{2}{\pi} \int_0^\pi \left(1 - \frac{x^2}{3!} + \frac{x^4}{5!} - \frac{x^6}{7!}\right) dx + \cdots$$

$$= \frac{2}{\pi}\left(\pi - \frac{\pi^3}{18} + \frac{\pi^5}{600} - \frac{\pi^7}{35,280}\right) + \cdots$$

$$= 2 - \frac{\pi^2}{9} + \frac{\pi^4}{300} - \frac{\pi^6}{17,640} + \cdots$$

$$= 2 - 1.11 + 0.33 - 0.04 + \cdots$$

$$= 1.18 \text{ to two decimal places}$$

so that to two decimal places, $\lim_M S_M f(1/2M) = 1.09$, demonstrating the Gibbs overshoot.

2.3.4 Dini Convergence Theorem

Returning to the theory, we now develop a basic convergence theorem.

Theorem 2.3.4. *Suppose that $f(x)$, $-\infty < x < \infty$ is a complex-valued integrable function that satisfies a Dini condition at x: for some $S \in \mathbb{C}$, $\delta > 0$,*

(2.3.3)
$$\int_0^\delta \frac{|f(x+t) + f(x-t) - 2S|}{t} \, dt < \infty.$$

Then

$$\lim_{M \to \infty} \int_{-M}^M \hat{f}(\xi) e^{2\pi i \xi x} \, d\xi = S.$$

(It is not asserted that $S = f(x)$).

Proof. From the Dirichlet kernel representation, we have the Fourier partial sum

$$S_M f(x) = \int_{-M}^M \hat{f}(\xi) e^{2\pi i \xi x} \, d\xi = \int_{-\infty}^\infty \frac{\sin 2\pi M z}{\pi z} f(x+z) \, dz.$$

Having proved Fourier inversion for the function $e^{-\pi x^2}$, we can replace $f(x)$ by $f(x) - Se^{-\pi x^2}$. The new choice of f is also in L^1 and satisfies the Dini condition with $S = 0$. The function $z \to [f(x+z) + f(x-z)]/z$ is integrable, since the Dini condition takes care of $\int_{|z| \le \delta}$ whereas

$$\int_{|z| > \delta} \frac{|f(x+z) + f(x-z)|}{z} \, dz \le \frac{1}{\delta} \int_{|z| > \delta} |f(x+z) + f(x-z)| \, dz < \infty.$$

Using the Riemann-Lebesgue lemma, it follows that the $S_M f(x) \to 0$, as required. ∎

Corollary 2.3.5. *Suppose that f satisfies a local Hölder condition with exponent $\alpha > 0$:*

$$|f(x) - f(y)| \le C|x - y|^\alpha, \qquad |y - x| < \delta.$$

Then Fourier inversion holds with $S = f(x)$.

Proof. Taking $S = f(x)$, we have for $0 < t < \delta$

$$|f(x+t) + f(x-t) - 2f(x)| = |(f(x+t) - f(x)) + (f(x-t) - f(x))| \le 2Ct^\alpha.$$

But the integral $\int_0^\delta t^{\alpha-1} dt$ is convergent, hence the Dini condition is satisfied. ∎

Corollary 2.3.6. *Suppose that f has right and left limits $f(x \pm 0)$ and satisfies a one-sided local Hölder condition with exponent $\alpha > 0$:*

$$|f(y) - f(x+0)| \le C|x - y|^\alpha, \qquad x < y < x + \delta$$
$$|f(y) - f(x-0)| \le C|x - y|^\alpha, \qquad x - \delta < y < x.$$

Then Fourier inversion holds with $S = [f(x+0) + f(x-0)]/2$.

Proof. We have for $0 < t < \delta$

$$|f(x+t) + f(x-t) - 2S| = |(f(x+t) - f(x+0)) + (f(x-t) - f(x-0))| \le 2Ct^\alpha.$$

But the integral $\int_0^\delta t^{\alpha-1} dt$ is convergent, hence the Dini condition is satisfied. ∎

The theorem of Dirichlet-Jordan also has a counterpart for Fourier transforms.

Theorem 2.3.7. *Suppose that $f \in L^1(\mathbb{R})$ is of finite total variation on the real line. Then $\lim_M S_M f(x) = \frac{1}{2} f(x+0) + \frac{1}{2} f(x-0)$.*

Proof. Letting $F(u) = \frac{1}{2}[f(x+u) + f(x-u)]$, we have

$$S_M f(x) = \int_0^\infty F(u)\, d\,\mathrm{Si}(2\pi M u)\, du.$$

For any $K > 0$, the contribution to the integral from $u \geq K$ is bounded by $\|f\|_1/K$, which can be made small by taking K sufficiently large. On the interval $(0, K)$ we can integrate by parts:

$$\int_{(0,K)} F(u)\, d\,\mathrm{Si}(2\pi M u) = F(K-0)\,\mathrm{Si}(2\pi MK) - \int_{(0,K)} \mathrm{Si}(2\pi M u)\, dF(u).$$

The integrand is bounded and tends to 1, so that we can apply dominated convergence to conclude

$$\lim_M \int_{(0,K)} F(u)\, d\,\mathrm{Si}(2\pi M u) = F(K-0) - [F(K-0) - F(0+0)] = F(0+0),$$

completing the proof. ∎

Exercise 2.3.8. *Prove that if f is of finite total variation, then the partial sums $S_M f$ are uniformly bounded: $\sup_{x \in \mathbb{R}, M > 0} |S_M f(x)| < \infty$.*

2.3.4.1 Extension to Fourier's single integral

We have proved the convergence of the Fourier inversion for functions in $L^1(\mathbb{R})$, which satisfy a Dini condition or have finite total variation. The operator $S_M f$ can be extended to a wider class of functions, if we note that the formula (2.3.2) is well-defined under the sole condition that $\int_\mathbb{R} |f(x)|/(1+|x|)\, dx < \infty$. This extended operator is called *Fourier's single integral* by Zygmund (1959). If we use the extended definition of the operator $f \to S_M f$ in formula (2.3.2), then we can extend each of the above theorems. We leave the details as exercises.

Exercise 2.3.9. *Suppose that $\int_\mathbb{R} |f(x)|/(1+|x|)\, dx < \infty$ and that f satisfies a Dini condition at x. Prove that $\lim_M S_M f(x) = S$.*

Exercise 2.3.10. *Suppose that $\int_\mathbb{R} |f(x)|/(1+|x|)\, dx < \infty$ and that f is of bounded variation in a neighborhood of x. Prove that $\lim_M S_M f(x) = \frac{1}{2} f(x+0) + \frac{1}{2} f(x-0)$.*

2.3.5 Smoothing Operations in $\mathbb{R}^1$–Averaging and Summability

The problem of pointwise convergence of Fourier series and integrals is beset with numerous pathologies, of which we recall two, in the context of Fourier series:

- There exists a continuous function whose Fourier series diverges at a point. The first such example was found by du Bois-Reymond and later simplified by Fejér.

By superposing these at different points, one can construct a continuous function whose Fourier series diverges at an infinite set of points.
- There exists an integrable function whose Fourier series diverges at *every* point. The first such example was found by Kolmogorov, thereby answering in the strongest negative sense the possibility of a general theorem on pointwise Fourier inversion for integrable functions.

These examples, which can be replicated in the context of Fourier transforms, suggest the difficulty of finding general sufficient conditions for convergence. In the other direction, it is impossible in general to obtain a necessary condition for convergence: if f is an odd function [$f(-x) = -f(x)$], its Fourier transform reduces to a sine transform: $F(\xi) = \int f(x) \sin 2\pi \xi x \, dx$, which is also an odd function; hence $S_M f(0) = 0$ identically and $\lim_M S_M f(0) = 0 = f(0)$. Thus pointwise Fourier inversion holds at $x = 0$ without any further regularity conditions.

In a positive direction, it was proved by Carleson (1966) that for an L^2 function, the Fourier series converges at almost every point. This was later generalized by Hunt (1968) to all of L^p if $p > 1$.

The Carleson-Hunt result is deep and difficult, beyond the scope of this work. Instead we shall be content with theorems that replace pointwise convergence by a weaker notion. If the sequence of numbers $S_M f(x)$ fail to converge, it is natural to form averages and hope that the averages behave better than the original sequence of partial sums. There are two possible ways to average:

- Average with respect to x: for example, form $1/(b-a) \int_a^b f(x) \, dx$.
- Average with respect to M: for example, the arithmetic mean $(1/M) \int_0^M S_m f(x) \, dm$. (Fejér mean). Another choice is the Abel mean $\epsilon \int_0^\infty e^{-\epsilon m} S_m f(x) \, dm$.

2.3.6 Averaging and Weak Convergence

These smoothing operations lead immediately to general convergence theorems without additional smoothness conditions.

Theorem 2.3.11. *Suppose that f is any integrable function on $\mathbb{R}$. Then for every $a < b$*

$$\lim_{M \to \infty} \int_a^b S_M f(x) \, dx = \int_a^b f(x) \, dx.$$

Proof. To see this, we first recall that the Dirichlet kernel is even, hence the operator $f \to S_M f$ is self-adjoint, so that we can write

$$\int_a^b S_M f(x) \, dx = \int_\mathbb{R} 1_{(a,b)} S_M f(x) \, dx = \int_\mathbb{R} f(x) S_M 1_{(a,b)} \, dx.$$

Since $1_{(a,b)}$ is of bounded variation, the partial sums $S_M 1_{(a,b)}$ converge boundedly so that we can write

$$\lim_M \int_a^b S_M f(x) \, dx = \int_\mathbb{R} 1_{(a,b)} f(x) \, dx = \int_a^b f(x) \, dx. \blacksquare$$

This type of averaging is natural in applications, where f may represent a density function of mass, charge or probability. We are interested only in the mass/charge/probability of an interval, which is defined by the integral $\int_a^b f(x)\,dx$. The previous theorem states for any integrable function f, we can always recover the mass/charge/probability of an interval as the limit of the Fourier partial sums.

Exercise 2.3.12. *Suppose that μ is a finite Borel measure on the real line. Modify the above proof to show that $\lim_M \int_a^b S_M \mu(\xi)\,d\xi = \mu((a,b)) + \tfrac{1}{2}\mu(\{a\}) + \tfrac{1}{2}\mu(\{b\})$ where $S_M\mu(\xi) = \int_{-M}^M \hat{\mu}(\xi) e^{2\pi i \xi x}\,dx$.*

We can use the above arguments to produce a continuous function vanishing at infinity, which is not the Fourier transform of any integrable function.[1] Let

(2.3.4) $$F(\xi) = \frac{\xi}{(1+|\xi|)\log(2+|\xi|)}.$$

Suppose that $F = \hat{f}$ for some $f \in L^1(\mathbb{R})$. Let $g = 1_{[0,1]}$ be the indicator function of the unit interval, with $\hat{g}(\xi) = (1 - e^{-2\pi i \xi})/2\pi i \xi$ for $\xi \neq 0$. Since g is of bounded variation, the Fourier partial sums $S_M g$ converge boundedly to g and we can write

$$\int_\mathbb{R} fg = \lim_M \int_\mathbb{R} f\, S_M g = \lim_M \int_\mathbb{R} g\, S_M f = \lim_M \int_{-M}^M \hat{f}(-\xi)\hat{g}(\xi)\,d\xi.$$

But a direct calculation shows that the real part of the last integral is

$$\int_{-M}^M \frac{1 - \cos 2\pi\xi}{(1+|\xi|)\log(2+|\xi|)}\,d\xi,$$

which diverges when $M \to \infty$.

2.3.7 Cesàro Summability

We now turn to the question of Cesàro summability. To study this, we rewrite the Fejér mean in terms of the original f, as follows:

$$\begin{aligned}
\frac{1}{M}\int_0^M S_m f(x)\,dm &= \frac{1}{M}\int_0^M \left(\int_{-M}^M \hat{f}(\xi) e^{2\pi i \xi x}\,d\xi\right) dm \\
&= \frac{1}{M}\int_0^M \left(\int_{-m}^m \left(\int_{-\infty}^\infty e^{2\pi i \xi(x-y)} f(y)\,dy\right) d\xi\right) dm \\
&= \frac{1}{M}\int_{-\infty}^\infty \left(\int_0^M \frac{\sin 2m\pi(x-y)}{\pi(x-y)}\,dm\right) f(y) \\
&= \int_{-\infty}^\infty \frac{1 - \cos(2M\pi(x-y))}{2M\pi^2(x-y)^2} f(y)\,dy.
\end{aligned}$$

---

[1] Igari, 1996, p. 172

The Fejér kernel is defined by $K_M(0) = M$ and

(2.3.5) $$K_M(x) = \frac{1 - \cos(2M\pi x)}{2M\pi^2 x^2}, \quad x \neq 0.$$

It has the properties of an approximate identity, expressed by

- $K_M(x) \geq 0$, $\int_{-\infty}^{\infty} K_M(x)\,dx = 1$.
- For any $\delta > 0$, $\int_{|x|>\delta} K_M(x)\,dx \to 0$, when $M \to \infty$.

The nonnegativity is obvious. The normalization can be found from Fourier inversion, as follows: The Fourier transform of $(1 - |\xi|)1_{[0,1]}(|\xi|)$ is $(1 - \cos 2\pi y)/2\pi^2 y^2$, which is an L^1 function. Therefore we can apply Fourier inversion at $y = 0$ to conclude that $\int_{\mathbb{R}} (1 - \cos 2\pi y)/2\pi^2 y^2\,dy = 1$, which transforms into $\int_{\mathbb{R}} K_M(x)\,dx = 1$ when we let $y = Mx$.

To estimate the integral for $|x| > \delta$, we replace the sine by 1, to obtain

$$\int_{|x|>\delta} K_M(x)\,dx \leq \frac{1}{M\pi^2} \int_{|x|>\delta} \frac{dx}{x^2} = \frac{2}{M\delta\pi^2} \to 0$$

when $M \to \infty$.

In order to minimize the new notation, we write

$$K_M f(x) = (K_M * f)(x) = \int_{-\infty}^{\infty} K_M(y) f(x - y)\,dy = \int_{-\infty}^{\infty} K_M(y) f(x + y)\,dy.$$

It will always be clear from context whether we are operating on a function or simply considering the kernel.

Theorem 2.3.13. *The Fejér means have the following properties:*

- *If f is integrable on $\mathbb{R}$ and continuous at x, then the Fejér means converge to $f(x)$: $K_M f(x) \to f(x)$ when $M \to \infty$.*
- *If f is integrable on $\mathbb{R}$ then the Fejér means converge in L^1: $\|K_M f - f\|_1 \to 0$, when $M \to \infty$.*
- *If f is bounded and uniformly continuous on $\mathbb{R}$, then the Fejér means converge uniformly to f: $\sup_{x \in \mathbb{R}} |K_m f(x) - f(x)| \to 0$ when $M \to \infty$.*

Proof. Since $\int_{\mathbb{R}} K_M = 1$, we can write

$$K_M f(x) - f(x) = \int_{-\infty}^{\infty} K_M(y)[f(x + y) - f(x)]\,dy,$$

$$|K_M f(x) - f(x)| \leq \left(\int_{|y|<\delta} + \int_{|y|\geq\delta} \right) K_M(y)[f(x + y) - f(x)]\,dy.$$

If f is continuous at x, then given $\epsilon > 0$ we can choose $\delta > 0$ so that $|f(x + y) - f(x)| < \epsilon$ for $|y| < \delta$. Therefore the first integral is bounded by ϵ. On the other hand, the second term is less than $2\sup_{\mathbb{R}} |f| \int_{|y|>\delta} K_M(y)\,dy$, which tends to zero, proving the first statement.

To estimate the L^1 norm, we write

$$\int_{-\infty}^{\infty} |K_M f(x) - f(x)|\, dx \leq \int_{-\infty}^{\infty} \left(\int_{-\infty}^{\infty} K_M(y) |f(x+y) - f(x)|\, dy \right) dx$$

$$= \int_{-\infty}^{\infty} K_M(y) \left(\int_{-\infty}^{\infty} |f(x+y) - f(x)|\, dx \right) dy$$

$$= \int_{-\infty}^{\infty} K_M(y) \|f_y - f\|_1\, dy.$$

But we can apply the reasoning of the previous step with $x = 0$, and f replaced by $\|f_y - f\|_1$, which is bounded by $2\|f\|_1$ and which is continuous at $y = 0$.

Now if f is bounded and uniformly continuous, we can choose $\delta > 0$ so that $|f(x+y) - f(x)| < \epsilon$ simultaneously for all x, when $|y| < \delta$. Thus

$$|K_M f(x) - f(x)| \leq \epsilon + 2 \sup_{\mathbb{R}} |f| \int_{|y| > \delta} K_M(y)\, dy.$$

Hence $\limsup_M \sup_{x \in \mathbb{R}} |K_M f(x) - f(x)| \leq \epsilon$ for any ϵ, which completes the proof. ∎

We used above, without proof, the fact that $\|f_y - f\|_1 \to 0$ when $y \to 0$. This can be proved in the same spirit as the Riemann-Lebesgue lemma. First prove it for indicator functions of an interval (a, b), then extend to finite linear combinations, and then to bounded measurable functions with compact support. But any L^1 function can be approximated in L^1 by a bounded measurable function with compact support.

Exercise 2.3.14. *Carry out the details of the proof that $\|f_y - f\|_1 \to 0$.*

2.3.7.1 Approximation properties of the Fejér kernel

As with Fourier series, we can find a universal bound for the accuracy of the Fejér approximation, as follows.

Theorem 2.3.15. *Let $f \in L^1(\mathbb{R})$ have the property that $\|K_M f - f\|_1 = o(1/M)$, $M \to \infty$. Then $f = 0$ a.e.*

Proof. The Fejér means are represented as

(2.3.6) $$K_M f(x) = \int_{-M}^{M} \left(1 - \frac{|\xi|}{M}\right) e^{2\pi i \xi x} \hat{f}(\xi)\, d\xi$$

(2.3.7) $$= \int_{\mathbb{R}} \frac{\sin^2 \pi M y}{\pi^2 M y^2} f(x - y)\, dy.$$

In particular $K_M f \in L^1(\mathbb{R})$ and its Fourier transform is $(1 - |\xi|/M)\hat{f}(\xi)1_{[-M,M]}(\xi)$. Hence for any $M > |\xi|$ we have

$$\left(1 - \frac{|\xi|}{M}\right)\hat{f}(\xi) = \int_{\mathbb{R}} e^{-2\pi i \xi x}(K_M f)(x)\, dx$$

$$\frac{|\xi|}{M}\hat{f}(\xi) = \int_{\mathbb{R}} e^{-2\pi i \xi x}[f(x) - (K_M f)(x)]\, dx$$

$$|\xi \hat{f}(\xi)| \leq M \int_{\mathbb{R}} |f(x) - (K_M f)(x)|\, dx$$

$$= M\|f - K_M f\|_1.$$

Letting $M \to \infty$ gives the conclusion that $\hat{f}(\xi) = 0$ for $\xi \neq 0$. But $\hat{f}$ is a continuous function, hence $\hat{f}(\xi) \equiv 0$, which implies that $f = 0$ a.e. ∎

If f has some additional regularity properties, one may obtain the first term in the asymptotic expansion of $K_M f$ when $M \to \infty$.

Exercise 2.3.16. *Suppose that $f \in L^1(\mathbb{R})$ and that $\hat{f} \in L^1(\mathbb{R})$, $\xi\hat{f} \in L^1(\mathbb{R})$. Then*

(2.3.8) $$\lim_{M \to \infty} M[f(x) - K_M f(x)] = \int_{\mathbb{R}} |\xi| e^{2\pi i \xi x} \hat{f}(\xi)\, d\xi.$$

Hint: Begin with the Fourier representation (2.3.6) of $K_M f$, noting that $f(x) = \int_{\mathbb{R}} e^{2\pi i \xi x}\hat{f}(\xi)\, d\xi$ and estimate each of the integral terms $\int_{|\xi|>M}$ and $\int_{|\xi|\leq M}$ separately.

2.3.8 Bernstein's Inequality

The Fejér means can be used to give a proof of an important inequality in approximation theory, originally due to Serge N. Bernstein (1912). Suppose that a function is represented in the form

(2.3.9) $$\boxed{f(x) = \int_{-M}^{M} e^{2\pi i t x} \mu(dt)}$$

where μ is a finite measure supported by the interval $[-M, M]$. This includes a finite trigonometric sum when we specialize μ to be a discrete measure on an arithmetic sequence with $2\pi t = 0, \pm 1, \pm 2, \ldots$. In general, f is an infinitely differentiable function whose derivatives are bounded. Bernstein's inequality gives an upper bound for these derivatives in terms of the upper bound of f.

Theorem 2.3.17. *Under the above conditions, we have for all $x \in \mathbb{R}$*

(2.3.10) $$\boxed{|f'(x)| \leq 4\pi M \sup_{x \in \mathbb{R}} |f(x)|.}$$

Proof. It is no loss of generality to prove the inequality at $x_0 = 0$, since we can change the measure μ by replacing it by $e^{2\pi i t x_0} \mu(dt)$. We begin with the Fejér kernel

$$\frac{1 - \cos 2M\pi x}{2\pi^2 x^2} = \int_{-M}^{M} (M - |t|) e^{2\pi i t x} dt := \int_{-\infty}^{\infty} \Delta_M(t) e^{2\pi i t x} dt$$

where $\Delta_M(t)$ is the triangular function defined by $\Delta_M(t) = M - |t|$ if $|t| < M$ and zero otherwise. From this it follows that

$$\int_{-\infty}^{\infty} \frac{1 - \cos 2M\pi x}{2\pi^2 x^2} e^{2\pi i x t} dx = \Delta_M(t)$$

$$\int_{-\infty}^{\infty} \sin 2M\pi x \frac{1 - \cos 2M\pi x}{2\pi^2 x^2} e^{2\pi i x t} dx = \frac{1}{2i}[\Delta_M(t + M) - \Delta_M(t - M)].$$

The right side is equal to $-t$ on the interval $[-M, M]$. Integrating both sides with respect to the measure $\mu(dt)$ on the interval $[-M, M]$ and applying Fubini, we obtain from (2.3.9)

$$\int_{-\infty}^{\infty} \sin 2M\pi x \frac{1 - \cos 2M\pi x}{2\pi^2 x^2} f(x) dx = \frac{1}{2i} \int_{-M}^{M} -t\mu(dt) = \frac{f'(0)}{4\pi}.$$

Therefore

$$\frac{|f'(0)|}{4\pi} \leq \sup_{-\infty < x < \infty} |f(x)| \times \int_{-\infty}^{\infty} \frac{1 - \cos 2M\pi x}{2\pi^2 x^2} dx = M \sup_{-\infty < x < \infty} |f(x)|,$$

completing the proof. ■

By applying this repeatedly, we obtain estimates for the higher derivatives.

Corollary 2.3.18. *Under the above conditions, we have for all $x \in \mathbb{R}$ and any $k = 1, 2, \ldots$*

$$|f^{(k)}(x)| \leq (4\pi M)^k \sup_{x \in \mathbb{R}} |f(x)|.$$

Bernstein's inequality can be used to characterize the smoothness of functions on the real line in terms of the speed of convergence of their approximation by Fourier integrals on finite intervals. In complex analysis, it is shown that these approximants are entire functions of exponential type. The following proposition is an immediate corollary of Bernstein's inequality.

Proposition 2.3.19. *Suppose that $f \in C(\mathbb{R})$ is a bounded continuous function with the property that there exist approximants of the form*

(2.3.11)
$$f_T(x) = \int_{|\xi| \leq T} e^{2\pi i x \xi} \mu_T(d\xi), \quad \text{with} \quad \|f_T - f\|_\infty = O\left(\frac{1}{T^{k+1}}\right), \quad T \to \infty.$$

Then $f \in C^k(\mathbb{R})$.

Proof. For any $N \in \mathbb{Z}^+$ we can write

$$f = f_{2^N} + \sum_{j=N}^{\infty} (f_{2^{j+1}} - f_{2^j}).$$

Clearly f_{2^N} is differentiable to any order. Now note that

$$f_{2^{j+1}} - f_{2^j} = (f_{2^{j+1}} - f) + (f - f_{2^j})$$
$$= O\left(\left(\frac{1}{2^j}\right)^{k+1}\right) + O\left(\left(\frac{1}{2^j}\right)^{k+1}\right)$$
$$= O\left(\left(\frac{1}{2^j}\right)^{k+1}\right).$$

But $f_{2^{j+1}} - f_{2^j}$ is of the form (2.3.9) with $M = 2^{j+1}$. Therefore by Bernstein's inequality, we have for any n

$$|(f_{2^{j+1}} - f_{2^j})^{(n)}(x)| \le [4\pi 2^{j+1}]^n \sup_{x \in \mathbb{R}} |f_{2^{j+1}}(x) - f_{2^j}(x)|$$
$$\le [4\pi 2^{j+1}]^n \times \left(\frac{C}{2^j}\right)^{k+1},$$

which is the general term of a convergent geometric series provided that $n \le k$. Hence the differentiated series $\sum_j (f_{2^{j+1}} - f_{2^j})^{(n)}$ are convergent for $n = 1, 2, \ldots, k$, which proves that the limit function f is k-times differentiable. ∎

In closing, we remark that in the best known version of Bernstein's inequality, the constant $4\pi M$ is replaced by $2\pi M$ and that this is sharp. Indeed, by considering the example $f(x) = e^{2\pi i M x}$, we see that $|f'(x)| = 2\pi M |f(x)|$. For details, consult Zygmund (1959), vol. 2, p. 276.

2.3.9 *One-Sided Fourier Integral Representation

Sometimes we have to deal with functions that are defined on the half line $0 \le x < \infty$. We can obtain several inequivalent representations by trigonometric integrals by extending the function to the entire real line $-\infty < x < \infty$ in different ways.

2.3.9.1 *Fourier cosine transform*
We can extend f as an even function by setting $f_{\text{even}}(x) = f(x)$ for $x \ge 0$ and $f_{\text{even}}(x) = f(-x)$ for $x < 0$. If $f \in L^1(0, \infty)$, then $f_{\text{even}} \in L^1(\mathbb{R})$. In order to have a symmetrical theory we define the Fourier cosine transform as

$$F_c(\xi) = \int_0^\infty f(x) \cos(\pi \xi x / 2) \, dx.$$

The Fourier transform of $f_{\text{even}}(x)$ is computed as

$$\mathcal{F}(f_{\text{even}})(\xi) = \int_{-\infty}^\infty f_{\text{even}}(x) e^{-2\pi i \xi x} \, dx = 2 \int_0^\infty f(x) \cos(2\pi \xi x) \, dx = 2 F_c(4\xi).$$

If in addition f satisfies a Dini condition at x, with $S = f(x)$, then we have the Fourier inversion as the improper integral

$$f(x) = \int_{-\infty}^{\infty} \mathcal{F}(f_{\text{even}})(\xi) e^{2\pi i \xi x} \, d\xi$$

$$= 2 \int_0^{\infty} 2F_c(4\xi) \cos(2\pi \xi x) \, d\xi$$

$$= \int_0^{\infty} F_c(\nu) \cos(\pi \nu x/2) \, d\nu, \qquad x > 0.$$

The partial sums of the Fourier cosine transform satisfy $(S_M f_{\text{even}})'(0) = 0$, which suggests the relation to the "boundary condition" $f'(0) = 0$.

2.3.9.2 Fourier sine transform

We can extend f as an odd function by setting $f_{\text{odd}}(x) = f(x)$ for $x > 0$ and $f_{\text{odd}}(x) = -f(-x)$ for $x < 0$. If $f \in L^1(0, \infty)$, then $f_{\text{odd}} \in L^1(\mathbb{R})$. In order to have a symmetrical theory, we define the *Fourier sine transform*

$$F_s(\xi) = \int_0^{\infty} f(x) \sin(\pi \xi x/2) \, dx.$$

The Fourier transform of $f_{\text{odd}}(x)$ is an odd function, written

$$\mathcal{F}(f_{\text{odd}}) = 2i \int_0^{\infty} f(x) \sin(2\pi \xi x) \, dx = 2i F_s(4\xi).$$

If f satisfies a Dini condition at $S = f(x)$, then we have the Fourier inversion in the form

$$f(x) = \int_0^{\infty} F_s(\xi) \sin(\pi \xi x/2) \, d\xi, \qquad x > 0.$$

Exercise 2.3.20. *Check this directly from the Fourier inversion theorem.*

The partial sums of the Fourier sine transform satisfy $(S_M f_{\text{odd}})(0) = 0$, which suggests the relation to the "boundary condition" $f(0) = 0$.

2.3.9.3 Generalized h-transform

It is also natural to consider Fourier integral representation for functions that satisfy a boundary condition of the form

$$f'(0) = h f(0).$$

The case $h = 0$ corresponds to the cosine transform, while the limiting case $h \to \infty$ corresponds to the sine transform.

In order to motivate the proper integral transform, we look for the combinations of $\sin \xi x$, $\cos \xi x$ that satisfy the boundary conditions. It is immediately verified that the function $f(x) = \xi \cos \xi x + h \sin \xi x$ satisfies the boundary condition. This function also has the property that $hf - f' = (\xi^2 + h^2) \sin \xi x$, an odd function. This immediately suggests a new recipe for extension of an arbitrary f to the half line $x < 0$, namely to require that $f'(x) - hf(x)$ be an odd function. This leads to a first order differential equation, which is solved in detail by writing

(2.3.12) $$\tilde{f}(x) = f(-x) - 2h \int_0^{-x} e^{h(y+x)} f(y)\, dy, \qquad x < 0,$$

while $\tilde{f}(x) = f(x)$ for $x > 0$. To proceed further we consider separately two cases.

Case 1. $h > 0$: In this case we verify directly that if $f \in L^1(0, \infty)$ then $\tilde{f} \in L^1(\mathbb{R})$. The first term of (2.3.12) is clearly integrable, while for the second term we have

$$\int_{-\infty}^0 |\tilde{f}(x)|\, dx \le 2h \int_{-\infty}^0 \left(\int_0^{-x} e^{h(y+x)} |f(y)|\, dy \right) dx$$

$$= 2h \int_0^\infty |f(y)| \left(\int_{-\infty}^{-y} e^{h(x+y)}\, dx \right) dy$$

$$= 2 \int_0^\infty |f(y)|\, dy < \infty.$$

Therefore $\tilde{f}$ is integrable. To compute the Fourier transform, we write

$$\tilde{f}(x) = f_{\text{even}}(x) - 2h(f \cdot 1_{(0,\infty)}) * (e^{-h\cdot} 1_{(0,\infty)})(-x), \qquad -\infty < x < \infty.$$

The Fourier transform of the convolution is the product of the Fourier transforms while the Fourier transform commutes with the reflection $x \to -x$. For the individual terms, we have

$$\mathcal{F}(f_{\text{even}}) = 2F_c(4\xi),$$

$$\mathcal{F}(f \cdot 1_{(0,\infty)}) = \int_0^\infty f(x) e^{-2\pi i \xi x}\, dx = F_c(4\xi) - iF_s(4\xi),$$

$$\mathcal{F}(e^{-h\cdot} 1_{(0,\infty)}) = \int_0^\infty e^{-hx} e^{-2\pi i \xi x}\, dx = \frac{1}{h + 2\pi i \xi}.$$

Therefore

$$\mathcal{F}(\tilde{f}) = 2F_c(4\xi) - 2h \frac{F_c(4\xi) + iF_s(4\xi)}{h - 2\pi i \xi} = -2i \frac{hF_s(4\xi) + 2\pi \xi F_c(4\xi)}{h - 2\pi i \xi}.$$

This is the h-transform of the given function $f(x)$, $0 < x < \infty$:

(2.3.13) $$F_h(\xi) = -2i \frac{hF_s(4\xi) + 2\pi \xi F_c(4\xi)}{h - 2\pi i \xi}.$$

The numerator is an odd function. In case $h = 0$ the h transform is twice the Fourier cosine transform, whereas in the case $h \to \infty$ it reduces to $(-2i) \times$ the Fourier sine transform. The Fourier inversion formula gives the improper integral

(2.3.14)
$$f(x) = \int_{-\infty}^{\infty} e^{2\pi i \xi x} F_h(\xi)\, d\xi$$
$$= -2i \int_{-\infty}^{\infty} \frac{e^{2\pi i \xi x}}{h - 2\pi i \xi} (hF_s(4\xi) + 2\pi \xi F_c(4\xi))\, d\xi$$
$$= 4 \int_{0}^{\infty} \frac{h \sin(2\pi \xi x) + 2\pi \xi \cos(2\pi \xi x)}{h^2 + 4\pi^2 \xi^2} (hF_s(4\xi) + 2\pi \xi F_c(4\xi))\, d\xi$$
$$= \int_{0}^{\infty} \frac{h \sin(\pi v x/2) + (\pi v/2) \cos(\pi v x/2)}{h^2 + (\pi^2 v^2/4)} (hF_s(v) + (\pi v/2) F_c(v))\, dv.$$

In this form we see more clearly the limiting cases $h \to 0$ and $h \to \infty$. The results are summarized as follows.

Proposition 2.3.21. *Suppose that $f \in L^1(0, \infty)$ with $h > 0$. Then $\tilde{f}$, defined by (2.3.12) is integrable on $\mathbb{R}$. If f satisfies a Dini condition at x, then f may be recovered from its h-transform, defined by formula (2.3.13) and (2.3.14).*

Case 2. $h < 0$: In this case we have the additional complication of a nonzero integrable function whose h transform is identically zero. The function $f(x) = e^{hx}$ is directly computed to have $F_h(\xi) \equiv 0$, hence f cannot be recovered from its h-transform. This is the only obstruction, however.

Proposition 2.3.22. *Suppose that $f \in L^1(0, \infty)$ with*

(2.3.15)
$$\int_0^\infty f(x) e^{hx}\, dx = 0.$$

Then $\tilde{f}$ defined by (2.3.12) is in $L^1(\mathbb{R})$ and the Fourier inversion holds at every point where the Dini condition is satisfied.

Proof. If we combine (2.3.15) with (2.3.12), we can write
$$\tilde{f}(x) = f(-x) - 2h \int_{-x}^{\infty} e^{h(y+x)} f(y)\, dy.$$

Now we can estimate the L^1 norm as before:
$$\left| h \int_{-\infty}^{0} \left(\int_{-x}^{\infty} e^{h(y+x)} |f(y)| dy \right) dx \right| = \int_{0}^{\infty} \left(\int_{-y}^{0} |h| e^{h(x+y)}\, dx \right) |f(y)|\, dy$$
$$= \int_{0}^{\infty} [1 - e^{hy}] |f(y)|\, dy < \infty$$

since $h < 0$. Therefore we can apply the Fourier inversion theorem to $\tilde{f}$. ∎

The new behavior for $h < 0$ is related to the existence of a *point spectrum* in the Laplace operator with these boundary conditions. The function $f(x) = e^{hx}$ satisfies the differential equation $f'' = h^2 f$ and also satisfies the boundary condition $f'(0) = hf(0)$.

If $h > 0$ this function is not in L^1 and hence does not figure in the Fourier analysis. But when $h < 0$ it is integrable and must be dealt with.

In case f does not satisfy the orthogonality condition, we simply subtract a multiple of e^{hx} and then apply Proposition 2.3.22. The function $f(x) + he^{hx} \int_0^\infty f(y) e^{hy} dy$ will satisfy the orthogonality condition, leading to the representation

$$f(x) = Ce^{hx} + \int_{-\infty}^{\infty} F_h(\xi) e^{2\pi i \xi x} d\xi, \qquad C = \int_0^\infty f(y) h e^{hy}\, dy.$$

This illustrates the role of the point spectrum in the Fourier analysis.

2.4 L^2 THEORY IN $\mathbb{R}^n$

The Fourier transform is well-adapted to the space of square-integrable functions, denoted $L^2(\mathbb{R}^n)$. Since L^1 and L^2 are not properly contained in one another, we first restrict to the common dense subspace $\mathcal{S}$ and extend by continuity.

2.4.1 Plancherel's Theorem

The Plancherel theorem serves as a replacement for the Riesz-Fischer theorem that appeared in the L^2 theory of Fourier series in Chapter 1. In contrast with the theory of $L^2(\mathbb{T})$, in the present context we have an isometric bijective correspondence on the space $L^2(\mathbb{R}^n)$. The precise statement follows.

Theorem 2.4.1. *Plancherel: The Fourier transform can be extended to the entire space $L^2(\mathbb{R}^n)$ so that the map $f \to \mathcal{F}f$ preserves the L^2 norm. Furthermore the extended mapping is 1:1 onto all of $L^2(\mathbb{R}^n)$.*

The key to the proof is to establish the isometry of the Fourier transform on the space $\mathcal{S}$, as follows.

Theorem 2.4.2. *Parseval: For any $f \in \mathcal{S}$, we have*

(2.4.1) $$\int_{\mathbb{R}^n} |f(x)|^2\, dx = \int_{\mathbb{R}^n} |\hat{f}(\xi)|^2\, d\xi.$$

Proof. This depends on three simple facts, valid for $u, v \in \mathcal{S}$:

(2.4.2) $$\int_{\mathbb{R}^n} u\hat{v} = \int_{\mathbb{R}^n} \hat{u} v$$

(2.4.3) $$\hat{\hat{u}}(x) = u(-x) := \mathcal{R}u(x)$$

(2.4.4) $$\hat{\bar{u}}(x) = \overline{\hat{u}(-x)}.$$

Formula (2.4.2) is a restatement of the Fourier reciprocity lemma, (2.4.3) is an expression of Fourier inversion for Schwartz functions, and (2.4.4) is from the definition of the Fourier transform. Letting $u = f$, $v = \bar{\hat{f}}$, we have $\hat{v} = \bar{f}$ and

$$\int_{\mathbb{R}^n} f\bar{f} = \int_{\mathbb{R}^n} u\hat{v} = \int_{\mathbb{R}^n} \hat{u}v = \int_{\mathbb{R}^n} \hat{u}\bar{\hat{u}},$$

which was to be proved. ∎

Proof of Plancherel's Theorem. By Proposition 2.2.42, any $f \in L^2(\mathbb{R}^n)$ can be approximated in the L^2 norm by a sequence $f_j \in \mathcal{S}$. Having done this, we apply Parseval to obtain

$$\int_{\mathbb{R}^n} |f_j - f_k|^2(x)\, dx = \int_{\mathbb{R}^n} |\hat{f}_j - \hat{f}_k|^2(\xi)\, d\xi.$$

By hypothesis the left side tends to zero when $j, k \to \infty$, hence $\hat{f}_j$ is a Cauchy sequence. By the completeness of the space L^2, there is a well-defined limit $\mathcal{F}(f)$ in the L^2 norm so that $\|\mathcal{F}(f) - \hat{f}_j\|_2 \to 0$, which defines the required extension $f \to \mathcal{F}f$. In particular,

$$\|\mathcal{F}f\|_2 = \lim_j \|\hat{f}_j\|_2 = \lim_n \|f_j\|_2 = \|f\|_2.$$

Clearly this definition is independent of the approximating sequence, since if g_j is another approximating sequence, another application of Parseval shows that $\|\hat{g}_j - \hat{f}_j\|_2 \to 0$.

To prove the onto property, we first note the properties of the reflection operator $\mathcal{R}$, defined by $\mathcal{R}f(x) = f(-x)$. Applied to the Fourier transform, we have for any $f \in \mathcal{S}$,

$$\mathcal{R}\hat{f}(\xi) = \int_{\mathbb{R}^n} e^{2\pi i x \cdot \xi} f(x)\, dx = \int_{\mathbb{R}^n} e^{-2\pi i x \cdot \xi} f(-x)\, dx = \widehat{(\mathcal{R}f)}(\xi).$$

Clearly $\|\mathcal{R}f\|_2 = \|f\|_2$ for all $f \in L^2(\mathbb{R}^n)$.

Now if $\psi \in L^2(\mathbb{R}^n)$, we approximate by $\psi_j \in \mathcal{S}$. From (2.4.3)

$$\psi_j = \mathcal{R}\hat{\hat{\psi}}_j = \widehat{(\mathcal{R}\hat{\psi}_j)} = \hat{\phi}_j$$

where $\phi_j = \mathcal{R}\hat{\psi}_j$. When $j \to \infty$ we have $\phi_j \to \mathcal{R}(\mathcal{F}\psi)$, $\hat{\phi}_j \to \mathcal{F}(\mathcal{R}\mathcal{F}\psi)$. Hence

$$\psi = \lim_j \psi_j = \lim_j \hat{\phi}_j = \mathcal{F}(\mathcal{R}\mathcal{F}\psi),$$

which proves that the operator $\mathcal{F}$ is onto all of $L^2(\mathbb{R}^n)$. ∎

Exercise 2.4.3. *Prove that the Fourier-Plancherel transform on $L^2(\mathbb{R}^n)$ satisfies the properties of linearity, translation, and phase factor that were proved for the Fourier transform on $L^1(\mathbb{R}^n)$, described in Proposition 2.2.1.*

2.4.2 *Bernstein's Theorem for Fourier Transforms

In Chapter 1 we proved that any Hölder continuous function with exponent $\alpha > \frac{1}{2}$ has an absolutely convergent Fourier series. Here is the corresponding result for Fourier transforms.

Proposition 2.4.4. *Suppose that $f \in L^2(\mathbb{R})$ satisfies*

$$\int_{\mathbb{R}} |f(x-h) - f(x)|^2 dx \le C^2 h^{2\alpha}$$

for some $C > 0$, $\alpha > \frac{1}{2}$ and all h sufficiently small. Then $\hat{f} = \mathcal{F}f \in L^1(\mathbb{R})$.

Proof. The Fourier-Plancherel transform of the translate f_h satisfies $\hat{f}_h(\xi) = e^{-2\pi i h \xi}\hat{f}(\xi)$ so that from the Parseval relation, we have

$$C^2 h^{2\alpha} \geq \int_{\mathbb{R}} |f(x-h) - f(x)|^2 \, dx = \int_{\mathbb{R}} |e^{-2\pi i \xi h} - 1|^2 |\hat{f}(\xi)|^2 \, d\xi.$$

Taking $h = (1/6)2^{-k}$, we estimate the L^1 norm of $\hat{f}$ for $|\xi| \in [2^k, 2^{k+1}]$ by Cauchy-Schwarz, noting that in this range $|e^{-2\pi i \xi h} - 1| \geq 1$:

$$\left(\int_{2^k \leq |\xi| \leq 2^{k+1}} |\hat{f}(\xi)| \, d\xi\right)^2 \leq 2^{k+1} \int_{2^k \leq |\xi| \leq 2^{k+1}} |\hat{f}(\xi)|^2 \, d\xi$$

$$\leq 2^{k+1} \int_{2^k \leq |\xi| \leq 2^{k+1}} |e^{-2\pi i \xi h} - 1|^2 |\hat{f}(\xi)|^2 \, d\xi$$

$$\leq 2^{k+1}(C^2/36) 2^{-2k\alpha}$$

$$= (C^2/18) 2^{k(1-2\alpha)}.$$

Therefore if $\alpha > \frac{1}{2}$

$$\sum_{k=0}^{\infty} \int_{2^k \leq |\xi| \leq 2^{k+1}} |\hat{f}(\xi)| \, d\xi \leq C \sum_{k=0}^{\infty} 2^{k(1-2\alpha)/2} < \infty,$$

and the proof is complete. ∎

In parallel with the theory of Fourier series, one can characterize the L^2 smoothness of a function in terms of the speed of L^2 convergence of the Fourier transform. These are listed in the following exercises.

Exercise 2.4.5. Suppose that for some $0 < \alpha < 1$ and some $C > 0$, we have $\|f - f_h\|_2 \leq Ch^\alpha$ for all $h > 0$. Prove that $\int_{|\xi|>M} |\hat{f}(\xi)|^2 d\xi \leq C_1 M^{-2\alpha}$ for all $M > 0$ and some $C_1 > 0$.

Hint: Mimick the proof of Theorem 1.3.3.

Exercise 2.4.6. Suppose that for some $0 < \alpha < 1$ and some $C_1 > 0$, we have $\int_{|\xi|>M} |\hat{f}(\xi)|^2 d\xi \leq C_1 M^{-2\alpha}$ for all $M > 0$. Prove that $\|f - f_h\|_2 \leq Ch^\alpha$ for all $h > 0$ and some $C > 0$.

Hint: Mimick the proof of Theorem 1.3.3.

Exercise 2.4.7. Suppose that for some $C_1 > 0$ we have $\int_{|\xi|>M} |\hat{f}(\xi)|^2 d\xi \leq C_1 M^{-2}$ for all $M > 0$. Prove that $\|f - f_h\|_2 \leq Ch \log(1/h)$ for $0 < h < \frac{1}{2}$, for some constant $C > 0$.

Exercise 2.4.8. Suppose that for some $k \in \mathbb{Z}^+$ and for some $0 < \alpha < 1$, $C > 0$ we have $\int_{|\xi|>M} |\hat{f}(\xi)|^2 d\xi \leq C_1 M^{-2k-2\alpha}$. Prove that $f, \ldots, f^{(k-1)}$ are absolutely continuous with $f' \in L^2(\mathbb{R}), \ldots, f^{(k)} \in L^2(\mathbb{R})$ and that $\|f_h^{(k)} - f^{(k)}\|_2 \leq Ch^\alpha$ for all $h > 0$, for some $C > 0$.

2.4.3 The Uncertainty Principle

The L^2 theory of the Fourier transform can be used to discuss the Heisenberg uncertainty principle, as follows. We restrict attention to the case of one dimension.

If $f \in L^2(\mathbb{R})$, a quantitative measure of the spread about $x = 0$ is given by the *dispersion about zero* and defined by the formula

$$D_0(f) = \frac{\int_{-\infty}^{\infty} x^2 |f(x)|^2 \, dx}{\int_{-\infty}^{\infty} |f(x)|^2 \, dx}.$$

This is defined whenever the relevant integrals are finite. The name is justified by Chebyshev's inequality, namely for any $M > 0$

$$\frac{\int_{|x|>M} |f(x)|^2 \, dx}{\int_{-\infty}^{\infty} |f(x)|^2 \, dx} \leq \frac{1}{M^2} \frac{\int_{-\infty}^{\infty} x^2 |f(x)|^2 \, dx}{\int_{-\infty}^{\infty} |f(x)|^2 \, dx} = \frac{D_0(f)}{M^2}.$$

The fraction of the L^2 norm due to $|x| > M$ is controlled by $D_0(f)$.

Exercise 2.4.9. *Suppose that f is a Gaussian density function:*

$$f(x) = \frac{1}{\sqrt{4\pi t}} e^{-x^2/4t}, \quad \text{with} \quad \hat{f}(\xi) = e^{-4\pi^2 t \xi^2}.$$

Check that $D_0(f) = t$ and $D_0(\hat{f}) = 1/(16\pi^2 t)$.

In this example the product is $D_0(f) D_0(\hat{f}) = 1/16\pi^2$. This can be paraphrased as the statement that if a Gaussian density function is highly concentrated about its midpoint, then the Fourier transform will be widely spread about its midpoint. A remarkable statement holds in the general case, where we have the following inequality:

Proposition 2.4.10. *Uncertainty principle: Let $f \in L^2(\mathbb{R})$, be a complex-valued function that is absolutely continuous and for which $x f \in L^2(\mathbb{R})$ and $f' \in L^2(\mathbb{R})$. Then we have the inequality*

(2.4.5) $$\boxed{D_0(f) D_0(\hat{f}) \geq \frac{1}{16\pi^2}.}$$

Equality holds if and only if f is a Gaussian density function centered at $x = 0$; in detail, $f(x) = C_1 e^{-x^2/\sigma^2}$, $\hat{f}(\xi) = C_2 e^{-\pi^2 \sigma^2 \xi^2}$ for suitable constants C_1, C_2, σ^2.

Remark. The term "uncertainty principle" comes from the interpretation that we cannot localize both $f(x)$ and $\hat{f}(\xi)$ in their respective spaces. If $f(x)$ is localized about $x = 0$, then $D_0(f)$ will be small; the uncertainty principle then asserts that $D_0(\hat{f})$ will be correspondingly large, indicating a lack of localization about $\xi = 0$.

Proof. For notational clarity, write $F = \hat{f}$. Unless otherwise noted, all integrals are taken over the entire real line. First we apply the Cauchy-Schwarz inequality to $|f(x)| = \sqrt{(1+x^2)} |f(x)| \times \sqrt{1/(1+x^2)}$ to conclude that $f \in L^1$, similarly $F = \hat{f} \in L^1$, hence they are Fourier transforms of one another, and in particular both f and F are equal a.e. to continuous functions vanishing at infinity. In addition, F' exists as an L^2 function, since its

Fourier transform is assumed to be L^2. Both the numerator and denominator of the expressions defining $D_0(f)$ and $D_0(F)$ will be transformed by Parseval's theorem. In this way one is led to examine a corresponding integral involving $F'(\xi)$. Writing $F = \hat{f}$, we write the real part of the integral of $\xi \bar{F} F'$ in two different ways. On the one hand,

$$2\,\mathrm{Re} \int \xi \bar{F} F' \, d\xi = \int \xi (F'\bar{F} + F\bar{F}') \, d\xi = \int \xi (F\bar{F})' \, d\xi = \int \xi (|F|^2)' \, d\xi = -\int |F|^2 \, d\xi$$

where we have integrated by parts in the last step. To justify discarding the term at the limits, note that since all of the integrals are absolutely convergent, they are also convergent improper Riemann integrals, so that $M|F(M)|^2$ and $N|F(-N)|^2$ tend to limits when either $M, N \to \infty$ in any order. If either limit is nonzero, then $F(\xi) \sim \mathrm{const}/|\xi|^{1/2}$, which contradicts the L^2 integrability of F. Now on the other hand,

(2.4.6)
$$-\mathrm{Re} \int \xi \bar{F} F' \, d\xi \le \left| \int \xi \bar{F}(\xi) F'(\xi) \, d\xi \right| \le \left(\int |\xi \bar{F}(\xi)|^2 \, d\xi \right)^{1/2} \left(\int |F'(\xi)|^2 \, d\xi \right)^{1/2}$$

where we have applied the Schwarz inequality to the functions $F'(\xi)$ and $\xi F(\xi)$. Now we apply Parseval's theorem twice, recalling that the Fourier transform of $xf(x)$ is $F'(\xi)/(-2\pi i)$:

$$\int |F(\xi)|^2 \, d\xi = \int |f(x)|^2 \, dx, \qquad \int |F'(\xi)|^2 \, d\xi = 4\pi^2 \int x^2 |f(x)|^2 \, dx.$$

Squaring both sides of (2.4.6) and making these substitutions gives the desired result, in the form $(1/16\pi^2) \int |f|^2 \int |F|^2 \le \int |xf|^2 \int |\xi F|^2$.

In case equality occurs in (2.4.6), we obtain two conditions: (i) Schwarz's inequality implies that $F'(\xi)$ and $\xi F(\xi)$ must be proportional a.e. thus F must satisfy a.e. the differential equation $F'(\xi) = -A\xi F(\xi)$ for some complex constant A; (ii) the imaginary part of $\int \xi \bar{F} F' d\xi$ must be zero. From (i) it follows that the derivative of the function $G(\xi) = F(\xi) e^{A\xi^2/2}$ is zero a.e., hence $G(\xi) = C$ a.e. for some complex constant C, which proves that $F(\xi) = C e^{-A\xi^2/2}$ a.e. But we noted above that F is continuous, hence the equality holds everywhere.

This function will yield a finite value of $D_0(F)$ if and only if $\mathrm{Re}\, A > 0$. To show that $\mathrm{Im}\, A = 0$, we write $A = \alpha + i\beta$ and compute $F'(\xi) = -CA\xi e^{-A\xi^2/2}$, $\bar{F}(\xi) = \bar{C} e^{-\bar{A}\xi^2/2}$:

$$\int \xi \bar{F}(\xi) F'(\xi) \, d\xi = |C|^2 A \int \xi^2 e^{-\alpha\xi^2} \, d\xi.$$

The imaginary part of the integral is zero if and only if $\beta = \mathrm{Im}\, A = 0$, which was to be proved. The constants can be identified by setting $A = 2\pi^2 \sigma^2$. ∎

In order to have a more flexible form of the uncertainty principle, we define the dispersion about a of a complex valued function f by

$$D_a(f) = \frac{\int_{-\infty}^{\infty} (x-a)^2 |f(x)|^2 \, dx}{\int_{-\infty}^{\infty} |f(x)|^2 \, dx}.$$

This can be reduced to the above case by defining

$$f_{a,b}(x) = e^{2\pi i b x} f(x-a), \qquad F_{b,-a}(\xi) = e^{-2\pi i a \xi} F(\xi - b).$$

It is immediately verified that the Fourier transform of $f_{a,b}$ is $e^{2\pi i ab} F_{-b,a}$ and that $D_a(f) = D_0(f_{a,b})$, $D_b(F) = D_0(F_{-b,a})$. Applying the uncertainty principle we see that $D_a(f) D_b(F) \geq 1/16\pi^2$ with equality if and only if $f(x) = C e^{2\pi i bx} e^{-(x-a)^2/2\sigma^2}$ some complex number C.

Exercise 2.4.11. Show that the uncertainty principle can be generalized to $\mathbb{R}^n$ in the form
$$D_0^n(f) D_0^n(\hat{f}) \geq \frac{n^2}{16\pi^2}$$
with equality if and only if $f(x) = C e^{-|x|^2/\sigma^2}$, $F(\xi) = C_2 e^{-\pi^2 |\xi|^2 \sigma^2}$ for suitable constants C_1, C_2, σ^2. Here we use the notation
$$D_0^n(f) = \frac{\int_{\mathbb{R}^n} |x|^2 |f(x)|^2 dx}{\int_{\mathbb{R}^n} |f(x)|^2 dx}.$$

Hint: Apply the proof of Proposition 2.4.10 in each coordinate and then apply the Cauchy-Schwarz inequality.

2.4.3.1 Uncertainty principle on the circle

Heisenberg's inequality has no direct analogue for Fourier series on the circle. This is related to the fact that there is no direct counterpart of the Gaussian density on the circle. The following modified form of Heisenberg's inequality was discovered by Grunbaum (1990).

Proposition 2.4.12. Suppose that $0 \neq f \in L^2(\mathbb{T})$ is absolutely continuous with $f' \in L^2(\mathbb{T})$ and $f(x_0) = 0$ for some $x_0 \in \mathbb{T}$. Then
$$\frac{\sum_{n \in \mathbb{Z}} |n|^2 |\hat{f}(n)|^2}{\sum_{n \in \mathbb{Z}} |\hat{f}(n)|^2} \frac{\int_{\mathbb{T}} (x - x_0)^2 |f(x)|^2 \, dx}{\int_{\mathbb{T}} |f(x)|^2 \, dx} > \frac{1}{4}.$$

Proof. By changing x to $x - x_0$, we may suppose that $x_0 = \pi$, so that $f(\pi) = f(-\pi) = 0$. Then
$$2 \operatorname{Re} \left(\int_{-\pi}^{\pi} x \bar{f} f' \right) = \int_{-\pi}^{\pi} x(\bar{f} f' + \bar{f}' f)$$
$$= \int_{-\pi}^{\pi} x(|f|^2)'$$
$$= -\int_{-\pi}^{\pi} |f|^2$$
so that we can apply the Cauchy-Schwarz inequality to obtain
$$\frac{1}{2} \int_{-\pi}^{\pi} |f|^2 = -\operatorname{Re} \int_{-\pi}^{\pi} x \bar{f} f' \leq \sqrt{\int_{-\pi}^{\pi} x^2 |f|^2} \sqrt{\int_{-\pi}^{\pi} |f'|^2}.$$

Now square both sides and use the Parseval identity for f and f' to obtain
$$\frac{1}{4} \int_{\mathbb{T}} |f|^2 \sum_{n \in \mathbb{Z}} |\hat{f}(n)|^2 \leq \int_{\mathbb{T}} x^2 |f|^2 \sum_{n \in \mathbb{Z}} n^2 |\hat{f}(n)|^2,$$

which proves the inequality with $\geq$. If equality holds, then we must have $f' = Axf$ for some A and also $f(\pi) = 0$, which implies that f is identically zero, a contradiction. ∎

2.4.4 Spectral Analysis of the Fourier Transform

In this section we show that the Fourier transform on $\mathbb{R}^1$ has a complete orthonormal system of eigenfunctions. To get started, we note that the Gaussian density function with $\sigma^2 = 1/2\pi$ is its own Fourier transform, since

$$\int_{-\infty}^{\infty} e^{-\pi x^2} e^{-2\pi i \xi x} \, dx = e^{-\pi \xi^2}.$$

If we differentiate both the sides of this identity, we see that the Fourier transform of $xe^{-\pi x^2}$ is $-i\xi e^{-\pi \xi^2}$.

2.4.4.1 Hermite polynomials

To proceed more generally, we introduce the generating function

$$(2.4.7) \qquad e^{tx - t^2/2} = \sum_{k=0}^{\infty} \frac{t^k}{k!} H_k(x) = H_0(x) + tH_1(x) + \frac{t^2}{2} H_2(x) + \cdots.$$

This power series converges for all t, real and complex; the coefficients $H_k(x)$ are the *Hermite polynomials*. Since the generating function is a Taylor series in the variable t, the coefficients can be obtained by successive difffentiation as

$$(2.4.8) \qquad H_k(x) = \left(\frac{d}{dt}\right)^k (e^{tx - t^2/2})|_{t=0} \qquad k = 0, 1, 2, \ldots$$

Equivalently, we can write

$$e^{-x^2/2} H_k(x) = \left(\frac{d}{dt}\right)^k \left(e^{-(t-x)^2/2}\right)\Big|_{t=0} = (-1)^k \left(\frac{d}{dx}\right)^k (e^{-x^2/2}).$$

The first few are written as follows:

$$H_0(x) = 1$$
$$H_1(x) = x$$
$$H_2(x) = x^2 - 1$$
$$H_3(x) = x^3 - 3x$$
$$H_4(x) = x^4 - 6x^2 + 3.$$

From the generating function, it follows that $H_{2k-1}(0) = 0$ and that $H_{2k}(0) = (2k)!/k!2^k$ for $k = 1, 2, \ldots$.

Lemma 2.4.13. $H'_k(x) = kH_{k-1}(x)$.

Proof. From (2.4.8), we write the derivatives and use Leibnitz's rule to write

$$H'_k(x) = \left(\frac{d}{dt}\right)^k (te^{tx-t^2/2})|_{t=0}$$

$$= t\left(\frac{d}{dt}\right)^k (e^{tx-t^2/2})|_{t=0} + k\left(\frac{d}{dt}\right)^{k-1} (e^{tx-t^2/2})|_{t=0}$$

$$= kH_{k-1}(x). \blacksquare$$

Lemma 2.4.14. For $x > 0$

$$\int_x^\infty e^{-y^2/2} \, dy \leq \sqrt{\frac{\pi}{2}} e^{-x^2/2}.$$

Proof. Let $y = x + t$. Then

$$\int_x^\infty e^{-y^2/2} \, dy = e^{-x^2/2} \int_0^\infty e^{-tx} e^{-t^2/2} \, dt$$

$$\leq e^{-x^2/2} \int_0^\infty e^{-t^2/2} \, dt$$

$$= \sqrt{\frac{\pi}{2}} e^{-x^2/2}. \blacksquare$$

Lemma 2.4.15. For $0 < a < 2\sqrt{2/\pi}$

$$A_k := \int_{-\infty}^\infty e^{-x^2/2} |H_k(ax)| \, dx \leq 3\sqrt{2\pi} 2^k k!.$$

Proof. We have for $x > 0$

$$H_k(ax) = H_k(0) + ak \int_0^x H_{k-1}(ay) \, dy,$$

$$|H_k(ax)| \leq |H_k(0)| + ak \int_0^x |H_{k-1}(ay)| \, dy.$$

Therefore

$$\int_0^\infty e^{-x^2/2} |H_k(ax)| \, dx \leq \sqrt{\frac{\pi}{2}} |H_k(0)| + ak \int_0^\infty |H_{k-1}(ay)| \left(\int_y^\infty e^{-x^2/2} dx\right) dy$$

$$\leq \sqrt{\frac{\pi}{2}} |H_k(0)| + ak\sqrt{\frac{\pi}{2}} \int_0^\infty |H_{k-1}(ay)| e^{-y^2/2} \, dy.$$

We perform the corresponding computation for $x < 0$. Combining the two, we see that the sequence A_k satisfies the system of inequalities.

$$A_k \leq \sqrt{2\pi} |H_k(0)| + 2kA_{k-1} \quad k \geq 1$$

$$\frac{A_k}{2^k k!} \leq \sqrt{2\pi} \frac{|H_k(0)|}{2^k k!} + \frac{A_{k-1}}{(k-1)! 2^{k-1}}$$

$$\frac{A_k}{2^k k!} - \frac{A_{k-1}}{2^{k-1}(k-1)!} \leq \sqrt{2\pi} \frac{|H_k(0)|}{2^k k!}$$

which telescopes to

$$\frac{A_n}{2^n n!} \leq A_0 + \sqrt{2\pi} \sum_{k=0}^{n} \frac{|H_k(0)|}{2^k k!}$$

$$\leq 3\sqrt{2\pi},$$

which was to be proved. ∎

Exercise 2.4.16. *Prove that $|H_k(x)| \leq c_k(1+|x|)^k$, $c_k := \int_{\mathbb{R}} |x|^k e^{-x^2/2} dx/\sqrt{2\pi}$.*

Hint: Use mathematical induction, applied to $H'_k = kH_{k-1}$.

2.4.4.2 Eigenfunctions of the Fourier transform

With this preparation, we can now list the eigenfunctions of the Fourier transform.

Proposition 2.4.17.

(2.4.9)
$$\frac{1}{\sqrt{2\pi}} \int_{-\infty}^{\infty} e^{-x^2/2} H_k(x\sqrt{2}) e^{-i\xi x} dx = (-i)^k e^{-\xi^2/2} H_k(\xi\sqrt{2}), \quad k = 0, 1, 2, \ldots.$$

Proof. To prove (2.4.9), we can use the generating function (2.4.7) with t real and x replaced by $x\sqrt{2}$ to write

$$\sum_{k=0}^{\infty} \frac{t^k}{k!} H_k(x\sqrt{2}) e^{-x^2/2} e^{-i\xi x} = e^{tx\sqrt{2} - t^2/2 - i\xi x - x^2/2}.$$

We apply Lemma 2.4.15 with $a = \sqrt{2}$, to see that the series

$$\sum_{k=0}^{\infty} \frac{t^k}{k!} \int_{-\infty}^{\infty} e^{-x^2/2} |H_k(x\sqrt{2})| dx$$

converges for $|t| < 1/2$. Hence we can integrate term-by-term to find that

$$\sum_{k=0}^{\infty} \frac{t^k}{k!} \int_{-\infty}^{\infty} H_k(x\sqrt{2}) e^{-x^2/2} e^{-i\xi x} dx = \int_{-\infty}^{\infty} e^{tx\sqrt{2} - t^2/2} e^{-i\xi x} e^{-x^2/2} dx.$$

This integral can be evaluated by completing the square in the exponent and making the substitution $y = x - t\sqrt{2}$ to obtain

$$e^{t^2/2} \int_{-\infty}^{\infty} e^{-y^2/2} e^{-i\xi(y+t\sqrt{2})} dy = \sqrt{2\pi} e^{-\xi^2/2} e^{-i\xi t\sqrt{2}} e^{t^2/2}.$$

When we compare this with the original generating function, we see that the only difference is the replacement of t by $-it$. But the series defining the generating function converges for all complex t, from which we conclude that for $|t| < \frac{1}{2}$,

$$\sum_{k=0}^{\infty} \frac{t^k}{k!} \int_{-\infty}^{\infty} H_k(x\sqrt{2}) e^{-x^2/2} e^{-i\xi x} dx = \sum_{k=0}^{\infty} \frac{(-it)^k}{k!} H_k(\xi\sqrt{2}) e^{-\xi^2/2}.$$

Identifying the coefficients of t^k completes the proof. ∎

If we make the substitution $x = y\sqrt{2\pi}$, $\xi = v\sqrt{2\pi}$, this can be written in terms of the usual notations as

(2.4.10)
$$\int_{-\infty}^{\infty} e^{-\pi y^2} e^{-2\pi i v y} H_k(2y\sqrt{\pi})\, dy = (-i)^k H_k(2v\sqrt{\pi}) e^{-\pi v^2}, \qquad k = 0, 1, 2, \ldots.$$

One can reinterpret the above result as providing a basis of functions in which the Fourier transform has a simple structure. For example, if a function is written as a finite sum:

$$f(x) = \sum_{k=0}^{N} a_k e^{-\pi x^2} H_k(2x\sqrt{\pi}),$$

then the Fourier transform is

$$F(\xi) = \sum_{k=0}^{N} (-i)^k a_k e^{-\pi \xi^2} H_k(2\xi\sqrt{\pi}).$$

2.4.4.3 Orthogonality properties

The orthogonality properties of the Hermite functions are obtained from a second-order differential equation which will be proved. Computing as above, we find

$$H_k'(x) = \left(\frac{d}{dt}\right)^k (t e^{tx - t^2/2})|_{t=0}$$

$$H_k''(x) = \left(\frac{d}{dt}\right)^k (t^2 e^{tx - t^2/2})|_{t=0}$$

$$x H_k'(x) = \left(\frac{d}{dt}\right)^k (xt e^{tx - t^2/2})|_{t=0}$$

$$H_k''(x) - x H_k'(x) = \left(\frac{d}{dt}\right)^k (t(t-x) e^{tx - t^2/2})|_{t=0}$$

$$= \left(\frac{d}{dt}\right)\left(t\frac{d}{dt}(e^{tx - t^2/2})\right)\bigg|_{t=0}$$

$$= t\left(\frac{d}{dt}\right)^{k+1} (e^{tx - t^2/2}) + k\left(\frac{d}{dt}\right)^k (e^{tx - t^2/2})\,|_{t=0}$$

$$= k H_k(x).$$

We now prove the orthogonality of these functions with respect to the measure with density $e^{-x^2/2}$. To do this we introduce the differential operator

$$Lf := f''(x) - x f'(x) = e^{x^2/2} [f'(x) e^{-x^2/2}]'.$$

Thus if f, g are polynomials, we can integrate by parts as follows:

$$\int_{\mathbb{R}} g(x)Lf(x)e^{-x^2/2}\,dx = \int_{\mathbb{R}} g(x)[f'e^{-x^2/2}]'\,dx$$

$$= -\int_{\mathbb{R}} g'(x)[f'e^{-x^2/2}]'\,dx$$

$$= +\int_{\mathbb{R}} f(x)[g'e^{-x^2/2}]'\,dx$$

$$= \int_{\mathbb{R}} f(x)Lg(x)e^{-x^2/2}\,dx.$$

Applying this with $f = H_m$, $g = H_m$, we see that

$$(n-m)\int_{\mathbb{R}} f(x)g(x)e^{-x^2/2}\,dx = 0,$$

which proves the orthogonality. To obtain the normalization, we write

$$\int_{\mathbb{R}} H_n(x)^2 e^{-x^2/2}\,dx = (-1)^n \int_{\mathbb{R}} H_n(x) D_x^n(e^{-x^2/2})\,dx$$

$$= (-1)^{n-1} \int_{\mathbb{R}} D_x H_n(x) D_x^{n-1}(e^{-x^2/2})\,dx$$

$$= n(-1)^{n-1} \int_{\mathbb{R}} H_{n-1}(x) D_x^{n-1}(e^{-x^2/2})$$

$$= n \int_{\mathbb{R}} H_{n-1}(x)^2 e^{-x^2/2}\,dx.$$

Proceeding inductively, we see that $\int_{\mathbb{R}} H_n(x)^2 e^{-x^2/2}\,dx = n! \int_{\mathbb{R}} e^{-x^2/2} = n!\sqrt{2\pi}$.
The orthogonality properties may be concisely written

(2.4.11) $$\int_{-\infty}^{\infty} H_k(x) H_j(x) e^{-x^2/2}\,dx = \begin{cases} k!\sqrt{2\pi} & k = j \\ 0 & k \neq j \end{cases}.$$

We also introduce the *normalized Hermite functions*

$$h_k(x) = (2\pi)^{-1/4} \frac{H_k(x)}{\sqrt{k!}} e^{-x^2/4}, \quad k = 0, 1, 2\ldots$$

which satisfy

$$\int_{-\infty}^{\infty} h_k(x)\,h_j(x)\,dx = \begin{cases} 1 & k = j \\ 0 & k \neq j \end{cases}.$$

2.4.4.4 Completeness

Finally we discuss the question of completeness of the Hermite functions. We want to show that the closed linear span of finite linear combinations of the Hermite functions is the entire space $L^2(\mathbb{R})$. If not, there would exist $f \in L^2(\mathbb{R})$ which is orthogonal to all of the eigenfunctions: $\int_{\mathbb{R}} f(x) h_k(x)\,dx = 0$ for $k = 0, 1, 2, \ldots$. Since the Hermite polynomial $= x^n +$ lower-order terms, we conclude that $\int_{\mathbb{R}} f(x) x^n e^{-x^2/4}\, dx = 0$ for $n = 0, 1, 2, \ldots$. Now we can compute the Fourier transform of $f(x) e^{-x^2/4}$ by integrating term-by-term:

$$\int_{-\infty}^{\infty} f(x) e^{-x^2/4} e^{-2\pi i \xi x}\, dx = \sum_{n=0}^{\infty} \frac{(-2\pi i \xi x)^n}{n!} \int_{-\infty}^{\infty} x^n f(x) e^{-x^2/4}\, dx = 0$$

where the interchange of sum and integral is justified by noting that the modulus of the integrand is bounded by

$$\sum_{n=0}^{\infty} \frac{|2\pi \xi x|^n}{n!} e^{-x^2/4} = e^{|2\pi \xi x|} e^{-x^2/4},$$

which is an integrable function, for any ξ. Hence the Fourier transform of $f(x) e^{-x^2/4}$ is zero, therefore $f(x) = 0$ a.e.

This immediately shows that any L^2 function has an L^2 convergent Hermite series. Indeed, we define the Fourier-Hermite coefficients of $f \in L^2(\mathbb{R})$ by

$$a_k = \int_{-\infty}^{\infty} f(x) h_k(x)\, dx.$$

Any finite linear combination $f_N = \sum_{k=0}^{N} a_k h_k$ is orthogonal to $f - f_N$, thus

$$\|f\|^2 = \|f - f_N\|_2^2 + \|f_N\|_2^2 \geq \|f_N\|_2^2 = \sum_{k=0}^{N} |c_k|^2$$

which proves Bessel's inequality: $\sum_{k=0}^{\infty} |c_k|^2 \leq \|f\|_2^2$. In particular $\tilde{f} := \sum_{k=0}^{\infty} a_k h_k$ is an L^2 convergent series and the difference $f - \tilde{f}$ is orthogonal to $h_k(x)$, for $k = 0, 1, 2, \ldots$, hence by the above argument $f - \tilde{f} = 0$ a.e. thus $f = \sum_{k=0}^{\infty} a_k h_k$, in the sense of L^2.

2.5 SPHERICAL FOURIER INVERSION IN $\mathbb{R}^n$

Bochner (1931) studied the pointwise convergence of the spherical partial sums of the Fourier integral in Euclidean space. The purpose of this work is to determine the minimal smoothness assumptions necessary for pointwise Fourier inversion in $\mathbb{R}^n$. In this section we will give an up-to-date treatment of this material, based on Pinsky (1994) and Pinsky and Taylor (1997).

2.5.1 Bochner's Approach

The spherical partial sums are defined by

(2.5.1) $$S_M f(x) = \int_{|\xi| \leq M} \hat{f}(\xi) e^{2\pi i x \cdot \xi}\, d\xi.$$

This integral may be rewritten as an integral transform on f by applying Fubini as follows:

$$S_M f(x) = \int_{|\xi| \leq M} \left(\int_{\mathbb{R}^n} f(y) e^{-2\pi i y \cdot \xi} \, dy \right) e^{2\pi i x \cdot \xi} \, d\xi$$

$$= \int_{\mathbb{R}^n} f(y) \left(\int_{|\xi| \leq M} e^{-2\pi i (x-y) \cdot \xi} \, d\xi \right) dy$$

$$= \int_{\mathbb{R}^n} D_M^n(x-y) f(y) \, dy$$

$$= \int_{\mathbb{R}^n} D_M^n(z) f(x-z) \, dz$$

where the n-dimensional spherical Dirichlet kernel is defined by

(2.5.2) $$\boxed{D_M^n(z) = \int_{|\xi| \leq M} e^{-2\pi i z \cdot \xi} \, d\xi = D_M^n(|z|)}$$

where we abuse the notation and identify a radial function on $\mathbb{R}^n$ with a function on the positive real line. Noting that D_M^n is a radial function, we may further reduce $S_M f$ in terms of the *spherical mean value*, defined by an integral over the surface of the unit sphere:

(2.5.3) $$\bar{f}_x(r) := \frac{1}{\omega_{n-1}} \int_{S^{n-1}} f(x + r\omega) \, dS_\omega$$

resulting in

(2.5.4) $$\boxed{S_M f(x) = \omega_{n-1} \int_0^\infty D_M^n(r) \bar{f}_x(r) r^{n-1} \, dr.}$$

Lemma 2.5.1. *The spherical Dirichlet kernel may be computed in terms of Bessel functions according to*

(2.5.5) $$\boxed{D_M^n(r) = M^n \frac{J_{n/2}(2\pi M r)}{(Mr)^{n/2}}.}$$

Proof. Taking a system of spherical polar coordinates with $\xi_1 = \mu \cos \theta$, we have

$$D_M^n(r) = C_n \int_0^M \int_0^\pi e^{2\pi i r \mu \cos \theta} \mu^{n-1} (\sin \theta)^{n-2} \, d\mu \, d\theta.$$

From equation (2.6.6) in the appendix to this chapter, the θ integral is recognized as the Bessel function $(r\mu)^{(2-n)/2} J_{(n-2)/2}(2\pi r \mu)$. When we perform the r-integration and use the differentation formula (2.6.2) for Bessel functions, we find the Bessel function $J_{n/2}$ as written. The dimensional constant can be identified by setting $r = 0$. Thus $D_M^n(0) = \text{vol}(\xi \in R^n : |\xi| \leq M) = (\pi M^2)^{n/2}/(n/2)!$ ∎

The useful properties of D_M^n are summarized as follows, where C_n denotes a dimension constant.

Proposition 2.5.2.

(i) $D_M^n(0) = \dfrac{(\pi M^2)^{n/2}}{(n/2)!}, \quad n \geq 1$

(ii) $D_M^n(r) = C_n M^{(n-1)/2} \left(\cos(2\pi M r - \theta_n) + O\left(\tfrac{1}{M}\right) \right), \ n \geq 1, \ r > 0, \ M \to \infty$

(iii) $|D_M^n(r)| \leq \dfrac{C_n M^n}{(1 + Mr)^{(n+1)/2}}, \quad n \geq 1, r > 0$

(iv) $D_M^n(r) = \dfrac{-1}{2\pi r} \dfrac{\partial}{\partial r} D_M^{n-2}(r), \quad n \geq 3, r > 0.$

Example 2.5.3. *If $n = 1$ we have the one-dimensional Dirichlet kernel*

$$D_M^1(r) = \frac{\sin(2\pi M r)}{\pi r}.$$

Example 2.5.4. *If $n = 2$ we have the two-dimensional Dirichlet kernel*

$$D_M^2(r) = \frac{M J_1(2\pi M r)}{r}.$$

The higher-dimensional Dirichlet kernels can be obtained by differentiation, using the recurrence formula (iv) from Proposition 2.5.2.

Theorem 2.5.5. *Suppose that $f \in L^1(\mathbb{R}^n)$ where $n = 2k + 1$ and the spherical mean $r \to \bar{f}_x(r)$ is absolutely continuous, together with its derivatives of order $k - 1$ and that $\int_0^\infty r^{j-1} |\bar{f}_x^{(j)}| \, dr < \infty$ for $j = 1, \ldots, k$. Then the spherical partial sum converges: $\lim_M S_M f(x) = \bar{f}_x(0 + 0)$.*

Proof. Let $\eta_M(r)$ be a C^∞ function with $0 \leq \eta_M(r) \leq 1$, $\eta_M(r) = 1$ for $r \leq M$, $\eta_M(r) = 0$ for $r > M + 1$, and $|\eta_M^{(j)}(r)| \leq C_j$ for $j = 1, \ldots, k$. We write (2.5.4) as $S_M f(x) = I_M(x) + II_M(x)$ where

$$I_M(x) = \omega_{n-1} \int_0^\infty \eta_M(r) \bar{f}_x(r) D_M^n(r) r^{n-1} \, dr$$

$$II_M(x) = \omega_{n-1} \int_0^\infty (1 - \eta_M(r)) \bar{f}_x(r) D_M^n(r) r^{n-1} \, dr.$$

The second term can be estimated by

$$II_M(x) \leq \omega_{n-1} \int_M^\infty |\bar{f}_x(r) D_M^n(r)| r^{n-1} \, dr$$

$$\leq C_n \int_M^\infty |\bar{f}_x(r)| \frac{M^n}{(1 + Mr)^{(n+1)/2}} r^{n-1} \, dr$$

$$\leq \frac{C_n}{M} \int_M^\infty r^{n-1} |\bar{f}_x(r)| \, dr = o\left(\frac{1}{M}\right)$$

where we have used the fact that $\int_0^\infty r^{n-1} |\bar{f}_x(r)| \, dr < \infty$. The term $I_M(x)$ is estimated by repeated integration-by-parts, where we exploit the fact that the integrand has compact support to write

$$\int_0^\infty \eta_M(r)\bar{f}_x(r)D_M^n(r)r^{n-1}\,dr = \int_0^\infty \eta_M(r)r^{n-1}\bar{f}_x(r)\frac{-1}{2\pi r}\frac{d}{dr}D_M^{n-2}(r)\,dr$$

$$= \int_0^\infty \frac{d}{dr}\left[r^{n-2}\eta_M(r)\bar{f}_x(r)\right]D_M^{n-2}(r)\,dr.$$

We repeat the partial integration k times to obtain

$$I_M(x) = \frac{\omega_{n-1}}{(2\pi)^k}\int_0^\infty r\left(\frac{1}{r}\frac{d}{dr}\right)^k\left[r^{n-2}\eta_M(r)\bar{f}_x(r)\right]\frac{\sin(2\pi Mr)}{\pi r}\,dr.$$

Writing out the derivative by Leibnitz, we find that

$$r\left(\frac{1}{r}\frac{d}{dr}\right)^k\left[r^{n-2}\eta_M(r)\bar{f}_x(r)\right] = \sum_{j=0}^k C_{jk}r^j\left(\frac{d}{dr}\right)^j[\eta_M(r)\bar{f}_x(r)]$$

for suitable constants C_{jk}. For each $j \geq 1$ we have $(d/dr)^j[\eta_M(r)\bar{f}_x(r)] \to (d/dr)^j\bar{f}_x(r)$ in $L^1(0,\infty)$ when $M \to \infty$. Therefore by the one-dimensional Riemann-Lebesgue lemma we have for $M \to \infty$

$$\int_0^\infty \frac{\sin(2\pi Mr)}{\pi r}r^j\left(\frac{d}{dr}\right)^j[\eta_M(r)\bar{f}_x(r)]\,dr \to 0 \qquad j = 1,\ldots,k.$$

For $j = 0$ we have the one-dimensional Fourier inversion of the absolutely continuous function $\eta_M(r)\bar{f}_x(r)$ at $r = 0$, which gives

$$\lim_{M \to \infty} I_M(x) = \frac{\omega_{n-1}}{(2\pi)^k}C_{0k}\bar{f}_x(0+0).$$

The constant is identified by choosing a function for which we have already proved the Fourier inversion, e.g. $f(x) = e^{-\pi|x|^2}$. ∎

The sharpness of the conditions is revealed by the following basic example in three dimensions.

Example 2.5.6. Let $n = 3$ and let $f(x) = 1$ for $0 \leq |x| < a$ and $f(x) = 0$ otherwise.

To compute the spherical mean value, we note that $\bar{f}_x(r)$ is the fractional area of the sphere $S(x;r)$ which is contained in the ball $B(0;a)$. This is zero if $r > a + |x|$ and is one if $r < a - |x|$. Otherwise it is computed as

$$\bar{f}_x(r) = \frac{1}{4\pi}\int_{\{\omega:|x+r\omega|\leq a\}} d\omega$$

$$= \frac{1}{2}\int_{\{|x|^2+2r|x|\cos\theta+r^2\leq a^2\}} \sin\theta\,d\theta$$

$$= \frac{1}{2}\left(\frac{a^2 - r^2 - |x|^2}{2r|x|} + 1\right)$$

$$= \frac{a^2 - (r - |x|)^2}{4r|x|}.$$

Therefore if $0 < |x| < a$, the spherical mean value $\bar{f}_x(r)$ is

$$\bar{f}_x(r) = \begin{cases} 1 & \text{if } 0 \le r \le a - |x| \\ \dfrac{a^2 - (r - |x|)^2}{4r|x|} & \text{if } a - |x| \le r \le a + |x| \\ 0 & \text{if } r \ge a + |x|. \end{cases}$$

The function $r \to \bar{f}_x(r)$ is Lipschitz continuous, in particular absolutely continuous, hence Theorem 2.5.5 applies to prove convergence at x. However if $x = 0$ we have $\bar{f}_x(r) = 1$ for $r \le a$ and zero otherwise, a discontinuous function. The Fourier inversion fails in a very simple way in this case, since we can use (2.5.4) to write

$$S_M f(0) = 4\pi \int_0^a D_M^3(r) r^2 \, dr$$

$$= -2 \int_0^a r \frac{d}{dr} D_M^1(r) \, dr$$

$$= -2a D_M^1(a) + 2 \int_0^a D_M^1(r) \, dr$$

$$= -\frac{2}{\pi} \sin(2M\pi a) + 2 \int_0^a \frac{\sin(2\pi M r)}{\pi r} \, dr.$$

The second term tends to 1 when $M \to \infty$, whereas the first term oscillates between $\pm 2/\pi$; in detail

$$\liminf_M S_M f(0) = 1 - \frac{2}{\pi}, \quad \limsup_M S_M f(0) = 1 + \frac{2}{\pi}.$$

This example provides a concrete illustration of the nonlocal dependence of Fourier inversion in three dimensions. The function is smooth in a neighborhood of $x = 0$ but has a jump at $|x| = a$. The jump effects the impossibility of Fourier inversion at $x = 0$. Figure 2.5.1 gives the profile of the spherical partial sum for this example.

Kahane (1995) generalized this example to the setting of a bounded region in $\mathbb{R}^3$ bounded by a smooth surface. If the surface is analytic and if the spherical Fourier inversion fails at a single point, then the surface must be a sphere and the point must be the center. For more general smooth surfaces, one may have divergence of the spherical Fourier partial sum at any preassigned finite set of points.

This example can be modified to provide a concrete example of nonlocalization.

Example 2.5.7. Let $f(x) = 1$ for $0 < a < |x| < b$ and zero otherwise. Then by applying the previous example twice and subtracting, we have

$$S_M f(0) = \frac{2}{\pi}[\sin(2M\pi a) - \sin(2M\pi b)] + o(1).$$

Clearly $f(x) = 0$ for $|x| < a$, but $\lim_M S_M f(0)$ fails to exist.

Exercise 2.5.8. Let $n = 2$ and let $f(x) = 1$ for $0 \le |x| < a$ and $f(x) = 0$ otherwise. Prove that if $x \ne 0$, then $r \to \bar{f}_x(r)$ is absolutely continuous and Hölder continuous with exponent $\frac{1}{2}$, but that if $x = 0$, $r \to \bar{f}_x(r)$ is discontinuous.

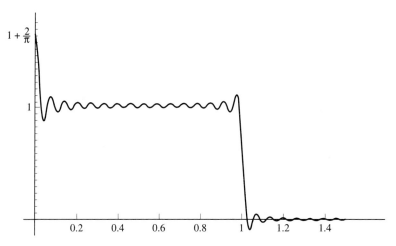

FIGURE 2.5.1
The spherical partial sum of the indicator function of the unit ball in $\mathbb{R}^3$ with $M = 99/2\pi$.

Theorem 2.5.5 can also be formulated in the case of even dimensions. The basic case of two dimensions is dealt with as follows.

Proposition 2.5.9. *Suppose that $f \in L^1(\mathbb{R}^2)$ and that the spherical mean value $r \to \bar{f}_x(r)$ is absolutely continuous with $\int_0^\infty |\bar{f}'_x(r)|\, dr < \infty$. Then $\lim_M S_M f(x) = \bar{f}_x(0+0)$.*

Proof. Appealing to (2.5.4) and (2.5.5) and the identity $(d/dr)J_0 = -J_1$, we have

$$S_M(x) = 2\pi \int_0^\infty MJ_1(2M\pi r)\bar{f}_x(r)\, dr$$

$$= -\int_0^\infty \bar{f}_x(r)\frac{d}{dr}J_0(2\pi Mr)\, dr$$

$$= \bar{f}_x(0) + \int_0^\infty J_0(2\pi Mr)\frac{d}{dr}\bar{f}_x(r)\, dr.$$

The final integral tends to zero by the dominated convergence theorem. ∎

Exercise 2.5.10. *Show that the conclusion of Proposition 2.5.9 holds true if f has compact support and the spherical mean $r \to \bar{f}_x(r)$ is assumed to be piecewise absolutely continuous. In particular for $f = 1_{[0,a]}(|x|)$, we have spherical Fourier inversion at $x = 0$.*

Now we can formulate Theorem 2.5.5 in even dimensions.

Theorem 2.5.11. *Suppose that $n = 2k$ and the spherical mean $r \to \bar{f}_x(r)$ is absolutely continuous, together with its derivatives of order $j = 1, \ldots, k-1$ and* that $\int_0^\infty r^{j-1}|\bar{f}_x^{(j)}(r)|\,dr < \infty$ for $j = 1,\ldots,k$. Then the spherical partial sum converges: $\lim_M S_M f(x) = \bar{f}_x(0+0)$.

Exercise 2.5.12. *Complete the details of the proof of Theorem 2.5.11, following the corresponding reduction in the odd-dimensional case.*

2.5.2 Piecewise Smooth Viewpoint

As in the case of one-dimensional Fourier inversion, it is not possible to obtain any necessary conditions for spherical Fourier inversion in $\mathbb{R}^n$. However we can isolate a class of functions for which we can obtain some simple necessary and sufficient conditions for convergence of $S_M f(x)$, $M \to \infty$. The original reference is Pinsky (1994).

Definition 2.5.13. $f \in L^1(\mathbb{R}^n)$ *is piecewise smooth of degree k with respect to $x \in \mathbb{R}^n$ if there exists a subdivision $0 = a_0 < a_1 < \cdots < a_K$ such that $r \to \bar{f}_x(r)$ is absolutely continuous on each subinterval, together with its derivatives of order $k-1$ and that $\int_0^\infty r^{j-1}|\bar{f}_x^{(j)}(r)|\,dr < \infty$ for $0 \leq j \leq k$. At each subdivision point we assume that there exist the one-sided limits $\bar{f}_x^{(j)}(a_i \pm 0)$ for $0 \leq j \leq k$. The jumps are denoted $\delta \bar{f}_x^{(j)}(a_i) = \bar{f}_x^{(j)}(a_i+0) - \bar{f}_x^{(j)}(a_i-0)$.*

Theorem 2.5.14. *Suppose that $n = 2k + 1$ and that $f \in L^1(\mathbb{R}^n)$ is piecewise smooth of degree k with respect to $x \in \mathbb{R}^n$. Then $\lim_M S_M f(x)$ exists if and only if $r \to \bar{f}_x(r)$ is of class C^{k-1}; in detail $\delta \bar{f}_x^{(j)}(a_i) = 0$ for $1 \leq i \leq K$, $0 \leq j \leq k-1$.*

Proof. The computations in the previous section can be repeated in this context on each subinterval (a_{i-1}, a_i). Each time we integrate by parts on (a_{i-1}, a_i), we obtain a contribution from the endpoints. When these are summed, the resultant contribution can be expressed in terms of the jumps. In detail, we have,

$$(2.5.6) \quad I_M f(x) = \omega_{n-1} \sum_{j=1}^{k} \sum_{i=1}^{K} \left(\frac{1}{2\pi}\right)^j a_i^{n-2j} \eta_M(a_i) D_M^{n-2j}(a_i) \delta \bar{f}_x^{(j-1)}(a_i)$$

$$+ \frac{\omega_{n-1}}{(2\pi)^k} \int_0^\infty r \left(\frac{1}{r}\frac{d}{dr}\right)^k \left[r^{n-2}\eta_M(r)\bar{f}_x(r)\right] \frac{\sin(2\pi M r)}{\pi r}\,dr.$$

If each of the jump terms is zero, then we obtain the desired convergence: $\lim_M S_M f(x) = \bar{f}_x(0+0)$. Conversely, suppose that $\lim_M S_M f(x)$ exists. Then it must be equal to $\bar{f}_x(0+0)$, by Gaussian summability. The final integral also converges to $\bar{f}_x(0+0)$, by one-dimensional Fourier inversion. Therefore the sum involving the jump terms must converge to zero. If we now divide by $M^{(n-3)/2}$ and apply the asymptotic formula for $D_M^n(r)$, only the term with $j = 1$ survives and we obtain

$$\lim_{M\to\infty} \sum_{i=1}^{K} \delta \bar{f}_x(a_i) \cos(Ma_i - \theta_n) = 0. \quad \blacksquare$$

To complete the proof, we state and prove a lemma on finite trigonometric sums.

Lemma 2.5.15. *If $\{C_i\}_{i=1}^K$ are complex numbers and $\theta \in \mathbb{R}$ is such that $\lim_{x\to\infty} \sum_{i=1}^{K} C_i \cos(xa_i - \theta) = 0$, then $C_i \equiv 0$.*

Proof. Multiply the sum by $\cos(xa_j - \theta)$, integrate over $[0, M]$ and divide by M. Each term in the sum tends to zero, save for the jth term, which tends to $\frac{1}{2}C_j$, which proves the result.

Applying this lemma we first see that $\delta \bar{f}_x(a_i) = 0$ for $i = 1, \ldots, K$. Now we return to (2.5.6) and divide by $R^{(n-5)/2}$ to obtain

$$\lim_{M \to \infty} \sum_{i=1}^{K} \delta \bar{f}_x^{(1)}(a_i) \cos(Ma_i - \theta_{n-2}) = 0.$$

Applying the lemma once more we see that $\delta \bar{f}_x^{(1)} x(a_i) = 0$ for $i = 1, \ldots, K$. Continuing inductively proves the result. ∎

The results of this section can be reformulated in terms of the *smoothness index*, which is defined as follows: If the spherical mean value $r \to \bar{f}_x(r)$ is discontinuous, we set $J(f; x) = -1$. Otherwise $r \to \bar{f}_x(r)$ is continuous with a certain number of continuous derivatives, denoted $J(f; x)$. The convergence theorem for piecewise smooth functions can be rephrased as follows, where [ ] denotes the integral part.

Theorem 2.5.16. *Suppose that $f \in L^1(\mathbb{R}^n)$ is piecewise smooth with respect to $x \in \mathbb{R}^n$ with smoothness index $J(f; x)$. Then the spherical partial sum $S_M f(x)$ converges when $M \to \infty$ if and only if $J(f; x) \geq [(n-3)/2]$, in which case the limit is $\bar{f}_x(0+0)$. If $J(f; x) < [(n-3)/2]$, then we have*

$$-\infty < \liminf_{M} M^{-k}(S_M f(x) - \bar{f}_x(0+0)) < \limsup_{M} M^{-k}(S_M f(x) - \bar{f}_x(0+0)) < \infty$$

where $k = (n - 5 - 2J(f; x))/2 \geq 0$

2.5.3 Relations with the Wave Equation

We have already seen the close relation between Fourier analysis and the partial differential equation of heat flow, whose steady-state limit is the Laplace equation. In each case the solution is defined by integration of an approximate identity applied to the initial-boundary data, as we have seen in detail.

When we come to the wave equation the situation is different, since the solution is no longer expressed as an integral with respect to a positive kernel, but rather a *Schwartz distribution* on the surface or interior of a sphere. To see this in detail, consider the initial-value problem for the wave equation

$$(2.5.7) \qquad \frac{\partial^2 u}{\partial t^2} = \Delta u := \sum_{i=1}^{n} \frac{\partial^2 u}{\partial x_i^2}$$

$$(2.5.8) \qquad u(x; 0) = f(x), \qquad \frac{\partial u}{\partial t}(x; 0) = 0$$

where $f \in \mathcal{S}$ is a rapidly decreasing function.

This initial-value problem can be solved in terms of the Fourier transform $\hat{f}$ by the formula

$$(2.5.9) \qquad u(x; t) = \int_{\mathbb{R}^n} \cos(2\pi t |\xi|) \hat{f}(\xi) e^{2\pi i \xi \cdot x} \, d\xi.$$

It is immediately verified that u solves the wave equation with the given initial conditions, in the sense that $\lim_{t \to 0} u(x; t) = f(x)$ and $\lim_{t \to 0} \partial u / \partial t(x; t) = 0$. A corresponding formula can also be developed for the more general initial conditions, which is left as an exercise.

In order to proceed further, we write (2.5.9) as

$$u(x; t) = \int_0^\infty \cos(2\pi t \mu) S'(\mu) \, d\mu$$

where

$$S(\mu) = S_\mu f(x) := \int_{|\xi| < \mu} e^{2\pi i \xi \cdot x} \hat{f}(\xi) \, d\xi, \qquad S'(\mu) = \int_{|\xi| = \mu} e^{2\pi i \xi \cdot x} \hat{f}(\xi) \omega_\mu(d\xi),$$

where $\omega_\mu(d\xi)$ is the surface measure on the indicated sphere. Since $f \in \mathcal{S}$ it follows that $\hat{f} \in \mathcal{S}$ and that both $\mu \to S'(\mu)$ and $t \to u(x, t)$ are rapidly decreasing functions. Hence we can apply one-dimensional Fourier inversion to obtain

$$S'(\mu) = \int_0^\infty \cos(2\pi t \mu) u(x; t) \, dt.$$

Integrating once more on $[0, M]$ and applying Fubini, we obtain the following proposition.

Proposition 2.5.17. *Suppose that $f \in \mathcal{S}$ and that $u(x; t)$ is the solution of the initial-value problem (2.5.7) and (2.5.8) for the wave equation. Then the Fourier partial sum can be retrieved through the formula*

(2.5.10)
$$\boxed{S_M f(x) = \int_0^\infty \frac{\sin(2\pi M t)}{\pi t} u(x; t) \, dt.}$$

Formula (2.5.10) was developed and applied by Pinsky and Taylor (1997) to study pointwise Fourier inversion on Euclidean space and other classical spaces on which the wave equation has a known solution.

Proposition 2.5.17 will now be used to find an explicit representation for $u(x; t)$, not involving the Fourier transform. To do this, recall the representation of $S_M f(x)$ in terms of the Dirichlet kernel

$$S_M f(x) = \omega_{n-1} \int_0^\infty D_M^n(r) \bar{f}_x(r) r^{n-1} \, dr.$$

Since $f \in \mathcal{S}$, the function $r \to \bar{f}_x(r)$ is also smooth and rapidly decreasing, so that we can integrate-by-parts and use the properties of the Dirichlet kernel to write $D_M^n(r) = -(1/2\pi r)(\partial/\partial r) D_M^{n-2}(r)$ and obtain if $n = 2k + 1$

$$S_M f(x) = \frac{\omega_{n-1}}{(2\pi)^k} \int_0^\infty \frac{\sin 2\pi M r}{\pi r} r \left(\frac{\partial}{r \partial r} \right)^k [r^{n-2} \bar{f}_x(r)] \, dr.$$

But Proposition 2.5.17 also provides a representation in terms of the same one-dimensional Dirichlet kernel. Therefore we obtain the following proposition.

Proposition 2.5.18. *The solution of the initial-value problem (2.5.7) and (2.5.8) is given by the explicit formula*

(2.5.11)
$$u(x;t) = \frac{\omega_{n-1}}{(2\pi)^k} t \left(\frac{\partial}{t\partial t}\right)^k [t^{n-2}\bar{f}_x(t)].$$

Example 2.5.19. *If $n = 3$, then the solution is $u(x;t) = (d/dt)[t\bar{f}_x(t)] = t\bar{f}'_x(t) + \bar{f}_x(t)$. The second term corresponds to a measure on the sphere $\{y:|y-x|=t\}$, while the first term corresponds to a (dipole) distribution, which is the first derivative of a measure—in the sense of Schwartz distributions.*

The representation formula (2.5.11) can be used to exhibit the finite speed of propagation of the wave equation, as contrasted with the "instantaneous speed of propagation" of the heat equation.

Proposition 2.5.20. *Suppose that $f(y) = 0$ in a ball of radius a centered at x. Then $u(x;t) = 0$ for $t < a$.*

Proof. If $f = 0$ in the ball, then $\bar{f}_x(t) = 0$ for $t < a$, likewise for all higher time derivatives. ∎

The odd-dimensional wave equation also exhibits Huygens' principle, stated as follows.

Proposition 2.5.21. *Suppose that f is supported in a ball of radius R centered at x. Then $u(x;t) = 0$ for $t > R$.*

Proof. In this case the surface integral $\bar{f}_x(t) = 0$ for $t > R$, likewise for the higher time derivatives. ∎

The wave equation in even-dimensional space can be solved by the *method of descent*, which we illustrate in the three-dimensional case. If $f \in \mathcal{S}(\mathbb{R}^2)$, we define $F(x,y,z) = f(x,y)$. If we parametrize the upper and lower halves of the sphere by writing $w_3 = \pm\sqrt{t^2 - w_1^2 - w_2^2}$, the surface integral over the sphere of radius t in three dimensions is transformed into the two-dimensional integral

$$F_{x,y,0}(t) = \frac{1}{2\pi t} \int_{|w|<t} \frac{f(x+w_1, y+w_2)}{\sqrt{t^2 - w_1^2 - w_2^2}} \, dw_1 \, dw_2.$$

From this formula we see that the finite speed of propagation still holds: if $f = 0$ in a disk of radius a centered at $x \in \mathbb{R}^2$, then $u(x;t) = 0$ for $t < a$. However Huygens' principle is not valid in two dimensions, since if f is supported in a ball of radius R, the solution $u(x;t)$ is written as an integral over the interior of the disc of radius t about x, hence will be nonzero for all t.

Figure 2.5.2 illustrates Huygens' principle; $f = 0$ outside the sphere of radius R about $(0, 0, 0)$. If $ct > R + |x|$, then $f = 0$ on the surface of outer sphere (labelled III) where the integration is performed.

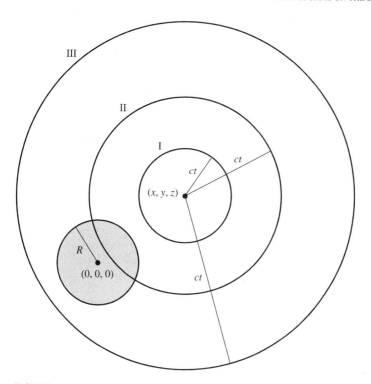

FIGURE 2.5.2
Illustrating Huygen's Principle.
From M. Pinsky, *Partial Differential Equations and Boundary-Value Problems with Applications*. Used by permission of The McGraw-Hill Companies.

2.5.3.1 The method of Brandolini and Colzani

The wave equation can be effectively used to study pointwise Fourier inversion and localization. In this section we treat the case of the Fourier integral. Brandolini and Colzani (1999) have also treated the corresponding problem for multiple Fourier series.

We begin with the spherical partial sum operator

$$S_M f(x) = \int_{|\xi| \leq M} \hat{f}(\xi) e^{2\pi i \xi \cdot x} d\xi$$

$$= \int_{\mathbb{R}^n} 1_{[0,M]}(|\xi|) \hat{f}(\xi) e^{2\pi i \xi \cdot x} d\xi.$$

Let $\psi \in C^N(\mathbb{R})$ for some $N > n/2$ be a nonnegative even function with $\int_{\mathbb{R}} \psi = 1$ and $\hat{\psi}(t) = 0$ for $|t| > \epsilon$, for some $\epsilon > 0$. For example we can take an iterated Fejér kernel

(2.5.12) $$\psi(t) = c_N \delta \left(\frac{1 - \cos 2\pi \delta t}{(\pi \delta t)^2} \right)^N$$

where c_N is a positive constant. Explicit computation shows that $\hat{\psi}(t) = 0$ for $|t| > N\delta$, so that we can take $\delta = \epsilon/N$. Without loss of generality, we take $\delta = 1$ in what follows.

Now $1_{[0,M]}$ is the restriction to the positive real axis of the even function $1_{[-M,M]}$. Writing $1_{[-M,M]} = 1_{[-M,M]} * \psi + (1_{[-M,M]} - 1_{[-M,M]} * \psi)$, we have

$$S_M f(x) = S_M^1 f(x) + S_M^2 f(x)$$

where

$$S_M^1 f(x) := \int_{\mathbb{R}^n} (1_{[-M,M]} * \psi)(|\xi|) \hat{f}(\xi) e^{2\pi i \xi \cdot x} d\xi$$

$$S_M^2 f(x) := \int_{\mathbb{R}^n} (1_{[-M,M]} - 1_{[-M,M]} * \psi)(|\xi|) \hat{f}(\xi) e^{2\pi i \xi \cdot x} d\xi.$$

The success of the method depends on two fundamental points: (i) $S_M^1 f(x)$ depends only on the values of f in a ball of radius ϵ about x. (ii) $S_M^2 f(x)$ is essentially bounded by a constant multiple of $\int_{M \leq |\xi| \leq M+1} |\hat{f}(\xi)| d\xi$. To see this, first write the convolution

$$(1_{[-M,M]} * \psi)(t) = \int_{\mathbb{R}} e^{2\pi i s t} \hat{1}_{[-M,M]}(s) \hat{\psi}(s) ds$$

$$= 2 \int_0^\infty \cos(2\pi s t) \frac{\sin 2\pi M s}{\pi s} \hat{\psi}(s) ds$$

$$S_M^1 f(x) = 2 \int_0^\infty \hat{\psi}(s) \frac{\sin 2\pi M s}{\pi s} \left(\int_{\mathbb{R}^n} \hat{f}(|\xi|) e^{2\pi i \xi \cdot x} \cos(2\pi s |\xi|) d\xi \right) ds$$

$$= 2 \int_0^\infty \hat{\psi}(s) \frac{\sin 2\pi M s}{\pi s} u(s, x) ds$$

where u is the solution of the wave equation $u_{tt} = \Delta u$ with $u(0, x) = f(x)$, $u_t(0, x) = 0$. Indeed, this equivalence has been demonstrated for the class $\mathcal{S}$, and the solution formula (2.5.11) easily extends to $f \in L^1 + L^2$. This formula shows two important properties.

- If $f = 0$ in a ball of radius ϵ about x, then $S_M^1 f(x) = 0$.
- If $s \to u(s, x)$ satisfies a Dini condition at $s = 0$, then $\lim_M S_M^1 f(x) = f(x)$.

To estimate $S_M^2 f(x)$, we note that $1_{[-M,M]} * \psi(t) = \int_{t-M}^{t+M} \psi(y) dy$. Assuming the iterated Fejér kernel (2.5.12) with $\delta = 1$, we see that for $M \geq 1$

$$\int_{|t|>M} |\psi(t)| dt \leq \frac{c_N}{M^{2N-1}}, \qquad M \geq 1$$

$$\int_{\mathbb{R}} \psi(t) dt = 1$$

which can be combined into the overall bound

$$\int_{|t|>M} \psi(t) dt \leq \frac{C_N}{(1+M)^{2N-1}}, \qquad 0 < M < \infty.$$

We use this to estimate the difference $(1_{[-M,M]} * \psi)(t) - 1_{[-M,M]}(t)$ separately for $|t| > M$ and $|t| \leq M$. In the first case $1_{[-M,M]}(t) = 0$ and we have

$$1_{[-M,M]} * \psi(t) = \int_{t-M}^{t+M} \psi(y)\, dy \leq \int_{t-M}^{\infty} \psi(y)\, dy \leq \frac{C_N}{(1+|t-M|)^{2N-1}},$$

while in the second case $1_{[-M,M]}(t) = 1$ so that

$$1_{[-M,M]} * \psi(t) - 1 = \int_{t-M}^{t+M} \psi(y)\, dy - \int_{\mathbb{R}} \psi(y)\, dy$$
$$\leq \left(\int_{t+M}^{\infty} + \int_{-\infty}^{t-M} \right) \psi(y)\, dy$$
$$\leq \frac{C_N}{(1+|t-M|)^{2N-1}}.$$

Proposition 2.5.22. *If* $f \in L^1(\mathbb{R}^n) + L^2(\mathbb{R}^n)$, *and the Fourier transform satisfies the Tauberian condition*

$$\lim_{M} \int_{M \leq 2\pi|\xi| \leq M+1} |\hat{f}(\xi)|\, d\xi = 0,$$

then $S_M^2 f(x) \to 0$; *if in addition* $s \to u(s; x)$ *satisfies a Dini condition at* $s = 0$, *then we have pointwise Fourier inversion:* $\lim_M S_M f(x) = f(x)$.

Proof. From the above computations,

$$S_M^2 f(x) = \int_{\mathbb{R}^n} [1_{[-M,M]} * \psi(|\xi|)] \hat{f}(\xi) e^{2\pi i x \cdot \xi}\, dx$$
$$\leq C_N \int_{\mathbb{R}^n} \frac{1}{(1+||\xi|-M|)^{2N-1}} |\hat{f}(\xi)|\, d\xi.$$

If $f \in L^2(\mathbb{R}^n)$ this can be estimated by Cauchy-Schwarz and applying the dominated convergence theorem, whereas for $f \in L^1(\mathbb{R}^n)$ it can be estimated by removing the supremum of $\hat{f}(\xi)$ and applying the dominated convergence theorem. ∎

Example 2.5.23. *Suppose that* $f \in L^1(\mathbb{R}^2)$ *is defined by a smooth function on the interior of a convex region with a smooth boundary with nonzero curvature, and defined to be zero outside. Then the Fourier transform satisfies the asymptotic estimate* $|\hat{f}(\xi)| \leq C/|\xi|^{3/2}$. *Hence*

$$\int_{M \leq 2\pi|\xi| \leq M+1} |\hat{f}(\xi)|\, d\xi \leq \frac{C}{M^{3/2}} M = O\left(\frac{1}{\sqrt{M}}\right) \to 0, \quad M \to \infty.$$

This example satisfies the Tauberian condition, hence we have Fourier inversion and localization at every point.

2.5.4 Bochner-Riesz Summability

Closely related to the techniques of spherical Fourier inversion is the notion of *Bochner-Riesz summability* of the Fourier integral. This is a natural substitute for the Fejér means, which have no direct counterpart in $\mathbb{R}^n$ if $n > 1$.

The Bochner-Riesz means of order $\alpha > 0$ are defined by

$$(2.5.13) \qquad B_M^\alpha f(x) = \int_{|\xi| \leq M} \left(1 - \frac{|\xi|^2}{M^2}\right)^\alpha \hat{f}(\xi) e^{2\pi i x \cdot \xi} \, d\xi.$$

By Fourier reciprocity, this operator can also be represented as the convolution with K_M^α, the inverse Fourier transform of the function $\xi \to (1 - |\xi|^2/M^2)^\alpha 1_{(0,M)}(|\xi|)$. In detail, the Bochner-Riesz kernel is

$$K_M^\alpha(x) = \int_{|\xi| \leq M} e^{2\pi i x \cdot \xi} \left(1 - \frac{|\xi|^2}{M^2}\right)^\alpha d\xi$$

$$= C_n \int_0^M \left(1 - \frac{\mu^2}{M^2}\right)^\alpha \mu^{n-1} \frac{J_{(n-2)/2}(2\pi \mu x)}{(2\pi \mu x)^{(n-2)/2}} \, d\mu$$

$$= 2\pi |x|^{-(n-2)/2} \int_0^1 (1 - s^2)^\alpha J_{(n-2)/2}(2\pi |x| s) s^{n/2} \, ds.$$

This integral is evaluated in the appendix to this chapter, with the result

$$(2.5.14) \qquad \boxed{K_M^\alpha(x) = \frac{\alpha!}{\pi^\alpha} M^n \frac{J_{\alpha+n/2}(2\pi M |x|)}{|Mx|^{\alpha+n/2}}.}$$

Proposition 2.5.24. *If $\alpha > (n-1)/2$, then K_M^α is an approximate identity. In particular for every $f \in L^p(\mathbb{R}^n)$, $1 \leq p < \infty$, $\|B_M^\alpha f - f\|_p \to 0$ when $M \to \infty$.*

Proof. From the properties of Bessel functions, we have the bound

$$|K_M^\alpha(r)| \leq C_{n\alpha} \frac{M^n}{(1 + Mr)^{\alpha + (n+1)/2}}.$$

Therefore

$$\int_{\mathbb{R}^n} |K_M^\alpha(x)| \, dx \leq C_{n\alpha} \int_0^\infty \frac{s^{n-1}}{(1+s)^{\alpha+(n+1)/2}} < \infty,$$

which proves that the L^1 norms remain bounded when $M \to \infty$. In particular the integral of K_M can be computed from the Fourier inversion at $\xi = 0$:

$$\left(1 - \frac{|\xi|^2}{M^2}\right)^\alpha = \int_{\mathbb{R}^n} K_M^\alpha(x) e^{2\pi i \xi x} \, dx, \qquad |\xi| \leq M$$

$$1 = \int_{\mathbb{R}^n} K_M^\alpha(x) \, dx.$$

Finally for any $\delta > 0$

$$\int_{|x| > \delta} |K_M^\alpha(x)| \, dx = \int_{|y| > \delta/M} K_1^\alpha(y) \, dy \to 0, \qquad M \to \infty,$$

which proves that K_M^α is an approximate identity. ∎

One can also prove a.e. summability, as follows.

Proposition 2.5.25. *If $\alpha > (n-1)/2$ and $f \in L^p(\mathbb{R}^n)$ with $1 \le p \le \infty$, then for every Lebesgue point, $\lim_M B_M^\alpha f(x) = f(x)$. In particular this holds at almost every $x \in \mathbb{R}^n$.*

Proof. Without loss of generality we can suppose that $x = 0$. Furthermore we may replace f by $f(x) - f(0)e^{-|x|^2}$ to reduce to the case $f(0) = 0$. Define

$$\Phi(r) := \int_{|y| \le r} |f(y)|\, dy, \qquad \epsilon(r) := \frac{\Phi(r)}{r^n}$$

From the hypothesis of Lebesgue point we have $\epsilon(r) \to 0$ when $r \to 0$. Now if $1 \le p < \infty$, for large r we can use Hölder's inequality to write

$$\Phi(r) \le C_{n,p} \|f\|_p\, r^{n/p'}, \qquad \epsilon(r) \le C_{n,p} r^{-n(1-1/p')}, \qquad p' = p/(p-1)$$

so that ϵ is a bounded function. In case $p = \infty$ it is immediate that ϵ is a bounded function. Now we can write

$$B_M^\alpha f(0) = \int_{\mathbb{R}^n} K_M^\alpha(x) f(x)\, dx$$

$$|B_M^\alpha f(0)| \le c_n M^n \int_{\mathbb{R}^n} \frac{1}{(1+M|x|)^{\alpha+(n+1)/2}} |f(x)|\, dx$$

$$= c_n M^n \int_0^\infty \frac{1}{(1+Mr)^{\alpha+(n+1)/2}}\, d\Phi(r)$$

$$= c_{n,\alpha} M^{n+1} \int_0^\infty \frac{1}{(1+Mr)^{\alpha+(n+3)/2}} \Phi(r)\, dr$$

$$= c_{n,\alpha} M^n \int_0^\infty \frac{\Phi(s/M)}{(1+s)^{\alpha+(n+3)/2}}\, ds$$

$$= c_{n,\alpha} \int_0^\infty \frac{s^n \epsilon(s/M)}{(1+s)^{\alpha+(n+3)/2}}\, ds$$

where we have integrated-by-parts and used $\Phi(0) = 0$ in the fourth line. The final integrand tends to zero when $M \to \infty$ and is dominated by an integable function since $\alpha > (n-1)/2$ and ϵ is a bounded function. Hence $B_M^\alpha f(0) \to 0$ when $M \to \infty$, which was to be proved. ∎

2.5.4.1 A general theorem on almost-everywhere summability

It is possible to abstract the features of the Bochner-Riesz kernel to prove a general theorem on almost-everywhere summability, originally due to Calderón and Zygmund (1952).

Theorem 2.5.26. *Suppose that $k_T(x)$, $T \to \infty$ is an approximate identity on $\mathbb{R}^n$ which is majorized in the form*

$$|k_T(x)| \le T^n K(T|x|), \qquad x \in \mathbb{R}^n, T > 0$$

where $K : [0, \infty) \to [0, \infty)$ is a decreasing function with $K(0) < \infty$, $\|K\|_1 := \int_{\mathbb{R}^n} K(|x|)\, dx < \infty$. Suppose that $f \in L^1(\mathbb{R}^n)$ and that 0 is a Lebesgue point of f.

Then

$$\lim_{T \to \infty} \int_{\mathbb{R}^n} k_T(x) f(x) \, dx = f(0).$$

Proof. The approximate identity can be applied directly to the bounded continuous function $x \to e^{-|x|^2}$; hence we may replace f by the integrable function $f(x) - f(0) e^{-|x|^2}$, and reduce attention to the case $f(0) = 0$. Having done this, we let

$$\Phi(r) := \int_{|x| \leq r} |f(x)| \, dx, \qquad \epsilon(r) := \frac{\Phi(r)}{r^n}.$$

Clearly $\Phi(r) \leq \|f\|_1$ and, since 0 is a Lebesgue point, ϵ is a bounded function with $\lim_{r \to 0} \epsilon(r) = 0$. Now we write

$$\left| \int_{\mathbb{R}^n} k_T(x) f(x) \, dx \right| \leq \int_{\mathbb{R}^n} T^n K(T|x|) |f(x)| \, dx = \int_0^\infty T^n K(Tr) \, d\Phi(r).$$

This will be integrated-by-parts, following the remarks that (i) since $\|K\|_1 < \infty$, we must have $\lim_{r \to \infty} K(r) = 0$ and (ii) since $\Phi(0) = 0$ and Φ is bounded, both terms at the limits vanish and we can write

$$\int_0^\infty T^n K(Tr) \, d\Phi(r) = - \int_0^\infty T^n \Phi(r) \, dK(Tr)$$

$$= - \int_0^\infty (Tr)^n \epsilon(r) \, dK(Tr)$$

$$= - \int_0^\infty s^n \epsilon\left(\frac{s}{T}\right) dK(s).$$

But another partial integration shows that

$$-\int_0^M s^n \, dK(s) = -s^M K(M) + \int_0^M K(s) s^{n-1} \, ds \leq C \|K\|_1 < \infty$$

so that $s^n dK(s)$ is a finite measure on $[0, \infty)$. Meanwhile, the function $\epsilon(s/T)$ is uniformly bounded and tends to zero when $T \to \infty$. Therefore by the Lebesgue dominated convergence theorem, the last integral tends to zero when $T \to \infty$, completing the proof. ∎

Exercise 2.5.27. *Extend the previous result to $f \in L^p(\mathbb{R}^n)$, $1 < p \leq \infty$ with a Lebesgue point at $x = 0$.*

2.6 BESSEL FUNCTIONS

Here we give a self-contained development of the necessary facts about Bessel functions. In Chapter 1 we encountered the so-called *modified Bessel function* $I_m(t)$, defined by a power series with positive coefficients. The standard Bessel function $J_m(t)$ is defined by

(2.6.1)
$$J_m(t) = \sum_{j=0}^\infty (-1)^j \frac{(t/2)^{2j+m}}{j!(j+m)!} \qquad m > -1.$$

This power series converges in the entire complex plane. If m is not an integer, we define the factorial as the Gamma function

$$m! = \int_0^\infty t^m e^{-t}\,dt = \Gamma(m+1).$$

Proposition 2.6.1. *If $m, \nu \geq 0$, the Bessel functions satisfy the relations*

(2.6.2) $\quad\quad\quad \dfrac{d}{dt}[t^m J_m(t)] = t^m J_{m-1}(t) \quad\quad\quad t > 0$

(2.6.3) $\quad\quad\quad \dfrac{d}{dt}[t^{-m} J_m(t)] = -t^{-m} J_{m+1}(t) \quad\quad\quad t > 0$

(2.6.4)

$$J_m''(t) + \frac{1}{t}J_m'(t) + \left(1 - \frac{m^2}{t^2}\right)J_m(t) = 0 \quad \left(' = \frac{d}{dt}\right) \quad t > 0$$

(2.6.5) $\quad \displaystyle\int_0^1 J_m(Rs) s^{m+1}(1-s^2)^\nu\,ds = 2^\nu \nu! \dfrac{J_{m+\nu+1}(R)}{R^{\nu+1}} \quad\quad R > 0$

(2.6.6) $\quad \displaystyle\int_{-1}^1 e^{2its}(1-s^2)^{m-1/2}\,ds = \dfrac{J_m(2t)}{t^m}\Gamma\left(m+\dfrac{1}{2}\right)\Gamma\left(\dfrac{1}{2}\right).$

The first three are obtained by termwise differentiation of the power series. For the fourth, we substitute the definition of J_m into the integral and integrate term-by-term:

$$\int_0^1 J_m(Rs)s^{m+1}(1-s^2)^\nu\,ds = \int_0^1 \left(\sum_{j=0}^\infty (-1)^j \frac{(Rs/2)^{m+2j}}{j!(j+m)!}\right) s^{m+1}(1-s^2)^\nu\,ds$$

$$= \frac{1}{2}\sum_{j=0}^\infty (-1)^j \frac{(R/2)^{m+2j}}{j!(j+m)!}\int_0^1 r^{m+j}(1-r)^\nu\,dr$$

$$= \frac{2^\nu \nu!}{R^{\nu+1}}\sum_{j=0}^\infty (-1)^j \frac{(R/2)^{m+\nu+1+2j}}{j!(m+\nu+j+1)!}$$

$$= \frac{2^\nu \nu!}{R^{\nu+1}} J_{m+\nu+1}(R)$$

The final integral formula is obtained by making the substitution $s = \sin\theta$ and recognizing the power series coefficients from Chapter 1.

Exercise 2.6.2. *Prove (2.6.2) by termwise differentiation of the power series definition (2.6.1).*

Exercise 2.6.3. *Prove (2.6.3) by termwise differentiation of the power series definition (2.6.1).*

Exercise 2.6.4. *Prove (2.6.4) by termwise differentiation of the power series definition (2.6.1).*

Exercise 2.6.5. *Complete the details of the proof of (2.6.6).*

The asymptotic behavior of $J_m(t), t \to \infty$ is most efficiently deduced from a differential equation. Let $y(t) = \sqrt{t} J_m(t)$. From (2.6.4), we have by successive differentiation

$$(2.6.7) \qquad y'' + y = \frac{Cy}{t^2}, \qquad 0 < t < \infty, \qquad C := m^2 - \frac{1}{4}.$$

This implies that y and y' remain bounded when $t \to \infty$, since

$$\frac{d}{dt}(y^2 + y'^2) = 2yy' + 2y'y'' = \frac{2Cyy'}{t^2} \leq \frac{|C|}{t^2}(y^2 + y'^2).$$

Hence for $t \geq t_0$,

$$y(t)^2 + y'(t)^2 \leq [y(t_0)^2 + y'(t_0)] \exp\left(|C| \int_{t_0}^{t} \frac{ds}{s^2}\right) \leq [y(t_0)^2 + y'(t_0)] e^{|C|/t_0},$$

which proves the required boundedness. From this we have the representation

$$(2.6.8) \qquad y(t) = \int_t^\infty \sin(t-s) \frac{Cy(s)}{s^2} ds + A_1 \cos t + A_2 \sin t.$$

Indeed, the first term on the right of (2.6.8) is a solution of the differential equation (2.6.7), so that it differs from $y(t)$ by a solution of the homogeneous equation $z'' + z = 0$, whose general solution is $A_1 \cos t + A_2 \sin t$.

From this follows an asymptotic formula for $y(t)$, since the integral term is bounded in the form

$$\left| \int_t^\infty \sin(t-s) \frac{Cy(s)}{s^2} ds \right| \leq \int_t^\infty \frac{|Cy(s)|}{s^2} ds = O\left(\frac{1}{t}\right), \qquad t \to \infty.$$

We summarize the above work as follows.

Proposition 2.6.6. *The Bessel function $J_m(t)$ satisfies the asymptotic relation*

$$\sqrt{t} J_m(t) = A_1 \cos t + A_2 \sin t + O\left(\frac{1}{t}\right), \qquad t \to \infty$$

for suitable constants A_1, A_2. Equivalently, we may write

$$J_m(t) = \frac{A}{\sqrt{t}} \cos(t - \theta) + O\left(\frac{1}{t^{3/2}}\right), \qquad t \to \infty$$

for suitable constants A, θ.

The constants A, θ can be explicitly identified in the case of integer m by the method of stationary phase, described in the next section. In the case of half integer m, it is often possible to identify the constants from elementary formulas, beginning with $J_{1/2}(x) = \sqrt{2/\pi x} \sin x$.

Exercise 2.6.7. *Prove that the derivative of the Bessel function satisfies the asymptotic formula $(d/dt)(\sqrt{t} J_m(t)) = -A_1 \sin t + A_2 \cos t + O(1/t), t \to \infty$.*

Hint: Compute $y'(t)$ from (2.6.8).

2.6.1 Fourier Transforms of Radial Functions

If $f \in L^1(\mathbb{R}^n)$, the Fourier transform $\hat{f}$ is a continuous function vanishing at infinity, whereas if $f \in L^2(\mathbb{R}^n)$, we can only say that $\hat{f} \in L^2(\mathbb{R}^n)$, in general. In Chapter 3 we will prove that the Fourier transform can be extended as a bounded operator from $L^p(\mathbb{R}^n)$ to $L^{p'}(\mathbb{R}^n)$ if $1 < p < 2$. However, if f depends only on $r = |x|$, then $\hat{f}$ will also be continuous and will vanish at infinity when $f \in L^p(\mathbb{R}^n)$ for restricted values of p.

We proved in Section 2.1 that the Fourier transform of a radial function is again a radial function. By writing this explicitly in terms of a suitable kernel, we can establish useful properties of the Fourier transforms of radial functions. Specifically, we have the following.

Proposition 2.6.8. *If $\varphi \in \mathcal{S}(0, \infty)$, and we set $f(x) = \varphi(|x|)$, then*

$$(2.6.9) \qquad \hat{f}(\xi) = \int_0^\infty \varphi(r) \frac{J_{(n-2)/2}(2\pi r|\xi|)}{(r|\xi|)^{(n-2)/2}} r^{n-1}\, dr.$$

If $f \in L^p(\mathbb{R}^n)$, $1 < p < 2$, then this integral is interpreted as a limit in $L^{p'}(\mathbb{R}^n)$ of $\int_0^M$ when $M \to \infty$.

This follows from the representation of Bessel functions, specifically (2.6.6) with $m = (n/2) - 1$, $t = \pi r|\xi|$. For details, see Stein and Weiss, 1971, p. 154.

We can use the representation (2.6.9) to prove additional properties of Fourier transforms of radial functions for restricted values of $p \in (1, 2)$.

Proposition 2.6.9. *Suppose that $1 < p < 2n/(n+1)$. Then the Fourier transform (2.6.9) is a continuous function for $\xi \neq 0$, which vanishes at infinity and satisfies*

$$|\hat{f}(\xi)|^p \le C_{np} \int_0^\infty |\varphi(r)|^p r^{n-1}\, dr = C'_{np} \int_{\mathbb{R}^n} |f(x)|^p\, dx.$$

Proof. Recalling the bounds on the Bessel function and applying Hölder's inequality, we have

$$\left| \int_0^M \varphi(r) \frac{J_{(n-2)/2}(2\pi r|\xi|)}{(r|\xi|)^{(n-2)/2}} r^{n-1}\, dr \right| \le C_n \int_0^M \frac{|\varphi(r)| r^{n-1}}{(1+r|\xi|)^{(n-1)/2}}\, dr$$

$$= \int_0^M |\varphi(r)| r^{(n-1)/p} \frac{r^{(n-1)/p'}}{(1+r|\xi|)^{(n-1)/2}}\, dr$$

$$\le \left(\int_0^M |\varphi(r)|^p r^{n-1}\, dr \right)^{1/p} \left(\int_0^M \frac{r^{n-1}\, dr}{(1+r|\xi|)^{((n-1)p')/2}} \right)^{1/p'}.$$

The final integral converges when $M \to \infty$ if and only if $(n-1) - p'(n-1)/2 < -1$, which is equivalent to $1/p' < (n-1)/2n$ or $1/p > (n+1)/2n$. The continuity for $\xi \neq 0$ now follows from the dominated convergence theorem. To prove that $\hat{f}$ vanishes at infinity, we write for any $\epsilon > 0$,

$$\hat{f}(\xi) = \left(\int_0^\epsilon + \int_\epsilon^\infty\right) \varphi(r) \frac{J_{(n-2)/2}(2\pi r|\xi|)}{(r|\xi|)^{(n-2)/2}} r^{n-1} dr$$

$$\left|\int_0^\epsilon \varphi(r) \frac{J_{(n-2)/2}(2\pi r|\xi|)}{(r|\xi|)^{(n-2)/2}} r^{n-1} dr\right| \leq C_n \int_0^\epsilon |\varphi(r)| r^{n-1} dr$$

$$= C_n \int_0^\epsilon |\varphi(r)| r^{(n-1)/p} r^{(n-1)/p'} dr$$

$$\leq C_n \left(\int_0^\epsilon |\varphi(r)|^p r^{n-1} dr\right)^{1/p} \left(\int_0^\epsilon r^{n-1} dr\right)^{1/p'}$$

$$= C_n \epsilon^{\frac{n}{p'}} \left(\int_0^\epsilon |\varphi(r)|^p r^{n-1} dr\right)^{1/p'}$$

$$\left|\int_\epsilon^\infty \varphi(r) \frac{J_{(n-2)/2}(2\pi r|\xi|)}{(r|\xi|)^{(n-2)/2}} r^{n-1} dr\right| \leq C_n \int_\epsilon^\infty \frac{|\varphi(r)|}{(r|\xi|)^{(n-1)/2}} r^{n-1} dr$$

$$= \frac{C_n}{|\xi|^{(n-1)/2}} \int_\epsilon^\infty |\varphi(r)| r^{(n-1)/p} r^{(n-1)/p'+(1-n)/2} dr$$

$$\leq \frac{C_n}{|\xi|^{(n-1)/2}} \left(\int_\epsilon^\infty |\varphi(r)|^p r^{n-1} dr\right)^{1/p} \left(\int_\epsilon^\infty r^q dr\right)^{1/p'}$$

where $q = n - 1 - p'(n-1)/2$. Taking $\epsilon = 1$ shows that $|\hat{f}(\xi)|^p \leq C_{np} \int_0^\infty |\varphi(r)|^p r^{n-1} dr$ provided that $p' > 2n/(n-1)$. On the other hand, for this same range of p we have for any $\epsilon > 0$,

$$\limsup_{|\xi|\to\infty} |\hat{f}(\xi)| \leq C_n \epsilon^{n/p'} \left(\int_0^\epsilon |\varphi(r)|^p r^{n-1} dr\right)^{1/p'}.$$

Taking $\epsilon \to 0$ shows that the limit is zero, as required.

To see that this range of p is sharp, let

$$\varphi(r) = a^n \frac{J_{n/2}(2\pi ra)}{(ra)^{n/2}},$$

which is the Fourier transform of the indicator function of a ball of radius a. From the asymptotics of Bessel functions, we have $\varphi(r) = [C + O(1/r)] \cos(r - \theta)/r^{(n+1)/2}$ when $r \to \infty$ so that $\varphi(|\cdot|) \in L^p(\mathbb{R}^n)$ if and only if $\infty > \int_1^\infty r^{(n-1)-p(n+1)/2} dr$, which happens if and only if $p > 2n/(n+1)$. But $\hat{f}$ is discontinuous at the sphere of radius a, which provides the required counterexample. ∎

The above results can be extended, in a suitable form, to nonradial functions by means of the *restriction theorems*, described below. These results depend in part on the complex interpolation method, to be developed in Chapter 3.

2.6.2 L^2-Restriction Theorems for the Fourier Transform

The Fourier transform of $f \in L^1(\mathbb{R}^n)$ is a continuous function that satisfies the pointwise bound $|\hat{f}(\xi)| \leq \|f\|_1$. However if $f \in L^p(\mathbb{R}^n)$ with $p > 1$ we cannot expect any such pointwise bound in general. However if f is a radial function, we have shown in the previous subsection that $|\hat{f}(\xi)| \leq C_{np}\|f\|_p$ for $1 \leq p < 2n/(n+1)$, hence this same estimate applies to the average over a sphere. The restriction theorems generalize this, by bounding the L^2 norm of the Fourier transform on a sphere in terms of suitable L^p norms. We have the following proposition, attributed to Tomas (1975).

Proposition 2.6.10. *Let $\omega(d\theta)$ be the uniform surface measure on the unit sphere. Let $f \in \mathcal{S}$, and $1 \leq p < 4n/(3n+1)$. Then for some constant $A = A_p$, we have*

$$(2.6.10) \qquad \left(\int_{S^{n-1}} |\hat{f}(\theta)|^2 \omega(d\theta) \right)^{1/2} \leq A_p \|f\|_p.$$

Proof. The Fourier transform of the surface measure is given by

$$\hat{\omega}(x) = \int_{S^{n-1}} e^{-2\pi i x \cdot \xi} \omega(d\xi) = C_n \frac{J_{(n-2)/2}(2\pi |x|)}{|x|^{(n-2)/2}}$$

and satisfies $\hat{\omega}(x) = O(|x|^{(1-n)/2})$, $|x| \to \infty$. Hence $\hat{\omega} \in L^q(\mathbb{R}^n)$, provided that $q > 2n/(n-1)$. Now by Fourier reciprocity, Hölder's inequality and the Young convolution estimate, we have

$$\int_{\mathbb{R}^n} |\hat{f}(\xi)|^2 \omega(d\xi) = \int_{\mathbb{R}^n} (f * f)(x) \hat{\omega}(x)\, dx$$
$$\leq \|f * f\|_{q'} \|\hat{\omega}\|_q$$
$$\leq \|f\|_p^2 \|\hat{\omega}\|_q$$

where $(2/p) = 1 + 1/q'$. But $q > 2n/(n-1)$ implies that $1/q' > (n+1)/2n$ so that $2/p > (3n+1)/2n$, which was to be proved. ∎

2.6.2.1 An improved result

We can extend the range of p-values in the previous result by the use of Stein's *complex interpolation method*, which will be developed in Chapter 3. In this method, we imbed the given problem in a one-parameter family of operators. A family of convolution operators is defined by the kernels

$$(2.6.11) \qquad K_z(x) = C_z \frac{J_{(n-2)/2 + z}(2\pi |\xi|)}{|\xi|^{(n-2)/2 + z}} \qquad \frac{1-n}{2} \leq \operatorname{Re} z \leq 1.$$

At the endpoints we have

$$z_0 = \frac{1-n}{2}, K_z(x) = \cos(2\pi |x|), \quad \text{a bounded function,}$$

$$z_1 = 1, K_z(x) = \frac{J_{n/2}(2\pi |\xi|)}{|\xi|^{n/2}} \quad \text{with a bounded F.T.,} \quad \hat{K}_1 = 1_{[0,1]}(|x|).$$

Therefore K_{z_0} maps $L^1 \to L^\infty$, and K_{z_1} maps $L^2 \to L^2$, so we can use the complex interpolation method with

$$p_0 = 1, q_0 = \infty, z_0 = \frac{1-n}{2},$$
$$p_1 = 2, q_1 = 2, z_1 = 1.$$

Thus
$$\frac{1}{p} = \frac{1-t}{1} + \frac{t}{2} = 1 - \frac{t}{2},$$
$$\frac{1}{q} = \frac{1-t}{\infty} + \frac{t}{2} = \frac{t}{2},$$
$$z = (1-t)\frac{1-n}{2} + t.$$

In particular p and q are conjugate exponents, $1/p + 1/q = 1$. In order to have $z = 0$ we must take $t = (n-1)/(n+1)$, hence $1/p = 1 - (t/2) = (n+3)/(2n+2)$, $p = (2n+2)/(n+3)$. We now apply this to the Fourier reciprocity formula and apply Hölder's inequality to write

$$\int_{S^{n-1}} |\hat{f}(\theta)|^2 \omega(d\theta) = \int_{\mathbb{R}^n} (f * f)(x)\hat{\omega}(x)\,dx$$
$$\leq \|f\|_p \|f * \hat{\omega}\|_{p'}.$$

The Fourier transform of ω is given by $\hat{\omega} = K_0$. Each member of the one-parameter family maps the Lebesgue space into its dual, in particular

$$\|f * \hat{\omega}\|_{p'} \leq C\|f\|_p,$$

from which we conclude the following theorem.

Theorem 2.6.11. *Let $\omega(d\theta)$ be the uniform surface measure on the unit sphere. Let $f \in \mathcal{S}$, and $1 \leq p \leq (2n+2)/(n+3)$. Then for some constant $A = A_p$, we have*

(2.6.12) $$\left(\int_{S^{n-1}} |\hat{f}(\theta)|^2 \omega(d\theta)\right)^{1/2} \leq C\|f\|_p.$$

In Chapter 3 it will be proved that the Fourier transform can be extended as a bounded operator from $L^p(\mathbb{R}^n)$ to $L^{p'}(\mathbb{R}^n)$ if $1 < p < 2$. In this setting the restriction theorem is also valid, by taking limits in the space $\mathcal{S}$. Many of the properties of the Fourier transform on $L^1(\mathbb{R}^n)$ have counterparts in this wider setting. In particular, the restriction theorem may be used to prove the following result on the *average decay* of the Fourier transform of a function in L^p for a suitable range of exponents.

Proposition 2.6.12. *Suppose that $f \in L^p(\mathbb{R}^n)$ where $1 < p \leq (2n+2)/(n+3)$. Then $\int_{|\xi|=1} |\hat{f}(r\xi)|^2 \omega(d\xi) \to 0$ when $r \to \infty$.*

Proof. Define $f_r(x) = r^{-n}f(x/r)$ for $r > 0$, $x \in \mathbb{R}^n$. Then

$$\hat{f}_r(\xi) = r^{-n}\int_{\mathbb{R}^n} e^{-2\pi i x\cdot\xi} f(x/r)\,dx = \int_{\mathbb{R}^n} e^{-2\pi i r y\cdot\xi} f(y)\,dy = \hat{f}(r\xi)$$

$$\|f_r\|_p^p = \int_{\mathbb{R}^n} |r^{-n}f(y/r)|^p\,dy = r^{-np}\int_{\mathbb{R}^n} |f(y/r)|^p\,dy = r^{n-np}\|f\|_p^p$$

Applying the L^2 restriction theorem, we have

$$\int_{|\xi|=1} |\hat{f}(r\xi)|^2 \omega(d\xi) \leq C_p \|f_r\|^2 = C_p r^{2n(1-p)/p} \|f\|_p^2 \to 0$$

since $p > 1$. ∎

Corollary 2.6.13. *Suppose that $f \in L^p(\mathbb{R}^n)$ where $1 < p \leq (2n+2)/(n+3)$. Then $\int_{|\xi|=1} \hat{f}(r\xi) \omega(d\xi) \to 0$ when $r \to \infty$.*

One may note that the results on radial transforms are valid on a wider set of L^p spaces than the L^2-restriction theorem, since

$$n > 1 \implies \frac{2n+2}{n+3} < \frac{2n}{n+1}.$$

Thus we may extend the previous corollary, as follows:

Exercise 2.6.14. *Suppose that $f \in L^p(\mathbb{R}^n)$ with $1 < p < 2n/(n+1)$. Show that $\int_{|\xi|=1} \hat{f}(r\xi)\omega(d\xi) \to 0$ when $r \to \infty$.*

Hint: Interchange the orders of integration to reduce to the case of a radial function.

2.6.2.2 Limitations on the range of p

The sufficient conditions on the range of p-values for the restriction theorem are also necessary, as revealed by the following example.

Example 2.6.15. *Let f be the Fourier transform of the indicator function of the rectangular region R defined by the inequalities*

$$1 - \epsilon \leq x_1 \leq 1, \qquad |x_j| \leq \sqrt{\epsilon} \qquad 2 \leq j \leq n$$

where $0 < \epsilon < 1$.

In detail

$$f(x) = \int_R e^{-2\pi i \xi \cdot x} d\xi$$

$$= \int_{1-\epsilon}^1 e^{-2\pi i x_1 \xi_1} d\xi_1 \prod_{j=2}^n \left(\int_{-\sqrt{\epsilon}}^{\sqrt{\epsilon}} e^{-2\pi i x_j \xi_j} d\xi_j \right)$$

$$= \left(\frac{e^{-2\pi i x_1} - e^{-2\pi i x_1 (1-\epsilon)}}{2\pi x_1} \right) \prod_{j=2}^n \left(\frac{\sin 2\pi x_j \sqrt{\epsilon}}{\pi x_j} \right)$$

$$|f(x)| = \left| \frac{\sin \pi x_1 \epsilon}{\pi x_1} \right| \prod_{j=2}^n \left| \frac{\sin 2\pi x_j \sqrt{\epsilon}}{\pi x_j} \right|.$$

To compute the L^p norm, we note that by direct computation

$$\int_\mathbb{R} \left| \frac{\sin \pi x \epsilon}{\pi x} \right|^p dx = \epsilon^{p-1} \int_\mathbb{R} \left| \frac{\sin \pi y}{\pi y} \right|^p dy = C_p \epsilon^{p-1}.$$

Applying this to both factors above yields

$$\|f\|_p^p = C_p \epsilon^{p-1} \sqrt{\epsilon}^{(n-1)(p-1)} = C_p \epsilon^{(p-1)(n+1)/2},$$

$$\|f\|_p = C_p \epsilon^{(n+1)/2p'}.$$

On the other hand, the L^2-norm of the Fourier transform on the unit circle is

$$\int_{|\xi|=1} |\hat{f}(\xi)|^2 \omega(d\xi) = \int_{|x_j| \leq \sqrt{\epsilon}, x_1^2 + \cdots + x_n^2 = 1} \omega(d\xi) = c_n \int_{|x_j| \leq \sqrt{\epsilon}} \frac{dx_2 \ldots dx_n}{\sqrt{x_1}} \sim c_p \epsilon^{(n-1)/2}.$$

Hence

$$\|f\|_{L^2(S^{n-1})} \sim A \epsilon^{(n-1)/4}.$$

If the restriction theorem (2.6.12) were valid, then we would have

$$\epsilon^{(n-1)/4} \leq A \epsilon^{(n+1)/2p'}.$$

This can be true for all $\epsilon > 0$ if and only if we have $(n+1)/2p' \leq (n-1)/4$, or equivalently

$$\frac{n-1}{2(n+1)} \geq \frac{1}{p'} = 1 - \frac{1}{p} \iff \frac{1}{p} \geq \frac{n+3}{2n+2},$$

which is precisely the range of admissible exponents in the L^2-restriction theorem (2.6.12).

2.7 THE METHOD OF STATIONARY PHASE

Often we have occasion to make an asymptotic evaluation of a one-dimensional Fourier transform, when $|\xi| \to \infty$. This can often be done by a simple integration-by-parts, for example if $f \in L^1[a, b]$ is absolutely continuous with $f' \in L^1[a, b]$, we can write

$$\hat{f}(\xi) = \int_a^b f(x) e^{-2\pi i \xi x} dx$$

$$= \int_a^b f(x) d\left(\frac{e^{-2\pi i \xi x}}{-2\pi i \xi}\right)$$

$$= \frac{f(b) e^{-2\pi i \xi b} - f(a) e^{-2\pi i \xi a}}{-2\pi i \xi} + \frac{1}{2\pi i \xi} \int_a^b f'(x) e^{-2\pi i \xi x} dx.$$

The new integral is $o(1/\xi)$ when $|\xi| \to \infty$ and we have obtained an asymptotic formula. If f has higher derivatives in L^1, then this may be iterated to obtain an asymptotic expansion.

In many problems of interest, we encounter Fourier transforms when $f' \notin L^1$, for example the Bessel function

$$J_0(t) = \frac{1}{\pi} \int_{-1}^{1} \frac{e^{itx}}{\sqrt{1-x^2}} \, dx.$$

If we make the change of variable $x = \cos\theta$, $0 < \theta < \pi$, we obtain the formula

$$J_0(t) = \frac{1}{\pi} \int_0^{\pi} e^{it\cos\theta} \, d\theta,$$

which looks less forbidding, however it is no longer written as a Fourier integral. If we try to integrate-by-parts, we find that the contribution to the integral from any interior interval tends to zero like $1/t$, but this analysis no longer applies near the endpoints $\theta = 0$, $\theta = \pi$.

To handle this and more general "oscillatory integrals," we develop the method of stationary phase. Specifically, we consider complex-valued functions of the form

(2.7.1) $$f(t) = \int_a^b e^{it\varphi(x)} g(x) \, dx,$$

where φ is a real-valued function called the *phase function*. The function $g(x)$ may be either real- or complex-valued. If $\varphi'(x) \neq 0$, then we may integrate-by-parts and conclude that $f(t) = O(1/t)$, $t \to \infty$. However, if $\varphi'(x) = 0$ for some x, then this conclusion is no longer valid. In order to find the correct result, we focus attention upon those points x_j where $\varphi'(x) = 0$, the so-called *stationary points*.

2.7.1 Statement of the Result

The complete statement of the result is given as follows:

Theorem 2.7.1. *The method of stationary phase:* Suppose that $g(x), \varphi(x)$ have two continuous derivatives for $a \leq x \leq b$, that $\varphi(x)$ is real-valued, and that $\varphi'(x) \neq 0$ except for a finite number of stationary points x_j, where $\varphi''(x_j) \neq 0$. Let these be labeled so that $\varphi''(x_j) > 0$ for $1 \leq j \leq K$ and $\varphi''(x_j) < 0$ for $K+1 \leq j \leq K+L$. Then when $t \to \infty$,

(2.7.2) $$\boxed{\int_a^b e^{it\varphi(x)} g(x) \, dx = I^+(t) + I^-(t) + O\left(\frac{1}{t}\right)}$$

where

(2.7.3) $$\boxed{I^+(t) = \sum_{j=1}^{K} \left(\frac{2\pi}{t\varphi''(x_j)}\right)^{1/2} e^{it\varphi(x_j)} e^{i\pi/4} g(x_j)}$$

(2.7.4) $$\boxed{I^-(t) = \sum_{j=K+1}^{K+L} \left(\frac{2\pi}{-t\varphi''(x_j)}\right)^{1/2} e^{it\varphi(x_j)} e^{-i\pi/4} g(x_j).}$$

If either of the endpoints $x = a$, $x = b$ are also stationary points, then they contribute to (2.7.3) and (2.7.4) with a factor of $\frac{1}{2}$.

A simple tool to remember this complicated formula is to observe that the result is identical to what is obtained by replacing $\varphi(x)$ by its two-term Taylor expansion and replacing $g(x)$ by its value at each stationary point, then doing the resultant integrals (one for each stationary point), and then summing the results.

We illustrate with a typical example.

Example 2.7.2. *Apply the method of stationary phase to find an asymptotic formula for the integral*

$$\int_{-\pi/2}^{\pi/2} (2x+3)e^{-it\cos x}\,dx.$$

In this case we have $g(x) = 2x + 3$, $\varphi(x) = -\cos x$, $\varphi'(x) = \sin x$, $\varphi''(x) = \cos x$. The only stationary point is $x = 0$, where $\varphi''(0) = +1$, $g(0) = 3$. Applying (2.7.2), we have

$$\int_{-\pi/2}^{\pi/2} e^{-it\cos x}\,dx = 3\sqrt{\frac{2\pi}{t}}\,e^{-it}e^{i\pi/4} + O\left(\frac{1}{t}\right), \qquad t \to \infty$$

2.7.2 Application to Bessel Functions

As a primary application of the method of stationary phase, we propose to identify the constants in the asymptotic behavior when $t \to \infty$ of the Bessel function $J_m(t)$, which is represented by the integral

(2.7.5) $$J_m(t) = \frac{i^{-m}}{2\pi}\int_{-\pi}^{\pi} e^{it\cos\theta}e^{-im\theta}\,d\theta \qquad m = 0, 1, 2, \ldots.$$

Proposition 2.7.3. *The Bessel function has the asymptotic behavior*

(2.7.6) $$J_m(t) = \sqrt{\frac{2}{\pi t}}\cos(t - \pi/4 - m\pi/2) + O\left(\frac{1}{t}\right), \qquad t \to \infty.$$

Proof. From the integral representation (2.7.5) we have (2.7.1), where $\varphi(x) = \cos x$, $g(x) = (1/2\pi)e^{-imx}e^{-im\pi/2}$. Since $\varphi'(x) = -\sin x$, $\varphi''(x) = -\cos x$, there are three stationary points: $x = 0$, $x = \pi$, $x = -\pi$, with $\varphi''(0) = -1$, $\varphi''(\pi) = 1 = \varphi''(-\pi)$; also $g(0) = (1/2\pi)e^{-im\pi/2}$, $g(\pi) = (1/2\pi)e^{-3im\pi/2} = g(-\pi)$. We apply the method of stationary phase, noting that the endpoints contribute with a factor of $\frac{1}{2}$. Hence

$$\begin{aligned}J_m(t) &= \sqrt{\frac{2\pi}{t}}e^{it}e^{-i\pi/4}\frac{e^{-im\pi/2}}{2\pi} + \left(\frac{1}{2}+\frac{1}{2}\right)\sqrt{\frac{2\pi}{t}}e^{-it}e^{i\pi/4}\frac{e^{-3im\pi/2}}{2\pi} + O\left(\frac{1}{t}\right) \\ &= \sqrt{\frac{2\pi}{t}}e^{it}e^{-i\pi/4}\frac{e^{-im\pi/2}}{2\pi} + \sqrt{\frac{2\pi}{t}}e^{-it}e^{i\pi/4}\frac{e^{im\pi/2}}{2\pi} + O\left(\frac{1}{t}\right) \\ &= \sqrt{\frac{1}{2\pi t}}\left(e^{i(t-\pi/4-m\pi/2)} + e^{-i(t-\pi/4-m\pi/2)}\right) + O\left(\frac{1}{t}\right) \\ &= \sqrt{\frac{2}{\pi t}}\cos(t-\pi/4-m\pi/2) + O\left(\frac{1}{t}\right). \quad \blacksquare\end{aligned}$$

2.7.3 Proof of the Method of Stationary Phase

We now outline the steps used to prove (2.7.2). The idea is to reduce the study to each stationary point, where we can approximate using with the Taylor expansions with an error of $O(1/t)$.

Step 1. If the interval $c \le x \le d$ does not contain any stationary points, then

$$\int_c^d e^{it\varphi(x)} g(x)\, dx = O\left(\frac{1}{t}\right), \quad t \to \infty.$$

Proof. We multiply and divide by $\varphi'(x)$ and integrate-by-parts as follows:

$$\begin{aligned}\int_c^d g(x) e^{it\varphi(x)}\, dx &= \int_c^d \frac{g(x)}{it\varphi'(x)}\, d\left(e^{it\varphi(x)}\right) dx \\ &= \frac{g(x)}{it\varphi'(x)} e^{it\varphi(x)}\Big|_{x=c}^{x=d} - \frac{1}{it}\int_c^d e^{it\varphi(x)} \frac{d}{dx}\left(\frac{g(x)}{\varphi'(x)}\right) dx.\end{aligned}$$

Both terms are $O(1/t)$, $t \to \infty$, and can therefore be included in the remainder term.

Therefore we can restrict attention to contributions from intervals containing the stationary points. Assume that x_1 is a stationary point for which $\varphi''(x_1) > 0$, and let $\delta > 0$ be chosen so that $\varphi(x) - \varphi(x_1) > 0$ in the interval $x_1 - \delta < x < x_1 + \delta$. We introduce a new variable of integration v through the equation

$$v = (x-x_1)\sqrt{\frac{\varphi(x)-\varphi(x_1)}{(x-x_1)^2}} \quad x_1 - \delta < x < x-1+\delta.$$

The function $x \to v(x)$ vanishes at $x = x_1$, with $v'(x_1) = \sqrt{\varphi''(x_1)/2} > 0$. Therefore there exists an inverse function $x = X(v)$ with $X(0) = x_1$, $X'(0) = \sqrt{2/\varphi''(x_1)}$. $\blacksquare$

Step 2.

$$\int_{x_1-\delta}^{x_1+\delta} g(x) e^{it\varphi(x)}\, dx = e^{it\varphi(x_1)} \int_{-\bar{\delta}_1}^{\bar{\delta}_2} G(v) e^{itv^2}\, dv$$

where $\bar{\delta}_2 = v(x_1+\delta)$, $-\bar{\delta}_1 = v(x_1-\delta)$ and $G(v) = g(X(v))/v'(X(v))$, $G(0) = g(x_1)\sqrt{2/\varphi''(x_1)}$.

Step 3.

$$\int_{-\tilde{\delta}_1}^{\tilde{\delta}_2} G(v)e^{itv^2}dv = G(0)\int_{-\tilde{\delta}_1}^{\tilde{\delta}_2} e^{itv^2}dv + O(1/t), \qquad t \to \infty.$$

Proof. We write $G(v) = G(0) + vh(v)$, which defines the differentiable function $h(v)$. The second term contributes to the integral

$$\int_{-\tilde{\delta}_1}^{\tilde{\delta}_2} vh(v)e^{itv^2}dv = \frac{1}{2it}\int_{-\tilde{\delta}_1}^{\tilde{\delta}_2} h(v)d\left(e^{itv^2}\right)$$

$$= \frac{1}{2it}\left(h(v)e^{itv^2}\Big|_{-\tilde{\delta}_1}^{\tilde{\delta}_2} - \int_{-\tilde{\delta}_1}^{\tilde{\delta}_2} h'(v)e^{itv^2}dv\right)$$

$$= O\left(\frac{1}{t}\right)$$

as required. ∎

Step 4.

$$\int_{-\tilde{\delta}_1}^{\tilde{\delta}_2} e^{itv^2}dv = \sqrt{\frac{\pi}{t}}e^{i\pi/4} + O\left(\frac{1}{t}\right), \qquad t \to \infty.$$

Proof. This is the *Fresnel integral*. Readers familiar with complex-variable methods will find this a one-liner: apply Cauchy's theorem to the function $f(z) = e^{iz^2}$ on the crescent-shaped contour formed by the ray $z = re^{i\pi/4}$, $0 \leq r \leq R$, the arc of the circle $|z| = R$, and the real axis from $(0,0)$ to $(R,0)$, when $R \to \infty$. We now outline a proof that does not use complex-variable methods.

The qualitative fact of convergence of the improper integral is established by the following partial integration:

$$\int_M^N e^{ix^2}dx = \int_M^N \frac{1}{2ix}d(e^{ix^2})$$

$$= \frac{e^{iN^2}}{2iN} - \frac{e^{iM^2}}{2iM} + \frac{1}{2i}\int_M^N \frac{e^{ix^2}}{x^2}dx.$$

The final integral is less than or equal to $1/N$, so that the right side tends to zero when $M, N \to \infty$. This proves that the improper integral $\int_0^\infty e^{ix^2}dx$ is convergent. Letting $N \to \infty$ shows furthermore that

$$\int_M^\infty e^{ix^2}dx = -\frac{e^{iM^2}}{2iM} + \frac{1}{2i}\int_M^\infty \frac{e^{ix^2}}{x^2}dx.$$

Both terms on the right are $O(1/M)$, so that we have the required speed of convergence:

$$\int_0^M e^{ix^2}dx = \int_0^\infty e^{ix^2}dx + O(1/M), \qquad M \to \infty.$$

It remains to compute the numerical value of the improper integral. To do this, we let $p > 0$ and examine the double integral

$$J_p = \int_0^\infty \int_0^\infty e^{-p(x^2+y^2)} e^{i(x^2+y^2)} dx\, dy.$$

On the one hand, we can take polar coordinates $x = r\cos\theta$, $y = r\sin\theta$ and compute

$$J_p = \int_0^\infty \int_0^{\pi/2} e^{-pr^2} e^{ir^2} r\, dr\, d\theta$$

$$= \frac{\pi}{2} \int_0^\infty r e^{-r^2(p-i)} dr$$

$$= \frac{\pi}{2} \frac{1}{2(p-i)}.$$

On the other hand, the double integral is the square of a single integral:

$$J_p = \left(\int_0^\infty e^{-px^2} e^{ix^2} dx \right)^2.$$

Letting $I = \int_0^\infty e^{ix^2} dx$, we conclude that

$$I^2 = \lim_{p \to 0} J_p = \frac{\pi i}{4}.$$

But the complex number I has both positive real and imaginary parts, so that the appropriate square root is

$$I = \sqrt{\frac{\pi}{4}} e^{i\pi/2}.$$

To apply this to Step 4, write

$$\int_0^\delta e^{itv^2} dv = \frac{1}{\sqrt{t}} \int_0^{\delta\sqrt{t}} e^{ix^2} dx = \frac{1}{\sqrt{t}} \left[I + O\left(\frac{1}{\sqrt{t}}\right) \right], \quad t \to \infty.$$

The equality of the two limits follows from Abel's lemma (see below), which completes the proof. ∎

2.7.4 Abel's Lemma

We give two forms of Abel's lemma for integrals, the second of which is applied above.

Proposition 2.7.4.

- *Suppose that $f(t)$, $t > 0$, is a locally integrable function and $\lim_{t \to \infty} f(t) = L$. Then $\lim_{p \downarrow 0} \int_0^\infty f(t) p e^{-pt} dt = L$.*
- *Suppose that $g(s)$, $s > 0$ is a locally integrable function and that the improper integral $\int_0^\infty g(s)\, ds$ converges to some L. Then*

$$\lim_{p \downarrow 0} \int_0^\infty g(s) e^{-ps} ds = L.$$

Proof. The proof consists of writing

$$\int_0^\infty f(t)pe^{-pt}\,dt - L = \int_0^\infty (f(t) - L)pe^{-pt}\,dt.$$

Given $\epsilon > 0$, we split the region of integration into the two regions $0 < t < T$ and $T < t < \infty$, where T is chosen such that $|f(t) - L| < \epsilon$ for $t > T$, so that the second integral is less than ϵ. The first integral is less than $p \int_0^T |f(t) - L|\,dt$ which tends to zero when $p \downarrow 0$ and the first statement follows.

To prove the second statement, we set $f(t) = \int_0^t g(s)\,ds$. By hypothesis $f(t) \to L$ when $t \to \infty$. Interchanging the orders of integration yields

$$\int_0^\infty f(t)pe^{-pt}\,dt = \int_0^\infty pe^{-pt}\left(\int_0^t g(s)\,ds\right)dt$$

$$= \int_0^\infty g(s)\left(\int_s^\infty pe^{-pt}\,dt\right)ds$$

$$= \int_0^\infty g(s)e^{-ps}\,ds.$$

Applying the first statement gives the result: if $\lim_{t\to\infty}\int_0^t g(s)\,ds = L$, then $\lim_{p\downarrow 0}\int_0^\infty g(s)e^{-ps}\,ds = L$. ∎

CHAPTER 3

FOURIER ANALYSIS IN L^p SPACES

3.1 MOTIVATION AND HEURISTICS

Much of modern Fourier analysis is concerned with bounded linear operators on the Lebesgue spaces $L^p(\mathbb{T})$ and $L^p(\mathbb{R}^n)$. This chapter is devoted to the development of systematic methods for proving the boundedness of relevant operators by the method of *interpolation*. Following M. Riesz, if we can first prove that an operator is bounded on two different pairs of Lebesgue spaces, then we can often deduce boundedness on the intermediate spaces. A more general concept, that of *weak boundedness*, can often be used in place of strict boundedness, following the work of Marcinkiewicz. These techniques are applied to prove the L^p boundedness of the classical Hilbert transform, both on the circle and on the real line. This yields the M. Riesz theorem on L^p convergence of Fourier series and integrals in one dimension. This chapter also includes the Hardy-Littlewood maximal inequality, which proves the weak L^1-boundedness of a fundamental operator that underlies the Lebesgue differentiation theorem and many other almost-everywhere convergence results in Fourier analysis.

3.2 THE M. RIESZ-THORIN INTERPOLATION THEOREM

In order to introduce the ideas, we first develop some elementary properties of L^p spaces.

Lemma 3.2.1. *Suppose that* $f \in L^{p_0}(\mathbb{R}^n) \cap L^\infty(\mathbb{R}^n)$. *Then* $f \in L^{p_1}(\mathbb{R}^n)$ *for any* $p_1 > p_0$.

Proof. Letting $M = \|f\|_\infty$, we write
$$\int_{\mathbb{R}^n} |f(x)|^{p_1}\, dx \leq M^{p_1 - p_0} \int_{\mathbb{R}^n} |f(x)|^{p_0}\, dx < \infty.$$
∎

We say that f *lives on a set of finite measure* if $|\{x : f(x) \neq 0\}| < \infty$.

Lemma 3.2.2. *Suppose that* $f \in L^{p_1}(\mathbb{R}^n)$ *lives on a set B of finite measure $|B|$. Then* $f \in L^{p_0}(\mathbb{R}^n)$ *for any* $p_0 < p_1$.

Proof. From Hölder's inequality we can write

$$\int_{\mathbb{R}^n} |f(x)|^{p_0}\, dx = \int_{\mathbb{R}^n} |f(x)|^{p_0} 1_B(x)\, dx \leq \left(\int_{\mathbb{R}^n} |f(x)|^{p_1}\, dx \right)^{p_0/p_1} (|B|)^{(p_1-p_0)/p_1} < \infty.$$

∎

Lemma 3.2.3. *Let* $0 < p_1 < p < p_1$ *and suppose that* $f \in L^{p_0}(\mathbb{R}^n) \cap L^{p_1}(\mathbb{R}^n)$. *Then* $f \in L^p(\mathbb{R}^n)$.

Proof. We write $f = f 1_{\{|f| \leq 1\}} + f 1_{\{|f| > 1\}} = f_1 + f_2$. Both f_1 and f_2 are dominated by f, in particular $f_1 \in L^{p_0}$ and $f_2 \in L^{p_1}$. But f_1 is bounded and f_2 lives on a set of finite measure, since

$$|\{x : f_2(x) \neq 0\}| = |\{x : |f(x)| > 1\}| \leq \int_{\mathbb{R}^n} |f(x)|^{p_0}\, dx < \infty.$$

Therefore by the preceding lemmas, $f_1 \in L^p(\mathbb{R}^n)$ and $f_2 \in L^p(\mathbb{R}^n)$. But $L^p(\mathbb{R}^n)$ is a linear space, hence $f \in L^p(\mathbb{R}^n)$. ∎

Lemma 3.2.3 has a sort of converse, as follows.

Lemma 3.2.4. *Let* $0 < p_0 < p < p_1$ *and suppose that* $f \in L^p(\mathbb{R}^n)$. *Then there exist* $f_0 \in L^{p_0}$ *and* $f_1 \in L^{p_1}$ *such that* $f = f_0 + f_1$.

Proof. It suffices to take $f_0 = f 1_{|f| \leq 1}$ and $f_1 = f 1_{|f| > 1}$. ∎

The last two lemmas can be restated as follows: if $p_0 < p < p_1$, then

(3.2.1) $\quad\boxed{L^{p_0}(\mathbb{R}^n) \cap L^{p_1}(\mathbb{R}^n) \subset L^p(\mathbb{R}^n) \subset L^{p_0}(\mathbb{R}^n) + L^{p_1}(\mathbb{R}^n).}$

The above lemmas show that for any measurable function f, the set $\{p : \|f\|_p < \infty\}$ is a connected subset of the real line. The theory of M. Riesz-Thorin quantifies this by showing, for example, that the mapping $p \to \log \|f\|_p$ is a *convex function* of $1/p$. At the same time we deal with linear operators that are simultaneously defined on two different L^p spaces, developing the interpolation properties of these linear operators.

Exercise 3.2.5. Use Hölder's inequality to show directly that if $1 \leq p_0 < p_1 < \infty$ and $0 < \alpha < 1$, then

$$\int_{\mathbb{R}^n} |f|^{\alpha p_1 + (1-\alpha) p_0} \leq \left(\int_{\mathbb{R}^n} |f|^{p_1} \right)^\alpha \left(\int_{\mathbb{R}^n} |f|^{p_0} \right)^{1-\alpha}.$$

Conclude that $p \to \log \|f\|_p^p$ *is a convex function.*

The basic convexity result of M. Riesz will be developed in the following general setting. Suppose we have two measure spaces (M, μ), (N, ν) and two pairs of indices

(p_0, q_0), (p_1, q_1) where $1 \leq p_0, p_1, q_0, q_1 \leq \infty$ with $p_0 \neq p_1$, $q_0 \neq q_1$. Further we assume given linear operators $A_0 : L^{p_0}(M) \to L^{q_0}(N)$ and $A_1 : L^{p_1}(M) \to L^{q_1}(N)$ so that $A_0 = A_1$ on the common domain $L^{p_0}(M) \cap L^{p_1}(M)$. Let $k_i = \|A\|_{p_i, q_i}$ be the respective operator norms, $i = 1, 2$. An interpolation is defined by a real number $t \in (0, 1)$ giving rise to indices (p_t, q_t) defined by the convex combinations

$$\frac{1}{p_t} = \frac{t}{p_1} + \frac{1-t}{p_0}, \qquad \frac{1}{q_t} = \frac{t}{q_1} + \frac{1-t}{q_0}.$$

Theorem 3.2.6. M. Riesz-Thorin: *Under the above hypotheses there exists a linear operator $A_t : L^{p_t}(N) \to L^{q_t}(N)$ that coincides with A_i on $L^{p_0}(M) \cap L^{p_1}(M)$ and whose operator norm satisfies*

$$\|A_t\|_{p_t, q_t} \leq k_0^{1-t} k_1^t.$$

The M. Riesz-Thorin theorem can be applied to prove the Hausdorff-Young inequalities for Fourier series in one dimension. This is illustrated in the following two examples.

Example 3.2.7. *Suppose that $M = \mathbb{T}$ with Lebesgue measure and that $N = \mathbb{Z}$ with counting measure. Let $Af(n) = 1/2\pi \int_\mathbb{T} f(\theta) e^{-in\theta} d\theta$ be the discrete Fourier transform. Letting $(p_0, q_0) = (1, \infty)$ we see from the contraction property that A_0 is bounded with $k_0 = 1$. Letting $(p_1, q_1) = (2, 2)$, we see from the Parseval identity that A_1 is bounded with $k_1 = 1$. Applying the theorem, we take $1/p_t = t + (1-t)/2 = (1+t)/2$ and $1/q_t = (1-t)/2$. These are conjugate exponents, satisfying $1/p_t + 1/q_t = 1$. Since $L^p(\mathbb{T}) \subset L^1(\mathbb{T})$ the operator A must agree with the discrete Fourier transform, so that we conclude for any $1 \leq p \leq 2$ $A : L^p(\mathbb{T}) \to L^{p'}(\mathbb{Z})$ where $1/p + 1/p' = 1$. Equivalently*

$$\left(\sum_{n \in \mathbb{Z}} |\hat{f}(n)|^{p'} \right)^{1/p'} \leq \left(\frac{1}{2\pi} \int_\mathbb{T} |f(\theta)|^p d\theta \right)^{1/p}.$$

The roles of $\mathbb{T}$ and $\mathbb{Z}$ can be reversed to obtain another example.

Example 3.2.8. *Suppose that $M = \mathbb{Z}$ with counting measure and that $N = \mathbb{T}$ with Lebesgue measure. For a bilateral sequence $\{c_n\}, n \in \mathbb{Z}$, define $Af(\theta) = \sum_{n \in \mathbb{Z}} c_n e^{in\theta}$. From the first properties of absolutely convergent series, this is bounded from $l^1(\mathbb{Z})$ to $L^\infty(\mathbb{T})$; also from the basic properties of L^2 it is also bounded from $l^2(\mathbb{Z})$ to $L^2(\mathbb{T})$. Therefore we can take $p_0 = 1$, $q_0 = \infty$, $p_1 = 2$, $q_1 = 2$ to obtain the same conjugate exponents $1/p_t = (1+t)/2$, $1/q_t = (1-t)/2$ and the conclusion that for $1 \leq p \leq 2$*

$$\left(\frac{1}{2\pi} \int_\mathbb{T} \left| \sum_{n \in \mathbb{Z}} c_n e^{in\theta} \right|^{p'} \right)^{1/p'} \leq \left(\sum_{n \in \mathbb{Z}} |c_n|^p \right)^{1/p} \qquad p' := \frac{p}{p-1}.$$

The next exercise treats the convexity of the p norm by M. Riesz-Thorin.

Exercise 3.2.9. Let $f \in L^p(E)$, where E is a measurable subset of $\mathbb{R}^n$ and $1 < p < \infty$, with $\|f\|_p := (\int_E |f|^p)^{1/p}$.

(i) Prove that $\|f\|_p = \sup_{\|\phi\|_{p'} \leq 1} \int_E |f\phi|$, where $p' = p/(p-1)$.

(ii) Use the M. Riesz-Thorin theorem to prove that if $f \in L_{p_0}(E) \cap L_{p_1}(E)$ with $1 < p_0 < p_1 < \infty$, then $\|f\|_{p_t} \leq \|f\|_{p_0}^{1-t} \|f\|_{p_1}^t$, where $1/p_t = (1-t)/p_0 + t/p_1$.

(iii) Conclude that the mapping $p \to \log \|f\|_{1/p}$ is a convex function.

To prove the theorem, we first prove the maximum-modulus theorem.

Lemma 3.2.10. *Suppose that $f(z)$ is an analytic function in an open connected subset S of the complex plane. This means that about every point z_0 there is a disk $|z - z_0| < r$ so that f is the sum of a convergent power series $f(z) = \sum_{n=0}^{\infty} a_n(z - z_0)^n$. Then f cannot have a local maximum at any interior point unless f is constant. In particular, $|f(z)|$ attains its maximum on the boundary.*

Proof. If f is nonconstant, then there is a smallest value of $n \geq 1$ so that $a_n \neq 0$. By translation, we may suppose that $z_0 = 0$. Near $z = 0$, $f(z) = a_0 + a_n z^n + O(z^{n+1})$, $|f(re^{i\theta})|^2 = |a_0|^2 + 2|a_0 a_n| r^n \cos(n\theta - \phi) + O(r^{n+1})$. Clearly $|f(z)|^2 - f(z_0)^2$ takes both positive and negative values in any neighborhood of $r = 0$, which is a contradiction. ∎

Now we can prove the three lines theorem.

Lemma 3.2.11. *Suppose that F is an analytic function defined in the strip $S = \{z : 0 \leq \text{Re}(z) \leq 1\}$ and such that $|F(iy)| \leq m_0$, $|F(1+iy)| \leq m_1$ for $-\infty < y < \infty$ where $m_0 > 0$, $m_1 > 0$. Then $|F(x+iy)| \leq m_0^{1-x} m_1^x$ for $0 \leq x \leq 1$, $-\infty < y < \infty$.*

Proof. Let $F_1(z) = F(z)/m_0^{1-z} m_1^z$, which is an analytic function with $|F(iy)| \leq 1$, $|F(1+iy)| \leq 1$, so that we can assume $m_0 = m_1 = 1$. First we prove the lemma under the added condition that $F(x+iy) \to 0$ uniformly when $|y| \to \infty$, $0 \leq x \leq 1$. Then we must have $|F(x \pm iM)| \leq \frac{1}{2}$ for M large enough. Therefore by the maximum principle, we must have $|F(x+iy)| \leq 1$ for $0 \leq x \leq 1$, $-M \leq y \leq M$. But M was arbitrary, so the inequality holds in the entire strip $0 \leq x \leq 1$, $-\infty < y < \infty$.

In the general case, we let $F_n(z) = F(z)e^{(z^2-1)/n}$. Then

$$|F(x+iy)| \leq |F(z)|e^{-y^2/n} e^{(x^2-1)/n} \leq e^{-y^2/n},$$

which tends to zero when $|y| \to \infty$. Hence we can apply the previous paragraph to conclude that $|F_n(x+iy)| \leq 1$ in the entire strip. But $|F(z)| = \lim_n |F_n(z)| \leq \limsup 1 = 1$, which completes the proof. ∎

Proof of the theorem. Having made the necessary preparations, we first note that for any measure space (M, μ), A can be defined on the spaces $L^{p_t}(M)$, since we can write $f = f 1_{|f| \leq 1} + f 1_{|f| > 1}$. By Hölder's inequality, this is the sum of an L^{p_0} function and an L^{p_1} function, for which A is defined. Now the L^q norm can be computed as

$$\|h\|_q = \sup_{\|g\|_{q'} \leq 1} \int_N hg \, d\nu,$$

where the supremum is taken over all simple functions, i.e., finite linear combinations of indicator functions of sets of finite measure. At the same time we have

$$\|A\|_{p,q} = \sup_{\|f\|_p=1, \|g\|_{q'}=1} \int_N (Af) g \, dv.$$

We extend the interpolated exponents to the complex plane by defining

$$\frac{1}{p(z)} = \frac{z}{p_1} + \frac{1-z}{p_0}, \quad \frac{1}{q'(z)} = \frac{z}{q'_1} + \frac{1-z}{q'_0}, \quad 0 \le \mathrm{Re}(z) \le 1.$$

It is sufficient to prove the theorem for simple functions: $f = \sum_{1 \le j \le N} a_j e^{i\alpha_j} 1_{A_j}$, $g = \sum_{1 \le j \le N} b_j e^{i\beta_j} 1_{B_j}$, where A_j, B_j are measurable sets, $a_j, b_j \ge 0$, and $\alpha_j, \beta_j \in (0, 2\pi]$. The functions f, g are extended to the strip in the complex plane by first defining $p = p_t$, $q' = q'_t$ and setting

$$\phi(\cdot, z) := \sum_{j=1}^N a_j^{p/p(z)} e^{i\alpha_j} 1_{A_j} = \sum_{j=1}^N a_j^{p/p(z)} \Theta_j,$$

$$\psi(\cdot, z) := \sum_{j=1}^N b_j^{q'/q'(z)} e^{i\beta_j} 1_{B_j} = \sum_{j=1}^N b_j^{q'/q'(z)} \Phi_j,$$

where we have set

$$\Theta_j = e^{i\alpha_j} 1_{A_j}, \quad \Phi_j = e^{i\beta_j} 1_{B_j}.$$

It is immediate that $\phi(\cdot, z) \in L^{p_j}(M)$, $\psi(\cdot, z) \in L^{q'_j}(N)$, in particular $A\psi(\cdot, z) \in L^{q_j}(N)$. Therefore the integral

$$(3.2.2) \qquad F(z) = \int_N A\phi(\cdot, z) \psi(\cdot, z) \, dv = \sum_{j,k=1}^N a_j^{p/p(z)} b_k^{q'/q'(z)} \int_N (A\Theta_j) \Phi_k \, dv$$

is a finite sum of exponential functions, in particular an analytic function in the open strip $0 < \mathrm{Re}(z) < 1$ and is bounded and continuous on the closed strip $0 \le \mathrm{Re}(z) \le 1$. On the boundary we have by direct computation

$$\|\phi(\cdot, iy)\|_{p_0} = \| |f|^{p/p_0} \|_{p_0} = \|f\|_p^{p/p_0} = 1,$$

$$\|\phi(\cdot, 1+iy)\|_{p_1} = \| |f|^{p/p_1} \|_{p_1} = \|f\|_p^{p/p_1} = 1,$$

$$\|\psi(\cdot, iy)\|_{q'_0} = \| |g|^{q'/q'_0} \|_{q'_0} = \|g\|_{q'_0}^{q'/q'_0} = 1,$$

$$\|\psi(\cdot, 1+iy)\|_{q'_1} = \| |g|^{q'/q'_1} \|_{q'_1} = \|g\|_{q'_1}^{q'/q'_1} = 1.$$

By the definition of $F(z)$ above and Hölder's inequality, we have from (3.2.2)

$$|F(iy)| \le \|A\phi(\cdot, iy)\|_{q_0} \|\psi(iy)\|_{q'_0} \le k_0$$

$$|F(1+iy)| \le \|A\phi(\cdot, 1+iy)\|_{q_1} \|\psi(\cdot, 1+iy)\|_{q'_1} \le k_1.$$

On the other hand, for $t \in (0, 1)$, $\phi(x, t) = f(x)$, $\psi(y, t) = g(y)$, so that $F(t) = \int_N Afg \, dv$. Applying Lemma 3.2.11, we conclude $|F(t)| \le k_0^{1-t} k_1^t$, which completes the proof. ∎

Now we examine some important applications of the M. Riesz-Thorin interpolation theorem.

3.2.0.1 Generalized Young's inequality

The interpolation theorem of M. Riesz-Thorin can be used to prove the generalized Young's inequality for convolutions:

$$(3.2.3) \qquad \|f * g\|_r \leq \|f\|_p \|g\|_q \qquad \left(1 + \frac{1}{r} = \frac{1}{p} + \frac{1}{q}\right)$$

where $1 \leq p \leq \infty$, $1 \leq q \leq \infty$ with $1/p + 1/q \geq 1$. Here f, g are measurable functions on $\mathbb{R}^n$ and the integrals are taken over all of $\mathbb{R}^n$.

Proof. To prove (3.2.3), we begin with the elementary estimates from Lebesgue integration theory:

$$\|f * g\|_1 \leq \|f\|_1 \|g\|_1 \qquad f \in L^1, g \in L^1$$
$$\|f * g\|_\infty \leq \|f\|_\infty \|g\|_1 \qquad f \in L^\infty, g \in L^1.$$

This shows that, for a fixed $g \in L^1$, the map $f \to f * g$ defines a bounded operator on L^1 and L^∞, with norm less than or equal to one. Therefore, by the M. Riesz-Thorin theorem, this map can be extended to L^p, with the same operator norm, to yield

$$(3.2.4) \qquad \|f * g\|_p \leq \|f\|_p \|g\|_1 \qquad f \in L^p, g \in L^1.$$

In other words, the map $g \to f * g$ is bounded from L^1 to L^p. On the other hand, by Hölder's inequality, if $f \in L^p$, $g \in L^{p'}$,

$$(3.2.5) \qquad \|f * g\|_\infty \leq \|f\|_p \|g\|_{p'} \qquad p' = p/(p-1).$$

Therefore the map $g \to f * g$ is bounded from $L^{p'}$ to L^∞. Hence we can apply the M. Riesz-Thorin theorem with $p_0 = 1$, $q_0 = p$, $p_1 = p'$, $q_1 = \infty$. In detail,

$$(3.2.6) \qquad \frac{1}{q} = \frac{1}{p_t} = \frac{1-t}{1} + \frac{t}{p'}$$

$$(3.2.7) \qquad \frac{1}{r} = \frac{1}{q_t} = \frac{1-t}{p} + 0.$$

Solving (3.2.6) for $t \in [0, 1]$, we have $t = p/q'$, which is possible since $p \leq q'$. Finally, we have

$$\frac{1}{r} = \frac{1}{q_t} = \frac{1-t}{p} = \frac{1}{p} - \frac{1}{q'} = \frac{1}{p} + \frac{1}{q} - 1,$$

which proves (3.2.3). ∎

Beckner (1975) has shown that the value of the "best constant" in the generalized Young's inequality (3.2.3) has the precise value $M_{pq} = (p^{1/p}/q^{1/q})^{n/2}$. The proof is beyond the scope of this book.

3.2.0.2 The Hausdorff-Young inequality

The M. Riesz-Thorin theorem can be immediately applied to the setting of Fourier transforms on $\mathbb{R}^n$. Let $M = N = \mathbb{R}^n$ with Lebesgue measure and $p_0 = 1$, $q_0 = \infty$. The Fourier transform $f \to \hat{f}$ is a bounded operator from $L^1(M)$ to $L^\infty(N)$ with norm 1. Also from the Plancherel theorem the Fourier transform is a bounded operator from $L^2(M)$ to

$L^2(N)$ with norm 1. Therefore we conclude that

Theorem 3.2.12. Hausdorff-Young: *If $1 < p < 2$, the Fourier transform is defined and is a bounded operator from $L^p(\mathbb{R}^n)$ to $L^{p'}(\mathbb{R}^n)$ where $p' = p/(p-1)$ is the conjugate exponent.*

Beckner (1975) has shown that the best constant in the Hausdorff-Young inequality has the value $(p^{1/p}/q^{1/q})^{n/2}$ where $q = p'$.

Exercise 3.2.13. *Let $f(x) = e^{-\pi|x|^2}$, so that $\hat{f}(\xi) = e^{-\pi|\xi|^2}$. Show that $\|\hat{f}\|_q/\|f\|_p$ attains the Beckner bound, where $1 < p \leq 2$.*

3.2.1 Stein's Complex Interpolation Theorem

The M. Riesz-Thorin theorem deals with a single operator A, initially defined and bounded on $L^{p_0} \cap L^{p_1}$ and subsequently extended to L^p, $p_0 < p < p_1$. E.M. Stein discovered a remarkable extension to a family of operators A_z which depend analytically on a complex parameter z, where $0 \leq \operatorname{Re}(z) \leq 1$. For the complete theory, see Stein and Weiss (1971), Chapter 4. The following is a special case of the general theory.

Theorem 3.2.14. *Let (M, μ) and (N, ν) be measure spaces with a family of linear operators A_z defined on the class of simple functions $S(M)$ so that $z \to \int_N (A_z f) g \, d\nu$ is analytic and bounded for $0 \leq \operatorname{Re}(z) \leq 1$ whenever $f \in S(M)$ and $g \in S(N)$. Furthermore suppose that for some $1 \leq p_0, q_0, p_1, q_1 \leq \infty$ we have*

$$\|A_{iy} f\|_{q_0} \leq M_0 \|f\|_{p_0} \quad f \in S(M), \ y \in \mathbb{R}$$
$$\|A_{1+iy} f\|_{q_1} \leq M_1 \|f\|_{p_1} \quad f \in S(M), \ y \in \mathbb{R}.$$

Then for all $t \in (0, 1)$

$$\|A_t f\|_{q_t} \leq M_0^{1-t} M_1^t \|f\|_{p_t} \quad f \in S(M),$$

where

$$\frac{1}{p_t} = \frac{1-t}{p_0} + \frac{t}{p_0}$$
$$\frac{1}{q_t} = \frac{1-t}{q_0} + \frac{t}{q_1}.$$

Proof. We can follow the steps of the proof of the M. Riesz-Thorin theorem, defining $\phi(\cdot, z), \psi(\cdot, z)$ as above. Then for $f \in S(M), g \in S(N)$ with $\|f\|_{p_0} = 1, \|g\|_{q_0'} = 1$ we set

$$F(z) = \int_N (A_z \phi(\cdot, z)) \psi(\cdot, z) \, d\nu,$$

which is an analytic function in the strip $\{0 \leq \operatorname{Re}(z) \leq 1\}$ and for which $|F(iy)| \leq M_0$, $|F(1+iy)| \leq M_1$. Hence by Lemma 3.2.11, $|F(t)| \leq M_0^{1-t} M_1^t$ for $0 \leq t \leq 1$, which completes the proof. ∎

3.3 THE CONJUGATE FUNCTION OR DISCRETE HILBERT TRANSFORM

We now pass to a particular operator of central importance in harmonic analysis, the so-called Hilbert transform. In the setting of Fourier series this is defined on the set $\mathcal{P}$ of trigonometric polynomials by the formula

$$(3.3.1) \qquad H\left(\sum_{n \in \mathbb{Z}} c_n e^{in\theta}\right) = -i \sum_{n \geq 1} c_n e^{in\theta} + i \sum_{n \leq -1} c_n e^{in\theta}.$$

This gives a convenient way of expressing the projection operator

$$f = \sum_{n \in \mathbb{Z}} c_n e^{in\theta} \to \sum_{n \geq 1} c_n e^{in\theta} = Pf$$

by writing

$$f + iHf = c_0 + 2\sum_{n \geq 1} c_n e^{in\theta} = \hat{f}(0) + 2Pf.$$

If we restrict to the subspace of functions with $\int_\mathbb{T} f = 0$, then we can write $2Pf = f + iHf$. The operator H is more convenient to work with, since it is invertible on this subspace: $H^2 = -I$.

Exercise 3.3.1. Show that the Hilbert transform is skew-adjoint, in the sense that for any trigonometric polynomials f, g, we have the identity

$$\int_\mathbb{T} Hf(\theta)\bar{g}(\theta)\,d\theta = -\int_\mathbb{T} f(\theta)\bar{H}g(\theta)\,d\theta.$$

Now we will show that the Fourier partial sum can also be expressed in terms of H by writing for any trigonometric polynomial f

$$e^{iN\theta}H(e^{-iN\theta}f) = -i\sum_{n>N} c_n e^{in\theta} + i\sum_{n<N} c_n e^{in\theta},$$

$$e^{-iN\theta}H(e^{iN\theta}f) = -i\sum_{n>-N} c_n e^{in\theta} + i\sum_{n<-N} c_n e^{in\theta}.$$

When we subtract, we obtain

$$e^{iN\theta}H(e^{-iN\theta}f) - e^{-iN\theta}H(e^{iN\theta}f) = 2i\sum_{n=-N}^{N} c_n e^{in\theta} - ic_N e^{iN\theta} - ic_{-N} e^{-iN\theta}.$$

This allows us to represent the partial sums $S_N f$ in terms of the norm-preserving operators $f \to e^{\pm iN\theta} f$ and the fixed operator H.

(3.3.2)
$$\boxed{S_N f(\theta) = \frac{1}{2i}\left(e^{iN\theta}H(e^{-iN\theta}f) - e^{-iN\theta}H(e^{iN\theta}f)\right) + \frac{1}{2}\hat{f}(N)e^{iN\theta} + \frac{1}{2}\hat{f}(-N)e^{-iN\theta}}$$

The last two terms tend to zero when $N \to \infty$ and are bounded by the L^p norm of f for any $p \geq 1$.

The representation (3.3.2) shows that H cannot be bounded on $L^1(\mathbb{T})$.

Proposition 3.3.2.

(3.3.3) $$\sup \frac{\|Hf\|_1}{\|f\|_1} = +\infty$$

where the supremum is taken over all trigonometric polynomials $f \neq 0$.

Proof. Suppose not; then we would have the estimate $\|Hf\|_1 \le C\|f\|_1$ for some constant C and all trigonometric polynomials f. Referring to (3.3.2), this implies that $\|S_N f\|_1 \le (1+C)\|f\|_1$, which implies that the (1,1) operator norm of S_N is bounded by a constant, independent of N. But we showed in Section 1.6.3 of Chapter 1 that $\|S_N\|_{1,1} = L_N \sim (4\log N)/\pi^2$ when $N \to \infty$. Thus we have a contradiction, which proves that the supremum in (3.3.3) is infinite. ∎

3.3.1 L^p Theory of the Conjugate Function

In order to prove uniform boundedness of S_N on the space $L^p(\mathbb{T})$ for $1 < p < \infty$, it suffices to prove that the operator H can be extended to a bounded operator on the space $L^p(\mathbb{T})$. This will be accomplished by interpolation, as follows:

Lemma 3.3.3. *The operator H is bounded on $L^2(\mathbb{T})$.*

Proof. From Parseval's identity, for any trigonometric polynomial f

$$\|Hf\|_2^2 = \sum_{0 \neq n \in \mathbb{Z}} |c_n|^2 \le \sum_{n \in \mathbb{Z}} |c_n|^2 = \|f\|_2^2.$$ ∎

This allows us to extend the definition of H to the space $L^2(\mathbb{T})$ as a bounded operator with norm at most 1. But this bound is attained, since $\|H(e^{i\theta})\|_2 = \|e^{i\theta}\|_2 = 1$. Since $L^p(\mathbb{T}) \subset L^2(\mathbb{T})$ for $p > 2$, we also obtain the existence of Hf when $f \in L^p(\mathbb{T}), p > 2$. We will now show by several steps that H is a bounded operator on any L^p space for $p > 2$.

We first prove that H is bounded on $L^p(\mathbb{T})$ if $p = 2k$ is an even integer. The following lemma is attributed to M. Riesz (1927).

Lemma 3.3.4. *For any $k = 2, 3, \ldots$ there exists a constant C_{2k}, so that if f is any trigonometric polynomial, $\|Hf\|_{2k} \le C_{2k}\|f\|_{2k}$.*

Proof. It suffices to first prove this for real-valued functions. If f is real-valued ($\overline{\hat{f}}(-n) = \hat{f}(n)$), then so is Hf and we have $Pf = \frac{1}{2}(f + iHf)$. Expanding this by the binomial theorem we note that $(Pf)^k$ has no constant term, so that

$$0 = \int_{\mathbb{T}} (2Pf)^{2k} = \sum_{j=0}^{2k} \binom{2k}{j} \int_{\mathbb{T}} f^j (iHf)^{2k-j}.$$

Taking the real part and writing $j = 2r$ we have

$$0 = \sum_{r=0}^{k} \binom{2k}{2r}(-1)^{k-r} \int_{\mathbb{T}} f^{2r}(Hf)^{2k-2r}.$$

178 INTRODUCTION TO FOURIER ANALYSIS AND WAVELETS

We isolate the term with $r = 0$ and apply Hölder's inequality to the remaining terms:

$$\int_{\mathbb{T}} (Hf)^{2k} \leq \sum_{r=1}^{k} \binom{2k}{2r} \int_{\mathbb{T}} f^{2r} (Hf)^{2k-2r}$$

$$\leq \sum_{r=1}^{k} \binom{2k}{2r} \left(\int_{\mathbb{T}} f^{2k} \right)^{\frac{r}{k}} \left(\int_{\mathbb{T}} (Hf)^{2k} \right)^{\frac{k-r}{k}}.$$

Dividing both sides by $\|f\|_{2k}^{2k}$, we have the polynomial inequality

(3.3.4) $$X^{2k} \leq \sum_{r=1}^{k} \binom{2k}{2r} X^{2k-2r}$$

where

$$X := \frac{\|Hf\|_{2k}}{\|f\|_{2k}} = \left(\frac{\int_{\mathbb{T}} (Hf)^{2k}}{\int_{\mathbb{T}} f^{2k}} \right)^{\frac{1}{2k}}.$$

If $X \leq 1$ there is nothing to prove. If $X > 1$, then each term in (3.3.4) is bounded by $\binom{2k}{2r} X^{2k-2}$, so that the sum is bounded by

$$X^{2k} \leq X^{2k-2} \sum_{r=1}^{k} \binom{2k}{2r} = X^{2k-2}(2^{2k} - 1)$$

equivalently $X^2 \leq 2^{2k} - 1$. The lemma is proved with $C_{2k} = \sqrt{2^{2k} - 1}$. ∎

Since the space of trigonometric polynomials is dense in $L^p(\mathbb{T})$, we immediately obtain the extension of H as follows.

Corollary 3.3.5. *The map $f \to Hf$ is a bounded operator on $L^{2k}(\mathbb{T})$ for any $k = 2, 3, \ldots$.*

Proof. In order to prove the boundedness of H on the intermediate L^p spaces, we can apply the M. Riesz-Thorin theorem to conclude that H can be extended to a bounded operator from $L^p(\mathbb{T})$ to $L^p(\mathbb{T})$ for any $1 < p < 2k$. But k was arbitrary, so we conclude boundedness for any $p > 2$. In order to prove boundedness for $1 < p < 2$ we use the duality of the norms, namely

$$\|Hf\|_p = \sup_{0 \neq g \in L^{p'}(\mathbb{T})} \frac{\int_{\mathbb{T}} |fHg|}{\|g\|_{p'}}$$

where the supremum is taken over all trigonometric polynomials. We can apply Hölder's inequality and the boundedness result on $L^{p'}$, $p' > 2$ to see that the numerator is bounded by

$$\|f\|_p \|Hg\|_{p'} \leq C_{p'} \|f\|_p \|g\|_{p'}$$

and thus conclude that

$$\|Hf\|_p \leq C_{p'} \|f\|_p \qquad f \in L^{p'}(\mathbb{T}), \ 1 < p < 2$$

which completes the proof. ∎

We list the result as follows.

Proposition 3.3.6. *The mapping $f \to Hf$ is a bounded operator on $L^p(\mathbb{T})$ whenever $1 < p < \infty$.*

This is now used to deduce the following result on L^p convergence of one-dimensional Fourier series.

Theorem 3.3.7. M. Riesz: *Suppose that $1 < p < \infty$ and $f \in L^p(\mathbb{T})$. Then the Fourier series of f converges in the norm of $L^p(\mathbb{T})$: $\lim_N \|f - S_N f\|_p = 0$.*

Proof. It suffices to note that we have convergence on the (dense) set $\mathcal{P}$ of trigonometric polynomials and that the partial sum operators have uniformly bounded norms. But from (3.3.2), we have $\sup_{N \geq 1} \|S_N f\|_{p,p} \leq 1 + C_p < \infty$. Therefore we have norm convergence on the entire space $L^p(\mathbb{T})$. ∎

Exercise 3.3.8. *Let Q_r be the conjugate Poisson kernel, defined on trigonometric polynomials by*

$$(3.3.5) \qquad Q_r f(\theta) = -i \sum_{0 \neq n \in \mathbb{Z}} \operatorname{sgn}(n) \hat{f}(n) r^{|n|} e^{in\theta}.$$

Prove that for any $k \in \mathbb{Z}^+$, there exists a constant C_{2k} such that $\|Q_r f\|_{2k} \leq C_k \|f\|_{2k}$ for all $f \in \mathcal{P}$ and $0 \leq r < 1$.

Hint: $P_r f + i Q_r f = 2 \sum_{n \geq 1} \hat{f}(n) r^{|n|} e^{in\theta}$. Now copy the proof of Lemma 3.3.4 and use the L^p boundedness of the Poisson kernel from Chapter 1.

Exercise 3.3.9. *Prove that for any $1 < p < \infty$ there exists a constant C_p so that $\|Q_r f\|_p \leq C_p \|f\|_p$ for all $f \in \mathcal{P}$ and $0 \leq r < 1$.*

Hint: Use the result of exercise 3.3.8 and the M. Riesz-Thorin theorem for $2 < p < \infty$. The duality argument takes care of $1 < p < 2$.

3.3.2 L^1 Theory of the Conjugate Function

An alternative method for proving L^p boundedness of the conjugate function is to first prove the following inequality of Kolmogorov:

$$(3.3.6) \qquad \boxed{|\{\theta : |Hf(\theta)| > \alpha\}| \leq \frac{C \|f\|_1}{\alpha}.}$$

This is then combined with the Marcinkiewicz interpolation theorem, to be proved later in this chapter. For this purpose, we need to define the conjugate function Hf when $f \in L^1(\mathbb{T})$. In order to develop these ideas, we begin with the Poisson integral P_r and its harmonic conjugate Q_r:

$$P_r f(\theta) = \sum_{n \in \mathbb{Z}} r^{|n|} \hat{f}(n) e^{in\theta},$$

$$Q_r f(\theta) = -i \sum_{0 \neq n \in \mathbb{Z}} \operatorname{sgn}(n) r^{|n|} \hat{f}(n) e^{in\theta}.$$

Exercise 3.3.10. Prove that both $P_r f(\theta)$ and $Q_r f(\theta)$ are harmonic functions and that $P_r f + i Q_r f$ is an analytic function of $z = re^{i\theta}$.

Note that, if f is a trigonometric polynomial, then $\lim_{r \to 1} Q_r f(\theta) = Hf(\theta)$, as defined in (3.3.1). The next theorem extends this definition to any $f \in L^1(\mathbb{T})$. The proof, which is adapted from Katznelson (1976), uses the properties of harmonic functions, which are developed in an appendix to this chapter.

Theorem 3.3.11. Let $f \in L^1(\mathbb{T})$. Then $Hf(\theta) := \lim_{r \to 1} Q_r f(\theta)$ exists for almost every $\theta \in \mathbb{T}$. For any $\alpha > 0$ the weak (1,1) inequality (3.3.6) holds.

Proof. Since any L^1 function can be written in terms of four nonnegative functions, we first assume that $f \geq 0$. In particular

$$P_r f(0) = \hat{f}(0) = \frac{1}{2\pi} \int_{\mathbb{T}} f = \|f\|_1.$$

It is readily verified that

$$P_r f(\theta) + i Q_r f(\theta) = \hat{f}(0) + 2 \sum_{n=1}^{\infty} r^n \hat{f}(n) e^{in\theta},$$

which is a holomorphic function of $z = re^{i\theta}$. By the Abel summability of L^1 functions, we see that $\lim_{r \to 1} P_r f(\theta) = f(\theta)$ for almost all θ. Now define $G(z) = e^{-P_r f - i Q_r f}$, a holomorphic function in D. Since $f \geq 0$, then $|G(z)| \leq 1$ and we can assert the existence of a radial limit $g(\theta) = \lim_{r \to 1} G(re^{i\theta})$ with $|g(\theta)| = e^{-f(\theta)} > 0$ a.e. since f is finite a.e. If $Q_r f(\theta)$ were unbounded when $r \to 1$, then the set of accumulation points of $e^{-i Q_r f(\theta)}$ would fill out an interval, which contradicts the convergence of $G(z)$, hence $Q_r f(\theta)$ remains bounded—in particular has at least one point of accumulation when $r \to 1$. If $Q_r f(\theta)$ had two different accumulation points when $r \to 1$, then the set of accumulation points of $e^{-i Q_r f(\theta)}$ would fill out an interval, which contradicts the existence of the radial limit of $G(z)$. From this it follows that there exists a radial limit of $Q_r f(\theta)$ for a.e. θ. A general $f \in L^1(\mathbb{T})$ can be written $f = f_1 - f_2 + if_3 - if_4$ where $f_j \in L^1(\mathbb{T})$ is nonnegative. We define $Hf = Hf_1 - Hf_2 + iHf_3 - iHf_4$, which completes the proof of the a.e. existence of $Hf(\theta)$. Note that the exceptional set, which comes from the Fatou theorem, may be different from the set where $P_r f$ fails to converge when $r \to 1$.

To prove the weak (1,1) inequality (3.3.6), we first suppose that $f \geq 0$ and set

$$F(re^{i\theta}) = P_r f(\theta), \qquad \tilde{F}(re^{i\theta}) = Q_r f(\theta).$$

Then we have a holomorphic function $z \to w = F(z) + i\tilde{F}(z)$ which maps the disk D to the right half plane $\operatorname{Re} w \geq 0$. For any $\lambda > 0$, consider the harmonic function H_λ, which equals 0 on the segment of the imaginary axis from $-i\lambda$ to $i\lambda$ and equals 1 on the two complementary rays. Equivalently

$$H_\lambda(w) = \frac{\pi - \Phi}{\pi},$$

where Φ is the angle that the point w makes with the points $\pm i\lambda$. Note that H_λ has the constant value of $\frac{1}{2}$ on the half circle $w = \lambda e^{i\varphi}$, $-\pi/2 < \varphi < \pi/2$ and outside of this circle is strictly larger than $\frac{1}{2}$. At any point w of the positive real axis we have

$H_\lambda(w) = (2/\pi)\arctan(w/\lambda) < 2w/\pi\lambda$. The composed mapping $z \to H_\lambda(F + i\tilde{F})$ is a harmonic function in the disk D. From the mean-value property of harmonic functions, we have

$$\frac{1}{2\pi}\int_{\mathbb{T}} H_\lambda\left(F(re^{i\theta}) + i\tilde{F}(re^{i\theta})\right) d\theta = H_\lambda[F(0)] = H_\lambda(\|f\|_1) < \frac{2\|f\|_1}{\pi\lambda}.$$

On the other hand,

$$\frac{1}{2\pi}\int_{\mathbb{T}} H_\lambda\left(F(re^{i\theta}) + i\tilde{F}(re^{i\theta})\right) d\theta \geq \frac{1}{2\pi}\int_{\{\theta:|F+i\tilde{F}|\geq\lambda\}} H_\lambda\left((F(re^{i\theta}) + i\tilde{F}(re^{i\theta}))\right) d\theta$$

$$\geq \frac{1}{4\pi}|\{\theta : |F(re^{i\theta}) + i\tilde{F}(re^{i\theta})| \geq \lambda\}|$$

$$\geq \frac{1}{4\pi}|\{\theta : |\tilde{F}(re^{i\theta})| \geq \lambda\}|,$$

from which we conclude that

$$|\{\theta : |\tilde{F}(re^{i\theta})| \geq \lambda\}| \leq \frac{8\|f\|_1}{\lambda}.$$

But $Hf(\theta) = \lim_{r\to 1}\tilde{F}(re^{i\theta})$ exists a.e. Therefore we have

$$|\{\theta : |Hf(\theta)| \geq \lambda\}| \leq \frac{8\|f\|_1}{\lambda},$$

which proves (3.3.6) in case $f \geq 0$. A general $f \in L^1(\mathbb{T})$ is written $f = f_1 - f_2 + if_3 - if_4$, for which $Hf = Hf_1 - Hf_2 + iHf_3 - iHf_4$. Then

$$|\{\theta : |Hf(\theta)| \geq \lambda\}| \leq \sum_{j=1}^{4}\left|\left\{\theta : |Hf_j(\theta)| \geq \frac{\lambda}{4}\right\}\right|$$

to which (3.3.6) is applied four times to obtain the result in general. ∎

Theorem 3.3.11 provides an extension of the Hilbert transform to the entire space $L^1(\mathbb{T})$. If, in addition, $f \in L^p(\mathbb{T})$ for some $1 < p < \infty$, then we can also compute Hf as the L^p limit of trigonometric polynomials, from Proposition 3.3.6. By taking subsequences, we see that the two definitions of Hf agree for $f \in L^p(\mathbb{T})$, $1 < p < \infty$.

The discrete Hilbert transform is integrable under a slight additional condition, where we use the notation $\log^+ x = \max\{0, \log x\}$ for $x > 0$.

Theorem 3.3.12. *If $f\log^+|f| \in L^1(\mathbb{T})$, then $Hf \in L^1(\mathbb{T})$.*

Proof. To prove this result, we introduce the *distribution function* of a nonnegative measurable function f, defined by

(3.3.7) $$\lambda_f(\alpha) = |\{\theta : f(\theta) > \alpha\}|.$$

The map $\alpha \to \lambda_f(\alpha)$ is decreasing and can be used to express the L^p norm ($0 < p < \infty$) as follows:

$$\int_{\mathbb{T}} |f(\theta)|^p\, d\theta = \int_{\mathbb{T}} \left(\int_0^{|f(\theta)|} p\alpha^{p-1}\, d\alpha \right) d\theta$$

$$= \int_{\mathbb{T}} \left(\int_0^{\infty} p\alpha^{p-1} 1_{\{\alpha < |f(\theta)|\}}\, d\alpha \right) d\theta$$

$$= \int_0^{\infty} \left(\int_{\mathbb{T}} 1_{\{\alpha < |f(\theta)|\}}\, d\theta \right) p\alpha^{p-1}\, d\alpha$$

$$= \int_0^{\infty} p\alpha^{p-1} \lambda_f(\alpha)\, d\alpha.$$

We also use the observation that

$$f \le f_1 + f_2 \implies \lambda_f(\alpha) \le \lambda_{f_1}(\alpha/2) + \lambda_{f_2}(\alpha/2).$$

Having made these preparations, we make the α-dependent decomposition $f = f_\alpha + f^\alpha$, where

$$f_\alpha = f 1_{|f| \le \alpha}, \qquad f^\alpha = f 1_{|f| > \alpha}.$$

Since H is a linear operator we have

$$|Hf| \le |Hf_\alpha| + |Hf^\alpha|,$$

$$\lambda_{|Hf|}(\alpha) \le \lambda_{|Hf_\alpha|}(\alpha/2) + \lambda_{|Hf^\alpha|}(\alpha/2).$$

The L^1 norm of HF is expressed in terms of the distribution function as

$$\int_{\mathbb{T}} |Hf| = \int_0^\infty \lambda_{|Hf|}(\alpha)\, d\alpha \le \lambda_{|Hf|}(0) + \int_1^\infty \lambda_{|Hf|}(\alpha)\, d\alpha.$$

It remains to estimate the last integral in terms of the distribution functions of $|Hf_\alpha|$ and $|Hf^\alpha|$. Since $\mathbb{T}$ has finite measure and f_α is bounded, it follows that $f_\alpha \in L^2(\mathbb{T})$. From this it follows that $Hf_\alpha \in L^2(\mathbb{T})$ with $\|Hf_\alpha\|_2 \le \|f_\alpha\|_2$. Now we transform by the Fubini theorem as follows:

$$\lambda_{|Hf_\alpha|}(\alpha/2) \le \frac{4}{\alpha^2} \|f_\alpha\|_2^2$$

$$\int_1^\infty \lambda_{|Hf_\alpha|}(\alpha/2)\, d\alpha \le \int_1^\infty \frac{4}{\alpha^2} \left(\int_{\mathbb{T}} |f(\theta)|^2 I_{|f(\theta)| \le \alpha}\, d\theta \right) d\alpha$$

$$= 4 \int_{\mathbb{T}} |f(\theta)|^2 \left(\int_{|f(\theta)| \vee 1}^\infty \frac{d\alpha}{\alpha^2} \right) d\theta$$

$$\le 4 \int_{\mathbb{T}} |f(\theta)|\, d\theta$$

$$< \infty.$$

Similarly, from (3.3.6) and the Fubini theorem, we have

$$\lambda_{|Hf^\alpha|}(\alpha/2) \leq \frac{C}{\alpha} \|f^\alpha\|_1$$

$$\int_1^\infty \lambda_{|Hf^\alpha|}(\alpha/2)\, d\alpha \leq \int_1^\infty \frac{C}{\alpha} \left(\int_{\mathbb{T}} |f(\theta)| I_{|f(\theta)|>\alpha}\, d\theta \right) d\alpha$$

$$= C \int_{\mathbb{T}} |f(\theta)| \left(\int_1^{|f(\theta)|\wedge 1} \frac{d\alpha}{\alpha} \right) d\theta$$

$$= C \int_{\mathbb{T}} |f(\theta)| \log^+ |f(\theta)|\, d\theta$$

$$< \infty.$$

The proof is complete. ∎

Exercise 3.3.13. *If $f \in L^1(\mathbb{T})$, prove that $Hf \in L^p(\mathbb{T})$ for $0 < p < 1$.*

Hint: Begin with the representation

$$\|Hf\|_p^p = p \int_0^\infty \alpha^{p-1} \lambda_{Hf}(\alpha)\, d\alpha \leq 2\pi \int_1^\infty p\alpha^{p-1} \lambda_{Hf}(\alpha)\, d\alpha$$

and follow the steps of the proof of Theorem 3.3.12.

3.3.2.1 Identification as a singular integral

We close this section by proving the a.e. representation.

$$(3.3.8) \qquad Hf(\theta) = \frac{1}{2\pi} \lim_{\epsilon \to 0} \int_{|\phi|>\epsilon} f(\theta - \phi) \frac{\sin \phi}{1 - \cos \phi}\, d\phi, \qquad f \in L^1(\mathbb{T}).$$

Proof. To do this, we take $r = 1 - \epsilon$ and write

$$Q_r f(\theta) - \frac{1}{2\pi} \int_{|\phi|>\epsilon} f(\theta - \phi) \frac{\sin \phi}{1 - \cos \phi}\, d\phi = \frac{I_1 + I_2}{2\pi},$$

where

$$I_1 := \int_{|\phi| \leq \epsilon} [f(\theta - \phi) - f(\theta)] \frac{2r \sin \phi}{1 + r^2 - 2r \cos \phi}\, d\phi$$

$$I_2 := \int_{|\phi| > \epsilon} [f(\theta - \phi) - f(\theta)] \left(\frac{2r \sin \phi}{1 + r^2 - 2r \cos \phi} - \frac{\sin \phi}{1 - \cos \phi} \right) d\phi$$

where we have used the oddness of $\phi \to \sin \phi$. To estimate I_1, we write $1 + r^2 - 2r \cos \phi = (1-r)^2 + 2r(1 - \cos \phi) \geq (1-r)^2$, so that

$$|I_1| \leq \int_{|\phi| \leq \epsilon} \frac{2|\phi|}{(1-r)^2} |f(\theta - \phi) - f(\theta)|\, d\phi$$

$$\leq \frac{2}{\epsilon} \int_{|\phi| \leq \epsilon} |f(\theta - \phi) - f(\theta)|\, d\phi,$$

which tends to zero for almost every $\theta \in \mathbb{T}$, when $\epsilon \to 0$.

To estimate I_2, we use the inequality $1 - \cos\phi \geq \phi^2/\pi^2$ for $|\phi| \leq \pi$ to write

$$I_2 = -(1-r)^2 \int_{|\phi|>\epsilon} \frac{\sin\phi}{(1+r^2-2r\cos\phi)(1-\cos\phi)} [f(\theta-\phi) - f(\theta)] d\phi$$

$$|I_2| \leq (1-r)^2 \int_{|\phi|>\epsilon} \frac{|\sin\phi|}{2r(1-\cos\phi)^2} [f(\theta-\phi) - f(\theta)] d\phi$$

$$\leq \pi^4 (1-r)^2 \int_{|\phi|>\epsilon} \frac{|f(\theta-\phi)-f(\theta)|}{|\phi|^3} d\phi \quad \left(r \geq \frac{1}{2}\right)$$

$$= \pi^4 (1-r)^2 \int_{|\phi|>\epsilon} \frac{dF(\phi)}{|\phi|^3}$$

where we have set $F(\phi) = \int_0^\phi |f(\theta-u) - f(\theta)| du$. Integration-by-parts shows that

$$|I_2| \leq \pi^4 (1-r)^2 \left(\left.\frac{|F(\phi)|}{|\phi|^3}\right|_{|\phi|=\epsilon}^{|\phi|=\pi} + 3 \int_{|\phi|>\epsilon} \frac{|F(\phi)|}{\phi^4} d\phi \right).$$

But $F(\phi)/\phi := \eta(1/v) \to 0$ a.e. when $\phi \to 0$ and $1 - r = \epsilon$ shows that the first term tends to zero. Similarly, the second term $= \int_{1/\pi}^{1/\epsilon} u\eta(u) du = o(\epsilon^{-2})$ when $\epsilon \to 0$, which shows that $I_2 \to 0$. Recalling that $Hf(\theta) = \lim_{r \to 1} Q_r f(\theta)$ a.e. completes the proof of (3.3.8). ∎

Exercise 3.3.14. Prove that $1 - \cos\phi \geq \phi^2/\pi^2$ for $|\phi| \leq \pi$.

Exercise 3.3.15. Let $\eta \in L^1_{loc}(\mathbb{R})$ with $\eta(x) \to 0$ when $x \to \infty$. Prove that $\lim_{x \to \infty} x^{-2} \int_0^x u\eta(u) du \to 0$ when $x \to \infty$.

3.4 THE HILBERT TRANSFORM ON $\mathbb{R}$

On the circle we developed the conjugate function beginning with its Fourier representation $\hat{H}f(n) = -i\,\text{sgn}(n)\hat{f}(n)$ for trigonometric polynomials f, eventually leading to the singular integral representation (3.3.8). When we pass to the corresponding problem on the real line, the relevant operator is the *Hilbert transform*, defined formally as the singular integral

(3.4.1) $$Hf(x) = \frac{1}{\pi} \lim_{\epsilon \to 0, M \to \infty} \int_{\epsilon < |y| < M} \frac{f(x-y)}{y} dy.$$

Proposition 3.4.1. *If $f \in \mathcal{S}$, the limit (3.4.1) exists.*

Proof. Denoting the integral in (3.4.1) as $H_{\epsilon,M} f(x)$, we have for $f \in \mathcal{S}$

$$\pi H_{\epsilon,M} f(x) = \int_{\epsilon<|y|<M} \frac{f(x-y)}{y} dy$$

$$= -\int_{\epsilon<|y|<M} \frac{f(x+y)}{y} dy$$

$$2\pi H_{\epsilon,M} f(x) = \int_{\epsilon<|y|<M} \frac{f(x-y) - f(x+y)}{y} dy.$$

If $f \in \mathcal{S}$, this family of integrals converge when $\epsilon \to 0, M \to \infty$ and we have

$$(3.4.2) \qquad Hf(x) = \frac{1}{2\pi} \int_{\mathbb{R}} \frac{f(x-y) - f(x+y)}{y} dy. \qquad \blacksquare$$

Example 3.4.2. Let $f = 1_{[a,b]}$. Then $Hf(x) = (1/\pi) \log(|x-a|/|x-b|)$ for $x \neq a, b$.

Exercise 3.4.3. *Prove this.*

Explicit calculation reveals that in this example $Hf(x) \sim (1/\pi)(b-a)/x$ when $x \to \infty$, showing that $Hf \notin L^1(\mathbb{R})$. The same behavior is generically true whenever $\int_{\mathbb{R}} f \neq 0$.

Exercise 3.4.4. *Suppose that $f \in \mathcal{S}(\mathbb{R})$. Prove that $\lim_{x \to \infty} xHf(x) = 1/\pi \int_{\mathbb{R}} f$.*

Exercise 3.4.5. *Suppose that $\int_{\mathbb{R}} |f(x)|/(1+|x|) dx < \infty$ and that f satisfies a Dini condition at x. Prove that the integral in (3.4.2) is absolutely convergent.*

3.4.1 L^2 Theory of the Hilbert Transform

In order to define H on $L^2(\mathbb{R})$, we let $K_{\epsilon,M}(x) = (1/\pi x)1_{\epsilon < |x| < M}$. Then $H_{\epsilon,M} f = f * K_{\epsilon,M} \in L^2$ whenever $f \in L^2$, since $K_{\epsilon,M} \in L^1$. We now study the Fourier transform. Clearly $\hat{K}_{\epsilon,M}(0) = 0$. The Fourier transform for $\xi \neq 0$ is computed as

$$\hat{K}_{\epsilon,M}(\xi) = \int_{\epsilon < |x| < M} \frac{e^{-2\pi i \xi x}}{\pi x} dx$$

$$= -2i \int_{\epsilon}^{M} \frac{\sin 2\pi x \xi}{\pi x} dx$$

$$= -i(\text{Si}(2\pi M \xi) - \text{Si}(2\pi \epsilon \xi)).$$

From the properties of the Si function, we have

$$|\hat{K}_{\epsilon,M}(\xi)| \leq 2\,\text{Si}(\pi), \qquad \lim_{\epsilon \to 0, M \to \infty} \hat{K}_{\epsilon,M}(\xi) = -i\pi \,\text{sgn}(\xi).$$

We can use this to define H on the space $L^2(\mathbb{R})$ as follows.

Proposition 3.4.6. *For $f \in L^2(\mathbb{R})$ there exists the L^2 limit*

$$Hf = \lim_{\epsilon \to 0, M \to \infty} H_{\epsilon,M} f = \lim_{\epsilon \to 0, M \to \infty} \int_{\mathbb{R}} \hat{K}_{\epsilon,M}(\xi) \hat{f}(\xi) e^{2\pi i \xi x} d\xi.$$

Proof. For any $0 < \epsilon < M < \infty$

$$\hat{H}_{\epsilon,M} f(\xi) = \hat{K}_{\epsilon,M}(\xi) \hat{f}(\xi),$$

and by the dominated convergence theorem for any $f \in L^2$,

$$\|\hat{H}_{\epsilon,M} f(\xi) + i\,\text{sgn}(\xi) \hat{f}(\xi)\|_2^2 \to 0, \qquad \epsilon \to 0, \; M \to \infty.$$

Therefore the L^2 limit of $\hat{H}_{\epsilon,M}f$ exists, especially $\hat{H}_{\epsilon,M}f$ is a Cauchy sequence. By Plancherel's formula it follows that $H_{\epsilon,M}f$ must also be a Cauchy sequence in L^2. Hence there exists $g \in L^2$, $g = \lim_{\epsilon \to 0, M \to \infty} H_{\epsilon,M}f$, which was to be proved. ∎

Proposition 3.4.6 can be paraphrased as the representation formula

$$Hf(x) = -i \int_{\mathbb{R}} \operatorname{sgn}(\xi) \hat{f}(\xi) e^{2\pi i \xi x} \, d\xi$$

valid for $f \in L^2(\mathbb{R})$. This shows, in particular, that H is norm-preserving: $\|Hf\|_2 = \|f\|_2$ and that $H(Hf) = -f$ whenever $f \in L^2(\mathbb{R})$.

3.4.2 L^p Theory of the Hilbert Transform, $1 < p < \infty$

We now develop the tools to prove the L^p boundedness of the Hilbert transform. For any $\epsilon > 0$, let

$$\mathcal{S}_\epsilon = \{f \in \mathcal{S} : \hat{f}(\xi) = 0 \text{ for } |\xi| \leq \epsilon\}$$

and $\mathcal{S}_0 = \cup_{\epsilon > 0} \mathcal{S}_\epsilon$. A typical element of $\mathcal{S}_0$ is written

$$f_\epsilon(x) = \int_{|\xi| > \epsilon} \hat{f}(\xi) e^{2\pi i \xi x} \, d\xi,$$

and the Hilbert transform

$$Hf_\epsilon(x) = i \int_{-\infty}^{-\epsilon} \hat{f}(\xi) e^{2\pi i \xi x} \, d\xi - i \int_\epsilon^\infty \hat{f}(\xi) e^{2\pi i \xi x} \, d\xi \in \mathcal{S}_0.$$

Proposition 3.4.7. *For any $p \geq 2$, $\mathcal{S}_0$ is dense in $L^p(\mathbb{R})$.*

Proof. Since $\mathcal{S}$ is dense in L^p, it suffices to prove that $\mathcal{S}_0$ is dense in $\mathcal{S}$ in the L^p norm. For $p = 2$ this follows immediately from the Plancherel formula, since

$$\|f - f_\epsilon\|_2^2 = \int_{-\epsilon}^\epsilon |\hat{f}(\xi)|^2 \, d\xi,$$

which can be made arbitrarily small with ϵ. For any $p \geq 2$, we can write

$$|f(x) - f_\epsilon(x)| \leq \int_{-\epsilon}^\epsilon |\hat{f}(\xi)| \, d\xi \leq C\epsilon,$$

which proves that $\mathcal{S}_0$ is dense in the supremum norm, hence also in the L^p norm on compact sets. But if $p \geq 2$ we can write

$$|f(x) - f_\epsilon(x)|^p \leq \epsilon^{p-2} |f(x) - f_\epsilon(x)|^2$$

$$\|f - f_\epsilon\|_p^p \leq (C\epsilon^{p-2}) \|f - f_\epsilon\|^2 \to 0,$$

which completes the proof. ∎

To prove L^p boundedness, we can follow the proof in the case of the circle. For $f \in \mathcal{S}_0$, we write

$$f(x) = \int_0^\infty \hat{f}(\xi)\, e^{2\pi i \xi x}\, d\xi + \int_{-\infty}^0 \hat{f}(\xi)\, e^{2\pi i \xi x}\, d\xi,$$

$$Hf(x) = -i \int_0^\infty \hat{f}(\xi)\, e^{2\pi i \xi x}\, d\xi + i \int_{-\infty}^0 \hat{f}(\xi)\, e^{2\pi i \xi x}\, d\xi,$$

so that

$$f + iHf = 2 \int_0^\infty \hat{f}(\xi) e^{2\pi i \xi x}\, d\xi.$$

Now if $p = 2k$ is any even integer, we can write

$$(f + iHf)^{2k} = 2^{2k} \int_0^\infty \left(\hat{f} * \cdots * \hat{f}\right)(\xi) e^{2\pi i \xi x}\, d\xi,$$

$$\int_{-T}^T (f + iHf)^{2k}\, dx = \int_0^\infty \frac{\sin(2\pi T \xi)}{\pi \xi} \left(\hat{f} * \cdots * \hat{f}\right) d\xi.$$

The right side is the partial Fourier inversion at $\xi = 0$ of a smooth function that vanishes for small ξ, hence $\lim_{T \to \infty} \int_{-T}^T (f + iHf)^{2k}\, dx = 0$. On the other hand this integral is absolutely convergent, since $f \in \mathcal{S}_0$ and $Hf \in \mathcal{S}_0$. Expanding by the binomial theorem we have

$$0 = \sum_{j=0}^{2k} \binom{2k}{j} \int_{\mathbb{R}} (iHf)^j (f)^{2k-j}.$$

If f is real, then Hf is also real, so that we can restrict attention to the even powers and write

$$\int_{\mathbb{R}} (Hf)^{2k} = \sum_{j=0}^{k-1} \binom{2k}{2j} \int_{\mathbb{R}} (f)^{2k-2j} (Hf)^{2j}.$$

Applying the Cauchy-Schwarz inequality as before, we can estimate $\int_{\mathbb{R}} (Hf)^{2k}$ in terms of $\int_{\mathbb{R}} f^{2k}$. This completes the proof that the Hilbert transform is bounded on the space $L^{2k}(\mathbb{R})$ for any $k = 1, 2, \ldots$. As before, we can apply the M. Riesz-Thorin interpolation theorem to conclude that $f \to Hf$ is bounded on any intermediate space L^p for $2 \leq p \leq 2k$. But k was arbitrary. So we conclude boundedness on L^p for any $p \geq 2$. By duality, it also follows, as before, that $f \to Hf$ is bounded on L^p for any $1 < p \leq 2$. We have proved the following theorem.

Theorem 3.4.8. *For any $1 < p < \infty$, the Hilbert transform can be extended from the space $\mathcal{S}_0$ to $L^p(\mathbb{R})$ as a bounded operator.*

3.4.2.1 Applications to convergence of Fourier integrals

In parallel with the case of the circle, the Hilbert transform on $\mathbb{R}$ can be used to study the convergence of the partial sums of the Fourier integral in the space $L^p(\mathbb{R})$, $1 < p < \infty$. This development is described in the following exercises.

Exercise 3.4.9. Let $f \in \mathcal{S}(\mathbb{R})$ and let $M > 0$. Prove the identity

$$S_M f(x) = \frac{1}{2i} \left(e^{2\pi i M x} H(e^{-2\pi i M x} f) - e^{-2\pi i M x} H(e^{2\pi i M x} f) \right).$$

Exercise 3.4.10. Prove that for any $f \in \mathcal{S}(\mathbb{R})$, we have the bound $\|S_M f\|_p \leq C \|f\|_p$ for $1 < p < \infty$, where $C = C_p$. In particular the operator S_M has a unique extension as a bounded operator on $L^p(\mathbb{R})$, with operator norm independent of M.

Exercise 3.4.11. If $2 < p < \infty$ and $f \in \mathcal{S}(\mathbb{R})$, prove that $\|S_M f - f\|_p \to 0$ when $M \to \infty$.

Hint: Use the Hausdorff-Young Theorem 3.2.12 to estimate $\|S_M f - f\|_p$ in terms of its Fourier transform.

Exercise 3.4.12. Combine the previous exercises to show that for any $f \in L^p(\mathbb{R})$, $2 < p < \infty$, we have $\|S_M f - f\|_p \to 0$ when $M \to \infty$.

Exercise 3.4.13. Use the duality of L^p and $L^{p'}$ to prove that for any $f \in L_p(\mathbb{R})$, $1 < p < 2$, we have $\|S_M f - f\|_p \to 0$ when $M \to \infty$.

3.4.3 L^1 Theory of the Hilbert Transform and Extensions

It remains to discuss the Hilbert transform in case $p = 1$. At the same time we will identify the Hilbert transform with the limit of the conjugate Poisson kernel, equivalently as the imaginary part of the boundary value of an analytic function in the upper half plane. This extension will be carried out in the Banach space

$$(3.4.3) \qquad B_1 = \left\{ f : \int_{\mathbb{R}} \frac{|f(x)|}{1+|x|} \, dx < \infty \right\},$$

which contains all of the Lebesgue spaces $L^p(\mathbb{R})$, $1 \leq p < \infty$. To do this, we begin with the absolutely convergent Fourier integrals:

$$\frac{y}{\pi(x^2 + y^2)} = \int_{\mathbb{R}} e^{2\pi i \xi x} e^{-2\pi y |\xi|} \, d\xi, \qquad y > 0, \, x \in \mathbb{R},$$

$$\frac{x}{\pi(x^2 + y^2)} = -i \int_{\mathbb{R}} e^{2\pi i \xi x} \operatorname{sgn}(\xi) e^{-2\pi y |\xi|} \, d\xi, \qquad y > 0, \, x \in \mathbb{R}.$$

For $f \in L^1(\mathbb{R})$, the Poisson integral and conjugate Poisson integral are written

$$P_y f(x) := \frac{1}{\pi} \int_{\mathbb{R}} f(x-t) \frac{y}{t^2 + y^2} \, dt = \int_{\mathbb{R}} \hat{f}(\xi) e^{2\pi i \xi x} e^{-2\pi y |\xi|} \, d\xi,$$

$$Q_y f(x) := \frac{1}{\pi} \int_{\mathbb{R}} f(x-t) \frac{t}{t^2 + y^2} \, dt = -i \int_{\mathbb{R}} \hat{f}(\xi) e^{2\pi i \xi x} e^{-2\pi y |\xi|} \, d\xi.$$

The defining equations for $P_y f, Q_y f$ are also meaningful for $f \in B_1$. It is immediate that

$$P_y f + i Q_y f = \int_{\mathbb{R}} \frac{f(x-t)}{y - it} \, dt = i^{-1} \int_{\mathbb{R}} \frac{f(u)}{x + iy - u} \, du,$$

which is an analytic function of the complex variable $z = x + iy$ for $y > 0$; hence the name *conjugate Poisson kernel*. In particular, the defining integrals are absolutely convergent and can be differentiated repeatedly to show that P_y and Q_y are both harmonic functions for $y > 0, x \in \mathbb{R}$. In case $f \in L^p(\mathbb{R})$, the harmonic property follows from repeated differentiation of the Fourier integral representation.

In order to study the conjugate Poisson kernel, we first develop the necessary properties of the operator

(3.4.4) $$P_y f(x) = \frac{1}{\pi} \int_{\mathbb{R}} \frac{y f(x-t)}{t^2 + y^2} \, dt.$$

This operator is defined on the Banach space

(3.4.5) $$B_2 = \left\{ f : \|f\|_{B_2} := \frac{1}{\pi} \int_{\mathbb{R}} \frac{|f(x)|}{1 + x^2} \, dx < \infty \right\}.$$

Clearly $L^1(\mathbb{R}) \subset B_1 \subset B_2$. The Poisson kernel has the following properties.

Proposition 3.4.14. *Suppose that $f \in B_2$. Then $\|P_y f\|_{B_2} \leq 2\|f\|_{B_2}$ for $0 < y \leq 1$ and for any $f \in B_2$, $\lim_{y \to 0} \|P_y f - f\|_{B_2} = 0$. Furthermore if $f \in B_1$, then $\lim_{y \to 0} P_y f(x) = f(x)$ for almost every $x \in \mathbb{R}$.*

Proof. We have

$$\|P_y f\|_{B_2} \leq \frac{1}{\pi^2} \int_{\mathbb{R}} \left(\int_{\mathbb{R}} \frac{y |f(t)|}{y^2 + (x-t)^2} \, dt \right) \frac{dx}{1 + x^2}$$

$$= \frac{1}{\pi^2} \int_{\mathbb{R}} |f(t)| \left(\int_{\mathbb{R}} \frac{y}{y^2 + (x-t)^2} \frac{1}{1 + x^2} \, dx \right) dt$$

$$= \frac{1}{\pi} \int_{\mathbb{R}} |f(t)| \frac{1+y}{(1+y)^2 + t^2} \, dt$$

$$\leq \frac{1}{\pi} \int_{\mathbb{R}} |f(t)| \frac{2}{1 + t^2} \, dt \quad 0 < y < 1$$

$$= 2\|f\|_{B_2},$$

where we have used the semigroup property of the Poisson kernel in the form $P_y * P_1 = P_{1+y}$. To prove the norm convergence, we first note that if $f = 1_{[a,b]}$, then $\pi P_y f(x) = \arctan[(x-b)/y] - \arctan[(x-a)/y]$, which is bounded by 2 and tends pointwise to $1_{[a,b]}$ except at the endpoints $x = a, b$. By the dominated convergence theorem, we have $\|P_y f - f\|_{B_2} \to 0$ when $y \to 0$. Similarly for a finite linear combination $f = \sum_{j=1}^{N} c_j 1_{[a_j, b_j]}$ we have $\|P_y f - f\|_{B_1} \to 0$. But these functions are dense in the space B_1, and we already have proved that the operator norms $\|P_y\|_{B_2}$ are uniformly bounded for $0 < y \leq 1$, hence the result. To prove the almost-everywhere convergence, we write

$$P_y f(x) - f(x) = \frac{1}{\pi} \int_0^\infty [f(x+t) + f(x-t) - 2f(x)] \frac{y}{t^2 + y^2} dt,$$

$$|P_y f(x) - f(x)| \le \int_0^\infty \frac{y}{t^2 + y^2} d\Phi_x(t)$$

$$\Phi_x(t) := \frac{1}{\pi} \int_0^t |f(x+u) + f(x-u) - 2f(x)| \, du.$$

From Lebesgue's theorem, we have for almost every x, $\Phi_x(t)/t \to 0$ when $t \to 0$. On the other hand, $f \in B_1$ implies that $\Phi_x(t) \le Ct$ for all $t \ge 0$. Now we integrate-by-parts:

$$|P_y f(x) - f(x)| \le \int_0^\infty \frac{2ty}{(t^2 + y^2)^2} \Phi_x(t) \, dt,$$

where the estimate $|\Phi_x(t)| \le Ct$ allows one to discard the term at the limits. Setting $t = yz$ in the integration gives

$$|P_y f(x) - f(x)| \le \int_0^\infty \frac{2z}{(1+z^2)^2} \frac{\Phi_x(zy)}{y} \, dz.$$

But the integrand is bounded by an L^1 function and tends to zero pointwise when $y \to 0$, hence $P_y f(x) \to f(x)$ as required. ∎

We now introduce a norm on the space B_1, by defining

(3.4.6) $$\|f\|_{B_1} = \frac{1}{\pi} \int_\mathbb{R} \frac{|xf(x)|}{1+x^2} dx.$$

Theorem 3.4.15. *Suppose that $f \in B_1$. Then $Q_y f(x)$ converges when $y \downarrow 0$ almost everywhere to a limiting function $\tilde{f}(x)$ and we have for almost every $x \in \mathbb{R}$,*

(3.4.7) $$\tilde{f}(x) = Hf(x) := \frac{1}{\pi} \lim_{\epsilon \to 0} \int_{|y| > \epsilon} \frac{f(x-y)}{y} dy.$$

Proof. Any complex-valued function can be written as $f = f_1 - f_2 + i(f_3 - f_4)$ where $f_j \ge 0$. We begin with the conjugate Poisson kernel operator

(3.4.8) $$Q_y f(x) = \frac{1}{\pi} \int_\mathbb{R} \frac{tf(x-t)}{t^2 + y^2} dt, \qquad y > 0, \ x \in \mathbb{R}.$$

Clearly $|Q_1 f(0)| \le \|f\|_{B_1}$. Then

(3.4.9) $$P_y f(x) + i Q_y f(x) = \frac{i}{\pi} \int_\mathbb{R} \frac{f(t)}{x + iy - t} dt.$$

For any $f \in B_1$, (3.4.9) defines an analytic function in the upper half plane $y > 0$. The mapping

(3.4.10) $$(x, y) \longrightarrow \exp[-(P_y f(x) + i Q_y f(x))]$$

is a bounded analytic function in the upper half plane. By the Fatou theorem, it possesses a.e. limits when $y \downarrow 0$. But $P_y f(x)$ converges to a finite limit a.e. whenever $f \in B_1 \subset B_2$. Hence we deduce the existence of the a.e. limit of $\exp[-i Q_y f(x)]$ when $y \downarrow 0$. From this it follows that $Q_y f(x)$ can have only one accumulation point when $y \downarrow 0$, hence the existence of $Hf(x) = \lim_{y \downarrow 0} Q_y f(x)$. ∎

It remains to identify the Hilbert transform as defined in (3.4.7), with the boundary values of $Q_y f$, namely to show that for a.e. $x \in \mathbb{R}$,

(3.4.11) $$Q_y f(x) = \frac{1}{\pi} \int_{\mathbb{R}} \frac{tf(x-t)}{t^2+y^2} \, dt \to Hf(x), \qquad y \downarrow 0.$$

Lemma 3.4.16. *Suppose that* $f \in B_1$. *Then*

$$\lim_{y \to 0} \left(\int_{\mathbb{R}} \frac{tf(x-t)}{t^2+y^2} \, dt - \int_{|t|>y} \frac{f(x-t)}{t} \, dt \right) = 0$$

for almost every $x \in \mathbb{R}$.

Proof. We write the above difference as $I_1 + I_2$ where

$$I_1 = \int_{|t|<y} \frac{tf(x-t)}{t^2+y^2} \, dt$$

$$I_2 = \int_{|t|\geq y} \left(\frac{t}{t^2+y^2} - \frac{1}{t} \right) f(x-t) \, dt.$$

The function $t \to t/(t^2+y^2)$ is odd and increasing for $|t| < y$, so that we can write

$$I_1 = \int_{|t|<y} \frac{t}{t^2+y^2} [f(x-t) - f(x)] \, dt$$

$$|I_1| \leq \frac{1}{2y} \int_{|t|<y} |f(x-t) - f(x)| \, dt \to 0$$

at every Lebesgue point of f, especially almost everywhere.

To estimate I_2 we note that its kernel is odd, hence for any $\delta > 0$ and $y < \delta$,

$$-I_2 = \int_{|t|>y} \frac{y^2}{t(t^2+y^2)} [f(x-t) - f(x)] \, dt$$

$$|I_2| \leq \int_{|t|>y} \frac{y^2}{|t|^3} |f(x-t) - f(x)| \, dt$$

$$= y^2 \int_{|t|>y} \frac{dF(t)}{|t|^3}$$

where $F(t) = \int_0^t |f(x-s) - f(x)| \, ds$. Clearly $F(t)/t \to 0$ at every Lebesgue point when $t \to 0$, whereas $F(t) \leq Ct$ when for all t. Therefore we can integrate-by-parts to obtain

$$\int_{|t|>y} \frac{dF(t)}{|t|^3} = \frac{F(y)}{|y|^3} + 3 \int_{|t|>y} \frac{F(t)}{t^4} \, dt.$$

The term at the limits is clearly $o(y^{-2})$ when $y \to 0$. To analyze the new integral, write $F(t)/t = \eta(t)$, $v = 1/t$ to obtain

$$\int_{|t|>y} \frac{F(t)}{t^4} \, dt = \int_0^{1/y} v \eta(1/v) \, dv = o(y^{-2}), \qquad y \to 0,$$

which completes the proof that $I_2 \to 0$ when $y \to 0$ for almost every $x \in \mathbb{R}$. ∎

As a special case, we can deduce the properties of the Hilbert transform on $L^p(\mathbb{R})$, $1 \leq p < \infty$.

Corollary 3.4.17. *Suppose that $f \in L^p(\mathbb{R})$, $1 \le p < \infty$. Then $P_y f(x) \to f$ for almost every $x \in \mathbb{R}$ and the convergence takes place in $L^p(\mathbb{R})$. Furthermore $Q_y f(x)$ converges almost everywhere to a limiting function $\tilde{f}(x)$ and the convergence takes place in L^p if $p > 1$.*

The convergence properties of $P_y f$ follow from the properties of the Poisson kernel: It is readily verified that $t \to t/(t^2 + y^2)$ satisfies the properties of an approximate identity. Therefore if B is any homogeneous Banach space, we have $P_y f \to f$ in the B-norm when $y \to 0$. The almost-everywhere convergence of $Q_y f$ follows from Theorem 3.4.15. ∎

3.4.3.1 Kolmogorov's inequality for the Hilbert transform

Following the discussion of the conjugate function on the circle, we can establish a corresponding inequality for the distribution function of the Hilbert transform whenever $f \in L_1(\mathbb{R})$. This takes the form

(3.4.12) $$|\{x : |Hf(x)| \ge \alpha\}| \le \frac{C}{\alpha} \int_{\mathbb{R}} |f(x)|\, dx$$

where C is an absolute constant. This will be deduced as a limiting case of a corresponding inequality for functions in the space B_1. Define a weighted measure by

(3.4.13) $$\mu(A) = \frac{1}{\pi} \int_A \frac{dx}{1+x^2}.$$

Theorem 3.4.18. *Suppose that $f \in B_1$. If $f \ge 0$, then we have the weak inequality*

(3.4.14) $$\mu\{x : |Hf(x)| \ge \alpha\} \le \frac{2}{\pi}\left(\frac{\|f\|_{B_2}}{\alpha - \|f\|_{B_1}} + \frac{\|f\|_{B_2}}{\alpha + \|f\|_{B_1}}\right), \quad \alpha > \|f\|_1.$$

For any complex-valued $f \in B_1$, (3.4.14) holds with four terms on the right side and with α replaced by $\alpha/4$.

Proof. We consider the harmonic function $J_\alpha(w)$, defined for $\mathrm{Re}(w) > 0$ as the harmonic measure of the two rays $\{w = iv, v \ge \alpha\}$ and $\{w = iv, v \le -\alpha\}$. This is the harmonic function that takes the value 1 on these rays and takes the value zero on the segment $\{w = iv, -\alpha \le v \le \alpha\}$. Equivalently, it can be obtained as the imaginary part of $(1/\pi) \log[(w - i\alpha)/(w + i\alpha)]$ for a suitable branch of the logarithm. The set $\{w : J_\alpha(w) \ge \frac{1}{2}\}$ is the exterior of the semicircle described by $\{w : \mathrm{Re}(w) > 0, |w| = \alpha\}$. On the strip $|\mathrm{Im}(w)| < \alpha$, we have

(3.4.15) $$J_\alpha(u+iv) = \frac{1}{\pi}\left[\arctan\left(\frac{u}{\alpha - v}\right) + \arctan\left(\frac{u}{\alpha + v}\right)\right], \quad u > 0, |v| < \alpha.$$

We now consider the harmonic function

$$U_\alpha(x, y) = J_\alpha[P_y f(x) + iQ_y f(x)].$$

We first recall a basic fact on harmonic functions.

Lemma 3.4.19. *If $U(x, y)$ is any bounded harmonic function in $y > 0$, then for any $y_1, y_2 > 0$, we have*

$$(3.4.16) \qquad U(x, y_1 + y_2) = \frac{1}{\pi} \int_{\mathbb{R}} \frac{y_2 U(t, y_1)}{(x-t)^2 + y_2^2} \, dt.$$

Applying (3.4.16) with $x = 0$, $y_2 = 1$, $U = U_\alpha$, we have

$$(3.4.17) \qquad J_\alpha \left(P_{1+y} f(0) + i Q_{1+y} f(0) \right) = \frac{1}{\pi} \int_{\mathbb{R}} \frac{J_\alpha (P_y f(t) + i Q_y f(t))}{1 + t^2} \, dt.$$

The right side of (3.4.17) is underestimated by

$$\frac{1}{\pi} \int_{\mathbb{R}} \frac{J_\alpha(P_y f(t) + i Q_y f(t))}{1 + t^2} \, dt \geq \frac{1}{2\pi} \int_{\{t : |Q_y f(t)| \geq \alpha\}} \frac{dt}{1 + t^2}.$$

Using the inequality $|\arctan(x)| \leq |x|$, applied to (3.4.15), we can overestimate the left side of (3.4.17) by writing

$$J_\alpha(P_{1+y}f(0) + iQ_{1+y}f(0)) \leq \frac{1}{\pi} \left(\frac{P_{1+y}f(0)}{\alpha - |Q_{1+y}f(0)|} + \frac{P_{1+y}f(0)}{\alpha + |Q_{1+y}f(0)|} \right).$$

Therefore we have

$$\mu\{x : |Q_y f(x)| \geq \alpha\} = \frac{1}{\pi} \int_{\{x : |Q_y f(x)| \geq \alpha\}} \frac{dx}{1 + x^2}$$

$$(3.4.18) \qquad \leq \frac{2}{\pi} \left(\frac{P_{1+y}f(0)}{\alpha - |Q_{1+y}f(0)|} + \frac{P_{1+y}f(0)}{\alpha + |Q_{1+y}f(0)|} \right).$$

Recall that $Hf(x) = \lim_{y \to 0} Q_y f(x)$ a.e., in particular we have convergence in measure. Now from (3.4.6), $|Q_1 f(0)| \leq \|f\|_{B_1}$, $P_1 f(0) = \|f\|_{B_2}$ and the right side of (3.4.18) is only increased when we replace $Q_1 f(0)$ by its upper bound $\|f\|_{B_1}$. Hence

$$\mu\{x : |Hf(x)| \geq \alpha\} \leq \frac{2}{\pi} \left(\frac{\|f\|_{B_2}}{\alpha - \|f\|_{B_1}} + \frac{\|f\|_{B_2}}{\alpha + \|f\|_{B_1}} \right),$$

which proves the result in case $f \geq 0$. In the general case, we write

$$\mu\{x : |Hf(x)| \geq \alpha\} \leq \sum_{j=1}^{4} \mu\{x : |Hf(x)| \geq \alpha/4\}$$

and apply the result for nonnegative functions to each of the terms on the right. ∎

The upper bound assumes a more familiar form in case f is even, as follows.

Corollary 3.4.20. *Suppose that $0 \leq f \in B_1$ is even: $f(-x) = f(x)$, $\forall x \in \mathbb{R}$. Then for any $\alpha > 0$ we have*

$$\mu\{x : |Hf(x)| \geq \alpha\} \leq \frac{4\|f\|_{B_2}}{\pi \alpha}.$$

Proof. In this case we have $Q_y f(0) = 0$ for all $y > 0$. Thus the right side of (3.4.18) becomes $(4/\pi) P_{1+y} f(0) \to 4\|f\|_{B_2}/\pi$ when $y \to 0$.

The inequality (3.4.14) contains the classical Kolmogorov inequality (3.4.12) as a limiting case, when we introduce a scaling parameter Y.

In detail, define

(3.4.19) $$\mu_Y(A) = \frac{1}{\pi} \int_A \frac{Y}{t^2 + Y^2} \, dt.$$

Then we have the following scaled replacement for (3.4.14) when $\alpha > |Q_Y f(0)|$:

(3.4.20) $$\mu_Y\{x : |Hf(x)| \geq \alpha\} \leq \frac{2}{\pi} \left(\frac{P_Y f(0)}{\alpha - Q_Y f(0)} + \frac{P_Y f(0)}{\alpha + Q_Y f(0)} \right).$$

Now multiply (3.4.20) by Y and take $Y \to \infty$. For the left side, we note that for any Borel set of finite Lebesgue measure we have from the dominated convergence theorem

$$\lim_{Y \to \infty} Y \mu_Y(A) = \frac{1}{\pi} |A|.$$

For the right side, we see that when $Y \to \infty$, the dominated convergence theorem shows that for any $f \in L^1(\mathbb{R})$

$$\lim_{Y \to \infty} Y P_Y f(0) = \frac{1}{\pi} \int_{\mathbb{R}} f(x) \, dx$$

whereas

$$|Q_Y f(0)| \leq \|f\|_{L^1(\mathbb{R})} \times \sup_{t \in \mathbb{R}} \frac{|t|}{t^2 + Y^2} \to 0, \qquad Y \to \infty.$$

Hence when we multiply (3.4.20) by Y and take $Y \to \infty$, we obtain (3.4.12), the original form of Kolmogorov's inequality. ∎

3.4.4 Application to Singular Integrals with Odd Kernels

The theory of the Hilbert transform can be transplanted to study n-dimensional singular integrals of the form

(3.4.21) $$K_{\epsilon,M} f(x) \int_{\epsilon < |y| < M} k(y) f(x - y) \, dy.$$

Here k is supposed to be an odd function that is homogeneous of degree $-n$ and satisfies $\int_{|x|=1} |k(x)| \, dx < \infty$. We write $k(x) = |x|^{-n} \Omega(x)$ where Ω is odd and homogeneous of degree zero. This can be reduced to the Hilbert transform by the *method of rotations*, developed by Calderón, as follows: We take spherical polar coordinates $y = r\omega$, with $dy = r^{n-1} \, dr \, d\omega$. Then

$$\int_{\epsilon < |y| < M} k(y) f(x - y) \, dy = \int_\epsilon^M r^{-n} \left(\int_{S^{n-1}} \Omega(\omega) f(x - r\omega) r^{n-1} \, dr \right) d\omega$$

$$= \int_{S^{n-1}} \Omega(\omega) \left(\int_\epsilon^M \frac{f(x - r\omega)}{r} \, dr \right) d\omega$$

$$= \frac{1}{2} \int_{S^{n-1}} \Omega(\omega) \left(\int_{\epsilon < r < M} \frac{f(x - r\omega) - f(x + r\omega)}{r} \, dr \right) d\omega$$

$$= \frac{1}{2} \int_{S^{n-1}} \Omega(\omega) \left(\int_{\epsilon < |r| < M} \frac{f(x - r\omega)}{r} \, dr \right) d\omega$$

where we have used the oddness of Ω in the last step. Now if $f \in \mathcal{S}$, the inner integral is the truncated Hilbert transform of the function $r \to F_{z,\omega}(r) = f(z + r\omega)$, where we make the ω-dependent decomposition of $\mathbb{R}^n \ni x = z + s\omega$ where $-\infty < s < \infty$ and z is in the hyperplane defined by $z \cdot \omega = 0$. In detail, $s = x \cdot \omega$ and $z = x - (x \cdot \omega)\omega$. Taking $\epsilon \to 0$, $M \to \infty$, we have

$$Kf(x) = \frac{1}{2} \int_{S^{n-1}} \Omega(\omega) HF_{z,\omega}(s)\, d\omega.$$

We estimate the n-dimensional L^p norm by using Minkowski's integral inequality and writing $dx = ds\, dz'$ as follows:

$$\|Kf\|_{L^p(\mathbb{R}^n)} \leq \frac{1}{2} \int_{S^{n-1}} |\Omega(\omega)| \left(\int_{\mathbb{R}^n} |HF_{z,\omega}(s)|^p\, dz\, ds \right)^{1/p} d\omega$$

$$\leq C_p \int_{S^{n-1}} |\Omega(\omega)| \left(\int_{\mathbb{R}^{n-1}} \int_{\mathbb{R}} |f(z+\omega s)|^p\, ds\, dz \right)^{1/p} d\omega$$

$$= C_p \|f\|_{L^p(\mathbb{R}^n)} \int_{S^{n-1}} |\Omega(\omega)|\, d\omega.$$

We summarize the above computations as a theorem.

Theorem 3.4.21. *There exists an absolute constant C_p such that for each $f \in \mathcal{S}$ and $1 < p < \infty$, we have the estimate*

$$\|K_{\epsilon,M} f\|_{L^p(\mathbb{R}^n)} \leq C_p \|f\|_{L^p(\mathbb{R}^n)} \int_{S^{n-1}} |\Omega(\omega)|\, d\omega.$$

Using the oddness of the kernel, we can also write

$$K_{\epsilon,M} f(x) = \frac{1}{2} \int_{\epsilon < |y| < M} k(y)[f(x-y) - f(x+y)]\, dy, \qquad f \in \mathcal{S}.$$

Taking limits, we have the absolutely convergent representation of the operator:

$$Kf(x) = \frac{1}{2} \int_{\mathbb{R}^n} k(y)[f(x-y) - f(x+y)]\, dy, \qquad f \in \mathcal{S}.$$

Example 3.4.22. The Riesz kernels are defined by $k_j(x) = c_n x_j / |x|^{n+1}$ for $1 \leq j \leq n$ where c_n is a constant. Clearly they satisfy all of the above conditions of oddness, homogeneity, and integrability.

The associated singular integral operator is denoted R_j. From Theorem 3.4.21 we have the estimate $\|R_j f\|_p \leq C_p \|f\|_p$ for $f \in \mathcal{S}$.

We now show that, to within a constant, the Riesz kernels can be regarded as the formal first partial derivatives of the operator $f \to \int_{\mathbb{R}^n} f(x-y)/|y|^{n-1}\, dy$.

To see this we use the method of subordination. We begin with the heat kernel transform of $f \in \mathcal{S}$:

$$\int_{\mathbb{R}^n} f(x-y) \frac{e^{-|y|^2/4t}}{(4\pi t)^{n/2}}\, dy = \int_{\mathbb{R}^n} \hat{f}(\xi) e^{-4\pi^2 t |\xi|^2} e^{2\pi i \xi \cdot x}\, d\xi.$$

Taking the partial derivative with respect to x_j, we have for $1 \leq j \leq n$,

$$-\int_{\mathbb{R}^n} f(x-y) \frac{y_j}{2t} \frac{e^{-|y|^2/4t}}{(4\pi t)^{n/2}} \, dy = \int_{\mathbb{R}^n} \hat{f}(\xi) e^{-4\pi^2 t |\xi|^2} (2\pi i \xi_j) e^{2\pi i \xi \cdot x} \, d\xi.$$

Multiply both sides by $t^{-1/2}$ and integrate over $0 \leq t < \infty$. We recognize the elementary integrals

$$\int_0^\infty t^{-1/2} e^{-4\pi^2 t |\xi|^2} \, dt = \frac{\Gamma(1/2)}{(4\pi^2 |\xi|^2)^{1/2}},$$

$$\int_0^\infty t^{-3/2-n/2} e^{-|y|^2/4t} \, dt = \Gamma\left(\frac{n+1}{2}\right) \left(\frac{4}{|y|^2}\right)^{(n+1)/2}.$$

Therefore

$$2^n \Gamma\left(\frac{n+1}{2}\right) \int_{\mathbb{R}^n} [f(x+y) - f(x-y)] \frac{y_j}{|y|^{n+1}} \, dy = \frac{\Gamma(1/2)}{2\pi} \int_{\mathbb{R}^n} \hat{f}(\xi) \frac{\xi_j}{|\xi|} e^{2\pi i \xi \cdot x} \, d\xi,$$

which displays the Fourier transform of the kernel. We choose the constant c_n in the definition so that the Fourier transform is as simple as possible, thus we define the Riesz transform by

(3.4.22) $$R_j f(x) := \int_{\mathbb{R}^n} \frac{i\xi_j}{|\xi|} e^{2\pi i \xi \cdot x} \hat{f}(\xi) \, d\xi, \qquad 1 \leq j \leq n, \; f \in \mathcal{S}.$$

This leads to a famous application of Calderón and Zygmund, as follows:

Proposition 3.4.23. *Suppose that $f \in \mathcal{S}(\mathbb{R}^n)$ then we have the a priori bound*

(3.4.23) $$\left\| \frac{\partial^2 f}{\partial x_j \, \partial x_k} \right\|_{L^p(\mathbb{R}^n)} \leq A_p \|\Delta f\|_{L^p(\mathbb{R}^n)}, \qquad 1 \leq j, \, k \leq n$$

where $\Delta = \sum_{j=1}^n \partial^2/\partial x_j^2$ is the n-dimensional Laplace operator.

Proof. We first prove the identity

(3.4.24) $$\frac{\partial^2 f}{\partial x_j \, \partial x_k} = -R_j R_k \Delta f.$$

The Fourier transform of the left side is $-4\pi^2 \xi_j \xi_k \hat{f}(\xi)$, while the Fourier transform of the right side is

$$-\frac{i\xi_j}{|\xi|} \frac{i\xi_k}{|\xi|} \left(-4\pi^2 |\xi|^2 \hat{f}(\xi) \right),$$

which proves (3.4.24). The proof is completed by applying the L^p boundedness of the operators R_j. ∎

3.5 HARDY-LITTLEWOOD MAXIMAL FUNCTION

A powerful tool in harmonic analysis is the concept of a *maximal function*. Strictly speaking, one should speak of a *maximal operator*, transforming a function into a supremum over an appropriate indexed family of sets.

If f is a locally integrable function on $\mathbb{R}^n$, the Hardy-Littlewood maximal function is defined by

$$(3.5.1) \qquad Mf(x) := \sup_{B \ni x} \frac{\int_B |f|}{|B|}$$

where the supremum is taken over all balls containing the point $x \in \mathbb{R}^n$. The balls need not be centered at x. The operator M is *sublinear*, in the sense that for any two locally integrable functions f, g we have $M(f+g)(x) \leq Mf(x) + Mg(x)$. It is also clear that $Mf(x) = M(|f|)(x)$, so that we can always assume that f is positive when studying Mf.

To see that Mf is a measurable function, note that both numerator and denominator of (3.5.1) are continuous functions of the radius and center of the ball B. Hence if we restrict the supremum to balls with rational centers and rational radii, the supremum will be unchanged. But for any fixed ball B, the average in (3.5.1) is a two-valued, hence measurable function of x.

It is obvious from the definition (3.5.1) that M is bounded on $L^\infty(\mathbb{R}^n)$: $\|Mf\|_\infty \leq \|f\|_\infty$. The following theorem describes the boundedness properties on the other L^p spaces.

Theorem 3.5.1. *Hardy-Littlewood:* (i) There exists a positive constant C, depending only on n such that for each $\alpha > 0$, $f \in L^1(\mathbb{R}^n)$

$$|\{x : Mf(x) > \alpha\}| \leq \frac{C}{\alpha} \int_{\mathbb{R}^n} |f|.$$

In particular Mf is finite almost everywhere.

(ii) If $1 < p < \infty$, there exists a positive constant $C_{p,n}$ so that for $f \in L^p(\mathbb{R}^n)$

$$\int_{\mathbb{R}^n} |Mf|^p \leq C_{p,n} \int_{\mathbb{R}^n} |f|^p.$$

(iii) If f lives on a set B of finite measure and $|f| \log^+ |f| \in L^1(B)$, then $\int_B Mf < \infty$.

Proof. We begin by studying the set $E_\alpha = \{x : Mf(x) > \alpha\}$. If $x \in E_\alpha$, then there is a ball $B_x \ni x$ so that $\int_{B_x} |f| > \alpha |B_x|$. Thus we have a covering: $E_\alpha \subset \cup_{x \in E_\alpha} B_x$. Since $\mathbb{R}^n$ has a countable dense subset (of points with rational coordinates), we can choose a countable subcollection x_k so that $E_\alpha \subset \cup_{k=1}^\infty B_{x_k}$. Calling these B_k, we must show that $\alpha |(\cup_{k=1}^\infty B_k)| \leq C\|f\|_1$. From the countable additivity of Lebesgue measure, $|\cup_{k=1}^\infty B_k| = \lim_N |\cup_{k=1}^N B_k|$, so it's enough to show that for any N

$$(3.5.2) \qquad \alpha \left| \bigcup_{k=1}^N B_k \right| \leq C\|f\|_1.$$

If the balls were pairwise disjoint we would be done, since for each k, $\int_{B_k} |f| \geq \alpha |B_k|$, and summing on k we would have

$$\|f\|_1 \geq \int_{\cup_{k=1}^N B_k} |f| \geq \alpha \left|\bigcup_{k=1}^N B_k\right|.$$

To complete the proof, it suffices to find a subcollection of balls that are pairwise disjoint and whose union covers at least a fixed fraction of E_α. ∎

Lemma 3.5.2. *Wiener covering lemma:* Given any finite collection of balls $B_k : 1 \leq k \leq N$ in $\mathbb{R}^n$, there exists a subcollection $\tilde{B}_i : 1 \leq i \leq p$ so that

(i) $\left|\left(\bigcup_{i=1}^p \tilde{B}_i\right)\right| \geq 3^{-n} \left|\left(\bigcup_{k=1}^N B_k\right)\right|$

(ii) $\{\tilde{B}_i\}_{i=1}^p$ are pairwise disjoint.

Proof. We order the balls in order of decreasing radii, renaming them $B_1, B_2, \ldots$. Let $\tilde{B}_1 = B_1$. Assuming that $\tilde{B}_1, \ldots, \tilde{B}_m$ have been chosen, choose $\tilde{B}_{m+1} = B_k$ where k is the smallest index so that $B_k \cap \tilde{B}_j = \phi$ for $j = 1, \ldots, m$. If no such index exists then the process terminates. We now check condition (i): if B_k is a ball that was discarded, then by definition there is a ball $\tilde{B}_i = B_j$ such that $j < k$ and $B_k \cap B_j \neq \phi$. Since the radii are ordered, we have $\tilde{r}_i = r_j \geq r_k$. We claim that $B_k \subset 3\tilde{B}_i$. Indeed, there is some $z \in B_k \cap \tilde{B}_i$. Now if x is any point of B_k, we can use the triangle inequality to estimate the distance from x to the center of $\tilde{B}_i$ as

$$|x| = |(x - z) + z| \leq |x - z| + |z| \leq 2r_k + \tilde{r}_i \leq 3\tilde{r}_i.$$

Hence $B_k \subset 3\tilde{B}_i$, which proves that $\cup_{k=1}^N B_k \subset \cup_{i=1}^p 3\tilde{B}_i$. But the measure of a disjoint union is the sum of the measures, and for any B, the n-dimensional measure of $3B$ is 3^n times the measure of B. The proof is complete. ∎

Having proved the covering lemma, part (i) of the Hardy-Littlewood maximal theorem is complete, since

$$\|f\|_1 \geq \int_{\cup_{i=1}^p \tilde{B}_i} |f| > \alpha \sum_{i=1}^p |\tilde{B}_i| = \alpha \left|\bigcup_{i=1}^p \tilde{B}_i\right| \geq \alpha 3^{-n} \left|\left(\bigcup_{k=1}^N B_k\right)\right|.$$

This proves (3.5.2) and hence part (i). Taking $\alpha \to \infty$ shows that $Mf < \infty$ a.e.

Proof. To prove part (ii), we introduce the distribution function

(3.5.3) $\qquad \lambda_f(\alpha) := |\{x : |f(x)| > \alpha\}| = \int_{\mathbb{R}^n} 1_{(\alpha,\infty)}(|f(x)|)\, dx.$

The L^p norm can be written in terms of the distribution function by writing

$$|f|^p = \int_0^{|f|} p u^{p-1}\, du = \int_0^\infty p u^{p-1} 1_{(0,|f|)}(u)\, du$$

and applying Fubini's theorem to obtain

$$\int_{\mathbb{R}^n} |f(x)|^p \, dx = \int_{\mathbb{R}^n} \left(\int_0^\infty p u^{p-1} 1_{(0,|f(x)|)}(u) \right) dx$$

$$= \int_0^\infty p u^{p-1} \left(\int_{\mathbb{R}^n} 1_{(u,\infty)}(f(x)) \, dx \right) du$$

$$= \int_0^\infty p u^{p-1} \lambda_f(u) \, du.$$

We will apply this identity to estimate $\|Mf\|_p$.

To do this, we first decompose f into a bounded part and an unbounded part (without loss of generality, we can assume that $f \geq 0$).

(3.5.4) $$f = f 1_{f \leq \alpha} + f 1_{f > \alpha} := f_\alpha + f^\alpha$$

Since the operator M is sublinear, we have

$$Mf(x) \leq Mf_\alpha(x) + Mf^\alpha(x) \leq \alpha + Mf^\alpha(x).$$

Hence if $Mf > 2\alpha$, then $Mf^\alpha > \alpha$. In terms of the distribution function, we have

$$\lambda_{Mf}(2\alpha) = |\{x : Mf(x) > 2\alpha\}| \leq |\{x : Mf^\alpha(x) > \alpha\}|.$$

But f^α lives on a set of finite measure, since

$$|\{x : f^\alpha(x) > 0\}| = |\{x : f(x) > \alpha\}| \leq \frac{\|f\|_p}{\alpha^p}$$

while $|f^\alpha| \leq \alpha + |f|$ so that $f^\alpha \in L^p(\mathbb{R}^n)$ to which we can apply the Hardy-Littlewood maximal inequality; we change α to 2α and write

$$\|Mf\|_p^p = p \int_0^\infty (2\alpha)^{p-1} \lambda_{Mf}(2\alpha) d(2\alpha)$$

$$\leq p 2^p C \int_0^\infty \alpha^{p-2} \left(\int_{\mathbb{R}^n} f^\alpha(x) \, dx \right) d\alpha$$

$$= p 2^p C \int_0^\infty \alpha^{p-2} \left(\int_{\mathbb{R}^n} f(x) 1_{(\alpha,\infty)}(f(x)) \, dx \right) d\alpha$$

$$= p 2^p C \int_{\mathbb{R}^n} \left(\int_0^{f(x)} \alpha^{p-2} d\alpha \right) f(x) \, dx$$

$$= \frac{C p 2^p}{p-1} \int_{\mathbb{R}^n} f(x)^{p-1} f(x) \, dx$$

$$= \frac{C p 2^p}{p-1} \|f\|_p^p,$$

which was to be proved. ∎

Proof. To deal with (iii), we write

$$\int_B Mf = \int_B Mf \mathbf{1}_{Mf \leq 2} + \int_B Mf \mathbf{1}_{Mf > 2}$$

$$\leq 2|B| + \int_0^\infty \lambda_{Mf}(\max(\alpha, 2)) \, d\alpha$$

$$= 2|B| + 2\lambda_{Mf}(2) + \int_2^\infty \lambda_{Mf}(\alpha) \, d\alpha.$$

The last term is estimated from the Hardy-Littlewood maximal inequality and the decomposition $f = f_\alpha + f^\alpha$

$$2 \int_1^\infty \lambda_{Mf}(2\alpha) \, d\alpha \leq 2C \int_1^\infty \frac{1}{\alpha} \left(\int_{\mathbb{R}^n} f^\alpha(x) \, dx \right) d\alpha$$

$$= 2C \int_1^\infty \frac{1}{\alpha} \left(\int_{\mathbb{R}^n} f(x) \mathbf{1}_{(\alpha, \infty)}(f(x)) \, dx \right) d\alpha$$

$$= 2C \int_{\mathbb{R}^n} \left(\int_1^{f(x)} \frac{d\alpha}{\alpha} \right) f(x) \, dx$$

$$= 2C \int_{\mathbb{R}^n} f(x) \log^+ f(x) \, dx$$

$$= 2C \|f \log^+ f\|_1,$$

which completes the proof. ∎

Remark. It is important to note that $Mf \in L^1(\mathbb{R}^n)$ if and only if $f \equiv 0$. To see this, let $f \in L^1(\mathbb{R}^n)$ and choose a (large) ball B centered at 0, so that $\int_B |f| > \frac{1}{2}\|f\|_1$. Then for any $x \notin B$, we have

$$Mf(x) \geq \frac{\int_{B(0;2|x|)} |f|}{|B(0; 2|x|)|} \geq \frac{\int_B |f|}{|B(0; 2|x|)|} \geq \text{const} \frac{\|f\|_1}{|x|^n}$$

where the constant is half the reciprocal of the volume of the ball $B(0; 2)$ in $\mathbb{R}^n$. Hence $Mf(x) \geq C/|x|^n$ for large x, which contradicts $\int_{\mathbb{R}^n} |Mf| < \infty$.

The upshot of this last remark is that no matter what decay condition we impose on f, it is never possible to achieve $Mf \in L^1(\mathbb{R}^n)$.

3.5.1 Application to the Lebesgue Differentiation Theorem

The Hardy-Littlewood maximal inequality can be used to give an efficient proof of the differentiability of the integral, as follows:

Proposition 3.5.3. *If f is a locally integrable function, then for almost every $x \in \mathbb{R}^n$*

(3.5.5) $$\lim_{r \to 0} \frac{\int_{B(x;r)} f}{|B(x; r)|} = f(x).$$

Here $B(x; r)$ is the ball of radius r centered at $x \in \mathbb{R}^n$.

Proof. To apply the Hardy-Littlewood weak $(1, 1)$ estimate, we need to replace f by an integrable function. This is easily accomplished by writing the exceptional set in (3.5.5) as the union of the exceptional sets S_M in the ball $|x| \leq M$, $M = 1, 2, \ldots$. On each S_M we can replace the locally integrable function f by the integrable function $f(x) I_{[0, M+1]}(|x|)$, to which Hardy-Littlewood can be applied.

To prove (3.5.5), we first note that if g is a continuous function, then the indicated limit exists and $= g(x)$ for *every* $x \in \mathbb{R}^n$. For an arbitrary f, define

$$f_r(x) := \frac{\int_{B(x;r)} f}{|B(x;r)|}, \qquad \Omega f(x) := \limsup_{r \to 0} f_r(x) - \liminf_{r \to 0} f_r(x).$$

Given $\epsilon > 0$, there exists a continuous function g such that $\|f - g\|_1 < \epsilon$. Writing $f = g + h$ with $\|h\|_1 < \epsilon$, we have

$$\limsup_{r \to 0} f_r(x) = g(x) + \limsup_{r \to 0} h_r(x),$$

$$\liminf_{r \to 0} f_r(x) = g(x) + \liminf_{r \to 0} h_r(x),$$

$$\Omega f(x) = \Omega h(x) \leq 2 \sup_{r > 0} |h_r(x)| \leq 2M|h|(x).$$

Therefore for any $\delta > 0$,

$$|\{x : \Omega f(x) > \delta\}| \leq |\{x : M|h|(x) \geq \delta/2\}| \leq \frac{2\epsilon}{\delta}.$$

But the left side does not depend on ϵ, hence we conclude that for every $\delta > 0$,

$$|\{x : \Omega f(x) > \delta\}| = 0$$

which means that $\Omega f(x) = 0$ almost everywhere, which was to be proved. ∎

The above reasoning can be strenthened and clarified in terms of the *Lebesgue set* of the locally integrable function f. This is defined as

(3.5.6) $$\text{Leb}(f) = \left\{ x \in \mathbb{R}^n : \lim_{r \to 0} \frac{\int_{B(x;r)} |f(y) - f(x)| m(dy)}{|B(x;r)|} = 0 \right\}.$$

Proposition 3.5.4. *For any locally integrable f, $|(\text{Leb}(f))^c| = 0$.*

Proof. For each real number c we apply the previous proof to the function $|f - c|$. Thus we obtain for almost every $x \in \mathbb{R}^n$,

(3.5.7) $$\lim_{r \to 0} \frac{\int_{B(x;r)} |f(y) - c| m(dy)}{|B(x;r)|} = |f(x) - c|.$$

If c_j is an enumeration of the rational numbers, we obtain a countable collection of exceptional sets E_j. Then $E := \cup_j E_j$ also has Lebesgue measure zero. But the right side and left side of (3.5.7) are continuous functions of c; the right side is obvious, but so is the left since the triangle inequality gives $\| |f(y) - c_1| - |f(y) - c_2| \| \leq |c_1 - c_2|$. Hence if $x \notin E$ we let $c_j \to f(x)$ to conclude that

$$\lim_{r \to 0} \frac{\int_{B(x;r)} |f(y) - f(x)| m(dy)}{|B(x;r)|} = 0 \qquad x \notin E,$$

which was to be proved. ∎

The Hardy-Littlewood maximal function can also be used to investigate certain questions of *nontangential convergence*. Normally this is considered in the framework of convergence in the unit disk or a higher-dimensional space. But the basic ideas are already present in the above framework, in the context of convergence of the type

$$\lim_{m \to \infty} f_{1/m}(x_m) = f(x) \qquad \text{when } x_m \to x.$$

We do not expect that this will be true unrestrictedly, since $x \to f_{1/m}(x)$ is a continuous function whereas the limit f is not continuous in general—hence we cannot expect uniform convergence. Nevertheless we have the following proposition.

Proposition 3.5.5. *Suppose that $x_m \to x$ so that $|x_m - x| < 1/m$. Then for any $f \in L^1(\mathbb{R}^n)$, $\lim_m f_{1/m}(x_m) = f(x)$ at almost every x.*

Proof. Repeating the above steps, we write $f = g + h$ where g is continuous and $\|h\|_1 < \epsilon$. Define

$$\Omega f(x) := \limsup_m f_{1/m}(x_m) - \liminf_m f_{1/m}(x_m) \le 2 \sup_m h_{1/m}(x_m).$$

But

$$h_{1/m}(x_m) = \frac{\int_{|y-x_m|<1/m} h}{c_n m^{-n}}.$$

The hypothesis $|x - x_m| < 1/m$ ensures that $x \in \{y : |y - x_m| < 1/m\}$, hence

$$h_{1/m}(x_m) \le Mh(x), \qquad \Omega f(x) \le 2Mh(x)$$

from which the proof can be completed as above:

$$|\{x : \Omega f(x) > \delta\}| \le |\{x : Mh(x) \ge \delta/2\}| \le \frac{2\delta}{\epsilon}.$$

But the left side does not depend on ϵ, hence we conclude that for every $\delta > 0$,

$$|\{x : \Omega f(x) > \delta\}| = 0. \qquad \blacksquare$$

3.5.2 Application to Radial Convolution Operators

The Hardy-Littlewood maximal function can be used to estimate more general convolution operators of the form

(3.5.8) $$(k_t * f)(x) = \int_{\mathbb{R}^n} k_t(y) f(x-y) \, dy,$$

where the *radial kernel* $k_t \in L^1(\mathbb{R}^n)$ is obtained from a monotone decreasing nonnegative function $K : \mathbb{R}^+ \to \mathbb{R}^+$ by setting $k_t(y) = t^{-n} K(y/t)$. The example of the Hardy-Littlewood operator is obtained when we set $K(x) = 1_{[0,1]}(|x|)$, the indicator function of the closed unit ball.

The following lemma shows that the Hardy-Littlewood maximal function can be used as a universal bound for any radial convolution. The original form is attributed to K.T. Smith (1956), and appears in Stein (1970). The proof below is attributed to S. Saeki, reproduced by Banuelos and Moore (1999).

Lemma 3.5.6. *For any radial kernel, we have the estimate*
$$|k_t f(x)| \leq (Mf)(x)\|k_1\|_1, \qquad f \in L^1(\mathbb{R}^n).$$

Proof. Let μ_x be the measure defined by $\mu_x(A) = \int_A |f(x-y)|\,dy$, for any Borel set A. In particular if A is a ball, we have $\mu_x(A) \leq |A|(Mf)(x)$. Applying this to the ball defined by $B_\lambda = \{y \in \mathbb{R}^n : k_t(y) > \lambda\}$, we have

$$\begin{aligned}|(k_t * f)(x)| &\leq \int_{\mathbb{R}^n} k_t(y)|f(x-y)|\,dy \\ &= \int_{\mathbb{R}^n} k_t(y)\mu_x(dy) \\ &= \int_0^\infty \mu_x(B_\lambda)\,d\lambda \\ &\leq (Mf)(x)\int_0^\infty |B_\lambda|\,d\lambda \\ &= (Mf)(x)\|k_1\|_1,\end{aligned}$$

where we have used the definition of Mf and twice used the representation of the L^1-norm as the integral of the distribution function. ∎

A typical application is to the n-dimensional heat kernel, where $K(x) = e^{-|x|^2/4}/(4\pi)^{n/2}$. Lemma 3.5.6 can then be combined with the Hardy-Littlewood maximal inequality to give a new proof of Proposition 2.2.35, that for any $f \in L^1(\mathbb{R}^n)$, and for almost every $x \in \mathbb{R}^n$, $(k_t * f)(x) \to f(x)$ when $t \to 0$.

Exercise 3.5.7. *Complete the details of this argument.*

Hint: Define $\Omega f(x) = \limsup_{t \to 0}(k_t * f)(x) - \liminf_{t \to 0}(k_t * f)(x)$ and argue as in the proof of the Lebesgue differentiation theorem to prove that $\lim_{t \to 0}(k_t * f)(x)$ exists a.e. Then identify the limit by a density argument.

As a second application of Lemma 3.5.6, consider the case of the n-dimensional Poisson kernel—$K(x) = C_n/(1+|x|^2)^{(n+1)/2}$, from (2.2.25). Following the steps of the proof of the Lebesgue differentiation theorem, we can prove that for any $f \in L^1(\mathbb{R}^n)$, $\lim_{t \to 0}(k_t * f)(x) = f(x)$ for almost every $x \in \mathbb{R}^n$.

Exercise 3.5.8. *Complete the details of this argument.*

3.5.3 Maximal Inequalities for Spherical Averages

Stein (1976) has shown that there exist L^p maximal inequalities for the spherical maximal function

(3.5.9) $$A^*f(x) = \sup_{t>0}\left|\int_{|y|=1} f(x-ty)\,d\omega(y)\right|$$

where $d\omega$ is the normalized surface measure on the sphere $\mathbf{S}^{n-1} \subset \mathbb{R}^n$. Results of this type can be used to prove Fatou-type theorems for solutions of the wave equation $u_{tt} = \Delta u$ for suitable values of n, p. In the treatment below we restrict attention to the simplest case, where $p = 2$, $n \geq 4$. A complete account of the subject for more general values of n, p can be found in Stein (1993), Chapter XI.

We begin with the spherical averaging opeator

$$f \to (A_t f)(x) := \int_{|y|=1} f(x - ty)\, d\omega(y) = (f * d\omega_t)(x).$$

We will also make use of the *square function*, defined by

(3.5.10) $$Sf(x) := \left(\int_0^\infty \left| \frac{\partial}{\partial t}(A_t f)(x) \right|^2 t\, dt \right)^{1/2},$$

which is well-defined whenever $f \in C^1$ has compact support.

Lemma 3.5.9. *For any $f \in C^1(\mathbb{R}^n)$ with compact support, we have*

$$|A_t f(x)| \leq (Mf)(x) + \frac{Sf(x)}{\sqrt{2n}}$$

where Mf is the Hardy-Littlewood maximal function.

Proof. We write

$$A_t f(x) = t^{-n}(t^n A_t f(x))$$
$$= t^{-n} \int_0^t \frac{d}{ds}(s^n A_s f(x))\, ds$$
$$= I_1 + I_2$$

where

$$I_1 = t^{-n} \int_0^t n s^{n-1} (A_s f)(x)\, ds$$
$$I_2 = s^n \frac{d}{ds}(A_s f)(x)\, ds.$$

I_1 is majorized by the Hardy-Littlewood maximal function, since for any $t > 0$,

$$I_1 = \frac{C_n}{t^n} \int_{|y-x| \leq t} f(y)\, dy \leq (Mf)(x).$$

Meanwhile, I_2 is estimated in terms of Sf by using Cauchy-Schwarz:

$$|I_2| = t^n \int_0^t s^{n-(1/2)} \sqrt{s}\, \frac{d}{ds}(A_s f)(x)\, ds$$
$$\leq t^n \left(\int_0^t s^{2n-1}\, ds \right)^{1/2} \left(s \left[\frac{d}{ds}(A_s f)(x) \right]^2 ds \right)^{1/2}$$
$$\leq \frac{1}{\sqrt{2n}}(Sf)(x),$$

which completes the proof. ∎

Lemma 3.5.10. *If $n \geq 4$ and f is C^1 with compact support, then*

$$\|Sf\|_2 \leq C_n \|f\|_2$$

where the L^2 norm is taken over all of $\mathbb{R}^n$ and the constant C_n depends only upon the dimension.

Proof. The Fourier representation of the spherical average is

$$A_t f(x) = \int_{\mathbb{R}^n} e^{2\pi i \xi \cdot x} \hat{f}(\xi) \hat{\omega}(t\xi) \, d\xi$$

where $\hat{\omega}(\xi) = c_n J_{(n-2)/2}(|\xi|)/|\xi|^{(n-2)/2}$ is the Fourier transform of the normalized measure ω. Now by the chain rule

$$\frac{d}{dt} A_t f(x) = \int_{\mathbb{R}^n} e^{2\pi i \xi \cdot x} \hat{f}(\xi) \left(\sum_{j=1}^n \xi_j \frac{\partial \hat{\omega}}{\partial \xi_j}(t\xi) \, d\xi \right)$$

(3.5.11)
$$= \int_{\mathbb{R}^n} e^{2\pi i \xi \cdot x} \hat{f}(\xi) \frac{\mu(t\xi)}{t} \, d\xi$$

where we have set

$$\mu(\xi) := \sum_{j=1}^n \xi_j \frac{\partial \hat{\omega}}{\partial \xi_j}.$$

Now from the asymptotic behavior of the Bessel function, we have

$$\hat{\omega}(\xi) = O(|\xi|^{(1-n)/2}), \quad \frac{\partial \hat{\omega}}{\partial \xi_j} = O(|\xi|^{(3-n)/2}), \quad |\xi| \to \infty.$$

On the other hand, when $|\xi| \to 0$, we have $\partial \hat{\omega}/\partial \xi_j = O(|\xi|)$. Combining these estimates, we can write

(3.5.12) $$|\mu(\xi)| \leq C_n \min\{|\xi|, |\xi|^{(3-n)/2}\}.$$

Applying Plancherel's theorem to (3.5.11), we have

$$\int_{\mathbb{R}^n} \left| \frac{d}{ds} (A_s \hat{f})(x) \right|^2 dx = \int_{\mathbb{R}^n} |\hat{f}(\xi)|^2 \frac{|\mu(s\xi)|^2}{s^2} \, d\xi.$$

Integrating both sides with respect to the measure $s \, ds$ and applying Fubini, we have

(3.5.13) $$\int_{\mathbb{R}^n} |Sf(x)|^2 \, dx = \int_{\mathbb{R}^n} |\hat{f}(\xi)|^2 \left(\int_0^\infty \frac{|\mu(s\xi)|^2}{s} \, ds \right).$$

But the inner integral can be estimated using (3.5.12). For any M, we have

$$\int_0^\infty |\mu(s\xi)|^2 \frac{ds}{s} = \left(\int_0^M + \int_M^\infty \right) |\mu(s\xi)|^2 \frac{ds}{s}$$

$$\leq C_n |\xi|^2 \int_0^M s \, ds + C_n |\xi|^{3-n} \int_M^\infty \frac{ds}{s^{n-2}}$$

$$= C_n |\xi|^2 M^2 + C_n \frac{|\xi|^{3-n}}{(n-2)M^{n-3}}.$$

The two terms are comparable to one another by taking $M = 1/|\xi|$, leading to the estimate

$$\int_0^\infty |\mu(s\xi)|^2 \frac{ds}{s} \leq C_n \left(1 + \frac{1}{n-2}\right) = C_n \frac{n-1}{n-2}.$$

Referring to (3.5.13) and applying Plancherel's theorem once again, the proof is complete. ∎

Combining Lemmas 3.5.9 and 3.5.10, with the L^2 bound for the Hardy-Littlewood maximal function, we have proved the following result.

Theorem 3.5.11. *Suppose that $n \geq 4$. Then we have the following estimate for any $f \in C^1(\mathbb{R}^n)$ with compact support:*

(3.5.14) $$\|\sup_{t>0}(A_t f)(x)\|_2 \leq C\|f\|_2$$

where the constant depends only on the dimension.

The estimate (3.5.14) can be extended to any $f \in L^2(\mathbb{R}^n)$ by noting that the set of C^1 functions with compact support is dense in L^2.

Example 3.5.12. *Taking $n = 1$ and f an unbounded function shows that $\sup_t A_t f(x) = +\infty$ for every $x \in \mathbb{R}^1$, hence the estimate (3.5.14) cannot hold for general $f \in L^2(\mathbb{R})$.*

Example 3.5.13. *To obtain an example in higher dimensions, take $f(x) = |x|^{1-n}[\log(1/|x|)]^{-1}\mathbf{1}_{[0,1]}(|x|)$, which fails to be in $L^2(\mathbb{R}^n)$ and for which it is verified that $\sup_t A_t f(x) = +\infty$ for every $x \in \mathbb{R}^n$.*

3.6 THE MARCINKIEWICZ INTERPOLATION THEOREM

In this section we develop the notion of *weakly bounded operators* and apply it to prove the theorem of Marcinkiewicz. To orient the thinking, we first define the notion of *weak Lebesgue space*.

Definition 3.6.1.

$$wkL^p(\mathbb{R}^n) := \{f : |\{x : |f(x)| > \alpha\}| \leq C\alpha^{-p}\}$$

for some $C > 0$, where $1 \leq p < \infty$. In case $p = \infty$, we set $wkL^\infty(\mathbb{R}^n) = L^\infty(\mathbb{R}^n)$.

This definition can be rewritten in terms of the distribution function λ_f in the form

$$\lambda_f(\alpha) \leq C\alpha^{-p}.$$

Clearly $L^p(\mathbb{R}^n) \subset wkL^p(\mathbb{R}^n)$ since if $f \in L^p(\mathbb{R}^n)$, then by Chebyshev's inequality we have for any $\alpha > 0$

$$\|f\|_p^p = \int_{\mathbb{R}^n} |f(x)|^p \, dx \geq \int_{|f|>\alpha} |f(x)|^\alpha \, dx \geq \alpha^p |\{x : |f(x)| \geq \alpha\}|.$$

But the converse is not true. For example $f(x) = 1/(1+|x|)$ is $wkL^1(\mathbb{R}^1)$ but is not integrable.

In parallel with the discussion preceding the M. Riesz-Thorin theorem at the beginning of this chapter, we have the following elementary properties.

Lemma 3.6.2. *Suppose that $f \in wkL^{p_0}(\mathbb{R}^n)$ and that $|f| \le M$ for some M. Then $f \in L^{p_1}(\mathbb{R}^n)$ for any $p_1 > p_0$.*

Proof. In terms of the distribution functions, we can write

$$\int_{\mathbb{R}^n} |f(x)|^{p_1}\, dx = p_1 \int_0^M \alpha^{p_1-1}\lambda_f(\alpha)\, d\alpha \le Cp_1 \int_0^M \alpha^{p_1-1-p_0}\lambda_f(\alpha)\, d\alpha < \infty.$$

∎

Lemma 3.6.3. *Suppose that $f \in wkL^{p_1}(\mathbb{R}^n)$ lives on a set of finite measure B. Then $f \in L^{p_0}(\mathbb{R}^n)$ for any $p_0 < p_1$.*

Proof. If $p_1 < \infty$, we write

$$\int_{\mathbb{R}^n} |f(x)|^{p_0}\, dx = p_0 \int_0^\infty \alpha^{p_0-1}\lambda_f(\alpha)\, d\alpha$$

$$\le \lambda_f(1) + Cp_0 \int_1^\infty \alpha^{p_0-p_1-1}\, d\alpha < \infty.$$

In $p_1 = \infty$, then the result is immediate, since on a space of finite measure, any bounded function is in L^{p_0} for any p_0. ∎

Lemma 3.6.4. *Suppose that $f \in wkL^{p_0}(\mathbb{R}^n)$ and $f \in wkL^{p_1}(\mathbb{R}^n)$ where $p_0 < p < p_1$. Then $f \in L^p(\mathbb{R}^n)$.*

Proof. We write $f = f 1_{\{|f| \le 1\}} + f 1_{\{|f| > 1\}} = f_1 + f_2$. Both f_1 and f_2 are dominated by f. In particular $f_1 \in wkL^{p_0}$ and $f_2 \in wkL^{p_1}$. But f_1 is bounded and f_2 lives on a set of finite measure, since

$$|\{x : f_2(x) \ne 0\}| = |\{x : |f(x)| > 1\}| \le C < \infty.$$

Therefore by the preceding lemmas, $f_1 \in L^p(\mathbb{R}^n)$ and $f_2 \in L^p(\mathbb{R}^n)$. But $L_p(\mathbb{R}^n)$ is a linear space, hence $f \in L^p$. ∎

These lemmas will be applied in many ways. As an elementary instance, we note that the Hardy-Littlewood maximal function is in the space $wkL^1(\mathbb{R}^n)$ and also in the space $L^\infty(\mathbb{R}^n)$. Therefore we may conclude that $Mf \in L^p(\mathbb{R}^n)$ for any $1 < p < \infty$.

In order to include the Hardy-Littlewood maximal function and related operators, we first formulate the general notion of *sublinear operator* as follows.

Definition 3.6.5. *An operator T is said to be sublinear if, whenever Tf_1 and Tf_2 are defined, and c is any complex number, then*

$$|T(f_1+f_2)| \le |Tf_1| + |Tf_2|, \qquad |T(cf_1)| \le c|f_1|.$$

We also need the notion of *type (p, q)*.

Definition 3.6.6. *Let (M, μ) and (N, ν) be measure spaces with a sublinear operator $T: L^p(M, \mu) \to L^q(N, \nu)$. We say that T is of type (p, q) if there exists a positive constant C so that for every $f \in L^p(M, \mu)$,*

(3.6.1) $$\|Tf\|_q \leq C\|f\|_p.$$

As a consequence, it follows from Chebyshev's inequality that for each $\alpha > 0$,

(3.6.2) $$\alpha^q \mu\{x : |Tf(x)| > \alpha\} \leq C^p \|f\|_p^p.$$

The converse is not true. This leads us to the notion of *weak type (p, q)*.

Definition 3.6.7. *An operator for which (3.6.2) holds is said to be of weak type (p, q).*

For example, the Hardy-Littlewood maximal operator is of weak type $(1, 1)$ but does not satisfy (3.6.1) with $(p, q) = (1, 1)$. In case $q = \infty$, condition (3.6.2) implies that $Tf \in L^\infty$ so that the definition can be taken in the strong sense.

Theorem 3.6.8. *Let T be a sublinear operator that is defined on the space $L^{p_0} + L^{p_1}$ so that it is weakly bounded of type (p_0, p_0) and weakly bounded of type (p_1, p_1), where $1 \leq p_0 \leq p_1 \leq \infty$. Then for any $p_0 < p < p_1$, T is defined on L^p and T is of type (p, p).*

Proof. Any $f \in L^p$ can be written as $f = f 1_{|f| \leq 1} + f 1_{|f| > 1}$, where the first term is in L^{p_1} and the second term is in L^{p_0}. Hence $L^p \subset L^{p_0} + L^{p_1}$. We first prove the theorem in case $p_1 = +\infty$. Dividing T by a constant, we may assume that $\|Tf\|_\infty \leq \|f\|_\infty$. Now we make the α-dependent decomposition

$$f = f_0 + f_1$$

where

$$f_1 = f 1_{\{|f| \leq \alpha/2\}}, \qquad f_0 = f 1_{\{|f| > \alpha/2\}}.$$

Now $|Tf| \leq |Tf_0| + |Tf_1|$, so that the distribution functions satisfy

$$\lambda_{Tf}(\alpha) \leq \lambda_{Tf_0}(\alpha/2) + \lambda_{Tf_1}(\alpha/2).$$

But $|f_1| \leq \alpha/2$ and $\|T\|_{\infty, \infty} \leq 1$ so that the second term is zero. Now from the weak (p_0, p_0) hypothesis,

$$\lambda_{Tf_0}(\alpha/2) \leq C \left(\frac{2}{\alpha}\right)^{p_0} \|f_0\|_{p_0}^{p_0}$$

and

$$\|f_0\|_{p_0}^{p_0} = \int_M |f(t)|^{p_0} 1_{\{|f| > \alpha/2\}} \mu(dt).$$

To compute the norm of Tf, we write

$$\|Tf\|_p^p = p \int_0^\infty \lambda_{Tf}(\alpha)\alpha^{p-1}\, d\alpha$$

$$\leq p \int_0^\infty \lambda_{Tf_0}(\alpha/2)\alpha^{p-1}\, d\alpha$$

$$\leq p \int_0^\infty \alpha^{p-1-p_0}\left(\int_M |f(t)|^{p_0} 1_{\{|f|>\alpha/2\}}\mu(dt)\right) d\alpha$$

$$= p \int_M \left(\int_0^{2|f(t)|} \alpha^{p-p_0-1}\, d\alpha\right) |f(t)|^{p_0}\, \mu(dt)$$

$$= \frac{p2^{p-p_0}}{p-p_0} \int_M |f(t)|^{p-p_0}|f(t)|^{p_0}\, \mu(dt)$$

$$= \frac{p2^{p-p_0}}{p-p_0} \|f\|_p^p.$$

This completes the proof that $\|Tf\|_p \leq C(p, p_0)\|f\|_p$ in case $p_1 = \infty$. To discuss the case $p_1 < \infty$, we make the same decomposition of $f = f_0 + f_1$:

$$\|Tf\|_p^p = p \int_0^\infty \lambda_{Tf}(\alpha)\alpha^{p-1}\, d\alpha$$

$$\leq p \int_0^\infty \lambda_{Tf_0}(\alpha/2)\alpha^{p-1}\, d\alpha + \lambda_{Tf_1}(\alpha/2)\alpha^{p-1}\, d\alpha$$

The term involving f_0 can be estimated in precisely the same fashion as above. For the term involving f_1, we use the weak (p_1, p_1) hypothesis to write

$$\int_0^\infty \lambda_{Tf_1}(\alpha/2) p\alpha^{p-1}\, d\alpha \leq C \int_0^\infty (2/\alpha)^{p_1} \|f_1\|_{p_1}^{p_1} p\alpha^{p-1}\, d\alpha$$

and hence

$$\int_0^\infty p\alpha^{p-1}\alpha^{-p_1}\|f_1\|_{p_1}^{p_1}\, d\alpha \leq \int_0^\infty p\alpha^{p-1-p_1}\left(\int_M 1_{\{|f(t)|\leq \alpha/2\}} |f(t)|^{p_1}\, \mu(dt)\right) d\alpha$$

$$= \int_M \left(\int_{2|f(t)|}^\infty p\alpha^{p-1-p_1}\, d\alpha\right) |f(t)|^{p_1}\, \mu(dt)$$

$$= \frac{2^{p-p_1}}{p_1 - p} \|f\|_p^p. \qquad\blacksquare$$

3.7 CALDERÓN-ZYGMUND DECOMPOSITION

In previous sections we had to decompose a function into a bounded part and an unbounded part. This suffices to prove the Marcinkiewicz interpolation theorem, for example. In other problems one needs a more sophisticated decomposition of an integrable function, due to Alberto Calderón and Antoni Zygmund in their seminal work on singular integral operators.

Proposition 3.7.1. Let $f \in L^1(\mathbb{R}^n)$ be nonnegative and let $\alpha > 0$. Then there exist disjoint subsets F, Ω so that

(i) $|f(x)| \leq \alpha$ for $x \in F$.
(ii) $\Omega = \cup_k Q_k$, where Q_k are cubes with disjoint interiors so that
$$\alpha \leq \frac{1}{|Q_k|} \int_{Q_k} f(x)\,dx \leq 2^n \alpha.$$
(iii) $\mathbb{R}^n = \Omega \cup F$, $|\Omega| \leq \|f\|_1/\alpha$.

Proof. We begin by decomposing $\mathbb{R}^n$ into cubes Q whose sides are parallel to the coordinate axes with $|Q| \geq \|f\|_1/\alpha$. Clearly this is possible since $\|f\|_1 < \infty$. In particular, we have

(3.7.1) $$\frac{1}{|Q|}\int_Q f \leq \frac{1}{|Q|}\int_{\mathbb{R}^n} f \leq \alpha.$$

For each of these cubes, subdivide into 2^n congruent cubes Q'. For each of these cubes, it's clear that

(3.7.2) $$\frac{1}{|Q'|}\int_{Q'} f \leq 2^n \alpha$$

for otherwise we would violate (3.7.1). These cubes are of two types:

$$\text{Type}(i): \frac{1}{|Q'|}\int_{Q'} f \leq \alpha,$$

$$\text{Type}(ii): \frac{1}{|Q'|}\int_{Q'} f > \alpha.$$

If Q' is of type (ii), then one does not subdivide further and it is added to the list of cubes Q_k. If Q' is of type (i), then we subdivide it into 2^n congruent cubes whose sides are parallel to the axes and repeat the decision process. As before, the discarded cubes satisfy (3.7.2). Continuing inductively defines the cubes Q_k; we let $\Omega := \cup_k Q_k$, $F := \mathbb{R}^n \setminus \Omega$. If $x \in F$, then every cube containing x is of type (i). By the corollary to the Lebesgue differentiation theorem we have for almost every x,

$$f(x) = \lim_Q \frac{1}{|Q|}\int f \leq \alpha, \quad x \in F.$$

Finally we note that since $\alpha|Q_k| \leq \int_{Q_k} f$ for each cube Q_k, we can sum these over k to obtain the estimate that $\alpha|\Omega| \leq \int_\Omega f \leq \|f\|_1$, which was to be proved. ∎

This can be restated as a decomposition of the function

$$f = g + b$$

where the *good part* $g(x) = f(x)$ for $x \in F$ and $g(x) = \int_{Q_j} f/|Q_j|$ for $x \in \text{int}(Q_j)$. Hence the *bad part* b must satisfy $b(x) = 0$ for $x \in F$ and $\int_{Q_j} b = 0$ for each cube Q_j.

3.8 A CLASS OF SINGULAR INTEGRALS

The Calderón-Zygmund decomposition can be used to treat a class of singular convolution operators that generalize the *odd kernels* which were discussed in Section 3.4.4.

We will only state the results, referring to Stein (1970) for a complete treatment of the details.

We begin with a real function $K(x), x \in \mathbb{R}^n \setminus \{0\}$, which satisfies the following growth, smoothness, and cancellation properties:

$$|K(x)| \leq \frac{C}{|x|^n} \qquad 0 \neq x \in \mathbb{R}^n, \tag{3.8.1}$$

$$|\nabla K(x)| \leq \frac{C}{|x|^{n+1}} \qquad 0 \neq x \in \mathbb{R}^n, \tag{3.8.2}$$

$$\int_{R_1 < |x| < R_2} K(x)\, dx = 0 \qquad 0 < R_1 < R_2 < \infty. \tag{3.8.3}$$

A truncated convolution operator is defined by setting

$$T_\epsilon f(x) = \int_{|y| > \epsilon} K(y) f(x-y)\, dy. \tag{3.8.4}$$

If $f \in L^p(\mathbb{R}^n)$ for $1 < p < \infty$, then this integral is well-defined by Hölder's inequality, in light of (3.8.1). If $f \in L^1(\mathbb{R}^n)$, then $T_\epsilon f$ is defined almost everywhere as the convolution with a function of class $L^2(\mathbb{R}^n)$. However the naive bounds in these estimates depend on ϵ, when $\epsilon \to 0$. It is remarkable that, under hypotheses (3.8.1) through (3.8.3) the operators T_ϵ have uniformly bounded operator norms and are convergent in L^p for $1 < p < \infty$.

Theorem 3.8.1. *Suppose that the kernel $K(x)$ satisfies (3.8.1) through (3.8.3). Then for each $p \in (1, \infty)$, there exists $A_p < \infty$ such that for each $f \in L^p(\mathbb{R}^n)$,*

$$\|T_\epsilon f\|_p \leq A_p \|f\|_p. \tag{3.8.5}$$

In addition, there exists the L^p limit $Tf = \lim_{\epsilon \to 0} T_\epsilon f$ and the operator T also satisfies the inequality (3.8.5).

The conditions (3.8.1) through (3.8.3) are by no means necessary. In fact, (3.8.2) can be replaced by the weaker *Hörmander condition*:

$$\int_{|x| \geq 2|y|} |K(x-y) - K(x)|\, dx \leq C, \qquad 0 \neq y \in \mathbb{R}^n. \tag{3.8.6}$$

Exercise 3.8.2. *Prove that (3.8.2) implies (3.8.6).*

Hint: Apply the mean-value theorem to the integrand in (3.8.6), noting that the segment from x to $x - y$ is outside of a sphere of radius $|x|/2$.

We now discuss some of the broad outlines of the proof of Theorem 3.8.1. The basic strategy is to apply the Marcinkiewicz interpolation theorem, first proving the boundedness on $L^2(\mathbb{R}^n)$ and the weak boundedness on $L^1(\mathbb{R}^n)$. The first of these is proved by showing that the truncated Fourier transforms

$$\hat{K}_\epsilon(\xi) := \int_{|x| \geq \epsilon} K(x) e^{-2\pi i \xi \cdot x}\, dx$$

remain uniformly bounded: $|\hat{K}_\epsilon(\xi)| \leq M, \forall \epsilon > 0, \xi \in \mathbb{R}^n$. This is proved in Lemma 3.3 of Stein (1970), pp. 36–37. The weak L^1 boundedness is proved by applying the Calderón-Zygmund decomposition at level α: $f = g_\alpha + b_\alpha$ and doing each term separately. We have

$$|\{x : |Tf(x)| \geq \alpha\}| \leq |\{x : |Tg_\alpha(x)| \geq \alpha/2\}| + |\{x : |Tb_\alpha(x)| \geq \alpha/2\}|.$$

We outline the treatment of the first term. From the properties of the Calderón-Zygmund decomposition, we can write

$$\begin{aligned}\|g_\alpha\|_2^2 &= \int_{\mathbb{R}^n} |g_\alpha(x)|^2 \, dx \\ &= \int_F |g_\alpha(x)|^2 \, dx + \int_\Omega |g_\alpha(x)|^2 \, dx \\ &\leq \alpha \int_F |f(x)| \, dx + 2^{2n} \alpha^2 |Q| \\ &\leq \alpha \|f\|_1 + 2^{2n} \alpha \|f\|_1 \\ &= \alpha(1 + 2^{2n}) \|f\|_1. \end{aligned}$$

This is combined with Chebyshev's inequality and the L^2-boundedness to write

$$\begin{aligned} |\{x : |Tg_\alpha(x)| \geq \alpha/2\}| &\leq \frac{1}{(\alpha/2)^2} \|Tg_\alpha\|^2 \\ &\leq \frac{M^2}{(\alpha/2)^2} \|g_\alpha\|_2^2 \\ &\leq \frac{M^2}{(\alpha/2)^2} (1 + 2^{2n}) \|f\|_1, \end{aligned}$$

which proves the weak L^1 bound for g_α. A more lengthy argument using the *Marcinkiewicz integrals* shows that the corresponding estimate holds for Tb_ϵ and allows one to complete the proof of boundedness for $1 < p < 2$. Then a duality argument is applied to prove the boundedness in case $2 < p < \infty$. For details see Stein (1970), pp. 30–33.

3.9 PROPERTIES OF HARMONIC FUNCTIONS

For completeness we state and prove the basic properties of harmonic functions in the disk.

3.9.1 General Properties

A twice differentiable function $u(x, y)$ is called *harmonic* in an open set D if it satisfies Laplace's equation $u_{xx} + u_{yy} = 0$ for every $(x, y) \in D$. The prototype examples of harmonic functions are the real and imaginary parts of $f(x, y) = (x + iy)^n$. It is clear that f satisfies Laplace's equation, hence each of the real and imaginary parts do also. These are abbreviated in the polar form $u_n = r^n \cos n\theta$ and $v_n = r^n \sin n\theta$. Taking the derivatives of f and forming the real and imaginary parts shows that $|D_x u_n| \le nr^{n-1}$, $|D_x^2 u_n| \le n(n-1)r^{n-2}$ and similarly for v_n. Hence we can form more general harmonic functions by the series $u(x,y) = \sum_{n \ge 0} a_n u_n(x,y)$. If $|a_n| \le M$ and $R < 1$, then this series converges uniformly in the closed disk $x^2 + y^2 \le R^2$ together with the differentiated series, thus defining a harmonic function in the open disk $x^2 + y^2 < 1$.

In particular, a useful class of harmonic functions are provided by the Poisson integrals of integrable functions (or more generally of finite measures):

$$(3.9.1) \quad u(x,y) = \frac{1}{2\pi} \int_{\mathbb{T}} \frac{R^2 - r^2}{R^2 + r^2 - 2rR\cos(\theta - \phi)} f(\phi)\, d\phi = \sum_{n \in \mathbb{Z}} \left(\frac{r}{R}\right)^{|n|} \hat{f}(n) e^{in\theta}.$$

Proposition 3.9.1. *Suppose that $f \in L^1(\mathbb{T})$ and $u(x,y)$ is defined by (3.9.1). Then u is a harmonic function in the disk $x^2 + y^2 < R^2$ and $\lim_{r \to R} u(re^{i\theta}) = f(\theta)$ for almost every $\theta \in \mathbb{T}$. If $f \in C(\mathbb{T})$, the convergence is uniform in $\mathbb{T}$. If $f \in L^p(\mathbb{T})$, $1 \le p < \infty$, then the convergence is in the norm of $L^p(\mathbb{T})$.*

Proof. We simply note that $|\hat{f}(k)| \le \|f\|_1$ so that we can infer that the sum of the series in (3.9.1) is a harmonic function. The convergence to the boundary values was already proved in Chapter 1 as part of the Abel summability of Fourier series in the spaces $L^1(\mathbb{T})$, $C(\mathbb{T})$, and $L^p(\mathbb{T})$. ∎

Exercise 3.9.2. *Suppose that $f \in L^\infty(\mathbb{T})$ and that u is defined by (3.9.1). Prove that $|u(x,y)| \le \|f\|_\infty$ in the disk $x^2 + y^2 < R^2$.*

Exercise 3.9.3. *Suppose that $m(d\theta)$ is a finite Borel measure on $\mathbb{T}$. Defining*

$$u(x,y) := \frac{1}{2\pi} \int_{\mathbb{T}} \frac{R^2 - r^2}{R^2 + r^2 - 2rR\cos(\theta - \phi)} m(d\phi),$$

show that u is a harmonic function in the disk $x^2 + y^2 < R^2$. Show by example that u is not necessarily a bounded function.

Hint: First show that $u(x,y) = \sum_{n \in \mathbb{Z}} (r/R)^{|n|} m_n e^{in\theta}$ for a bounded sequence m_n. If m is a point measure, then u is unbounded near that point.

We now turn the picture around, assuming only that u is a given harmonic function in the disk. The first problem is to prove the uniqueness of harmonic functions with given boundary values.

Proposition 3.9.4. *Suppose that D is a bounded and connected region with a piecewise smooth boundary. Suppose that v, w are twice differentiable in D and the first derivatives have continuous extensions to $\bar{D}$. If v, w are harmonic in D and satisfy $v = w$ on the boundary. Then $v = w$ in D.*

Proof. Let $u = v - w$, so that u is harmonic in D with $u = 0$ on the boundary. We will apply Green's theorem to the vector identity

$$\operatorname{div}(u \operatorname{grad} u) = u(u_{xx} + u_{yy}) + (u_x^2 + u_y^2).$$

The first term on the right is zero since u is harmonic. Applying Green's theorem transforms the double integral of the divergence into a line integral on the boundary. Thus

$$\int_{\partial D} u \operatorname{grad} u \, dS = \int_D (u_x^2 + u_y^2) \, dx \, dy.$$

But the left side is also zero, since u is zero on the boundary. Hence we conclude that the continuous function $u_x^2 + u_y^2 = 0$ in D, meaning that u is constant on D. But the boundary value is zero, hence $u \equiv 0$ in D. ∎

Corollary 3.9.5. *Any harmonic function in the disk $x^2 + y^2 < 1$ can be represented on the closed disk $x^2 + y^2 \leq R^2 < 1$ by a Poisson integral, in the form*

(3.9.2)
$$u(x, y) = \frac{1}{2\pi} \int_{\mathbb{T}} \frac{R^2 - r^2}{R^2 + r^2 - 2rR \cos(\theta - \phi)} u(Re^{i\phi}) \, d\phi, \qquad x^2 + y^2 \leq R^2 < 1.$$

Proof. From the above discussion, the right side of (3.9.2) is a harmonic function in the disk $x^2 + y^2 < R^2$ and has the same boundary values as u. Hence by Proposition 3.9.4 the equality (3.9.2) follows. ∎

Corollary 3.9.6. *Any harmonic function in the disk $x^2 + y^2 < 1$ has the mean value property:*

(3.9.3) $$u(0) = \frac{1}{2\pi} \int_{\mathbb{T}} u(Re^{i\theta}) \, d\theta, \qquad R < 1.$$

Proof. It suffices to take $(x, y) = (0, 0)$ in (3.9.2). ∎

3.9.2 Representation Theorems in the Disk

The Poisson integral can be characterized in each of the spaces $L^\infty(\mathbb{T})$, $L^p(\mathbb{T})$, ($1 < p < \infty$) and the space of nonnegative measures. We enumerate these results separately, beginning with the classical Fatou theorem for bounded harmonic functions.

Theorem 3.9.7. *Fatou: Suppose that u is a harmonic function that is bounded in the unit disk $x^2 + y^2 < 1$: $|u(x, y)| \leq M$ for some M. Then there exists $u_1 \in L^\infty(\mathbb{T})$ so that*

(3.9.4) $$u(x, y) = \frac{1}{2\pi} \int_{\mathbb{T}} \frac{1 - r^2}{1 + r^2 - 2r \cos(\theta - \phi)} u_1(\phi) \, d\phi, \qquad x^2 + y^2 < 1.$$

In particular we have $\lim_{r \to 1} u(re^{i\theta}) = u_1(\theta)$ for almost every $\theta \in \mathbb{T}$.

Proof. We will make a compactness argument based on duality of $L^1(\mathbb{T})$ and $L^\infty(\mathbb{T})$. Consider the linear functionals on $L^1(\mathbb{T})$ defined by

(3.9.5) $$L_R f = \frac{1}{2\pi} \int_{\mathbb{T}} f(\theta) u(Re^{i\theta}) \, d\theta, \qquad 0 \leq R < 1.$$

Clearly $|L_R f| \leq M \|f\|_1$, so that this family of linear functionals is a bounded subset of the ball $\|L\| \leq M$ in the dual space of $L^1(\mathbb{T})$, namely $L^\infty(\mathbb{T})$. But this ball is compact in the weak* topology, which can be seen directly by taking a countable dense subset of $L^1(\mathbb{T})$ and applying Cantor's diagonal argument. Thus we obtain a weak* convergent subsequence L_{R_j} with $R_j \to 1$ with a weak* limit, call it u_1 with the property that for every $f \in L^1(\mathbb{T})$, $\lim_j L_{R_j} f = \frac{1}{2\pi} \int_{\mathbb{T}} f(\theta) u_1(\theta)\, d\theta$. We apply this to the Poisson integral representation of u:

(3.9.6)
$$u(x,y) = \frac{1}{2\pi} \int_{\mathbb{T}} \frac{R_j^2 - r^2}{R_j^2 + r^2 - 2rR_j \cos(\theta - \phi)} u(R_j e^{i\phi})\, d\phi, \qquad x^2 + y^2 \leq R_j^2 < 1.$$

For any fixed (x, y) the Poisson kernels P_{R_j} converge uniformly on $\mathbb{T}$ when $R_j \to 1$ to the Poisson kernel P_1; this follows from the series representation

$$|P_1(r, \theta) - P_R(r, \theta)| = 2 \left| \sum_{n \geq 1} r^n \frac{1 - R^n}{R^n} \cos(n\theta) \right|$$

$$\leq 2(1 - R) \sum_{n \geq 1} r^n \frac{n}{R^n}$$

$$\leq 2(1 - R) \sum_{n \geq 1} n \left(\frac{2r}{1 + r} \right)^n$$

provided that $R > (1 + r)/2$. For any fixed $r < 1$ the last sum is finite, which proves the uniform convergence of the Poisson kernel. Hence we have established (3.9.4). The convergence of $u(re^{i\theta})$ follows from the almost-everywhere Abel summability of the Fourier series. ∎

The next result concerns representation of nonnegative harmonic functions.

Theorem 3.9.8. *Suppose that u is a nonnegative harmonic function in the disk $x^2 + y^2 < 1$. Then there exists a nonnegative Borel measure m on $\mathbb{T}$ so that*

(3.9.7)
$$u(x, y) = \frac{1}{2\pi} \int_{\mathbb{T}} \frac{1 - r^2}{1 + r^2 - 2r \cos(\theta - \phi)} m(d\phi), \qquad x^2 + y^2 < 1.$$

For any $f \in C(\mathbb{T})$, we have

(3.9.8)
$$\lim_{r \to 1} \int_{\mathbb{T}} u(re^{i\theta}) f(\theta)\, d\theta = \int_{\mathbb{T}} f(\theta) m(d\theta).$$

Proof. Again we consider the linear functional (3.9.5), now defined on the space $C(\mathbb{T})$ whose dual space is the set of finite Borel measures on $\mathbb{T}$. We have

$$|L_R f| \leq \|f\|_\infty \frac{1}{2\pi} \int_{\mathbb{T}} u(re^{i\theta})\, d\theta.$$

But from the Poisson integral representation of u in the disk $x^2 + y^2 \leq R^2$, we have $u(0) = 1/2\pi \int_{\mathbb{T}} u(re^{i\theta})\, d\theta$, hence $|L_R f| \leq u(0) \|f\|_\infty$. Applying the weak compactness argument once again yields a measure m on $\mathbb{T}$ and a weak* convergent subsequence L_{R_j} so that $L_{R_j} f \to \int_{\mathbb{T}} f(\theta) m(d\theta)$. As in the previous proof we have (3.9.6) on which we can take the limit $R_j \to 1$ to conclude (3.9.7). To prove the convergence, we multiply (3.9.7) by a continuous $f(\theta)$ and integrate on the circle of radius r. The left side is $L_r f = \int_{\mathbb{T}} u(re^{i\theta}) f(\theta)\, d\theta$. The right side can be written as $\int_{\mathbb{T}} P_r f(\theta) m(d\theta)$. But $P_r f$ converges uniformly to f, so that the right side converges to $\int_{\mathbb{T}} f(\theta) m(d\theta)$, from which it follows that $\lim_{r \to 1} L_r f$ exists and is given by (3.9.8). ∎

To complete the picture, we state and prove the theorem on L^p boundedness.

Theorem 3.9.9. *Suppose that u is a harmonic function in the disk $x^2 + y^2 < 1$ with the property that*

$$\sup_{0 \leq r < 1} \int_{\mathbb{T}} |u(re^{i\theta})|^p \, d\theta \leq M < \infty$$

where $1 < p < \infty$. Then there exists $u_1 \in L^p(\mathbb{T})$ so that

$$(3.9.9) \quad u(x, y) = \frac{1}{2\pi} \int_{\mathbb{T}} \frac{1 - r^2}{1 + r^2 - 2r \cos(\theta - \phi)} u_1(\phi) \, d\phi, \qquad x^2 + y^2 < 1.$$

In particular

$$\lim_{r \to 1} \int_{\mathbb{T}} |u(re^{i\theta}) - u_1(\theta)|^p \, d\theta = 0$$

and $\lim_{r \to 1} u(re^{i\theta}) = u_1(\theta)$ for a.e. $\theta \in \mathbb{T}$.

Proof. Again we consider the linear functional (3.9.5), now defined on the space $L^{p'}(\mathbb{T})$ which is the dual space of $L^p(\mathbb{T})$ where $p' = p/(p-1)$. We have for any $f \in L^{p'}(\mathbb{T})$,

$$|L_R f| \leq M \|f\|_{p'}$$

so that the linear functionals L_R have bounded norms. Applying the weak compactness argument once again yields an L^p function f on $\mathbb{T}$ and a weak* convergent subsequence L_{R_j} so that $L_{R_j} f \to \int_{\mathbb{T}} f(\theta) u_1(\theta) \, d\theta$. As in the previous proof we have (3.9.6) to which we can apply take the limit $R_j \to 1$ to conclude (3.9.9). The convergence follows from the L^p and a.e. Abel summability of the Fourier series proved in Chapter 1. ∎

3.9.3 Representation Theorems in the Upper Half Plane

The results in the previous section can be transformed to obtain representation theorems for harmonic functions in the upper half plane $\mathbb{R}^{2+} = \{(x, y) : -\infty < x < \infty, y > 0\}$. To see this, we write $z = x + iy$ and introduce the fractional linear transformation

$$(3.9.10) \quad w = \frac{z - i}{z + i} = \frac{x + i(y - 1)}{x + i(y + 1)}$$

which maps i to 0 and maps the real axis $-\infty < x < \infty$ to the unit circle $|w| = 1$, deleted by the point $w = 1$, which corresponds to the point $z = \infty$. The upper half plane $\mathbb{R}^{2+}$ is mapped $1:1$ conformally onto the unit disk $D = \{w : |w| < 1\}$.

If $U(z)$ is a harmonic function defined for $z = x + iy \in \mathbb{R}^{2+}$, we obtain a harmonic function on D by setting $u(w) = U(z)$. This can be seen directly by computing the partial derivatives by the chain rule or by observing that u is the real part of the holomorphic function obtained by composition of a holomorphic function with the fractional linear transformation (3.9.10). If U is a bounded harmonic function on $\mathbb{R}^{2+}$, then u is a bounded harmonic function in D. If U is a nonnegative harmonic function in $\mathbb{R}^{2+}$, then u is a nonnegative harmonic function in D. We state and prove the corresponding Fatou theorems.

Theorem 3.9.10. *Suppose that U is a bounded harmonic function in $\mathbb{R}^{2+}$. Then there exists $U_1 \in L^\infty(\mathbb{R})$ so that*

(3.9.11) $$U(x,y) = \frac{y}{\pi} \int_\mathbb{R} \frac{U_1(t)}{(t-x)^2 + y^2} \, dt.$$

Proof. From the Fatou theorem in the disk, we set $w = re^{i\theta}$ and

(3.9.12) $$u(w) = \frac{1}{2\pi} \int_\mathbb{T} \frac{1-r^2}{1+r^2 - 2r\cos(\theta-\phi)} u_1(\phi) \, d\phi = \frac{1}{2\pi} \int_\mathbb{T} \frac{1-|w|^2}{|e^{i\phi} - w|^2} u_1(\phi) \, d\phi.$$

It remains to transform this to the $z = x + iy$ coordinates. Direct computation shows that the numerator of the Poisson kernel is computed from

$$|w|^2 = \frac{x^2 + (y-1)^2}{x^2 + (y+1)^2}, \qquad 1 - |w|^2 = \frac{4y}{x^2 + (y+1)^2}.$$

The denominator is computed by writing $e^{i\phi} = (t-i)/(t+i)$, $w = (z-i)/(z+i)$ to obtain

$$e^{i\phi} - w = \frac{t-i}{t+i} - \frac{z-i}{z+i}$$

$$= \frac{2y + 2i(t-x)}{(tx - y - 1) + i(ty + x + t)},$$

$$|e^{i\phi} - w|^2 = \frac{4y^2 + 4(t-x)^2}{(tx - y - 1)^2 + (ty + x + t)^2}$$

$$= \frac{4}{1+t^2} \frac{y^2 + (t-x)^2}{x^2 + (y+1)^2},$$

resulting in the identity

$$\frac{1 - |w|^2}{|e^{i\phi} - w|^2} = \frac{y(1+t^2)}{y^2 + (t-x)^2}.$$

It remains to compute the Jacobian of the mapping $t \to \phi$. This is computed directly by writing $ie^{i\phi} \, d\phi = 2i \, dt/(t+i)^2$ from which $d\phi = 2 \, dt/(1+t^2)$. Substituting this into (3.9.12), we conclude the representation formula (3.9.11), where $U_1(t) = u_1(\phi)$. ∎

The representation theorem for nonnegative harmonic functions contains a new term, which was not present in the disk.

Theorem 3.9.11. *Suppose that U is a nonnegative harmonic function in $\mathbb{R}^{2+}$. Then there exists a nonnegative finite Borel measure M on $\mathbb{R}$ and a nonnegative constant c so that*

(3.9.13) $$U(x,y) = cy + \frac{y}{\pi} \int_\mathbb{R} \frac{1}{(t-x)^2 + y^2} M(dt).$$

Proof. Again setting $u(w) = U(z)$, we apply Theorem 3.9.8 to obtain

(3.9.14) $$u(w) = \frac{1}{2\pi} \int_\mathbb{T} \frac{1-r^2}{1+r^2 - 2r\cos(\theta-\phi)} m(d\phi) = \frac{1}{2\pi} \int_\mathbb{T} \frac{1-|w|^2}{|e^{i\phi} - w|^2} m(d\phi).$$

If the measure m attributes no mass to the point $\phi = 0$, then we can transform this integral exactly as we did in the proof of the previous theorem, by setting $M(A) = m(\tilde{A})$ if the Borel set $\tilde{A}$ is the image of the Borel set A under the mapping (3.9.10). If $m(\{0\}) > 0$, we must compute the contribution as the Poisson kernel corresponding to $t = \infty$:

$$\frac{1 - |w|^2}{|1 - w|^2} = \frac{4y}{x^2 + (y+1)^2} \cdot \frac{x^2 + (y+1)^2}{4} = y,$$

which gives the additional term in (3.9.13), where $c = m(\{0\})$. ∎

The Fatou theorem for the L^p norm is also different from the case of the disk. To see this, we first note that the image of a horizontal line $y = \text{const.}$ under the map $z \to (z-i)/(z+i)$ is a circle whose center is on the line $\text{Re } w = 0$ and which passes through the point $w = 1$. When $y \to 0$ this circle tends to the circle $|w| = 1$. In detail, we write

$$\frac{x + iy - i}{x + iy + i} = \frac{y}{y+1} + \frac{1}{y+1} \cdot \frac{x - i(y+1)}{x + i(y+1)}$$

so that the center is at $y/(y+1)$ and the radius $= 1/(y+1)$. We parametrize the circle by writing

$$w = \frac{y}{y+1} + \frac{1}{y+1} e^{i\Psi}$$

from which we compute $d\Psi = 2(y+1)/(x^2 + (y+1)^2)\, dx$. Therefore the L^p norms transform according to

$$(3.9.15) \quad \int_{-\pi}^{\pi} |u(w)|^p\, d\Psi = 2(y+1) \int_{\mathbb{R}} |U(x,y)|^p \frac{1}{x^2 + (y+1)^2}\, dx.$$

Theorem 3.9.12. *Suppose that U is a harmonic function in $\mathbb{R}^{2+}$ such that for each $Y > 0$,*

$$\sup_{Y > y > 0} \int_{\mathbb{R}} \frac{|U(x,y)|^p}{1 + x^2}\, dt \leq M < \infty.$$

Then there exists $U_1 \in L^p(\mathbb{R}; dt/(1+t^2))$ so that

$$(3.9.16) \quad U(x,y) = \frac{y}{\pi} \int_{\mathbb{R}} \frac{1}{(t-x)^2 + y^2} U_1(t)\, dt.$$

Proof. We have $t^2 + (y+1)^2 > 1 + t^2$, thus

$$\frac{1}{t^2 + (y+1)^2} < \frac{1}{t^2 + 1}$$

so that

$$2(y+1)\int_{\mathbb{R}} \frac{|U(t,y)|^p}{t^2 + (y+1)^2}\, dt \leq 2(y+1) \int_{\mathbb{R}} \frac{|U(t,y)|^p}{t^2 + 1}\, dt \leq 4M(Y+1).$$

Transforming the integrals as in (3.9.15),

$$\sup_{Y > y > 0} \int_{-\pi}^{\pi} \left| u\left(\frac{y}{y+1} + \frac{1}{y+1} e^{i\Psi} \right) \right|^p d\Psi < \infty.$$

Now we can apply the compactness argument from the previous section, defining a sequence of linear functionals on $L^{p'}(\mathbb{T})$ by

$$L_R f = \int_{-\pi}^{\pi} f(\Psi) u\left(\frac{y}{y+1} + \frac{1}{y+1}e^{i\Psi}\right) d\Psi.$$

We see that their norms are uniformly bounded, hence we can choose a weak* convergent subsequence L_{y_j} where $y_j \to 0$ and $L_{y_j} \to u_1 \in L^p(\mathbb{T})$. Writing the Poisson integral representation of u with respect to a fixed but arbitrary point in the interior of the circle C_{y_j}, we note that the Poisson kernels converge uniformly in $\Psi \in (-\pi, \pi)$ so that we can take the limit $y_j \to 0$ and conclude that $u(w)$ is represented by the Poisson integral of u_1 as in (3.9.9). Finally we transform this into the (x, y) variables, exactly as in the proof of Theorem 3.9.9. ∎

3.9.4 Herglotz/Bochner Theorems and Positive Definite Functions

We can use the representation theorem for positive harmonic functions to characterize the Fourier coefficients of a nonnegative measure on the circle. A bilateral sequence of complex numbers $\{u_n\}_{n\in\mathbb{Z}}$ is called *positive definite* if for every finite set of complex numbers $\{c_n\}_{n=-N}^{N}$, we have

(3.9.17) $$\sum_{m,n=-N}^{N} c_m \bar{c}_n u_{m-n} \geq 0.$$

Exercise 3.9.13. *Prove that a positive definite sequence $\{u_n\}$ satisfies $u_0 \geq 0$ and $|u_n| \leq u_0$ for $n \in \mathbb{Z}$.*

Hint: Apply the definition with $c_0 = 1$, $c_n = re^{i\theta}$ and otherwise $c_k = 0$. By suitable choice of θ, first prove that u_n is hermitian symmetric: $\bar{u}_n = u_{-n}$. Then minimize a quadratic polynomial to obtain the inequality.

Theorem 3.9.14. *Herglotz: A sequence of complex numbers $\{u_n\}_{n\in\mathbb{Z}}$ is positive definite if and only if there exists a nonnegative Borel measure M on $\mathbb{T}$ such that*

(3.9.18) $$u_n = \int_{\mathbb{T}} e^{-in\theta} M(d\theta), \quad n \in \mathbb{Z}.$$

Proof. If (3.9.18) holds, then we have for any finite set of complex numbers $\{c_n\}_{n=-N}^{N}$,

$$\sum_{m,n=-N}^{N} c_m \bar{c}_n u_{m-n} = \int_{\mathbb{T}} \left|\sum_{m=-N}^{N} c_m e^{im\theta}\right|^2 M(d\theta)$$
$$\geq 0.$$

Conversely, suppose that (3.9.17) holds. Let $c_n = r^n e^{in\theta}$ for $n \geq 0$ and $c_n = 0$ for $n < 0$, where $0 \leq r < 1$ and set

$$F(r, \theta) := \sum_{m,n \geq 0} u_{m-n} r^m e^{im\theta} r^n e^{-in\theta}.$$

The double sum is majorized by $u_0 \sum_{m,n\geq 0} r^{n+m} = u_0/(1-r)^2 < \infty$. It is also the $(N \to \infty)$ limit of the finite sum for $0 \leq m, n \leq N$, hence $F(r, \theta) \geq 0$ by (3.9.17). We can rewrite

$F(r, \theta)$ as a sum along 45 degree lines in the first quadrant:

$$0 \le F(r,\theta) = \sum_{j \in \mathbb{Z}} \sum_{m,n \ge 0, m-n=j} u_{m-n} r^m e^{im\theta} r^n e^{-in\theta}$$

$$= \sum_{j \in \mathbb{Z}} u_j e^{ij\theta} \sum_{m,n \ge 0, m-n=j} r^{m+n}.$$

If $m - n = 0$ the inner sum is $\sum_{m=0}^{\infty} r^{2m} = 1/(1 - r^2)$. Otherwise we have $m - n = j \ne 0$ and we can remove the common factor $r^{|j|}$ from the inner sum to obtain

$$F(r,\theta) = \frac{1}{1-r^2} \sum_{j \in \mathbb{Z}} u_j r^{|j|} e^{ij\theta} \ge 0,$$

which proves that the harmonic function $u(r,\theta) := \sum_{j \in \mathbb{Z}} u_j r^{|j|} e^{ij\theta}$ is nonnegative in the unit disk. Hence by Theorem 3.9.8 there exists a nonnegative Borel measure M on $\mathbb{T}$ so that

(3.9.19) $$\sum_{j \in \mathbb{Z}} u_j r^{|j|} e^{ij\theta} = \int_{\mathbb{T}} \frac{1-r^2}{1+r^2 - 2r\cos(\theta - \phi)} M(d\phi).$$

Now we multiply both sides of (3.9.19) by $e^{-iN\theta}$ and integrate over $\mathbb{T}$ to conclude that $u_N = \int_{\mathbb{T}} e^{-iN\theta} M(d\theta)$ for any $N \in \mathbb{Z}$, which was to be proved. ∎

One can also consider positive definite functions $f(\xi)$ on the real line. These are defined by the statement that for every finite set of real numbers $\xi_1, \ldots, \xi_N$ and every finite set of complex numbers $\{c_n\}_{n=1}^N$ we have

(3.9.20) $$\sum_{m,n=1}^{N} c_m \bar{c}_n f(\xi_m - \xi_n) \ge 0.$$

Again it follows that $f(0) \ge 0$ and that $|f(\xi)| \le f(0)$.

Exercise 3.9.15. *Prove this.*

Theorem 3.9.16. *Bochner: A continuous function $f(\xi)$ is positive definite if and only if there exists a nonnegative Borel measure M on $\mathbb{R}$ such that*

(3.9.21) $$f(\xi) = \hat{M}(\xi) := \int_{\mathbb{R}} e^{-i\xi x} M(dx), \quad \xi \in \mathbb{R}.$$

Proof. Suppose that $f(\xi)$ is defined by (3.9.21). Then for any finite sets (ξ_n) and (c_n) we have

$$\sum_{m,n=1}^{N} c_m \bar{c}_n f(\xi_m - \xi_n) = \int_{\mathbb{R}} \left| \sum_{n=1}^{N} c_n e^{i\xi_n x} \right|^2 M(dx)$$

$$\ge 0.$$

Conversely, suppose that (3.9.20) holds. Now for any $x \in \mathbb{R}$, $y > 0$, define

$$F(x,y) = \int_0^\infty \int_0^\infty f(\xi - \eta) e^{-\xi y} e^{i\xi x} e^{-\eta y} e^{-i\eta x} \, d\xi \, d\eta.$$

This double integral is majorized by the convergent double integral $f(0) \int_0^\infty \int_0^\infty e^{-\xi y} e^{-\eta y} \, d\xi \, d\eta = f(0)/y^2$. It is also the limit of finite Riemann sums of the form $\sum_{m,n} f(\xi_m - \xi_n) e^{-\xi_m y} e^{i\xi_m x} e^{-\xi_n y} e^{-i\xi_n x}$ which are nonnegative, by hypothesis. Hence $F(x, y) \geq 0$. On the other hand we can write

$$0 \leq F(x, y) = \int_{\mathbb{R}} f(u) e^{iux} \left(\int_{\xi, \eta \geq 0, \xi - \eta = u} e^{-y(\xi + \eta)} d\eta \right) du$$

$$= \frac{1}{2y} \int_{\mathbb{R}} f(u) e^{iux} e^{-y|u|} du,$$

which proves that $2yF(x, y)$ is a nonnegative harmonic function in the upper half plane. Hence by Theorem 3.9.11 there exists a nonnegative finite Borel measure M on $\mathbb{R}$ and a nonnegative number c such that

(3.9.22) $$cy + 2 \int_{\mathbb{R}} \frac{y}{y^2 + (t - x)^2} M(dt) = \int_{\mathbb{R}} e^{iux} f(u) e^{-y|u|} du.$$

However both integrals are bounded functions of y when $y \to \infty$. Hence the constant term must be $c = 0$. To complete the identification, we recall the representation of the Poisson kernel:

$$\frac{2y}{y^2 + (t - x)^2} = \int_{\mathbb{R}} e^{i\xi x} e^{-i\xi t} e^{-y|\xi|} d\xi.$$

Integrate both sides with respect to $M(dt)$ over $\mathbb{R}$ to obtain

$$\int_{\mathbb{R}} f(u) e^{iux} e^{-y|u|} du = \int_{\mathbb{R}} \hat{M}(\xi) e^{i\xi x} e^{-y|\xi|} d\xi$$

from which we conclude, by the uniqueness of Fourier transforms, that for all $\xi \in \mathbb{R}, y > 0$, we have $f(\xi) e^{-y|\xi|} = \hat{M}(\xi) e^{-y|\xi|}$, hence $f(\xi) = \hat{M}(\xi)$, as required. ∎

CHAPTER 4

POISSON SUMMATION FORMULA AND MULTIPLE FOURIER SERIES

4.1 MOTIVATION AND HEURISTICS

Up until now we have treated Fourier series and Fourier integrals independently of one another. Given the strong parallels between the one-dimensional theories, we may ask if there is a systematic link for passing back and forth. This is supplied by the notion of *periodization* and the closely related *Poisson summation formula*. For the sake of clarity, we will first pursue these ideas in one dimension where the formulas are simpler. As applications, we will obtain the Shannon sampling formula for band-limited signals and the transformation formulas for Gaussian sums from number theory using these ideas. The higher-dimensional Poisson summation formula allows us to study multiple Fourier series. As a by-product we can derive the famous Landau asymptotic formula for the number of lattice points in a large sphere in Euclidean space.

In order to simplify the notations, we will consider Fourier series for functions of period 1; these are defined by their restriction to the interval $(0, 1)$. In this setting a Fourier series is written

$$f(x) \sim \sum_{n \in \mathbb{Z}} \hat{f}(n) e^{2\pi i n x}, \qquad \hat{f}(n) := \int_0^1 f(x) e^{-2\pi i n x}\, dx,$$

and the Dirichlet kernel of Fourier series is written

$$D_M^{FS}(x) = \sum_{k=-M}^{M} e^{2\pi i k x} = \frac{\sin((2M+1)\pi x)}{\sin \pi x}.$$

4.2 THE POISSON SUMMATION FORMULA IN $\mathbb{R}^1$

4.2.1 Periodization of a Function

Given a function on the line, one can construct a periodic function by summing over the integer translates, defining

(4.2.1) $$\boxed{\bar{f}(x) = \sum_{n \in \mathbb{Z}} f(x+n)}$$

where the series is supposed to converge in the sense of symmetric partial sums. This is called the *periodization* of f.

Example 4.2.1. *If $f \in L^1(0,1)$ and is defined to be zero elsewhere, then $\bar{f}$ is simply the periodic extension of f to the entire real line.*

Example 4.2.2. *If f is the heat kernel, defined by $f(x) = (4\pi t)^{-1/2} e^{-x^2/4t}$ for $t > 0, x \in \mathbb{R}$, then $\bar{f}(x) = (4\pi t)^{-1/2} \sum_{n \in \mathbb{Z}} e^{-(x+n)^2/4t}$ is the periodic heat kernel.*

Example 4.2.3. *Let f be the Dirichlet kernel relative to the Fourier transform, studied in Chapter 2:*

$$f(x) = D_M^{FT}(x) := \frac{\sin(2M\pi x)}{\pi x} = \int_{-M}^{M} e^{2\pi i \xi x} \, d\xi.$$

In this case the series defining $\bar{f}$ is not absolutely convergent, but we can compute the periodization $\bar{f}$ directly as follows:

(4.2.2) $$\sum_{n=-N}^{N} \frac{\sin 2\pi M(x+n)}{\pi(x+n)} = \sum_{n=-N}^{N} \left(\int_{-M}^{M} e^{2\pi i \xi (x+n)} \, d\xi \right)$$

$$= \int_{-M}^{M} e^{2\pi i \xi x} \left(\sum_{n=-N}^{N} e^{2\pi i \xi n} \right) d\xi$$

$$= \int_{-M}^{M} e^{2\pi i \xi x} \left(\frac{\sin(2N+1)\pi \xi}{\sin \pi \xi} \right) d\xi.$$

We apply the pointwise inversion of Fourier series on each interval $(k-1/2, k+1/2)$, each time obtaining a contribution from the center. If $M \notin \mathbb{Z}$, we obtain

(4.2.3) $$\lim_{N \to \infty} \sum_{n=-N}^{N} \frac{\sin 2\pi M(x+n)}{\pi(x+n)} = \sum_{k=-[M]}^{[M]} e^{2\pi i k x} = \frac{\sin((2[M]+1)\pi x)}{\sin \pi x}.$$

Furthermore the partial sums (4.2.2) converge boundedly for each $M < \infty$, since the right side consists of a finite number of Fourier partial sums for the function $\xi \to e^{2\pi i x \xi}$ at $\xi = 0, \pm 1, \ldots, \pm[M]$.

We have shown that *the periodization of the Dirichlet kernel D_M^{FT} is the periodic Dirichlet kernel $D_{[M]}^{FS}$.*

Exercise 4.2.4. *If $M \in \mathbb{Z}$, show that we must add an additional term of $\cos 2M\pi x$ to the right side of (4.2.3).*

We can further exploit the bounded convergence to represent the Fourier partial sum of an arbitrary $f \in L^1(0,1)$:

$$\int_{-N}^{N} \tilde{f}(x+t) D_M^{FT}(t)\, dt = \int_0^1 \tilde{f}(x+t) \sum_{n=-N}^{N} \frac{\sin 2\pi M(t+n)}{\pi(t+n)}\, dt \to \int_0^1 f(x+t) D_{[M]}^{FS}(t)\, dt.$$

When $N \to \infty$, the left side converges to the *Cauchy principal value*, or symmetric improper integral, which can be summarized by writing

$$PV \int_{\mathbb{R}} \tilde{f}(x+t) D_M^{FT}(t)\, dt = \int_0^1 \tilde{f}(x+t) D_{[M]}^{FS}(t)\, dt, \qquad M \notin \mathbb{Z}.$$

As a final example of periodization, we consider the kernel of the Hilbert transform, which was studied in Chapter 3.

Example 4.2.5. *Let $f(x) = 1/x$ for $x \neq 0$.*

The periodization is determined from the identity

(4.2.4) $$\sum_{n \in \mathbb{Z}} \frac{1}{x-n} = \pi \cot \pi x, \qquad x \notin \mathbb{Z}.$$

This can be proved by residue calculus, which the reader is invited to supply. We offer a Fourier-analytic proof as follows:

Proof. Let $f(x) = \sum_{n \in \mathbb{Z}} 1/(x-n) - \pi \cot \pi x$ for $x \notin \mathbb{Z}$. Then $x \to f(x)$ is an odd function of period 1. It is also a continuous function, since $\sum_{0 \neq n \in \mathbb{Z}} 1/(x-n)$ converges uniformly for $-1/2 \leq x \leq 1/2$, [since $|1/(x^2 - n^2)| \leq 4/(4n^2 - 1)$], hence to a continuous function. On the other hand the function $x \to \pi \cot \pi x - 1/x$ can be defined by continuity at $x = 0$ and thus is a continuous function for $-1/2 \leq x \leq 1/2$. To prove that $f \equiv 0$, we note that its Fourier cosine coefficients are all zero; it remains to show that its Fourier sine coefficients also vanish, i.e., $\int_{-1/2}^{1/2} f(x) \sin 2\pi Mx\, dx = 0$ for $M = 1, 2, \ldots$. Now the series $\sum_{n \in \mathbb{Z}} \sin(2\pi Mx)/(x-n)$ converges uniformly in $-1/2 \leq x \leq 1/2$, so we can integrate term-by-term: On the one hand,

(4.2.5) $$\lim_N \int_{-1/2}^{1/2} \sin(2\pi Mx) \left(\sum_{-N}^{N} \frac{1}{x-n} \right) dx = \lim_N \int_{-N-1/2}^{N+1/2} \frac{\sin 2\pi Mx}{x}\, dx = \pi.$$

On the other hand

$$(4.2.6) \qquad \int_{-1/2}^{1/2} \sin(2\pi Mx)\pi \cot \pi x \, dx = \int_{-1/2}^{1/2} \frac{\sin(2\pi Mx)}{\sin \pi x} \pi \cos \pi x \, dx.$$

But elementary trigonometric identities reveal that $\sum_{k=1}^{M} \cos(2k-1)\pi x = \sin(2M\pi x)/2\sin \pi x$, so that by orthogonality

$$(4.2.7) \qquad \int_{-1/2}^{1/2} \sin(2\pi Mx)\pi \cot \pi x \, dx = \int_{-1/2}^{1/2} 2\pi \cos^2 \pi x \, dx = \pi.$$

Therefore $\int_{-1/2}^{1/2} f(x) \sin(2\pi Mx) \, dx = 0$ for $M = 1, 2, \ldots$, hence $f(x) = 0$ almost everywhere. But we already noted that f is continuous, and a continuous function that is zero a.e. must be zero everywhere, which completes the proof of the required identity. ∎

Additional examples of periodization are obtained from the Fejér kernel and the Poisson kernel.

Exercise 4.2.6. *Consider the Fejér kernel on the line, $K_T(x) = (1 - \cos 2\pi Tx)/2\pi^2 Tx^2$ for $x \neq 0$. Show that if $T \in \mathbb{Z}^+$, the periodization of K_T is the Fejér kernel of Fourier series: $\tilde{K}_{T-1}(x) = \sin^2(T\pi x)/T \sin^2(\pi x)$.*

Hint: Begin with the representation of K_T as a Fourier integral: $K_T(x) = \int_{-T}^{T} e^{2\pi i x\xi}(1 - |\xi|/T) \, d\xi$.

Exercise 4.2.7. *Consider the Poisson kernel on the line, $P_y(x) = y/[\pi(x^2 + y^2)]$ where $y > 0$, $x \in \mathbb{R}$. Show that the periodization of P_y is the Poisson kernel of Fourier series.*

Hint: Begin with the representation of P_y as a Fourier integral: $P_y(x) = \int_{\mathbb{R}} e^{2\pi i x\xi} e^{-2\pi |\xi| y} \, d\xi$.

4.2.2 Statement and Proof

The Poisson summation formula allows us to compute the Fourier series of $\bar{f}$ in terms of the Fourier transform of f at the integers.

Theorem 4.2.8. *Suppose that $f \in L^1(\mathbb{R})$. Then $\bar{f}(x)$ is finite a.e., satisfies $\bar{f}(x+1) = \bar{f}(x)$ a.e. and is an integrable function on any period, e.g., $[0, 1]$. The Fourier coefficient is obtained as*

$$\int_0^1 \bar{f}(x) e^{-2\pi i mx} \, dx = \hat{f}(m) = \int_{-\infty}^{\infty} f(x) e^{-2\pi i mx} \, dx, \qquad m \in \mathbb{Z}.$$

If in addition $\sum_{n=-\infty}^{\infty} |\hat{f}(n)| < \infty$, then the Fourier series of $\bar{f}$ converges and we have the a.e. equality

$$(4.2.8) \qquad \boxed{\sum_{n \in \mathbb{Z}} f(x+n) = \bar{f}(x) = \sum_{m \in \mathbb{Z}} \hat{f}(m) e^{2\pi i mx}.}$$

In particular $\bar{f}$ is a.e. equal to a continuous function on $\mathbb{T} = \mathbb{R}/\mathbb{Z}$. Redefining f if necessary, then the equality holds everywhere and we have the Poisson identity

(4.2.9)
$$\boxed{\sum_{n \in \mathbb{Z}} f(n) = \bar{f}(0) = \sum_{m \in \mathbb{Z}} \hat{f}(m).}$$

Proof. We have

$$\int_0^1 \left(\sum_{n=-\infty}^{\infty} |f(x+n)| \right) dx = \sum_{n=-\infty}^{\infty} \int_n^{n+1} |f(x)|\, dx = \int_{-\infty}^{\infty} |f(x)|\, dx < \infty,$$

which shows that $\bar{f}$ is finite almost everywhere and integrable on $[0, 1]$. The same calculation applied to $\bar{f}(x)e^{-2\pi imx}$ allows us to integrate term-by-term:

$$\int_0^1 \bar{f}(x)e^{-2\pi imx}\, dx = \int_0^1 \left(\sum_{n=-\infty}^{\infty} f(x+n) \right) e^{-2\pi imx}\, dx$$

$$= \sum_{n=-\infty}^{\infty} \int_0^1 f(x+n)e^{-2\pi imx}\, dx$$

$$= \sum_{n=-\infty}^{\infty} \int_n^{n+1} f(y)e^{-2\pi im(y-n)}\, dy$$

$$= \int_{-\infty}^{\infty} f(y)e^{-2\pi imy}\, dy$$

$$= \hat{f}(m),$$

which proves the first statement. Now if the series $\sum_{m \in \mathbb{Z}} |\hat{f}(m)|$ converges, then the Fourier series of f converges uniformly and in L^1, in particular the Cesàro means converge to the same limit. But the Cesàro means converge in L^1 to f. Therefore f is almost everywhere equal to the sum of its Fourier series. In particular, for a dense set S. Finally if f is continuous, we let $y \to x$ through the set S to conclude that $f(x) = \sum_{n \in \mathbb{Z}} \hat{f}(n)e^{2\pi inx}$. Finally, set $x = 0$ to obtain the Poisson identity (4.2.9). ∎

Exercise 4.2.9. *Suppose that $f \in L^1(\mathbb{R})$, that f has finite total variation on $\mathbb{R}$ and is normalized so that $2f(x) = f(x+0) + f(x-0)$. Prove the Poisson identity (4.2.9).*

Hint: Show that $\bar{f}$ satisfies the hypotheses of Dirichlet's convergence theorem from Chapter 1. Check first the uniform convergence of the series (4.2.1).

Example 4.2.10. *If $f(x) = (4\pi t)^{-1/2} e^{-x^2/4t}$, then all of the conditions are satisfied and we have $\hat{f}(n) = e^{-4\pi^2 t n^2}$ and thus the identity*

$$\sum_{n \in \mathbb{Z}} \frac{e^{-(x-n)^2/4t}}{\sqrt{4\pi t}} = \sum_{n \in \mathbb{Z}} e^{2\pi inx} e^{-4\pi^2 t n^2}.$$

This example shows that the periodic heat kernel has a natural Fourier series representation in terms of separated solutions of the heat equation.

Remark. It is clear that some continuity restriction is necessary to obtain the Poisson summation formula. Indeed, if $f(x) = 0$ for $x = 0, \pm 1, \pm 2, \ldots$ and $f(x) = e^{-\pi x^2}$ otherwise, then the left side of Poisson's identity (4.2.9) is zero, but the right side is obtained as $\sum_{n \in \mathbb{Z}} e^{-\pi n^2}$, from the above example.

Closely related to the Poisson summation formula is a bilinear identity similar to the Fourier reciprocity studied previously.

Proposition 4.2.11. *Suppose that either*

(a) $f \in L^\infty(\mathbb{T})$ and $K \in L^1(\mathbb{R}^1)$ or
(b) $f \in L^1(\mathbb{T})$ and $\sum_{n \in \mathbb{Z}} K(x+n)$ converges boundedly on $\mathbb{T}$.

Then

$$\tag{4.2.10} \int_0^1 f(x) \bar{K}(x)\, dx = \int_\mathbb{R} \bar{f}(x) K(x)\, dx.$$

Proof. In case (a) we appeal to Theorem (4.2.8) to obtain the a.e. convergent series

$$\bar{K}(x) = \sum_{n \in \mathbb{Z}} K(x+n) \qquad x \in (0,1),$$

$$f(x)\bar{K}(x) = \sum_{n \in \mathbb{Z}} f(x) K(x+n) \qquad x \in (0,1).$$

The right side is dominated by the integrable function $\|f\|_\infty \sum_{n \in \mathbb{Z}} |K(x+n)|$ so that we can integrate term-by-term to obtain

$$\int_0^1 f(x) \bar{K}(x)\, dx = \sum_{n \in \mathbb{Z}} \int_0^1 f(x) K(x+n)\, dx$$

$$= \sum_{n \in \mathbb{Z}} \int_n^{n+1} f(x-n) K(x)\, dx$$

$$= \int_\mathbb{R} \bar{f}(x) K(x)\, dx.$$

The proof in case (b) is another application of the dominated convergence theorem. ∎

The previous proposition can be paraphrased in terms of *adjoint operators*. Let $\mathcal{P}$ be the operator that transforms $f \in L^1(\mathbb{R})$ into its periodization $\bar{f} = \mathcal{P}f \in L^1(\mathbb{T})$. In each case the dual space is L^∞ and (4.2.10) becomes the identity

$$\int_\mathbb{T} f\, (\mathcal{P}K)\, dx = \int_\mathbb{R} (\mathcal{P}^* f)\, K\, dx, \qquad K \in L^1(\mathbb{T}),\, f \in L^\infty(\mathbb{T}),$$

which shows that *the adjoint of the periodization operator $\mathcal{P}$ is the operator that forms the periodic extension of $f \in L^\infty(\mathbb{T})$ to $L^\infty(\mathbb{R})$.*

These ideas can be applied to "lift" computations from the circle to the real line where the formulas may be simpler. For example, if $f \in L^1(\mathbb{T})$ is extended periodically to $\mathbb{R}$ and K_M is the Fejér kernel with $K_M(t) = (\sin^2 M\pi t)/(\pi M t^2)$, then from Exercise 4.2.6, its periodization is the Fejér kernel for Fourier series, defined by $\bar{K}_{M-1}(t) = (\sin^2 M\pi t)/(M \sin^2 \pi t)$ so that we can transform the Fejér mean of the

Fourier series as follows:
$$\int_0^1 f(x+t)\frac{\sin^2(M\pi t)}{M\sin^2 \pi t}\,dt = \int_{\mathbb{R}} \bar{f}(x+t)\frac{\sin^2(M\pi t)}{\pi^2 M t^2}\,dt.$$

In the same fashion, the Poisson integral is transformed according to the identity
$$\int_0^1 f(x+t)\frac{1-r^2}{1+r^2-2r\cos 2\pi t}\,dt = \int_{\mathbb{R}} \bar{f}(x+t)\frac{y}{\pi(t^2+y^2)}\,dt \qquad (r=e^{-2\pi y}).$$

4.2.3 Shannon Sampling

As a first application of the Poisson summation formula, we consider the problem of reconstructing a *band-limited signal* from its values on the integers. By definition, these are functions of the form

(4.2.11) $$f(t) = \int_{-\lambda}^{\lambda} F(\xi) e^{2\pi i t \xi}\,d\xi$$

where $F \in L^1(-\lambda, \lambda)$ and where we set $F(\xi) = 0$ for $|\xi| > \lambda$. The number λ is the *bandwidth*. The Shannon sampling formula is the following identity:

(4.2.12) $$f(t) = \sum_{n \in \mathbb{Z}} f(n)\frac{\sin \pi(t-n)}{\pi(t-n)} \qquad t \in \mathbb{R}$$

where the series is taken as the limit of the symmetric partial sums and where the fraction is set equal to 1 when $n = t$.

The following example shows that one cannot expect (4.2.12) to hold for an arbitrary bandwidth.

Example 4.2.12.
$$f(t) = \left(\frac{\sin \pi t}{\pi t}\right)^2 = \int_{-1}^1 (1-|\xi|) e^{2\pi i t \xi}\,d\xi$$

is a band-limited signal with $\lambda = 1$. Clearly f is zero on the integers, so that we cannot retrieve f from $\{f(n)\}, n \in \mathbb{Z}$.

The following general theorem gives sufficient conditions for the validity of (4.2.12).

Theorem 4.2.13. Whittaker, Shannon, Boas. *Suppose that $F \in L^1(-\lambda, \lambda)$ defines a band-limited signal with $\lambda \leq \frac{1}{2}$. Then the Shannon sampling formula (4.2.12) holds. More generally, if $\lambda > \frac{1}{2}$ we have the estimate*

(4.2.13) $$\left| f(t) - \sum_{n \in \mathbb{Z}} f(n)\frac{\sin \pi(t-n)}{\pi(t-n)} \right| \leq 2\int_{|\xi| > \frac{1}{2}} |F(\xi)|\,d\xi.$$

Proof. Let $\bar{F}(\xi) = \sum_{n \in \mathbb{Z}} F(\xi + n)$ be the periodization of $F \in L^1(\mathbb{R})$. The Fourier transform of F is $x \to f(-x)$ so that by the Poisson summation formula, we have the Fourier series

$$(4.2.14) \qquad \bar{F}(\xi) \sim \sum_{n \in \mathbb{Z}} f(-n) e^{2\pi i n \xi} = \sum_{n \in \mathbb{Z}} f(n) e^{-2\pi i n \xi}.$$

From the results of Chapter 1, any L^1 Fourier series may be integrated term-by-term after multiplication by a function g of bounded variation. Applying this with $g(\xi) = e^{2\pi i t \xi}$, we have

$$\int_{-1/2}^{1/2} \bar{F}(\xi) e^{2\pi i t \xi} = \sum_{n \in \mathbb{Z}} f(n) \int_{-1/2}^{1/2} e^{2\pi i \xi (t-n)} d\xi$$

$$= \sum_{n \in \mathbb{Z}} f(n) \frac{\sin \pi (t-n)}{\pi (t-n)}.$$

If $\lambda \leq \frac{1}{2}$, then $\bar{F}$ is the periodic extension of F to $\mathbb{R}$, so that $\bar{F}(\xi) = F(\xi)$ for $|\xi| \leq \frac{1}{2}$, and the left side reduces to $\int_{-1/2}^{1/2} F(\xi) e^{2\pi i t \xi} = f(t)$, which proves (4.2.12).

Otherwise, we can rewrite the left side as

$$\int_{-1/2}^{1/2} \bar{F}(\xi) e^{2\pi i t \xi} d\xi = \sum_{n \in \mathbb{Z}} \int_{-1/2}^{1/2} F(\xi + n) e^{2\pi i t \xi} d\xi$$

$$= \sum_{n \in \mathbb{Z}} \int_{n-1/2}^{n+1/2} F(u) e^{2\pi i t (u-n)} du$$

$$= \sum_{n \in \mathbb{Z}} e^{-2\pi i n t} \int_{n-1/2}^{n+1/2} F(u) e^{2\pi i t u} du; \quad \text{but}$$

$$f(t) = \sum_{n \in \mathbb{Z}} \int_{n-1/2}^{n+1/2} F(u) e^{2\pi i t u} du; \quad \text{thus}$$

$$\left| f(t) - \int_{-1/2}^{1/2} \hat{F}(\xi) e^{2\pi i t \xi} d\xi \right| = \left| \sum_{n \neq 0} (1 - e^{-2\pi i n t}) \int_{n-1/2}^{n+1/2} F(u) e^{2\pi i t u} du \right|$$

$$\leq 2 \sum_{n \neq 0} \int_{n-1/2}^{n+1/2} |F(u)| du$$

$$= 2 \int_{|u| \geq 1/2} |F(u)| du,$$

which completes the proof. ∎

Band-limited signals have the further property that the total signal strength $\int_{\mathbb{R}} f(t) dt$ can be computed by sampling at the integers.

Proposition 4.2.14. *Suppose that $\lambda \leq \frac{1}{2}$ and that $F \in L^1(-\lambda, \lambda)$ satisfies a Dini condition at $\xi = 0$ with value S. Then the series $\sum_{n \in \mathbb{Z}} f(n)$ converges and we have the identity*

$$(4.2.15) \qquad \lim_{T \to \infty} \int_{-T}^{T} f(t) dt = S = \sum_{n \in \mathbb{Z}} f(n).$$

Proof. From (4.2.11), the left side of (4.2.15) is computed as

$$\int_{-T}^{T} f(t)\, dt = \int_{-\lambda}^{\lambda} F(\xi) \frac{\sin 2\pi \xi T}{\pi \xi}\, d\xi,$$

which converges to S by one-dimensional Fourier inversion. The convergence of the series on the right side of (4.2.15) can be seen directly by computing from (4.2.11):

$$\sum_{-N}^{N} f(n) = \int_{-\lambda}^{\lambda} F(\xi) \frac{\sin(2N+1)\pi \xi}{\sin \pi \xi}\, d\xi.$$

Since $\lambda \leq \frac{1}{2}$, this is an integral on a single copy of the basic period interval $(-\frac{1}{2}, \frac{1}{2})$ and converges to S by applying one-dimensional Fourier inversion. ∎

Exercise 4.2.15. Suppose that $f(t) = \int_{-\lambda}^{\lambda} e^{2\pi i t \xi} \mu(d\xi)$, where μ is a finite Borel measure and $\lambda \leq \frac{1}{2}$. Prove the identity

(4.2.16) $$\lim_{T \to \infty} \frac{1}{2T} \int_{-T}^{T} f(t)\, dt = \lim_{N} \frac{1}{2N+1} \sum_{n=-N}^{N} f(n).$$

Hint: Suitably apply the dominated convergence theorem and identify both sides with $\mu(\{0\})$.

Remark. The formula (4.2.12) for a band-limited signal with $\lambda \leq \frac{1}{2}$ is not canonical. If instead we integrate (4.2.14) on the interval $[-\lambda, \lambda]$, we obtain the alternative representation

(4.2.17) $$f(t) = \sum_{n \in \mathbb{Z}} f(n) \frac{\sin \pi \lambda (t-n)}{\pi (t-n)} \qquad t \in \mathbb{R}.$$

Equivalent formulas can be obtained for any $\nu \in [\lambda, \frac{1}{2}]$.

Exercise 4.2.16. Suppose that $f(t)$ is any band-limited signal. Show that we can reconstruct f from its values at the points $n/2\lambda$, $n \in \mathbb{Z}$ by means of the formula

$$f(t) = 2\lambda \sum_{n \in \mathbb{Z}} f\left(\frac{n}{2\lambda}\right) \frac{\sin \pi (2\lambda t - n)}{\pi (2\lambda t - n)}.$$

Hint: Apply the Shannon sampling formula to $F_\lambda(\xi) := F(2\lambda \xi)$.

4.3 MULTIPLE FOURIER SERIES

Multiple Fourier series are naturally associated with functions on the torus $\mathbb{T}^d = (0, 1)^d = \mathbb{R}^d/\mathbb{Z}^d$. There is a natural 1:1 correspondence between functions on $\mathbb{T}^d$ and functions on $\mathbb{R}^d$, which are periodic in each coordinate: $f(x_1, \ldots, x_i + 1, \ldots, x_d) = f(x_1, \ldots, x_i, \ldots, x_d)$ for $1 \leq i \leq d$, $(x_1, \ldots, x_d) \in \mathbb{R}^d$.

We begin with an integrable function on the d-dimensional torus $\mathbb{T}^d = (0, 1)^d$. The L^1 norm is denoted $\|f\|_1 = \int_{\mathbb{T}^d} |f(x)|\, dx$ and the Fourier coefficients of

$f \in L^1(\mathbb{T}^d)$ are

$$\hat{f}(n) := \int_{\mathbb{T}^d} f(x) e^{-2\pi i n \cdot x}\, dx, \qquad n \in \mathbb{Z}^d$$

and the Fourier series is written

(4.3.1) $$f(x) \sim \sum_{n \in \mathbb{Z}^d} \hat{f}(n) e^{2\pi i n \cdot x}.$$

4.3.1 Basic L^1 Theory

The elementary properties of multiple Fourier series may be obtained from the periodic heat kernel. We first develop the lemma of Fourier reciprocity in the following form.

Proposition 4.3.1. *Suppose that $K(x) := \sum_{n \in \mathbb{Z}^d} \hat{K}(n) e^{2\pi i n \cdot x}$ is an absolutely convergent trigonometric series: $\sum_{n \in \mathbb{Z}^d} |\hat{K}(n)| < \infty$. If $f \in L^1(\mathbb{T}^d)$, then*

(4.3.2) $$\int_{\mathbb{T}^d} f(y) K(x - y)\, dy = \sum_{n \in \mathbb{Z}^d} \hat{K}(n) \hat{f}(n) e^{2\pi i n \cdot x}.$$

Proof. Multiply the defining equation for $K(x - y)$ by $f(y)$ and integrate term-by-term. ∎

We apply this to the periodic heat kernel

(4.3.3) $$\boxed{K_t(x) := \sum_{n \in \mathbb{Z}^d} \frac{e^{-|x-n|^2/4t}}{(4\pi t)^{n/2}}.}$$

The Fourier representation of $K_t(x)$ is obtained by repeated application of the one-dimensional Poisson summation formula to obtain

(4.3.4) $$K_t(x) = \sum_{n \in \mathbb{Z}^d} e^{2\pi i n \cdot x} e^{-4\pi^2 t |n|^2},$$

since both (4.3.3) and (4.3.4) are absolutely convergent sums, and can be evaluated by multiplication of the corresponding one-dimensional sums. Applying Fourier reciprocity, we obtain

(4.3.5) $$\boxed{\int_{\mathbb{T}^d} f(y) K_t(x - y)\, dy = \sum_{n \in \mathbb{Z}^d} \hat{f}(n) e^{2\pi i n \cdot x} e^{-4\pi^2 t |n|^2} \qquad f \in L^1(\mathbb{T}^d).}$$

This can also be written in terms of $\bar{f}$, the periodic extension of f to $\mathbb{R}^d$ as

(4.3.6) $$\boxed{\int_{\mathbb{R}^d} \bar{f}(y) \frac{e^{-|x-y|^2/4t}}{(4\pi t)^{n/2}}\, dy = \sum_{n \in \mathbb{Z}^d} \hat{f}(n) e^{2\pi i n \cdot x} e^{-4\pi^2 t |n|^2} \qquad t > 0,\ x \in \mathbb{R}^d.}$$

Exercise 4.3.2. *Prove that the series $\sum_{n \in \mathbb{Z}^d} \int_{\mathbb{T}^d} |f(y)| e^{-|x-y-n|^2/4t} \, dy$ converges whenever $f \in L^1(\mathbb{T}^d)$ and use this to show that the left side of (4.3.5) is equal to the left side of (4.3.6).*

From this we conclude the following.

Proposition 4.3.3.
(i) *If $\hat{f}(n) \equiv 0$, then $f = 0$ a.e.*
(ii) *If $\sum_{n \in \mathbb{Z}^d} |\hat{f}(n)| < \infty$, then the Fourier series (4.3.1) converges to a continuous function and we have almost everywhere*

$$f(x) = \sum_{n \in \mathbb{Z}^d} \hat{f}(n) e^{2\pi i n \cdot x}.$$

Proof. Both statements are direct consequences of the pointwise almost everywhere summability associated with the Gauss kernel, proved in Chapter 2. ∎

The periodic heat kernel is a nonnegative approximate identity, meaning that

(4.3.7) $\quad K_t(x) \geq 0, \quad \int_{\mathbb{T}^d} K_t(x) \, dx = 1, \quad \lim_{t \to 0} \int_{|x|>\delta} K_t(x) \, dx = 0, \quad \forall \delta > 0.$

Exercise 4.3.4. *Prove the three properties (4.3.7) of K_t.*

A multiple Fourier series is said to be *Gauss-summable* if (4.3.5) converges to f when $t \to 0$. Since the periodic heat kernel is an approximate identity, we immediately obtain the following properties of Gauss summability.

Proposition 4.3.5. *Suppose that $f \in L^1(\mathbb{T}^d)$. Then the Fourier series is Gauss-summable in $L^1(\mathbb{T}^d)$ to f. If, in addition, f is continuous at $x \in \mathbb{T}^d$, then the Fourier series is Gauss-summable to $f(x)$.*

The Gauss-summability of Fourier series has the further consequence that the set of trigonometric polynomials is dense in $L^1(\mathbb{T}^d)$. A direct proof can be obtained by applying properties of one-dimensional Fourier series, as follows.

Lemma 4.3.6. *The trigonometric polynomials $\sum_{|n| \leq N} a_n e^{2\pi i n \cdot x}$ are dense in $L^1(\mathbb{T}^d)$.*

Proof. From Chapter 1, we know that the Fourier series of an indicator function converges boundedly and in the L^1 norm. Now if we have a product of indicator functions $1_{(a_i,b_i)}$ each approximated by a Fourier partial sum S_M^i, then we can write

$$\prod_{i=1}^d S_M^i - \prod_{i=1}^d 1_{(a_i,b_i)} = \sum_{i=1}^d \left[S_M^i - 1_{(a_i,b_i)} \right] C_i$$

where C_i contains $i - 1$ factors of S_M^j and $d - i$ factors of $1_{(a_j, b_j)}$. But these are uniformly bounded, so that we can write

$$\left\| \prod_{i=1}^{d} S_M^i - \prod_{i=1}^{d} 1_{(a_i, b_i)} \right\|_1 \leq C \sum_{i=1}^{d} \left\| S_M^i - 1_{(a_i, b_i)} \right\|_1 \to 0, \qquad M \to \infty.$$

But finite sums of indicators of rectangles are dense in $L^1(\mathbb{T}^d)$, which completes the proof. ∎

4.3.1.1 Pointwise convergence for smooth functions

The condition of absolute convergence: $\sum_{m \in \mathbb{Z}^d} |\hat{f}(m)| < \infty$ can be verified in case f is sufficiently smooth. To see this, we compute the Fourier coefficients of any mixed partial derivative corresponding to a multiindex $\alpha = (\alpha_1, \ldots, \alpha_d)$, obtaining

$$(2\pi i m)^\alpha \hat{f}(m) = \int_{\mathbb{T}^d} D^\alpha f(x) e^{-2\pi i m \cdot x} \, dx.$$

Applying this twice to each of the coordinate derivatives and summing, we obtain

$$(1 + |2\pi m|^2) \hat{f}(m) = \int_{\mathbb{T}^d} [1 - \Delta] f(x) e^{-2\pi i m \cdot x} \, dx,$$

where $\Delta = \sum_{j=1}^{d} \partial^2/\partial x_j^2$ is the Laplace operator. Applying this k times we obtain

$$(1 + |2\pi m|^2)^k \hat{f}(m) = \int_{\mathbb{T}^d} [1 - \Delta]^k f(x) e^{-2\pi i m \cdot x} \, dx.$$

If $f \in C^{2k}(\mathbb{T}^d)$, the right side is the Fourier coefficient of a continuous function, hence the estimate

$$|\hat{f}(m)| \leq \frac{C}{(1 + |2\pi m|^2)^k}, \qquad m \in \mathbb{Z}^d.$$

In particular, if $2k > d$, then the series $\sum_{m \in \mathbb{Z}^d} |\hat{f}(m)|$ converges and we have absolute and uniform convergence of the Fourier series. This is summarized as follows:

Proposition 4.3.7. *Suppose that $f \in C^{2k}(\mathbb{T}^d)$ with $2k > d$. Then the Fourier series converges absolutely and uniformly to f.*

In particular, the Fourier series of an infinitely differentiable function on $\mathbb{T}^d$ is uniformly convergent. If f has fewer than $d/2$ derivatives, examples show that one may have Fourier series that are divergent at a point. This will be discussed in Section 4.5.3 in the context of *radial functions*.

4.3.1.2 Representation of spherical partial sums

The partial sum of a one-dimensional Fourier series can be written in terms of a corresponding partial Fourier integral. Restricting attention to $\theta = 0$ and $f \in L^1(\mathbb{T})$, we can write

$$S_M f(0) := \sum_{k=-M}^{M} \hat{f}(k) = \int_{-1/2}^{1/2} \frac{\sin(2M+1)\pi\theta}{\sin\pi\theta} f(\theta)\, d\theta$$

$$= \int_{-1/2}^{1/2} \frac{\sin(2M+1)\pi\theta}{\pi\theta} \left(\frac{\pi\theta f(\theta)}{\sin\pi\theta}\right) d\theta.$$

The last integral is a Fourier partial integral of the associated function $\theta \to [\pi\theta f(\theta)/\sin\pi\theta]\,1_{(-1/2,1/2)}(\theta)$. This allows convergence questions for Fourier series to be reduced to corresponding convergence questions for Fourier integrals.

In higher dimensions the spherical partial sum of a multiple Fourier series bears no simple relation to the spherical partial sum of the corresponding Fourier integral, as it does in the case of one dimension. To obtain a suitable substitute for the latter, we consider the *quasispherical partial sum* of the Fourier integral, defined for $f \in L^1(\mathbb{R}^d)$ as

(4.3.8)
$$\tilde{S}_M f(x) := \int_{B_M} \hat{f}(\xi) e^{2\pi i \xi \cdot x}\, d\xi,$$

where B_M is the set of cubes S_k of side 1, centered at the integer points k with $|k| \leq M$. In one dimension B_M is the interval $[-M - \frac{1}{2}, M + \frac{1}{2}]$ if $M = 1, 2, \ldots$. The corresponding *quasispherical Dirichlet kernel* is

$$\tilde{D}_M(x) := \int_{B_M} e^{2\pi i \xi \cdot x}\, d\xi.$$

Clearly we have the representation formula that for any $f \in L^1(\mathbb{R}^d)$

(4.3.9)
$$\tilde{S}_M f(x) = \int_{\mathbb{R}^d} \tilde{D}_M(x-y) f(y)\, dy.$$

This is to be compared with the spherical Dirichlet kernel of Fourier series, defined as

$$\bar{D}_M(x) := \sum_{|k| \leq M} e^{2\pi i k \cdot x}.$$

To compute $\tilde{D}_M$ in general, we first compute the integral

$$\int_{S_k} e^{2\pi i \xi \cdot x}\, d\xi = e^{2\pi i k \cdot x} \prod_{j=1}^{d} \frac{\sin \pi x_j}{\pi x_j}, \qquad k \in \mathbb{Z}^d.$$

Summing these for $|k| \leq M$, we have

$$\int_{B_M} e^{2\pi i \xi \cdot x}\, d\xi = \left(\prod_{j=1}^{d} \frac{\sin \pi x_j}{\pi x_j}\right) \sum_{|k| \leq M} e^{2\pi i k \cdot x}$$

and the formula

$$\tilde{D}_M(x) = \left(\prod_{j=1}^{d} \frac{\pi x_j}{\sin \pi x_j}\right) \bar{D}_M(x).$$

The spherical Dirichlet kernel of Fourier series differs from the quasispherical Dirichlet kernel by a factor which is smooth and bounded above and below over the basic cube $[-\frac{1}{2}, \frac{1}{2}]^d$.

This can be immediately applied to write the spherical partial sum of the Fourier series.

$$S_M f(x) = \sum_{|k| \leq M} \hat{f}(k) e^{2\pi i k \cdot x}$$

$$= \int_{\mathbb{T}^d} f(y) \left(\sum_{|k| \leq M} e^{2\pi i k \cdot (x-y)} \right) dy$$

$$= \int_{\mathbb{T}^d} \tilde{D}_M(z) f(x+z) \, dz$$

$$= \int_{\mathbb{T}^d} \left(\prod_{j=1}^{d} \frac{\pi z_j}{\sin \pi z_j} \right) \tilde{D}_M(z) f(x+z) \, dz.$$

At $x = 0$ this is the quasispherical partial sum of the Fourier integral for the associated function defined by $f_0(z) = \prod_{j=1}^{d} (\pi z_j / \sin(\pi z_j)) f(z) 1_{\mathbb{T}^d}$. From (4.3.8) and (4.3.9) it follows that

$$S_M f(0) = \int_{\mathbb{R}^d} \tilde{D}_M(z) f_0(z) \, dz = \int_{B_M} \hat{f_0}(\xi) \, d\xi.$$

Proposition 4.3.8. *Suppose that* $f \in L^1(\mathbb{T}^d)$ *satisfies the condition that for the associated function* f_0, $\lim_{M \to \infty} \int_{\{M-\sqrt{d} \leq |\xi| \leq M+\sqrt{d}\}} |\hat{f_0}(\xi)| \, d\xi = 0$. *Then the spherical partial sum of the Fourier series is equiconvergent with the spherical partial sum of the corresponding Fourier integral.*

Proof. From the above computations, we have

$$\left| S_M f(0) - \int_{|\xi| \leq M} \hat{f_0}(\xi) \, d\xi \right| \leq \int_{\{M-\sqrt{d} \leq |\xi| \leq M+\sqrt{d}\}} |\hat{f_0}(\xi)| \, d\xi \to 0. \quad \blacksquare$$

This can be applied to certain two-dimensional Fourier series where the Fourier coefficients satisfy $\sum_{M \leq |n| \leq M+1} |\hat{f}(n)| \to 0$ when $M \to \infty$ but not the stronger condition that $\sum_{n \in \mathbb{Z}^2} |\hat{f}(n)| < \infty$.

Exercise 4.3.9. Let $d = 2$ and $f(x) = 1_{[0,a]}(|x|)$. Show that $\sum_{n \in \mathbb{Z}^2} |\hat{f}(n)| = +\infty$ but $\sum_{M \leq |n| \leq M+1} |\hat{f}(n)| \to 0$ when $M \to \infty$.

4.3.2 Basic L^2 Theory

As in the case of the circle, multiple Fourier series have a very satisfactory theory in the space $L^2(\mathbb{T}^d)$. For any finite set of complex numbers $c_1, \ldots, c_N$, and multiindices

$m_1, \ldots, m_N$, we have

$$\left\| f - \sum_{j=1}^{N} c_j e^{2\pi i m_j \cdot x} \right\|_2^2 = \sum_{j=1}^{N} |c_j - \hat{f}(j)|^2 + \|f\|^2 - \sum_{j=1}^{N} |\hat{f}(j)|^2.$$

Thus we see, as in the one-dimensional case, that the Fourier coefficients minimize the mean-square distance between f and the finite dimensional set spanned by $e^{2\pi i m_j \cdot x}$. In particular we have Bessel's inequality $\sum_{j \in \mathbb{Z}^d} |\hat{f}(j)|^2 \leq \|f\|^2$.

Proposition 4.3.10. *If $f \in L^2(\mathbb{T}^d)$, then the Fourier series converges to f in $L^2(\mathbb{T}^d)$ and we have the Parseval equality*

$$\sum_{n \in \mathbb{Z}^d} |\hat{f}(n)|^2 = \|f\|_2^2.$$

Proof. From Bessel's inequality, the series $\sum_{n \in \mathbb{Z}^d} |\hat{f}(n)|^2$ converges. Let $F = \sum_{n \in \mathbb{Z}^d} \hat{f}(n) e^{2\pi i n \cdot x}$, an L^2 convergent series. The function $f - F$ has all Fourier coefficients equal to zero, hence $f - F = 0$ a.e. Since $\sum_{n \in \mathbb{Z}^d} \hat{f}(n) e^{2\pi i n \cdot x}$ converges to f in $L^2(\mathbb{T}^d)$, the L^2 norms also converge, which yields Parseval's equality. ∎

4.3.3 Restriction Theorems for Fourier Coefficients

Zygmund (1974) discovered a universal bound for the L^2 norm of the Fourier coefficients of two-dimensional Fourier series in terms of the L^p norm of the original function, for some $p < 2$. This is closely related to the restriction theorems for Fourier transforms, which were treated in Chapter 2. To formulate the result, we begin with $f \in L^1(\mathbb{T}^d)$ and its Fourier coefficients

(4.3.10) $$\hat{f}(\xi) = \int_{\mathbb{T}^2} f(x) e^{-2\pi i \xi \cdot x} \, dx.$$

For any given $r > 0$, the set $\{\xi \in \mathbb{Z}^2 : |\xi| = r\}$ is a (possibly empty) finite collection of lattice points on the circle of radius r. Then we have the following theorem.

Theorem 4.3.11. *For any $f \in L^{4/3}(\mathbb{T}^2)$ and any $r > 0$, we have the bound*

(4.3.11) $$\left(\sum_{\xi \in \mathbb{Z}^2 : |\xi| = r} |\hat{f}(\xi)|^2 \right)^{1/2} \leq 5^{1/4} \|f\|_{4/3}.$$

Proof. If the left side of (4.3.11) is zero, then there is nothing to prove. Otherwise, let $c(\xi) = \bar{\hat{f}}(\xi) / \sqrt{\sum_{|\xi|=r} |\hat{f}(\xi)|^2}$. Then $\sum_{|\xi|=r} |c(\xi)|^2 = 1$ and we have

$$|\hat{f}(\xi)|^2 = \hat{f}(\xi)\overline{\hat{f}}(\xi)$$

$$= \hat{f}(\xi)c(\xi)\sqrt{\sum_{|\xi|=r}|\hat{f}(\xi)|^2}$$

$$\sum_{|\xi|=r}|\hat{f}(\xi)|^2 = \left(\sum_{|\xi|=r}\hat{f}(\xi)c(\xi)\right)\sqrt{\sum_{|\xi|=r}|\hat{f}(\xi)|^2}$$

$$\sqrt{\sum_{|\xi|=r}|\hat{f}(\xi)|^2} = \sum_{|\xi|=r}\hat{f}(\xi)c(\xi)$$

$$= \int_{\mathbb{T}^2} f(x)\left(\sum_{|\xi|=r}c(\xi)e^{-2\pi i\xi\cdot x}\right)dx.$$

We now apply Hölder's inequality with $p = \frac{4}{3}, p' = 4$, to obtain

(4.3.12) $$\sqrt{\sum_{|\xi|=r}|\hat{f}(\xi)|^2} \leq \|f\|_{4/3}\left\|\sum_{|\xi|=r}c(\xi)e^{2\pi i\xi\cdot}\right\|_4.$$

Therefore we need to show that if $\phi(x) = \sum_{|\xi|=r}c(\xi)e^{2\pi i\xi\cdot x}$ is a trigonometric sum with $\sum_{|\xi|=r}|c(\xi)|^2 = 1$, then $\|\phi\|_4 \leq 5^{1/4}$. To do this, we define

(4.3.13) $$\Gamma(x) := |\phi(x)|^2 = \sum_{\mu,\nu}c(\mu)\bar{c}(\nu)e^{2\pi i(\mu-\nu)\cdot x}$$

$$= \sum_{\rho\in\mathbb{Z}^2}\gamma(\rho)e^{2\pi i\rho\cdot x}$$

where $\gamma(\rho) := \sum_{\mu-\nu=\rho}c(\mu)\bar{c}(\nu),$

and where the final sum is over those pairs (μ, ν) with $|\mu| = |\nu| = r$ and $\mu - \nu = \rho$. From the complex orthonormality of $\{e^{2\pi i\rho\cdot x}\}$, we have

$$\|\phi\|_4^4 = \int_{\mathbb{T}^2}|\phi(x)|^4 dx = \int_{\mathbb{T}^2}|\Gamma(x)|^2 dx = \sum_{\rho\in\mathbb{Z}^2}|\gamma(\rho)|^2.$$

The nonzero terms in this sum are of three types: (i) $\rho = 0$, (ii) $|\rho| = 2r$ and (iii) $0 < |\rho| < 2r$. The contribution of the terms of type (i) is given by

$$\gamma(0) = \sum_{|\mu|=r}|c(\mu)|^2 = 1.$$

For the terms of type (ii), the pairs (μ, ν), which enter into the defining sum, are antipodal points $(\mu = -\nu)$ of the circle $|\xi| = r$, one for each admissible value of ρ. Therefore the sum of these pairs contributes

$$\sum_{|\rho|=2r}|\gamma(\rho)|^2 = \sum_{|\mu|=r}|c(\mu)|^2|\bar{c}(-\mu)|^2.$$

To study the terms of type (iii), note that for a given value of $\rho \in \mathbb{Z}^2$ with $0 < |\rho| < 2r$, we can have at most two pairs (μ, ν) and (μ', ν'), corresponding to two nondiametrical chords of the circle with diametrically opposite endpoints; in detail, either $\mu + \mu' = 0$ or $\mu + \nu' = 0$. Thus, if $0 < |\rho| < 2r$, we have

$$\gamma(\rho) = c(\mu)\bar{c}(\nu) + c(\mu')\bar{c}(\nu')$$

$$|\gamma(\rho)|^2 \leq 2|c(\mu)|^2|c(\nu)|^2 + 2|c(\mu')|^2|c(\nu')|^2.$$

The sum of these pairs contributes at most

$$\sum_{0<|\rho|<2r} |\gamma(\rho)|^2 \leq 2 \sum_{\mu,\nu:0<|\mu-\nu|<2r} |c(\mu)|^2|c(\nu)|^2 + 2 \sum_{\mu',\nu':0<|\mu'-\nu'|<2r} |c(\mu')|^2|c(\nu')|^2$$

$$= 4 \sum_{\mu,\nu:0<|\mu-\nu|<2r} |c(\mu)|^2|c(\nu)|^2.$$

On the other hand, we have from the normalization

$$1 = \left(\sum_{|\xi|=r} |c(\xi)|^2\right)^2 = \sum_{\mu,\nu} |c(\mu)|^2|c(\nu)|^2.$$

Therefore, the sum of terms of type (ii), (iii) can be bounded by

$$4 \sum_{\mu,\nu:0<|\mu-\nu|\leq 2r} |c(\mu)|^2|c(\nu)|^2 \leq 4,$$

leading to the final estimate

$$\int_{\mathbb{T}^2} |\Gamma(x)|^2\, dx \leq 1 + 4 = 5,$$

which was to be proved. ∎

4.4 POISSON SUMMATION FORMULA IN $\mathbb{R}^d$

The Poisson summation formula in $\mathbb{R}^d$ is entirely similar to the one-dimensional case. The periodization of $f \in L^1(\mathbb{R}^d)$ is defined by

$$\bar{f}(x) = \sum_{n \in \mathbb{Z}^d} f(x+n).$$

This is a periodic function, whose Fourier coefficients are computed as

$$\int_{\mathbb{T}^d} \bar{f}(x) e^{-2\pi i k \cdot x}\, dx = \int_{\mathbb{T}^d} \left(\sum_{n \in \mathbb{Z}^d} f(x+n)\right) e^{-2\pi i k \cdot x}\, dx$$

$$= \sum_{n \in \mathbb{Z}^d} \int_{\mathbb{T}^d} f(x+n) e^{-2\pi i k \cdot x}\, dx$$

$$= \sum_{n \in \mathbb{Z}^d} \int_{n+\mathbb{T}^d} f(y) e^{-2\pi i k \cdot (y-n)}\, dy$$

$$= \int_{\mathbb{R}^d} f(y) e^{-2\pi i k \cdot y}$$

$$= \hat{f}(k),$$

leading to the formal identity

(4.4.1)
$$\bar{f}(x) := \sum_{n \in \mathbb{Z}^d} f(x+n) \sim \sum_{k \in \mathbb{Z}^d} \hat{f}(k) e^{2\pi i k \cdot x}.$$

Without any additional conditions on f, we can only interpret this as a formal computation. In order to obtain a pointwise identity, we assume that f and its Fourier transform satisfy the decay estimates

(4.4.2)
$$|f(x)| \le \frac{C_1}{|x|^{d+\epsilon}}, \quad |\hat{f}(k)| \le \frac{C_2}{|k|^{d+\epsilon}}$$

for positive constants C_1, C_2, ϵ. Then both sides of (4.4.1) converge absolutely and uniformly on $\mathbb{T}^d$. This is seen from writing the sum as a Steiltjes integral with respect to the lattice point counting function $N(R) = \sum_{|k| \le R} 1$, which satisfies $N(R) \le c_d R^d$. Hence

$$\sum_{|k| \le R} |\hat{f}(k)| \le |\hat{f}(0)| + C_1 \int_1^R |x|^{-d-\epsilon} dN(x)$$

$$= |\hat{f}(0)| + R^{-d-\epsilon} N(R) + (d+\epsilon) \int_1^R |x|^{-d-\epsilon-1} N(x)\, dx.$$

The term at the limits tends to zero and the final integral is absolutely convergent, proving that the right side of (4.4.1) is absolutely convergent.

Exercise 4.4.1. *Place a cube of unit side with center at the lattice point $k \in \mathbb{Z}^d$ to prove that the lattice point counting function satisfies the two-sided estimate*

$$c_d \left(R - \tfrac{1}{2}\sqrt{d}\right)^d \le N(R) \le c_d \left(R + \tfrac{1}{2}\sqrt{d}\right)^d$$

where c_d is the volume of the unit ball in R^d.

We can summarize the above discussion as follows:

Theorem 4.4.2. *Suppose that $f \in L^1(\mathbb{R}^d)$. Then the Fourier series of the periodization $\bar{f}$ is given by*

(4.4.3)
$$\bar{f}(x) \sim \sum_{n \in \mathbb{Z}^d} \hat{f}(n) e^{2\pi i n \cdot x}.$$

If in addition we have the decay estimates (4.4.2), then both sides of (4.4.3) converge absolutely and uniformly on $\mathbb{T}^d$ and we have equality in (4.4.3) almost everywhere.

4.4.1 *Simultaneous Nonlocalization

When we discussed the uncertainty principle in Chapter 2, we remarked that one can prove that a function and its Fourier transform cannot both be nonzero on a bounded set. We now use the Poisson summation formula to prove a stronger proposition on simultaneous nonlocalization due to Benedicks (1985). The statement and proof are carried out in the d-dimensional setting, as follows:

Proposition 4.4.3. *Suppose that $f \in L^1(\mathbb{R}^d)$ is supported by a set of finite measure:*

(4.4.4) $$A := \{x : f(x) \neq 0\}, \qquad |A| < \infty$$

and the same for the Fourier transform:

(4.4.5) $$B := \{\xi : \hat{f}(\xi) \neq 0\}, \qquad |B| < \infty.$$

Then $f = 0$ almost everywhere.

Proof. By a scaling transformation $x \to ax$, we may assume that $|A| < 1$. Consider the periodization of the indicator function of B:

$$\sum_{k \in \mathbb{Z}^d} 1_B(\xi - k) \geq 0.$$

Since $1_B \in L^1(\mathbb{R}^d)$, it follows that except for a set U of measure zero, this sum is finite for all $\xi \in \mathbb{R}^d$. But the terms are natural numbers, therefore

$$\xi \notin U \quad \text{implies} \quad \text{card}\{k \in \mathbb{Z}^d : 1_B(\xi - k) \neq 0\} < \infty.$$

But the Fourier transform $\hat{f}(\xi)$ is nonzero iff $\xi \in B$, therefore

(4.4.6) $$\xi \notin U \quad \text{implies} \quad \text{card}\{k \in \mathbb{Z}^d : \hat{f}(\xi - k) \neq 0\} < \infty.$$

Now let $f_\xi(x) = e^{2\pi i x \cdot \xi} f(x)$ and let $\bar{f}_\xi$ be its periodization. The Fourier transform of f_ξ is $\hat{f}(\cdot - \xi)$ so that the Poisson summation formula gives

(4.4.7) $$\bar{f}_\xi(x) = \sum_{\nu \in \mathbb{Z}^d} e^{2\pi i (x-\nu) \cdot \xi} f(x - \nu) \sim \sum_{k \in \mathbb{Z}^d} \hat{f}_\xi(k) e^{2\pi i k x} = \sum_{k \in \mathbb{Z}^d} \hat{f}(k - \xi) e^{2\pi i k x}.$$

Now since $|f_\xi(x)| = |f(x)|$ for $x \in \mathbb{R}^d$, $|\bar{f}_\xi(x)| \leq \sum_{\nu \in \mathbb{Z}^d} |f(x - \nu)|$ so that

$$\{x \in \mathbb{T}^d : \bar{f}_\xi(x) \neq 0\} \subset \bigcup_{\nu \in \mathbb{Z}^d} \{x \in \mathbb{T}^d : |f(x + \nu)| \neq 0\}$$

$$|\{x \in \mathbb{T}^d : \bar{f}_\xi(x) \neq 0\}| \leq \sum_{\nu \in \mathbb{Z}^d} |\{x \in \mathbb{T}^d : |f(x + \nu)| \neq 0\}|$$

$$= \sum_{\nu \in \mathbb{Z}^d} |\{y \in \nu + \mathbb{T}^d : |f(y)| \neq 0\}|$$

$$= |A| < 1$$

so that

(4.4.8) $$\bar{f}_\xi \in L^1(\mathbb{T}^d), \qquad |\{x \in \mathbb{T}^d : \bar{f}_\xi(x) \neq 0\}| < 1.$$

Property (4.4.6) implies that for $\xi \notin U$, $\bar{f}_\xi$ is a trigonometric polynomial. But property (4.4.8) implies that this polynomial is zero on a set of positive measure. Hence we must have $\bar{f}_\xi = 0$ almost everywhere for $\xi \notin U$; furthermore (4.4.7) shows that the Fourier transform $\hat{f}(k - \xi) = 0$ for $\xi \notin U$, $k \in \mathbb{Z}$. But the set of translates $\{k - U; k \in \mathbb{Z}^d\}$ has measure zero. Hence we conclude that $\hat{f} = 0$ almost everywhere, hence $f = 0$ almost everywhere, which completes the proof. ∎

4.5 APPLICATION TO LATTICE POINTS

The number of integer lattice points in a ball centered at $x \in \mathbb{R}^d$ is defined by

$$N(x; R) = \sum_{n \in \mathbb{Z}^d} 1_{[0,R]}(|x - n|).$$

This is the periodization of the function $x \to 1_{[0,R]}(|x|)$; hence the Fourier series is found from the Poisson summation formula as

(4.5.1) $$N(x; R) \sim \sum_{k \in \mathbb{Z}^d} F_R(k) e^{2\pi i k \cdot x}$$

where $F_R(k)$ is the Fourier transform of $x \to 1_{[0,R]}(|x|)$:

$$F_R(k) = \int_{|x| \leq R} e^{-2\pi i k \cdot x} \, dx$$

$$= R^d \frac{J_{d/2}(2\pi |k| R)}{(|k| R)^{d/2}}.$$

For each $R > 0$, $x \to N(x; R)$ is a bounded function, in particular in $L^2(\mathbb{T}^d)$, so that the Fourier series (4.5.1) is convergent in $L^2(\mathbb{T}^d)$. But the series is divergent at $x = 0$ if $d > 2$, as we shall show below. We first do the L^2 theory of lattice points, due to Kendall (1948).

4.5.1 Kendall's Mean Square Error

The Fourier transform of $1_{[0,R]}$ at $k = 0$ is the volume of the ball, whereas the Fourier transform has slower growth for $k \neq 0$; in detail

$$F_R(0) = \frac{(\pi R^2)^{d/2}}{(d/2)!}, \qquad |F_R(k)| \leq C_d \frac{R^d}{(1 + kR)^{(1+d)/2}} \quad (k \neq 0)$$

from the asymptotic behavior of Bessel functions. This sequence is square-summable and we can apply Parseval's theorem to $N(x; R) - (\pi R^2)^{d/2}/(d/2)!$ to obtain

$$\int_{\mathbb{T}^d} \left| N(x; R) - \frac{(\pi R^2)^{d/2}}{(d/2)!} \right|^2 dx = \sum_{k \neq 0} |F_R(k)|^2$$

$$\leq C_d \sum_{k \neq 0} \left(\frac{R^d}{(1 + kR)^{(1+d)/2}} \right)^2$$

$$\leq C_d \int_{|k| \geq 1} \frac{R^{2d}}{(1 + kR)^{1+d}} k^{d-1} \, dk$$

$$= R^d \int_R^\infty \frac{s^{d-1}}{(1 + s)^{d+1}} \, ds$$

$$= O(R^{d-1}) \qquad R \to \infty.$$

We can summarize the result as follows.

Proposition 4.5.1. *The L^2 norm of the error term is bounded in the form*

$$\text{(4.5.2)} \qquad \left\| N(\cdot; R) - \frac{(\pi R^2)^{d/2}}{(d/2)!} \right\|_2 \leq C_d R^{(d-1)/2}.$$

This L^2 estimate can be transformed into an almost-everywhere result by using the Chebyshev inequality: for any $\delta > 0$

$$\int_{\mathbb{T}^d} \left| N(x; R) - \frac{(\pi R^2)^{d/2}}{(d/2)!} \right|^2 dx \geq \delta^2 \times \left| \left\{ x \in \mathbb{T}^d : \left| N(x; R) - \frac{(\pi R^2)^{d/2}}{(d/2)!} \right| \geq \delta \right\} \right|.$$

If we let $R \to \infty$ along integer values we can take $\delta = R^{(d/2)+\epsilon}$ and obtain a convergent series:

$$\sum_{R=1}^{\infty} \left| \left\{ x \in \mathbb{T}^d : \left| N(x; R) - \frac{(\pi R^2)^{d/2}}{(d/2)!} \right| \geq R^{(d/2)+\epsilon} \right\} \right| \leq C_d \sum_{R \in \mathbb{Z}^+} R^{-d-2\epsilon} R^{d-1} < \infty.$$

Therefore the set of points in infinitely many of these sets has measure zero. We summarize this as follows:

Proposition 4.5.2. *For each $\epsilon > 0$ and for almost every $x \in \mathbb{T}^d$, the number of integer lattice points in a ball centered at x satisfies the estimate*

$$\left| N(x; R) - \frac{(\pi R^2)^{d/2}}{(d/2)!} \right| \leq R^{(d/2)+\epsilon}, \qquad \mathbb{Z}^+ \ni R \to \infty.$$

Exercise 4.5.3. *Show that if $R \to \infty$ along the sequence of squares $R = j^2$, then we have the improved estimate: $\forall \epsilon > 0$ and almost every $x \in \mathbb{T}^d$,*

$$\left| N(x; R) - \frac{(\pi R^2)^{d/2}}{(d/2)!} \right| \leq R^{(d/2)-(1/4)+\epsilon}, \qquad R = j^2 \to \infty.$$

Generalize to any power law $R = j^\alpha$ with $\alpha > 0$.

Kendall also obtained a formula for the limiting average variance, defined as

$$\text{(4.5.3)} \qquad \sigma^2 = \lim_{T \to \infty} \frac{1}{T} \int_1^T dR \int_{\mathbb{T}^d} R^{1-d} \left| N(x; R) - \frac{(\pi R^2)^{d/2}}{(d/2)!} \right|^2 dx.$$

In case $d = 2$, this can be computed from the asymptotic expansion of the Bessel function

$$F_R(n) = \frac{\sqrt{R}}{\pi |n|^{3/2}} \left[\cos\left(2\pi |n| R - \frac{3\pi}{4}\right) + O\left(\frac{1}{R}\right) \right].$$

Thus

$$\frac{1}{T}\int_1^T \left(\int_{\mathbb{T}^d} \frac{|N(x;R) - \pi R^2|^2}{R} dx\right) dR$$
$$= \sum_{0 \neq n \in \mathbb{Z}^2} \frac{1}{\pi^2 |n|^3} \frac{1}{T} \int_1^T \left[\cos^2\left(2\pi |n| R - \frac{3\pi}{4}\right) dR + O\left(\frac{1}{T}\right)\right].$$

The average value of the trigonometric term is $\frac{1}{2}$, leading to the evaluation

$$\sigma^2 = \frac{1}{2\pi^2} \sum_{0 \neq n \in \mathbb{Z}^2} \frac{1}{|n|^3},$$

which can be expressed in terms of the Riemann zeta function and Dirichlet's L function (see Kendall, 1948).

Exercise 4.5.4. *Obtain a formula for the limiting average variance σ^2 in the general case $d \geq 3$.*

4.5.2 Landau's Asymptotic Formula

A more specific result is obtained if we fix attention on a single point, which we take to be the origin. This leads to the famous *Landau estimate*, as follows.

Proposition 4.5.5. *The lattice-point counting function satisfies the asymptotic estimate*

(4.5.4) $$N(0;R) = \frac{(\pi R^2)^{d/2}}{(d/2)!} + O(R^{d-2+2/(d+1)}), \qquad R \to \infty.$$

Proof. For $d = 1$ the estimate is exact, so we assume that $d > 1$. The lattice point counting function can be represented as

$$N(0;R) = \sum_{\nu \in \mathbb{Z}^d} f_R(\nu)$$

where $f_R(x) = 1_{[0,R]}(|x|)$ with Fourier transform $F_R(k) = R^d J_{d/2}(2\pi R|k|)/(R|k|)^{d/2}$. However we cannot apply the pointwise form of the Poisson summation formula directly. Instead we will apply the pointwise form of the Poisson summation formula to the regularized function $f_R * \rho_\epsilon$ where $\rho_\epsilon(x) = \epsilon^{-d} \rho(x/\epsilon)$ and ρ is a nonnegative C^∞ function supported in the ball $|x| \leq 1$ and of total integral 1. Both ρ_ϵ and its Fourier transform are rapidly decreasing, so that we can apply the pointwise form of Poisson's formula to obtain

(4.5.5) $$N_\epsilon(R) := \sum_{\nu \in \mathbb{Z}^d} (f_R * \rho_\epsilon)(\nu) = \sum_{k \in \mathbb{Z}^d} F_R(k) \hat{\rho}_\epsilon(k) = \sum_{k \in \mathbb{Z}^d} F_R(k) \hat{\rho}(\epsilon k).$$

On the other hand, the smoothing density can be chosen so that for any desired integer $N > (d+1)/2$,

$$|\hat{\rho}(k)| \leq C_{d,N} \left(\frac{1}{1+|k|}\right)^N$$

for a constant $C_{d,N}$. The term of the series (4.5.5) with $k=0$ is simply the volume $\pi^{d/2} R^d/(d/2)!$. Subtracting this, we estimate the remainder as

$$|N_\epsilon(R) - (\pi R^2)^{d/2}/(d/2)!| = \left|\sum_{k \neq 0} F_R(k)\hat{\rho}(\epsilon k)\right|$$

$$\leq \frac{C_{d,N} R^d}{R^{(d+1)/2}} \sum_{k \neq 0} \frac{1}{k^{(d+1)/2}} \left(\frac{1}{1+\epsilon|k|}\right)^N$$

$$\leq \frac{C_{d,N} R^d}{R^{(d+1)/2}} \int_{|\xi| \geq 1/2} \frac{1}{\xi^{(d+1)/2}} \left(\frac{1}{1+\epsilon|\xi|}\right)^N d\xi$$

$$\leq C_{d,N} R^{(d-1)/2} \int_{\mathbb{R}^d} \frac{\epsilon^{(d+1)/2}}{|y|^{(d+1)/2}} \left(\frac{1}{1+|y|}\right)^N \epsilon^{-d} dy$$

(4.5.6) $$= \tilde{C}_{d,N} \left(\frac{R}{\epsilon}\right)^{(d-1)/2}.$$

On the other hand, since ρ_ϵ is supported in the ball of radius ϵ, we have

(4.5.7) $$(f_{R-\epsilon} * \rho_\epsilon)(v) \leq f_R(v) \leq (f_{R+\epsilon} * \rho_\epsilon)(v), \quad v \in \mathbb{R}^d.$$

This follows from the fact that the middle term is either zero or one; in the first case $|v| > R$ and the ball $\{|z - v| \leq \epsilon\}$ does not intersect the ball $\{|z| \leq R - \epsilon\}$. In the second case $|v| \leq R$ and the ball $\{|z - v| \leq \epsilon\}$ is contained in the ball $\{|z| \leq R + \epsilon\}$.

Summing (4.5.7) over $v \in \mathbb{Z}^d$, we obtain

$$N_\epsilon(R - \epsilon) \leq N(0; R) \leq N_\epsilon(R + \epsilon).$$

Applying the estimate (4.5.6) gives the upper and lower bounds

(4.5.8) $$N(0; R) \leq c_d (R+\epsilon)^d + \tilde{C}_{d,N} \left(\frac{R+\epsilon}{\epsilon}\right)^{(d-1)/2},$$

(4.5.9) $$N(0; R) \geq c_d (R-\epsilon)^d - \tilde{C}_{d,N} \left(\frac{R-\epsilon}{\epsilon}\right)^{(d-1)/2}.$$

The two error terms are balanced when we choose $\epsilon = R^{(1-d)/(1+d)}$. Making the necessary substitutions produces the stated result. ∎

4.5.3 Application to Multiple Fourier Series

We can use the Landau lattice-point formula to estimate the partial sums of multiple Fourier series of radial functions at the center of the torus. We begin with a function F on the real line, which is supported in the interval $[0, a]$, obtaining a radial function on $\mathbb{R}^d$ through the formula $f(x) = F(|x|)$.

The Fourier transform is again a radial function: $\hat{f}(\xi) = A(|\xi|)$, where A is the Hankel transform

$$A(|\xi|) = \hat{f}(\xi) = \int_{|x| \leq a} F(|x|) e^{-2\pi i \xi \cdot x} dx = C_d \int_0^a \frac{J_{(d-2)/2}(2\pi |\xi| r)}{(\pi r)^{(d-2)/2}} F(r) r^{d-1} dr.$$

We now consider the Fourier series of the periodized function

(4.5.10) $$\bar{f}(x) = \sum_{n \in \mathbb{Z}^d} F(|x-n|) \sim \sum_{m \in \mathbb{Z}^d} A(|m|) e^{2\pi i m \cdot x}.$$

The partial sum of the Fourier series at $x = 0$ is written as a Steiltjes integral:

$$S_M \bar{f}(0) = \sum_{|n| \leq M} A(n) = A(0) + \int_1^M A(\mu) \, dN(\mu).$$

4.5.3.1 Three-dimensional case

We can obtain a simple necessary and sufficient condition for the convergence of the spherical partial sums of a periodized radial function in three dimensions, as follows. The case $a < \frac{1}{2}$ was treated in Pinsky, Stanton, and Trapa (1993).

Proposition 4.5.6. *Suppose that $d = 3$ and that F is a C^2 function on the interval $[0, a]$ for some $a > 0$. Then the spherical partial sums of the Fourier series (4.5.10) converge at $x = 0$ if and only if $F(a) = 0$.*

This will be proved by developing an asymptotic expansion for the Fourier transform.

Lemma 4.5.7. *We have the asymptotic estimates when $\mu \to \infty$*

$$A(\mu) = -\frac{a \cos 2\pi a\mu}{\pi \mu^2} F(a) + \frac{\sin 2\pi \mu a}{2\pi^2 \mu^3} \frac{d}{dr}(rF(r))(a-0) + O\left(\frac{1}{\mu^4}\right),$$

$$A'(\mu) = \frac{2a^2 \sin 2\pi a\mu}{\mu^2} F(a) + O\left(\frac{1}{\mu^3}\right).$$

Proof. We have $A(\mu) = \hat{f}(\xi)$ where $\mu = |\xi|$ and

$$\hat{f}(\xi) = \int_{-\infty}^{\infty} \int_{-\infty}^{\infty} \int_{-\infty}^{\infty} f(x_1, x_2, x_3) e^{-2\pi i(\xi_1 x_1 + \xi_2 x_2 + \xi_3 x_3)} \, dx_1 \, dx_2 \, dx_3$$

$$= \int_0^a \int_0^\pi \int_0^{2\pi} F(r) e^{-2\pi i \mu r \cos\theta} r^2 \sin\theta \, dr \, d\theta \, d\phi$$

$$= 2 \int_0^a r^2 F(r) \frac{\sin 2\pi r\mu}{r\mu} \, dr$$

$$= -\frac{2}{\mu} \int_0^a rF(r) \frac{d}{dr} \left[\frac{\cos(2\pi \mu r)}{2\pi \mu}\right] dr$$

$$= -\frac{2}{\mu} \left(\frac{aF(a) \cos(2\pi a\mu)}{(2\pi \mu)} - \int_0^a \frac{\cos(2\pi \mu r)}{2\pi \mu} \frac{d}{dr}(rF(r)) \, dr\right).$$

The final integral can be integrated by parts once again to obtain the indicated form, with the indicated remainder term for $A(\mu)$. Finally the last formula can be differentiated to obtain the asymptotic formula for $A'(\mu)$. ∎

Now we use the Landau formula (4.5.4). We must compare a sum of the form

$$\int_0^M A(\mu)\, dN(\mu) \quad \text{with} \quad \int_0^M A(r) 4\pi r^2 \, dr.$$

To do this we integrate both by parts:

$$\int_0^M A(\mu) dN(\mu) = A(M)N(M) - \int_0^M A'(\mu) N(\mu)\, d\mu,$$

$$\int_0^M A(\mu) 4\pi \mu^2 \, d\mu = A(M)\frac{4\pi M^3}{3} - \int_0^M A'(\mu) \frac{4\pi \mu^3}{3}\, d\mu.$$

Now we subtract these two expressions and analyze the two terms separately.

If $F(a) = 0$, then we have the asymptotic forms

$$A(\mu) = C_2 \frac{\sin a\mu}{\mu^3} + O\left(\frac{1}{\mu^4}\right), \qquad A'(\mu) = C_2 \frac{a \cos a\mu}{\mu^3} + O\left(\frac{1}{\mu^4}\right).$$

Therefore the product $A(M)(N(M) - 4\pi M^3/3) = O(M^{3/2} \times M^{-3}) = O(M^{-3/2})$. Similarly the integral is estimated by $\int_1^\infty r^{3/2} r^{-3}\, dr$, an absolutely convergent integral. Therefore the difference

$$\int_0^M A(\mu) dN(\mu) - \int_0^M 4\pi \mu^2 A(\mu)\, d\mu$$

has a limit when $M \to \infty$. But the integral is easily seen to be convergent from the form of $A(r)$.

To treat the case $F(a) \neq 0$, we examine the sum $\sum_{\mu_k \leq |n| \leq \mu_{k+1}} A(|n|)$ for large k, where $\mu_k = (2k+1)/4a$. If the series $\sum_n \hat{f}(n)$ converges, this sum must tend to zero when $k \to \infty$. But the above analysis allows one to compare this with the corresponding integral, with a smaller error. Thus

$$\sum_{\mu_k \leq |n| < \mu_{k+1}} A(|n|) - \int_{\mu_{k-1}}^{\mu_k} 4\pi r^2 A(r)\, dr = \int_{\mu_k}^{\mu_{k+1}} A(\mu) d\left[N(\mu) - \frac{4\pi \mu^3}{3}\right].$$

When we integrate-by-parts, we find that the terms at the limits yield

$$A(\mu_k)\left[N(\mu_k) - \frac{4\pi \mu_k^3}{3}\right] = O(k^{-2} \times k^{3/2}) = O(k^{-1/2})$$

while the new integral is

$$\int_{\mu_k}^{\mu_{k+1}} \left[N(\mu) - \frac{4\pi \mu^3}{3}\right] A'(\mu) d\mu = O(k^{3/2} \times k^{-2}) = O(k^{-1/2}).$$

Thus

$$\sum_{\mu_k \leq |n| < \mu_{k+1}} A(|n|) - \int_{\mu_k}^{\mu_{k+1}} 4\pi \mu^2 A(\mu)\, d\mu = O(k^{-1/2}) \qquad k \to \infty.$$

On the other hand, the explicit form of $A(\mu)$ with $F(a) \neq 0$ shows that the integral has the explicit asymptotic expression

$$\int_{\mu_k}^{\mu_{k+1}} 4\pi \mu^2 A(\mu) = 4\pi F(a) \int_{\mu_k}^{\mu_{k+1}} \cos 2\pi a\mu\, d\mu + O\left(\frac{1}{k}\right)$$

$$= 2F(a)(-1)^k + O\left(\frac{1}{k}\right),$$

which fails to tend to zero when $k \to \infty$, completing the proof. ∎

4.5.3.2 Higher-dimensional case

In higher dimensions, we can use the above method to prove that the Fourier series of the periodized indicator function of a ball diverges at $x = 0$.

Proposition 4.5.8. *Suppose that $d > 3$. Then the spherical partial sums*

$$S_M f(0) = \sum_{|k| \leq M} a^k \frac{J_{d/2}(2\pi |k| a)}{|ka|^{\frac{d}{2}}}$$

are unbounded when $M \to \infty$, where $f(x) = 1_{[0,a]}(|x|)$.

Proof. We can repeat the asymptotic analysis done above in case $d = 3$. The Bessel function $A(\mu) = a^d (J_{d/2}(2\pi \mu a))/(a\mu)^{d/2}$ has the asymptotic behavior

$$A(\mu) = \frac{C_d}{\mu^{(d+1)/2}} \left[\cos\left(2\pi a\mu - \frac{(d-3)\pi}{4}\right) + O\left(\frac{1}{\mu}\right)\right],$$

$$A'(\mu) = \frac{-2\pi a C_d}{\mu^{(d+1)/2}} \left[\sin\left(2\pi a\mu - \frac{(d-3)\pi}{4}\right) + O\left(\frac{1}{\mu}\right)\right].$$

Letting μ_k be the consecutive zeros of the above cosine function, we estimate as before:

$$\sum_{\mu_k \leq |n| < \mu_{k+1}} A(|n|) - \int_{\mu_k}^{\mu_{k+1}} dc_d \mu^{d-1} A(\mu)\, d\mu = \int_{\mu_k}^{\mu_{k+1}} A(\mu) d\left[N(\mu) - c_d \mu^d\right].$$

When we integrate-by-parts, we find that the terms at the limits yield

$$A(\mu_k)\left[N(\mu_k) - c_d \mu_k^d\right] = O(k^{-(d+1)/2} \times k^{d-2+2/(d+1)}) = O(k^{(d-3)/2-(d-1)/(d+1)})$$

while the new integral is

$$\int_{\mu_k}^{\mu_{k+1}} \left[N(\mu) - c_d \mu^d\right] A'(\mu)\, d\mu = O(k^{-(d+1)/2} \times k^{d-2+2/(d+1)}) = O(k^{(d-3)/2-(d-1)/(d+1)}).$$

Thus we have when $k \to \infty$,

$$\sum_{\mu_k \leq |n| < \mu_{k+1}} A(|n|) - \int_{\mu_k}^{\mu_{k+1}} dc_d \mu^{d-1} A(\mu)\, d\mu = O(k^{-(d+1)/2} \times k^{d-2+2/(d+1)}) = O(k^{(d-3)/2-(d-1)/(d+1)}).$$

On the other hand, the explicit form of $A(\mu)$ shows that the integral has the explicit asymptotic expression

$$\int_{\mu_k}^{\mu_{k+1}} \mu^{k-1} A(\mu)\, d\mu = \int_{\mu_k}^{\mu_{k+1}} \mu^{(k-3)/2} \left[\cos\left(2\pi a\mu - \pi\frac{d-3}{4}\right) + O\left(\frac{1}{k}\right)\right] d\mu$$

$$= \text{const } k^{(d-3)/2} (-1)^k \left[1 + O\left(\frac{1}{k}\right)\right],$$

which is unbounded, and of larger order than the error term, completing the proof. ∎

4.6 SCHRÖDINGER EQUATION AND GAUSS SUMS

In this section we will use the Poisson summation formula to evaluate some finite sums that occur in number theory. In order to explain the setting, we first formulate the notion of Fourier series of Schwartz distributions on the circle. This will be applied to the fundamental solution of the Schrödinger equation, which is a finite linear combination of δ-distributions at equally spaced points whenever $2\pi t$ is a rational number. The treatment follows Taylor (1999).

4.6.1 Distributions on the Circle

$C^\infty(\mathbb{T})$ is the space of infinitely differentiable functions on $\mathbb{R}$ that are periodic with period 1. Convergence is defined by requiring that all derivatives converge uniformly on any period interval. This can be defined by the metric

$$d(\phi, \psi) = \sum_{n=0}^{\infty} 2^{-n} \frac{d_n}{1 + d_n}, \qquad d_n := \sup_{0 \le x \le 1} |\phi^{(n)}(x) - \psi^{(n)}(x)|.$$

A *periodic distribution* is a continuous linear functional L on the space $C^\infty(\mathbb{T})$. The Fourier coefficients of a periodic distribution are defined by

(4.6.1) $$\hat{L}(n) = L(e^{-2\pi i n})$$

and the Fourier series is written $L \sim \sum_{n \in \mathbb{Z}} \hat{L}(n) e^{2\pi i n x}$.

Example 4.6.1. *The linear functional $L(\phi) = \phi(0)$ is the Dirac mass at zero, written $L = \delta_0$. Its Fourier coefficients are given by $\hat{L}(n) \equiv 1$ so that we have the Fourier series*

$$\delta_0 \sim \sum_{n \in \mathbb{Z}} e^{2\pi i n x}.$$

Example 4.6.2. *Any $f \in L^1(\mathbb{T})$ becomes a periodic distribution by setting $L_f(\phi) = \int_\mathbb{T} f(x)\phi(x)\, dx$. The Fourier coefficients are given by the usual integrals*

$$\hat{L}_f(n) = \int_\mathbb{T} f(x) e^{-2\pi i n x}\, dx = \hat{f}(n).$$

The *Fourier representation* of a periodic distribution is obtained from the Fourier series of $\phi \in C^\infty(\mathbb{T})$, which has the convergent Fourier series

$$\phi(x) = \sum_{n \in \mathbb{Z}} \hat{\phi}(n) e^{2\pi i n x}$$

so that we can apply L to both sides to obtain the Fourier representation

(4.6.2) $$L(\phi) = \sum_{n \in \mathbb{Z}} \hat{\phi}(n) \hat{L}(-n).$$

This allows us to identify a distribution in terms of its Fourier series, as follows.

Proposition 4.6.3. *If L_1, L_2 are periodic distributions with the same Fourier series, then $L_1 = L_2$.*

Proof. Applying (4.6.2) to $L = L_1 - L_2$ shows that $L(\phi) = 0, \forall \phi \in C^\infty(\mathbb{T})$, hence L is identically zero. ∎

The Fourier representation can also be used to define new distributions.

Exercise 4.6.4. *Suppose that L_n is a bilateral sequence of complex numbers of polynomial growth: $|L_n| \leq C(1+|n|)^N$ for some $C > 0, N > 0$. Prove that there exists a periodic distribution L such that $\hat{L}(n) = L_n$ for every $n \in \mathbb{Z}$.*

Hint: For every $\phi \in C^\infty(\mathbb{T})$, and every $k > 0$, there exists $C_{kn} > 0$ so that $|\hat{\phi}(n)| \leq C_{kn}(1+|n|)^{-k}$ for every $n \in \mathbb{Z}$.

Exercise 4.6.5. *Define the convolution of a periodic distribution L with $\phi \in C^\infty(\mathbb{T})$ by $L * \phi(x) = L(\phi(x - \cdot))$. Show that $L * \phi \in C^\infty(\mathbb{T})$ and that we have the convergent Fourier series*

$$L * \phi(x) = \sum_{n \in \mathbb{Z}} \hat{L}(n)\hat{\phi}(n)e^{2\pi inx}.$$

An important class of distributions are those that are sums of a finite number of delta measures at equally spaced points; in detail we write $L = \sum_{j=0}^{N-1} c_j \delta_{j/N}$ where

$$L(\phi) = \sum_{j=0}^{N-1} c_j \phi\left(\frac{j}{N}\right).$$

The Fourier coefficients are

$$\hat{L}(k) = \sum_{j=0}^{N-1} c_j e^{-2\pi i kj/N}.$$

This sequence is periodic with period N, since

$$\hat{L}(k+N) = \sum_{j=0}^{N-1} c_j e^{-2\pi i(k+N)j/N} = \sum_{j=0}^{N-1} c_j e^{-2\pi i kj/N} = \hat{L}(k).$$

Conversely, suppose that we are given a bilateral sequence L_k with the property that for some $N \in \mathbb{Z}^+$, $L_{k+N} = L_k$ for all $k \in \mathbb{Z}$. The smallest such value N is called the *period*. Then we can uniquely solve the system of linear equations

$$L_k = \sum_{0 \leq j \leq N-1} c_j e^{-2\pi i jk/N} \qquad 0 \leq k \leq N-1$$

in the form

$$c_j = \frac{1}{N} \sum_{0 \leq j \leq N-1} L_k e^{2\pi i jk/N} \qquad 0 \leq j \leq N-1$$

to obtain a periodic distribution $L = \sum_{j=0}^{N-1} c_j \delta_{j/N}$ and conclude the following.

Proposition 4.6.6. *Every periodic sequence is the set of Fourier coefficients of a unique distribution that is obtained as a finite sum of delta measures: $L = \sum_{j=0}^{N-1} c_j \delta_{j/N}$ where N is the period of the sequence.*

Example 4.6.7. *Let $L = \frac{1}{N} \sum_{j=0}^{N-1} \delta_{j/N}$. The Fourier coefficients are given by $\hat{L}(n) = \frac{1}{N} \sum_{j=0}^{N-1} e^{-2\pi inj/N}$, which is one if $n = 0, \pm N, \pm 2N, \ldots$, and zero* otherwise. Therefore we have the Fourier series

$$\frac{1}{N}\sum_{j=0}^{N-1}\delta_{j/N} \sim \sum_{k\in\mathbb{Z}} e^{2\pi i Nkx}.$$

4.6.2 The Schrödinger Equation on the Circle

The initial-value problem for the one-dimensional Schrödinger equation on the circle is to find $u(x,t)$ defined for $x \in \mathbb{T}$, $t > 0$ so that

$$-i\frac{\partial u}{\partial t} = \frac{\partial^2 u}{\partial x^2}$$
$$u(x,0) = \phi(x) \in C^\infty(\mathbb{T}).$$

Separation of variables produces the factored solutions $u = e^{2\pi i n x} e^{-4\pi^2 i t n^2}$ and the Fourier series

$$u(x,t) = \sum_{n\in\mathbb{Z}} \hat{\phi}(n) e^{2\pi i n x} e^{-4\pi^2 i t n^2}.$$

It is immediately verified that for $\phi \in C^\infty(\mathbb{T})$ this series converges in $C^\infty(\mathbb{T})$ and that $\lim_{t\to 0} u(x,t) = \phi(x)$. The *fundamental solution* is the distribution with Fourier series

$$L_t \sim \sum_{n\in\mathbb{Z}} e^{2\pi i n x} e^{-4\pi^2 i t n^2}, \qquad \hat{L}_t(n) = e^{-4\pi^2 i t n^2}.$$

From the definition of convolution, we have $u = L_t * \phi$.

Proposition 4.6.8. *Assume that $2\pi t = M/N$ for some $M, N \in \mathbb{Z}^+$. Then $\hat{L}_t$ is periodic with period N and L_t is a finite sum of delta measures at the equally spaced points $0, 1/N, \ldots, (N-1)/N$ where the measure of j/N is*

$$c_j = C_{MN}(j/N), \qquad C_{MN}(x) := \sum_{r=0}^{N-1} e^{2\pi i r x} e^{-2\pi i r^2 M/N}$$

for $j = 0, \ldots, (N-1)/N$.

Proof. It is immediate that if $2\pi t = M/N$, then

$$\hat{L}_t(k+N) = e^{-4\pi^2 i(k+N)^2 M/2N\pi}$$
$$= e^{-2\pi i(N^2 + 2Nk + k^2)(M/N)}$$
$$= e^{-2\pi i k^2 (M/N)},$$

which proves the periodicity of the sequence. In order to obtain an explicit evaluation, we write the Fourier series at $2\pi t = M/N$: $L_t = \lim_{n\to\infty} \sum_{k=-n}^{n} \hat{L}_t(k)$ in the sense that for each $\phi \in C^\infty(\mathbb{T})$, we obtain $L_t(\phi)$ by integrating and taking the limit of the partial sums. This limit can be computed by writing $k = Nj + r$ and considering an arithmetic sequence of sums as follows:

$$\sum_{k=-Nn}^{N(n+1)-1} e^{2\pi ikx} e^{-2\pi ik^2(M/N)} = \sum_{j=-n}^{n} \sum_{r=0}^{N-1} e^{2\pi i(Nj+r)x} e^{-2\pi i(Nj+r)^2(M/N)}$$

$$= \sum_{j=-n}^{n} \sum_{r=0}^{N-1} e^{2\pi i(Nj+r)x} e^{-2\pi ir^2(M/N)}$$

$$= \sum_{j=-n}^{n} e^{2\pi iNjx} \sum_{r=0}^{N-1} e^{2\pi irx} e^{-2\pi ir^2(M/N)}$$

$$= C_{MN}(x) \sum_{j=-n}^{n} e^{2\pi iNjx}.$$

When $n \to \infty$, the last sum converges to the distribution $\sum_{j=0}^{N-1} \delta_{j/N}$, which completes the proof. ∎

We now look for another formula for L_t, by means of the Poisson summation formula. The solution of the Schrödinger equation on the real line can be obtained from the Fourier transform as

$$u(x, t) = \int_{\mathbb{R}} \hat{f}(\xi) e^{2\pi i\xi x} e^{-4\pi^2 it\xi^2} d\xi$$

where $f \in \mathcal{S}$. Clearly $u(\cdot, t) \in \mathcal{S}$ for each $t > 0$. This can be represented as an integral in x if we identify

$$e^{-4\pi^2 it\xi^2} = \int_{\mathbb{R}} e^{2\pi i\xi x} \frac{e^{-x^2/4ti}}{\sqrt{4\pi it}} dx$$

where the integral is taken as a Cauchy principal value and the square root is taken with positive real part. This gives the explicit representation

$$u(x, t) = \int_{\mathbb{R}} \frac{e^{-(x-y)^2/4ti}}{\sqrt{4\pi it}} f(y) \, dy.$$

Applying the Poisson summation formula leads to the representation

$$\sum_{n \in \mathbb{Z}} \hat{f}(n) e^{2\pi inx} e^{-4\pi^2 itn^2} = \sum_{v \in \mathbb{Z}} u(x - v, t)$$

$$= \int_{\mathbb{R}} f(y) \left(\sum_{v \in \mathbb{Z}} \frac{e^{-(x-y-v)^2/4it}}{\sqrt{4\pi it}} \right) dy.$$

This allows us to represent the distribution L_t as

$$L_t = \sum_{k \in \mathbb{Z}} \frac{e^{-(x-k)^2/4it}}{\sqrt{4\pi it}} = \lim_{n \to \infty} \sum_{k=-n}^{n} \frac{e^{-(x-k)^2/4it}}{\sqrt{4\pi it}}.$$

Taking $2\pi t = M/N$, we have

$$\frac{1}{4it} = \frac{N\pi}{2Mi}, \quad \frac{1}{4\pi it} = \frac{N}{2Mi}.$$

The sum is evaluated by letting $k = 2Mj + r$, leading to

$$e^{-(x-k)^2/4it} = e^{-(x-2Mj-r)^2(N\pi/2Mi)}$$
$$= e^{-(x^2+r^2-2rx-4Mjx)(N\pi/2Mi)}$$
$$= e^{-x^2 N\pi/2Mi} e^{-4Mjx(N\pi/2Mi)} e^{iN\pi rx/M} e^{ir^2 N\pi/2M}.$$

Therefore

$$\frac{1}{\sqrt{4\pi it}} \sum_{k=-2Mn}^{2M(n+1)-1} e^{-(x-k)^2/4it} = \sqrt{\frac{N}{2Mi}} e^{N\pi ix^2/2M} D_{MN}(x) \sum_{j=-n}^{n} e^{2\pi i N x j}$$

where we have set

$$D_{MN}(x) = \sum_{r=0}^{2M-1} e^{iN\pi rx/M} e^{-r^2 N\pi/2Mi}.$$

The last sum converges to a sum of unit delta measures at the points $0, 1/N, \ldots, (N-1)/N$. Equating the two forms of the distribution, we obtain the identity

$$C_{MN}(x) = \sqrt{\frac{N}{2Mi}} e^{-N\pi ix^2/2M} D_{MN}(x) \qquad x = 0, \frac{1}{N}, \ldots, \frac{N-1}{N}.$$

In detail,

(4.6.3) $$\sum_{r=0}^{N-1} e^{2\pi ir(k/N)} e^{-2\pi ir^2(M/N)} = \sqrt{\frac{N}{2Mi}} e^{-i\pi k^2/2MN} \sum_{r=0}^{2M-1} e^{i\pi rk/M} e^{ir^2 N\pi/2M}$$

for each $N = 1, 2, \ldots$ and each $k \in \{0, 1, \ldots, N-1\}$. In particular for $k = 0$ we have

$$\sum_{r=0}^{N-1} e^{-2\pi ir^2(M/N)} = \sqrt{\frac{N}{2Mi}} \sum_{r=0}^{2M-1} e^{ir^2 N\pi/2M}.$$

Specializing this to the case $M = 1$ yields the most classical Gauss sum:

$$\sum_{r=0}^{N-1} e^{-2\pi ir^2/N} = \sqrt{\frac{N}{2i}} (1 + i^N).$$

4.7 RECURRENCE OF RANDOM WALK

The formalism of multiple Fourier series can be combined with Laplace's asymptotic method to study the recurrent behavior of a simple random walk on the integer lattice.

The integer lattice $\mathbb{Z}^d$ is the set of all d-dimensional vectors whose coordinates are integers: $\mathbb{Z}^d = \{(k_1, \ldots, k_d) : k_j = 0, \pm 1, \pm 2, \ldots\}$. The simple random walk begins at the origin and moves with equal probability $1/2d$ to one of its neighbors, with the successive steps being independent of one another. In detail, denoting the coordinate

vectors by e_j,
$$\text{Prob}[S_{n+1} - S_n = \pm e_j] = 1/2d \qquad 1 \leq j \leq d, \ n = 0, 1, 2, \ldots.$$
Fourier series are introduced through the *characteristic function*
$$f_n(t) = E[e^{i\langle t, S_n \rangle}] = \sum_{k \in \mathbb{Z}^d} e^{i\langle t, k \rangle} \text{Prob}[S_n = k]$$
where $\langle , \rangle$ denotes the standard inner product of $\mathbb{R}^d$. The hypothesis of independence allows us to write
$$f_n(t) = E\left[\prod_{j=1}^n e^{i\langle t, (S_j - S_{j-1}) \rangle}\right] = \prod_{j=1}^n E[e^{i\langle t, (S_j - S_{j-1}) \rangle}] = f(t)^n.$$
But
$$f(t) = \frac{e^{it_1} + e^{-it_1} + \cdots + e^{it_d} + e^{-it_d}}{2d} = \frac{\cos t_1 + \cdots + \cos t_d}{d}.$$
The recurrent behavior is determined by $\text{Prob}[S_n = 0]$, as follows.

Proposition 4.7.1.

- If n is odd, then $\text{Prob}[S_n = 0] = 0$.
- If $n = 2m$ is even, then
$$\lim_m m^{d/2} \text{Prob}[S_{2m} = 0] = 2^{1-d} d^{d/2} \pi^{-d/2}.$$

Proof. A return to zero occurs at time n if and only if in each coordinate the number of positive and negative steps are equal, hence the total number of steps in each coordinate must be even, in particular the total number of steps (in all coordinates) must be even. We have
$$f(t)^n = \sum_{k \in \mathbb{Z}^d} e^{i\langle t, k \rangle} \text{Prob}[S_n = k].$$
From the formulas for multiple Fourier series,
$$\text{Prob}[S_n = k] = \frac{1}{(2\pi)^d} \int_{(-\pi, \pi)^d} f(t)^n e^{-i\langle t, k \rangle} \, dt.$$
In particular
$$\text{Prob}[S_{2m} = 0] = \frac{1}{(2\pi)^d} \int_{(-\pi, \pi)^d} f(t)^{2m} \, dt.$$
This can be further simplified by noting that f is 2π-periodic in each variable and satisfies the oddness property $f(t_1 + \pi, \ldots, t_d + \pi) = -f(t_1, \ldots, t_d)$. The periodicity allows us to write
$$\int_{(-\pi, \pi)^d} f(t)^{2m} \, dt = \int_{(-\pi/2, 3\pi/2)^d} f(t)^{2m} \, dt.$$
The oddness property further allows us to write $\int_{|t| < \delta} f(t)^{2m} \, dt = \int_{|t - \pi| < \delta} f(t)^{2m} \, dt$. Now we can apply the asymptotic method of Laplace. On the cube $(-\pi/2, 3\pi/2)^d$, $|f(t)| \leq 1$ with

global maxima assumed at $t_j \equiv 0$, and $t_j \equiv \pi$ where $|f(0)| = 1$, $|f(\pi)| = 1$. On the sets $|t| > \delta$, $|t - \pi| > \delta$ we have $|f(t)| \le 1 - \eta_1 < 1$ so that

$$\int_{|t|>\delta, |t-\pi|>\delta} f(t)^{2m}\, dt \le (2\pi)^d (1-\eta_1)^{2m}.$$

Thus

$$m^{d/2} \int_{(-\pi/2, 3\pi/2)^d} f(t)^{2m}\, dt = 2m^{d/2} \int_{|t|<\delta} f(t)^{2m}\, dt + O\!\left(m^{d/2}(1-\eta_1)^{2m}\right)$$

$$= 2 \int_{|s|<\delta m^{1/2}} f(s/\sqrt{m})^{2m}\, ds + o(1).$$

In order to apply the dominated convergence theorem, we first note that the integrand is bounded by the integrable function $e^{-|s|^2/3d}$. To find the limit, we use the Taylor series expansion and the inequality $|A^{2m} - B^{2m}| \le 2m|A - B|$ to write

$$f(t) = 1 - \frac{|t|^2}{2d} + O(|t|^4), \qquad t \to 0$$

$$f\!\left(\frac{s}{m^{1/2}}\right)^{2m} - e^{-|s|^2/d} = f\!\left(\frac{s}{\sqrt{m}}\right)^{2m} - (e^{-|s|^2/(2md)})^{2m}$$

$$\le 2m \left| f\!\left(\frac{s}{\sqrt{m}}\right) - e^{-|s|^2/(2md)} \right|$$

$$\le 2m \left(1 - \frac{|s|^2}{2md} + O\!\left(\frac{1}{m^2}\right) - 1 + \frac{|s|^2}{2md} + O\!\left(\frac{1}{m^2}\right) \right) \qquad m \to \infty$$

$$= O\!\left(\frac{1}{m}\right) \qquad m \to \infty$$

from which we conclude that

$$\lim_{m \to \infty} m^{d/2} \int_{(-\pi/2, 3\pi/2)^d} f(t)^{2m}\, dt = 2 \int_{\mathbb{R}^d} e^{-|s|^2/d}\, ds = 2(\pi d)^{d/2},$$

which gives the required result. ∎

This can be used to estimate the *mean occupation time* of the origin, defined by

$$\text{Mean occupation time} = \sum_{n=0}^{\infty} \text{Prob}[S_n = 0].$$

If the dimension $d = 1$, resp. $d = 2$, then the $2m$th term of this series is asymptotic to the general term of a divergent series ($1/\sqrt{m}$ resp. $1/m$), thus we have an infinite mean occupation time. However if $d \ge 3$, then the general term of the series is asymptotic to the general term of a convergent series ($1/m^{d/2}$), thus a finite mean occupation time. In this sense we say that symmetric random walk is recurrent in one and two dimensions and transient in dimensions three and greater.

Exercise 4.7.2. *Suppose that a one-dimensional random walk is defined by independent steps with*

$$\text{Prob}[S_n - S_{n-1} = 1] = p, \qquad \text{Prob}[S_n - S_{n-1} = 1] = 1 - p$$

where $p \neq 1/2$. Show that $\text{Prob}[S_{2m} = 0] = \binom{2m}{m} p^m (1-p)^m$. *Find an asymptotic formula for* $\text{Prob}[S_{2m} = 0]$ *and conclude the transient behavior.*

Exercise 4.7.3. *Suppose that a one-dimensional random walk is defined by independent steps with*
$$\text{Prob}[S_n - S_{n-1} = k] = p_k \qquad k = 0, \pm 1, \pm 2, \ldots$$
where $\sum_{k \in \mathbb{Z}} p_k = 1, \sum_{k \in \mathbb{Z}} k p_k = 0, \sum_{k \in \mathbb{Z}} k^2 p_k = \sigma^2 < \infty$. *Find an asymptotic formula for* $\text{Prob}[S_n = 0]$ *and conclude the recurrent behavior.*

CHAPTER

5

APPLICATIONS TO PROBABILITY THEORY

5.1 MOTIVATION AND HEURISTICS

The previous chapters have dealt with the Fourier analysis of functions in one and several dimensions. While this is sufficient for many applications, it does not fully cover problems from probability theory, where we must deal with measures that do not have a density with respect to Lebesgue measure. Unlike the L^2 theory of the Fourier transform from Chapter 2, this theory is inherently nonsymmetric: The Fourier transform of a finite measure is a continuous function rather than another measure. For this reason we change slightly the definition of the Fourier transform for notational convenience.

5.2 BASIC DEFINITIONS

Let m be a finite Borel measure on $\mathbb{R}^d$. This is a nonnegative, countably additive set function defined on the Borel sets of $\mathbb{R}^d$ with $m(\mathbb{R}^d) < \infty$. The Euclidean inner product is denoted $\xi \cdot x = \sum_{i=1}^{d} \xi_i x_i$. The Fourier transform of the measure m is the function

(5.2.1) $$\hat{m}(\xi) = \int_{\mathbb{R}^d} e^{i\xi \cdot x} m(dx).$$

Note the change in the sign convention and the omission of the factor 2π in the definition. These adjustments are made in order to conform with the conventions of the theory of probability. If m is a probability measure ($m(\mathbb{R}^d) = 1$), we refer to $\hat{m}$ as the *characteristic function* of the measure m. The properties are listed below.

Proposition 5.2.1. *The Fourier transform has the following properties:*

1. $\hat{m}$ *is a continuous function with* $\hat{m}(0) = m(\mathbb{R}^d)$.

2. $\hat{m}$ is a positive definite function, meaning that for every set of complex numbers $(c_j)_{1\le j\le n}$ and vectors $(\xi_j)_{1\le j\le n}$

$$\sum_{j,k=1}^n c_j \bar{c}_k \hat{m}(\xi_j - \xi_k) \ge 0.$$

3. If m_1, m_2 are two measures, the Fourier transform of the convolution is the product of their Fourier transforms, where the convolution is defined by

$$(m_1 * m_2)(B) = \int_{\{(x,y):x+y\in B\}} m_1(dx) m_2(dy).$$

Equivalently, for any bounded continuous function g

$$\int_{\mathbb{R}^d} g(z)(m_1 * m_2)(dz) = \int_{\mathbb{R}^{2d}} g(x+y) m_1(dx) m_2(dy).$$

Proof. The continuity of $\hat{m}$ follows from the dominated convergence theorem: If $\xi_n \to \xi$, then the complex-valued functions $e^{i\xi_n \cdot x}$ are bounded by 1, and converge to $e^{i\xi \cdot x}$ when $n \to \infty$. The positive definite property is a direct computation:

$$\sum_{j,k=1}^n c_j \bar{c}_k \hat{m}(\xi_j - \xi_k) = \sum_{j,k=1}^n c_j \bar{c}_k \int_{\mathbb{R}^d} e^{i\xi_j \cdot x} e^{-\xi_k \cdot x} m(dx)$$

$$= \int_{\mathbb{R}^d} \left| \sum_{j=1}^n c_j e^{i\xi_j \cdot x} \right|^2 m(dx)$$

$$\ge 0.$$

To prove the convolution property, multiply the two transforms to obtain

$$\hat{m}_1(\xi) \hat{m}_2(\xi) = \int_{\mathbb{R}^{2d}} e^{i\xi \cdot (x+y)} m_1(dx) m_2(dy) = \int_{\mathbb{R}^d} e^{i\xi \cdot (x+y)} (m_1 * m_2)(dz),$$

which was to be proved. ∎

Example 5.2.2. The centered Gaussian distribution with variance parameter $\sigma > 0$ is the measure with density $e^{-|x|^2/2\sigma^2}$. Its Fourier transform can be computed in terms of a product of one-dimensional transforms as

$$\hat{m}(\xi) = \int_{\mathbb{R}^d} e^{i\xi \cdot x} e^{-|x|^2/2\sigma^2} dx$$

$$= \int_{\mathbb{R}^d} \prod_{j=1}^d \left(e^{i\xi_j x_j} e^{-x_j^2/2\sigma^2} dx_j \right)$$

$$= \prod_{j=1}^d \left(\int_{\mathbb{R}} e^{i\xi_j x_j} e^{-x_j^2/2\sigma^2} dx_j \right)$$

$$= \prod_{j=1}^d \left(\sqrt{2\pi\sigma^2} e^{-\xi_j^2 \sigma^2} \right)$$

$$= (2\pi\sigma^2)^{d/2} e^{-\sigma^2 |\xi|^2/2}.$$

Example 5.2.3. *The Fourier transform of the uniform measure on the rectangle $\prod_{j=1}^{d}(a_j, b_j)$ is computed as*

$$\hat{m}(\xi) = \int_{\prod_{j=1}^{d}(a_j,b_j)} e^{i\xi \cdot x}\, dx = \prod_{j=1}^{d} \int_{a_j}^{b_j} e^{i\xi_j x_j}\, dx_j = \prod_{j=1}^{d} \frac{e^{i\xi_j b_j} - e^{i\xi_j a_j}}{i\xi_j}, \qquad \xi_j \neq 0.$$

If $\xi_j = 0$ for some j, then the corresponding factor is replaced by $b_j - a_j$.

We now prove that the mapping $m \to \hat{m}$ is $1:1$.

Proposition 5.2.4. *The measure m can be retrieved from its Fourier transform by the inversion formula*

$$(2\pi)^d m\left(\prod_{j=1}^{d}(a_j, b_j)\right) = \lim_{\sigma \to 0} \int_{\mathbb{R}^d} \hat{m}(\xi) e^{-\sigma^2 |\xi|^2/2} \prod_{j=1}^{d} \frac{e^{-i\xi_j b_j} - e^{-i\xi_j a_j}}{-i\xi_j}\, d\xi$$

provided that the m-measure of the boundary of this rectangle is zero. In particular, m is uniquely determined by $\hat{m}$.

Proof. Multiply the defining equation (5.2.1) by $e^{-\sigma^2 |\xi|^2/2} e^{-i\xi \cdot y}$ to obtain

$$e^{-\sigma^2 |\xi|^2/2} e^{-i\xi \cdot y} \hat{m}(\xi) = \int_{\mathbb{R}^d} e^{i\xi \cdot (x-y)} e^{-\sigma^2 |\xi|^2/2}\, m(dx).$$

If we integrate this with respect to ξ and use the Gaussian Example 5.2.2, we obtain

$$\int_{\mathbb{R}^d} e^{-\sigma^2 \xi \cdot \xi /2} e^{-i\xi \cdot y} \hat{m}(\xi)\, d\xi = (2\pi)^d \int_{\mathbb{R}^d} \frac{e^{-|x-y|^2/2\sigma^2}}{(\sqrt{2\pi\sigma^2})^{d/2}} m(dx).$$

Now we integrate with respect to y over $\prod_j (a_j, b_j)$ to obtain

(5.2.2)
$$\int_{\mathbb{R}^d} e^{-\sigma^2 \xi \cdot \xi /2} \prod_{j=1}^{d} \frac{e^{-i\xi_j b_j} - e^{-i\xi_j a_j}}{-i\xi_j} \hat{m}(\xi)\, d\xi$$
$$= (2\pi)^d \int_{\mathbb{R}^d} \left(\int_{\prod_{j=1}^{d}(a_j,b_j)} \frac{e^{-|x-y|^2/2\sigma^2}}{(\sqrt{2\pi\sigma^2})^{d/2}}\, dy \right) m(dx).$$

The integrand on the right side is bounded by 1; it tends to 1 if $x \in \prod_{j=1}^{d}(a_j, b_j)$ and tends to zero if $x \notin \prod_{j=1}^{d}[a_j, b_j]$. The boundary of this rectangle is supposed to have m-measure zero. The conclusion now follows from the dominated convergence theorem. ■

If the Fourier transform of a finite measure is integrable, then the measure has a density that can be recovered by Fourier inversion, as follows.

Corollary 5.2.5. *If $\hat{m} \in L^1(\mathbb{R}^d)$ then the measure m has a density, given by*

$$\frac{dm}{dy} = \frac{1}{(2\pi)^d} \int_{\mathbb{R}^d} e^{-i\xi \cdot y} \hat{m}(\xi)\, d\xi.$$

Proof. With this extra hypothesis, we can take the limit under the integral in Proposition 5.2.4 to obtain

$$(2\pi)^d m\left(\prod_{j=1}^d (a_j, b_j)\right) = \int_{\mathbb{R}^d} \hat{m}(\xi) \prod_{j=1}^d \frac{e^{-i\xi_j b_j} - e^{-i\xi_j a_j}}{-i\xi_j}\, d\xi$$

$$= \int_{\prod_{j=1}^d (a_j, b_j)} \left(\int_{\mathbb{R}^d} e^{-i\xi \cdot y} \hat{m}(\xi)\, d\xi\right) dy,$$

which displays the measure of the rectangle as the integral of the required density function. ∎

These ideas allow us to generalize Maxwell's characterization of the Gaussian density in Chapter 2, Proposition 2.2.51, from density functions to the larger class of finite measures on $\mathbb{R}^d$. The precise statement is the following.

Proposition 5.2.6. *Suppose that m is a finite measure on $\mathbb{R}^d$, $d \geq 2$ with the following two properties:*

(5.2.3) $\qquad m(dx_1 \cdots dx_d) = m_1(dx_1) \cdots m_d(dx_d)$

where $m_1, \ldots, m_d$ are finite measures on $\mathbb{R}$.
For any orthogonal transformation T of $\mathbb{R}^d$,

(5.2.4) $\qquad \int_{\mathbb{R}^d} f(Tx) m(dx) = \int_{\mathbb{R}^d} f(x) m(dx), \qquad \forall f \in C(\mathbb{R}^d).$

Then either $m = A\,\delta_0$ or $m(dx) = A e^{-B|x|^2}\, dx$ where $A \geq 0$, $B > 0$.

Proof. The Fourier transform $\hat{m}(\xi) = \int_{\mathbb{R}^d} e^{i\xi \cdot x} m(dx)$ satisfies the corresponding factorization

(5.2.5) $\qquad \hat{m}(\xi_1, \ldots, \xi_d) = \hat{m}_1(\xi_1) \cdots \hat{m}_d(\xi_d).$

Taking $f(x) = e^{-i\xi \cdot x}$, (5.2.4) shows that $\hat{m}(T\xi) = \hat{m}(\xi)$, hence $\hat{m}(\xi) = G(|\xi|^2)$ for some continuous function G. Now we can follow the steps of the proof of Proposition 2.2.51 in Chapter 2 to conclude that, since G is bounded, $G(x) = A e^{-Bx}$, where $B \geq 0$. If $B = 0$, then m is a multiple of δ_0; otherwise $B > 0$, which completes the proof. ∎

Remark. We can render more transparent the computations in Proposition 5.2.4 and Corollary 5.2.5 following by using the notations of Chapter 2, beginning with the Fourier representation of the heat kernel H_t, a bounded function with $\hat{H}_t \in L^1(\mathbb{R}^d)$:

$$H_t(x-y) = \int_{\mathbb{R}^d} e^{2\pi i \xi \cdot (x-y)} \hat{H}_t(\xi)\, d\xi.$$

Multiply both sides by $1_A(y)m(dx)$ and integrate over $\mathbb{R}^d \times \mathbb{R}^d$:

$$\int_{\mathbb{R}^d} (H_t * 1_A)(x) m(dx) = \int_{\mathbb{R}^d} \hat{m}(-\xi) \hat{1}_A(\xi) \hat{H}_t(\xi) \, d\xi$$

$$m(A) = \lim_{t \to 0} \int_{\mathbb{R}^d} \hat{m}(-\xi) \hat{1}_A(\xi) \hat{H}_t(\xi) \, d\xi \quad \text{if } m(\partial A) = 0$$

$$= \int_{\mathbb{R}^d} \hat{m}(-\xi) \hat{1}_A(\xi) \, d\xi \quad \text{if } \hat{m} \in L^1(\mathbb{R}^d)$$

$$= \int_A \left(\int_{\mathbb{R}^d} e^{2\pi i \xi \cdot x} \hat{m}(\xi) \, d\xi \right) dx \quad \text{by Fubini.}$$

The above methods can also be used to prove the following *continuity theorem* for Fourier transforms of measures.

Proposition 5.2.7. *Suppose that $m_n, n = 0, 1, 2, \ldots$ is a sequence of finite measures whose Fourier transforms converge:*

$$\lim_n \hat{m}_n(\xi) = \hat{m}_0(\xi).$$

Then the measures converge on every rectangle whose boundary has m_0 measure zero.

Proof. For any $\phi \in \mathcal{S}$, let $\psi(\xi) = \hat{\phi}(-\xi)$, so that $\phi = \hat{\psi}$. Then by Fourier reciprocity we have

$$\int_{\mathbb{R}^n} \psi(\xi) \hat{m}_n(\xi) \, d\xi = \int_{\mathbb{R}^n} \phi(x) \, m_n(dx).$$

Letting $n \to \infty$, the dominated convergence theorem implies that

$$\lim_n \int_{\mathbb{R}^n} \phi(x) m_n(dx) = \lim_n \int_{\mathbb{R}^n} \psi(\xi) \hat{m}_n(\xi) \, d\xi = \int_{\mathbb{R}^n} \psi(\xi) \hat{m}_0(\xi) \, d\xi = \int_{\mathbb{R}^n} \phi(x) m_0(dx).$$

If $R = \Pi_{j=1}^n (a_j, b_j)$ is any rectangle, let $\phi^\pm \in \mathcal{S}$ so that $\phi^- \leq 1_R \leq \phi^+$. Thus

$$\limsup_n m_n(R) \leq \limsup_n \int_{\mathbb{R}^n} \phi^+(x) m_n(dx) = \int_{\mathbb{R}^n} \phi^+(x) m_0(dx)$$

$$\liminf_n m_n(R) \geq \liminf_n \int_{\mathbb{R}^n} \phi^-(x) m_n(dx) = \int_{\mathbb{R}^n} \phi^-(x) m_0(dx).$$

Now let $\phi^+ \downarrow 1_R$, $\phi^- \uparrow 1_R$ to conclude that

$$m_0(R^o) \leq \liminf_n m_n(R) \leq \limsup_n m_n(R) \leq m_0(\bar{R}).$$

If $m_0(\partial R) = 0$, then the extreme values are equal, so that the limit exists as required. ∎

5.2.1 The Central Limit Theorem

Fourier analysis of measures is particularly well suited to study the convolutions of a single probability measure. In the case of the measure $p\delta_1 + (1-p)\delta_0$ this was effectively studied in Chapter 1 in connection with the DeMoivre-Laplace local limit theorem. The central limit theorem extends this to an arbitrary probability measure with a finite second moment, which we now describe.

Theorem 5.2.8. *Suppose that m is a probability measure on the real line with*

$$\int_{\mathbb{R}} x \, m(dx) = 0 \qquad \int_{\mathbb{R}} x^2 \, m(dx) = \sigma^2 < \infty.$$

Then for any interval A,

$$\lim_n (m * m * \cdots * m)(A\sqrt{n}) = \frac{1}{\sqrt{2\pi\sigma^2}} \int_A e^{-x^2/\sigma^2} \, dx.$$

Proof. From Proposition 5.2.7 it suffices to compute the Fourier transform of the indicated convolution. This is $[\hat{m}(\xi/\sqrt{n})]^n$. The Taylor expansion at $\xi = 0$ is

$$\hat{m}(\xi) = 1 - \xi^2 \sigma^2/2 + o(\xi^2).$$

The characteristic function of the Gaussian measure with mean zero and variance σ^2 has the same Taylor expansion. Now we can write

$$\left| \hat{m}\left(\frac{\xi}{\sqrt{n}}\right)^n - e^{-\xi^2/2\sigma^2} \right| \leq n \left| \hat{m}\left(\frac{\xi}{\sqrt{n}}\right) - e^{-\xi^2/2n\sigma^2} \right|$$

where we have used the fact that if a, b are complex numbers with $|a| \leq 1, |b| \leq 1$, then $|a^n - b^n| \leq n|a - b|$. But from the Taylor expansions,

$$\hat{m}\left(\frac{\xi}{n^{1/2}}\right) - e^{-\xi^2/2n\sigma^2} = o(1/n),$$

which proves that $\lim_n \hat{m}(\xi/n^{1/2})^n = e^{-\xi^2/2\sigma^2}$. The conclusion now follows from the continuity theorem proved above. ∎

Exercise 5.2.9. *Central limit theorem for Abel sums.* Suppose that m is a probability measure on the real line with

$$\int_{\mathbb{R}} x \, m(dx) = 0, \qquad \int_{\mathbb{R}} x^2 \, m(dx) = 1.$$

For $0 < r < 1$ and $n = 1, 2, \ldots$, let $m_n^r(A) = m(A/(r^n\sqrt{1-r^2}))$. Prove that for any interval A

$$\lim_{r \to 1} (m_1^r * m_2^r * \cdots)(A) = \frac{1}{\sqrt{2\pi}} \int_A e^{-x^2/2} \, dx.$$

Hint: It suffices to show that the Fourier transform satisfies the limiting relation $\lim_{r \to 1} \prod_{n=0}^{\infty} \hat{m}(tr^n \sqrt{1-r^2}) = e^{-t^2/2}$. Use the fact that $\hat{m}(\xi) = e^{-\xi^2/2(1+o(1))}$ when $\xi \to 0$.

5.2.1.1 Restatement in terms of independent random variables

The central limit theorem is presented as a result on the convolution powers of a single probability measure. This can also be recast as a result about the measure induced by a *sum of independent random variables*, as follows:

Definition 5.2.10. *A set of real-valued functions $X_1(t), \ldots, X_n(t)$ on a probability measure space (Ω, μ) is* mutually independent *if for every choice of real numbers $x_1, \ldots, x_n$, we have*

$$\mu\{t : X_1(t) \leq x_1, \ldots, X_n(t) \leq x_n\} = \mu\{t : X_1(t) \leq x_1\} \cdots \mu\{t : X_n(t) \leq x_n\}.$$

Let m_i be the distribution of X_i, namely the measure induced on $\mathbb{R}$ by the equation $m_i(A) = \mu\{t : X_i(t) \in A\}$. Then the distribution of the sum $X_1(t) + \cdots + X_n(t)$ is the convolution $m_1 * \cdots * m_n$, so that $(m * \cdots * m)(A\sqrt{n})$ is the distribution of the sum $(X_1(t) + \cdots + X_n(t))/\sqrt{n}$. In probability theory, the term "random variable" is synonymous with "real-valued measurable function." The central limit theorem can now be recast as follows:

Theorem 5.2.11. *Suppose that $\{X_n(t)\}_{n=1,2,\ldots}$ is a sequence of mutually independent random variables with distribution m, where $\int_{\mathbb{R}} x\, m(dx) = 0$, $\int_{\mathbb{R}} x^2\, m(dx) = \sigma^2 < \infty$. Then the distribution of the normalized sum $[X_1(t) + \cdots + X_n(t)]/\sigma\sqrt{n}$ converges to a standard normal distribution when $n \to \infty$.*

Independent random variables may be constructed on the unit interval $\Omega = [0, 1]$ as follows. Let $\phi : \mathbb{N} \to \mathbb{N}^2$ be a bijective mapping. For example, this may be constructed by listing all of the integers in a doubly infinite array as follows:

$$\begin{array}{cccccc} 1 & 3 & 6 & 10 & 15 & 21\ldots \\ 2 & 5 & 9 & 14 & 20\ldots \\ 4 & 8 & 13 & 19\ldots \\ 7 & 12 & 18\ldots \\ 11 & 17\ldots \\ 16 & \ldots \end{array}$$

In this example we have, for example $\phi(8) = (2, 3)$, $\phi(18) = (3, 4)$ and so forth. Now we expand $t = \sum_{k=1}^{\infty} \omega_k/2^k$ and define

$$X_1(t) = \sum_{n=1}^{\infty} \frac{\omega_{\phi(1,n)}}{2^n} = \frac{\omega_1}{2} + \frac{\omega_3}{4} + \frac{\omega_6}{8} + \cdots$$

$$X_2(t) = \sum_{n=1}^{\infty} \frac{\omega_{\phi(2,n)}}{2^n} = \frac{\omega_2}{2} + \frac{\omega_5}{4} + \frac{\omega_9}{8} + \cdots$$

$$X_3(t) = \sum_{n=1}^{\infty} \frac{\omega_{\phi(3,n)}}{2^n} = \frac{\omega_4}{2} + \frac{\omega_8}{4} + \frac{\omega_{13}}{8} + \cdots$$

and so forth. Then for every n, $\{X_1(t), X_2(t), \ldots, X_n(t)\}$ are independent random variables, each of which is distributed according to Lebesgue measure on $[0, 1]$. To achieve more general distributions, it suffices to form Borel functions in the form $Y_i(t) = \phi_i(X_i(t))$ for $i = 1, 2, \ldots$.

5.3 EXTENSION TO GAP SERIES

The asymptotic normal distribution is not restricted to sums of independent random variables. In this section we consider a class of trigonometric series that are asymptotically normal. More general results are found in the book of Zygmund (1959, Volume 2,

Chapter XVI). Here we consider sums of the form

(5.3.1) $$S_k(t) = \sum_{j=1}^{k} a_j \cos n_j t$$

where (a_j) are real numbers and $n_1 < n_2 < \cdots$ are integers which satisfy

(5.3.2) $$n_{k+1} \geq q n_k \qquad (k = 1, 2, \ldots)$$

for some $q > 1$. The growth of the sum is measured by the L^2 norm, which is

(5.3.3) $$A_k = \left(\frac{1}{2} \sum_{j=1}^{k} a_j^2 \right)^{1/2}$$

where we assume that

(5.3.4) $$A_k \to \infty, \quad \frac{a_k}{A_k} \to 0 \qquad (k \to \infty).$$

As preparation for the theorem, we first prove a simple lemma.

Lemma 5.3.1. *Under the conditions (5.3.4), we have*

$$\frac{1}{A_k} \max_{1 \leq j \leq k} |a_j| \to 0 \qquad (k \to \infty).$$

Proof. Given $\epsilon > 0$, let K_ϵ be such that $|a_k|/A_k < \epsilon$ for $k > K_\epsilon$. On the one hand, since (A_k) is increasing, we have for $k > K_\epsilon$

$$\frac{1}{A_k} \max_{K_\epsilon \leq j \leq k} |a_j| \leq \frac{1}{A_k} \epsilon \max_{K_\epsilon \leq j \leq k} |A_j| = \epsilon,$$

while

$$\frac{1}{A_k} \max_{1 \leq j \leq K_\epsilon} |a_j| \to 0 \qquad (k \to \infty).$$

Hence $\limsup_{k \to \infty} (1/|A_k|) \max_{1 \leq j \leq k} |a_j| \leq \epsilon$. But ϵ was arbitrary, so the proof is complete. ∎

This lemma allows one to conclude, for example, that for any $p > 2$

$$\sum_{1 \leq i \leq k} \frac{|a_i|^p}{A_k^p} \leq 2 \max_{1 \leq i \leq k} \frac{|a_i|^{p-2}}{A_k^{p-2}} \to 0.$$

Theorem 5.3.2. *Suppose that the integers (n_k) and the real numbers (a_j) satisfy (5.3.2) and (5.3.4) with $q \geq 3$. Then we have for $k \to \infty$*

(5.3.5) $$\frac{1}{2\pi} \left| \left\{ t \in \mathbb{T} : y_1 \leq \frac{S_k(t)}{A_k} \leq y_2 \right\} \right| \to \frac{1}{\sqrt{2\pi}} \int_{y_1}^{y_2} e^{-u^2/2} \, du.$$

Proof. We compute the Fourier transform, defining

$$\Phi_k(\xi) = \frac{1}{2\pi} \int_{\mathbb{T}} e^{i\xi S_k(t)/A_k} \, dt$$

$$= \frac{1}{2\pi} \int_{\mathbb{T}} \prod_{j=1}^{k} e^{i\xi(a_j/A_k)\cos n_j t} \, dt.$$

It suffices to prove that $\Phi_k(\xi) \to e^{-\xi^2/2}$ when $k \to \infty$. From the power series of the exponential function, we have for small $|z|$,

$$(1+z)e^{-z} = (1+z)\left(1 - z + \frac{z^2}{2} + O(|z|^3)\right)$$

$$= 1 - \frac{z^2}{2} + O(|z|^3)$$

$$= e^{-z^2/2 + O(|z|^3)}$$

so that

$$e^z = (1+z)e^{z^2/2 + O(|z|^3)} \qquad (z \to 0)$$

and we can write

$$\Phi_k(\xi) = \frac{1}{2\pi} \int_{\mathbb{T}} \prod_{j=1}^{k} \left(1 + i\xi \frac{a_j}{A_k} \cos n_j t\right) \exp\left\{-\xi^2 \frac{a_j^2}{2A_k^2}[\cos^2 n_j t + o(1)]\right\} dt$$

where the $o(1)$ term is uniform in $t \in \mathbb{T}$. Now we write

$$\sum_{j=1}^{k} \frac{a_j^2}{A_k^2} \cos^2 n_j t = 1 + \frac{1}{2}\sum_{j=1}^{k} \frac{a_j^2}{A_k^2} \cos 2n_j t$$

$$:= 1 + T_k(t)$$

noting that

$$\frac{1}{2\pi}\int_{\mathbb{T}} T_k(t)^2 \, dt = \frac{1}{8} \sum_{j=1}^{k} \frac{a_j^4}{A_k^4} \to 0 \qquad (k \to \infty),$$

in particular $T_k(t) \to 0$ in measure. Meanwhile

$$\left|\prod_{j=1}^{k}\left(1 + i\xi \frac{a_j}{A_k}\cos n_j t\right)\right|^2 \le \prod_{j=1}^{k}\left(1 + \frac{\xi^2 a_j^2}{A_k^2}\right) \le e^{2\xi^2}$$

so that we can write

$$\Phi_k(\xi) = \frac{1}{2\pi}\int_{\mathbb{T}} \prod_{j=1}^{k}\left(1 + i\xi \frac{a_j}{A_k}\cos n_j t\right)\exp\left[-\frac{x^2}{2}(1 + \xi_k(t) + o(1))\right] dt$$

$$= o(1) + \frac{e^{-x^2/2}}{2\pi}\int_{\mathbb{T}} \prod_{j=1}^{k}\left(1 + ix\frac{a_j}{A_k}\cos n_j t\right) dt.$$

It remains to analyze the final integral. For this purpose we expand the cosine products using repeatedly the identity $2\cos a \cos b = \cos(a+b) + \cos(a-b)$ to obtain a finite sum

(5.3.6) $$\prod_{j=1}^{k}\left(1 + i\xi \frac{a_j}{A_k} \cos n_j t\right) = \sum_{\nu=0}^{n_1+\cdots+n_k} \alpha_\nu \cos \nu t$$

where the sum is over those indices of the form $\nu = n_{i_1} \pm n_{i_2} \pm \cdots$ with $n_{i_1} > n_{i_2} > \cdots$.

Lemma 5.3.3. *Suppose that $n_{k+1} \geq q n_k$ with $q \geq 3$. Suppose that an integer ν is represented in two (possibly) different ways*

$$n_{i_1} \pm n_{i_2} \pm \cdots = \nu = n_{j_1} \pm n_{j_2} \pm \cdots$$

where $i_1 > i_2 > \cdots, j_1 > j_2 > \cdots$. Then $i_1 = j_1, i_2 = j_3, \ldots$.

Proof. If all of the subscripts are equal, there is nothing to prove. Otherwise there is a first subscript that differs in the two representations. By relabeling the subscripts, we may assume without loss of generality that $i_1 > j_1$. By a further relabeling and moving all of the terms to one side, we may write

$$n_{i_1} = a_1 n_{k_1} + a_2 n_{k_2} + \cdots$$

where the coefficients $a_j \in \{0, \pm 1, \pm 2\}$ and $i_1 > k_1 > k_2 > \cdots$. Hence

$$n_{i_1} \leq 2\left(n_{k_1} + n_{k_2} + \cdots\right)$$
$$\leq 2\left(n_{k_1} + \frac{1}{3}n_{k_1} + \frac{1}{9}n_{k_1} + \cdots\right)$$
$$< 3 n_{k_1},$$

which is a contradiction.

To complete the proof of the theorem, we note that in the product (5.3.6) the only contribution to the term α_0 occurs when all of the frequencies are zero, hence $\alpha_0 = 1$. Applying the orthogonality of $\cos \nu t$, we conclude that

$$\frac{1}{2\pi}\int_{\mathbb{T}} \prod_{j=1}^{k}\left(1 + i\xi \frac{a_j}{A_k}\cos n_j t\right) dt = 1,$$

which completes the proof of the theorem. ∎

Exercise 5.3.4. *Suppose that the coefficients (a_n) satisfy (5.3.4). Prove that $n^{-1} \log|a_n| \to 0$ when $n \to \infty$.*

Hint: Write $a_n^2/A_n^2 = \epsilon_n \to 0$ and solve for $a_n^2 = a_N^2(\epsilon_n/\epsilon_N)\prod_{N+1}^{n}(1/(1-\epsilon_k))$ for $n > N$, where $\epsilon_N > 0$ and $\epsilon_k < \epsilon$ for $n > N$.

Exercise 5.3.5. *Suppose that the coefficients (a_n) satisfy (5.3.4) and the integers (n_k) satisfy (5.3.2). Let $S_k(t) = \sum_{j=1}^{k} a_j \cos(n_j t - \theta_j)$ where $\theta_j \in \mathbb{R}$. Prove that (5.3.5) holds for this wider class of series.*

5.3.1 Extension to Abel Sums

The central limit theorem for gap series can naturally be extended to obtain the limiting distribution of the harmonic function

$$(5.3.7) \qquad u(r,t) = \sum_{j=1}^{\infty} a_j r^{n_j} \cos n_j t$$

under the same conditions as in Theorem 5.3.2.

We define

$$(5.3.8) \qquad A(r)^2 := \frac{1}{2\pi} \int_{\mathbb{T}} u(r,t)^2 \, dt = \frac{1}{2} \sum_{j=1}^{\infty} a_j^2 r^{2n_j}.$$

Then Exercise 5.3.5 implies that $A(r) < \infty$ for $0 < r < 1$.

The following result was first proved by Kac (1939) and later extended by Salem and Zygmund (1948).

Theorem 5.3.6. *Suppose that the integers (n_k) and the real numbers (a_j) satisfy (5.3.2) and (5.3.4) with $q \geq 3$. Then for any interval $C \subset \mathbb{T}$, we have for $r \to 1$*

$$(5.3.9) \qquad \left| \left\{ t \in \mathbb{T} : \frac{u(r,t)}{A(r)} \in C \right\} \right| \to \int_C \frac{e^{-y^2/2}}{\sqrt{2\pi}} \, dy.$$

We first develop the Abelian counterpart of Lemma 5.3.1.

Lemma 5.3.7. *Let $b_k \geq 0$ for $k \geq 1$ and set $B_0 = 0$, $B_n = b_1 + \cdots + b_n$ for $n \geq 1$. Suppose that $B_n \to \infty$ and $b_n/B_n \to 0$ when $n \to \infty$. Furthermore let $B(r) = \sum_{n \geq 1} b_n r^n$ for $0 < r < 1$. Then*

$$\frac{1}{B(r)} \sup_{k \geq 1} b_k r^k \to 0$$

when $r \to 1$.

Proof. We set $e_n = b_n/B_n$ for $n \geq 2$. Without loss of generality we may assume that $b_1 = 1$ and $e_n < 1$ for all n. Then $B_{n-1}/B_n = 1 - e_n$, so that we can write for $n \geq 2$

$$B_n = \left(\frac{1}{1-e_2} \right) \cdots \left(\frac{1}{1-e_n} \right)$$

$$b_n = e_n \left(\frac{1}{1-e_2} \right) \cdots \left(\frac{1}{1-e_n} \right).$$

Since $B_n \to \infty$, we have $B(r) \to \infty$ when $r \to 1$. On the other hand we can write for any $N \in \mathbb{Z}^+$,

$$B(r) = \sum_{n=1}^{\infty} b_n r^n$$

$$\geq \sum_{n=1}^{N} (B_n - B_{n-1}) r^n$$

$$= r^N B_N + (1-r) \sum_{n=1}^{N} r^n B_n$$

$$\geq r^N B_N$$

$$= r^N \left(\frac{1}{1-e_2}\right) \cdots \left(\frac{1}{1-e_N}\right).$$

Given $\epsilon > 0$, let K_ϵ be such that $e_k < \epsilon$ for $k > K_\epsilon$. If $\sup_k b_k r^k$ is attained at some $k > K_\epsilon$, then

$$b_k r^k = \frac{e_k r^k}{(1-e_2) \cdots (1-e_k)} \leq e_k B(r) < \epsilon B(r).$$

On the other hand, if the supremum is attained at some $k \leq K_\epsilon$, then

$$\frac{b_k r^k}{B(r)} \leq \frac{\max_{1 \leq k \leq K_\epsilon} b_k r^k}{B(r)} \leq \epsilon.$$

But ϵ was arbitrary, which completes the proof. ∎

We can apply this lemma by taking $b_{n_k} = a_k^2$ and $b_n = 0$ if $n \notin \{n_{k_1}, n_{k_2}, \ldots\}$, to conclude that $r^{n_k} a_k / A(r) \to 0$ when $r \to 1$, uniformly in $k \in \mathbb{Z}^+$.

Proof of the Theorem. We compute the Fourier transform

$$\Phi_r(\xi) = \frac{1}{2\pi} \int_{\mathbb{T}} \exp\left(\frac{i\xi}{A(r)} \sum_{j=1}^{\infty} a_j r^{n_j} \cos n_j t\right) dt.$$

We proceed as in the proof of Theorem 5.3.2, beginning with the estimate $e^z = (1+z) \exp[z^2/2 + O(|z|^3)]$ applied to $z_j = (i\xi/A(r)) a_j r^{n_j} \cos n_j t$, noting that $z_j \to 0$ when $r \to 1$, uniformly in j. Then

$$\Phi_r(\xi) = \int_{\mathbb{T}} \prod_{k=1}^{\infty} \left(1 + \frac{i\xi}{A(r)} a_k r^{n_k} \cos n_k t\right) \exp\left(-\frac{\xi^2 a_k^2 r^{2n_k}}{2A(r)^2} \left[\cos^2(n_k t) + o(1)\right]\right) dt.$$

Write

$$\frac{1}{A(r)^2} \sum_{k=1}^{\infty} s_k^2 r^{n_k} \cos^2 n_k t = \frac{1}{2} + \frac{1}{2A(r)^2} \sum_{k=1}^{\infty} a_k^2 r^{n_k} \cos 2n_k t$$

$$:= \frac{1}{2} + \Psi_r(t).$$

The L^2 norm of Ψ_r is estimated as before

$$\frac{1}{2\pi}\int_{\mathbb{T}} \Psi_r(t)^2\,dt = \frac{1}{8A(r)^2}\sum_{k=1}^{\infty} r^{4n_k} a_k^4 \to 0, \qquad r \to 1$$

where we have used Lemma 5.3.7 to replace one factor by the supremum, the remaining sum being equal to 1. Meanwhile, we have the uniform bound

$$\prod_{k=1}^{\infty}\left|1 + i\frac{\xi a_k}{A(r)}\cos n_k t\right|^2 \le \prod_{k=1}^{\infty}\left(1 + \frac{\xi^2 a_k^2}{A(r)^2}\right) \le e^{\xi^2}$$

so that we can apply the dominated convergence theorem to conclude that

$$\Phi_r(\xi) = o(1) + \frac{e^{-\xi^2/2}}{2\pi}\int_{\mathbb{T}}\prod_{k=1}^{\infty}\left(1 + \frac{i\xi a_k r^{n_k}}{A(r)}\cos n_k t\right)dt.$$

But this infinite product can be expanded as a sum:

$$\prod_{k=1}^{\infty}\left(1 + \frac{i\xi a_k r^{n_k}}{A(r)}\cos n_k t\right) = \sum_{\nu=0}^{\infty} a_\nu(r)\cos \nu t.$$

From Lemma 5.3.3 and the condition (5.3.4) we have $\alpha_0(r) = o(1)$, so that we conclude

$$\frac{1}{2\pi}\int_{\mathbb{T}}\prod_{k=1}^{\infty}\left(1 + \frac{i\xi a_k r^{n_k}}{A(r)}\cos n_k t\right)dt = 1,$$

which completes the proof. ∎

5.4 WEAK CONVERGENCE OF MEASURES

To probe the deeper aspects of the convergence question, we develop the following notion of weak convergence of measures.

Definition 5.4.1. *A sequence of finite Borel measures (m_n) is said to converge weakly to a limit measure m if for every bounded continuous function g*

$$\lim_n \int_{\mathbb{R}^d} g(x)m_n(dx) = \int_{\mathbb{R}^d} g(x)m(dx).$$

If A is any set in $\mathbb{R}^d$, the interior A° is the set of points $x \in A$ such that A contains an open ball about x. The closure $\bar{A}$ is the complement of the interior of the complement; in symbols $(\bar{A})^c = (A^c)^\circ$ and we have $A^\circ \subset A \subset \bar{A}$. The boundary is defined as $\partial A = \bar{A}\setminus A^\circ$. The *portmanteau theorem* gives equivalent conditions for weak convergence.

Theorem 5.4.2. *The following conditions are equivalent*

1. *m_n converges weakly to m.*
2. *For every closed set A, $\limsup_n m_n(A) \le m(A)$.*

3. For every open set A, $\liminf_n m_n(A) \geq m(A)$.
4. For every Borel set A with $m(\partial A) = 0$, $\lim_n m_n(A) = m(A)$.

The proof of this theorem, which has nothing to do with Fourier analysis, can be found in Billingsley (1999).

It is also helpful to develop the appropriate notions of *compactness* in the context of weak convergence of measures. In this setting we refer to a *tight* family of measures, formalized as follows:

Definition 5.4.3. *A sequence of finite measures (m_n) is tight if*

$$\lim_{A \to \infty} \sup_n m_n(\{x : |x| > A\}) = 0.$$

Theorem 5.4.4. *Suppose that m_n is a tight sequence of finite measures with $\sup_n m_n(\mathbb{R}^d) < \infty$. Then there exists a weakly convergent subsequence.*

Again we refer to Billingsley (1999) for the details.

Exercise 5.4.5. *Prove that any weakly convergent sequence of finite Borel measures is tight.*

Example 5.4.6. *With $d = 1$, let $m_n = \delta_n$, a point mass at the point n. Then m_n is not a tight sequence, since for any n, $m_n\{x : |x| > n/2\} = 1$.*

5.4.1 An Improved Continuity Theorem

The theory of tightness can be used to formulate an improved version of the continuity theorem for sequences of characteristic functions of probability measures. In the previous version, Proposition 5.2.7, we required that the Fourier transforms $\hat{m}_n(\xi)$ converge to a limit $M(\xi)$, which is assumed to be a characteristic function. We now have the following improved version.

Theorem 5.4.7. *Suppose that m_n is a sequence of probability measures on $\mathbb{R}^d$ with Fourier transforms $\hat{m}_n(\xi)$, with the property that there exists*

(5.4.1) $$M(\xi) = \lim_n \hat{m}_n(\xi)$$

and that M is continuous at $\xi = 0$. Then there exists a probability measure m so that m_n converges weakly to m and $\hat{m}(\xi) = M(\xi)$.

Proof. From the hypotheses, we can write

$$1 - \hat{m}_n(\xi) = \int_{\mathbb{R}^d} \left(1 - e^{i\xi \cdot x}\right) m_n(dx).$$

This equation is integrated over the cube $C_a = \cap_{j=1}^{d}\{|\xi_j| \leq a\}$ to obtain

$$\int_{C_a} [1 - \hat{m}_n(\xi)] d\xi = \int_{\mathbb{R}^d} \left((2a)^d - \prod_{j=1}^{d} \frac{2\sin a x_j}{x_j} \right) m_n(dx)$$

$$\frac{1}{(2a)^d} \int_{C_a} [1 - \hat{m}_n(\xi)] d\xi = \int_{\mathbb{R}^d} \left(1 - \prod_{j=1}^{d} \frac{\sin a x_j}{a x_j} \right) m_n(dx)$$

$$\geq \int_{(C_{2a^{-1}})^c} \left(1 - \prod_{j=1}^{d} \frac{\sin a x_j}{a x_j} \right) m_n(dx),$$

since the integrand on the right side is nonnegative. If $(x_1, \ldots, x_d)$ lies outside the cube $C_{2a^{-1}}$, then the indicated product contains at least one factor that is less than $\frac{1}{2}$, the remaining factors are less than 1. Hence

$$\frac{1}{(2a)^d} \int_{C_a} [1 - \hat{m}_n(\xi)] d\xi \geq \frac{1}{2} \int_{(C_{2a^{-1}})^c} m_n(dx).$$

From the dominated convergence theorem we conclude that

$$\limsup_n \int_{(C_{2a^{-1}})^c} m_n(dx) \leq \frac{2}{(2a)^d} \int_{C_a} [1 - M(\xi)] d\xi.$$

From the continuity of M, the right side can be made arbitrarily small by taking a sufficiently small. This proves that the measures m_n are tight, hence we can extract a weakly convergent subsequence. If we had two different subsequential limits m_0 and n_0, then both of these measures must have the same characteristic function, hence they must be the same measure by the uniqueness of Fourier transforms, Proposition 5.2.4. Hence every subsequence has a subsubsequence that converges weakly to the same measure m_0. From this it follows that the original sequence converges weakly to m_0. ∎

Exercise 5.4.8. *Show that the hypothesis of continuity at $\xi = 0$ in Theorem 5.4.7 can be weakened to the hypothesis that $\xi = 0$ is a Lebesgue point for M, in the sense that $\lim_{a \to 0} a^{-d} \int_{C_a} |1 - M(\xi)| d\xi = 0$.*

5.4.1.1 Another proof of Bochner's theorem

In Chapter 3 we introduced the concept of *positive-definite function* and used a Fatou theorem for harmonic functions in the upper half plane to prove Bochner's theorem, which affirms that any continuous positive-definite function on $\mathbb{R}$ is the Fourier transform of a nonnegative measure. In this section we will give an independent proof of Bochner's theorem using the theory of weak convergence applied to a Gaussian convolution.

Proof. We begin with $f(\xi)$, a complex-valued positive-definite function that is assumed to be continuous. Define

(5.4.2) $$m_\epsilon(x) = \frac{1}{2\pi} \int_{\mathbb{R}} e^{-ixu} e^{-\epsilon u^2} f(u) du.$$

By hypothesis, we have for every $\psi \in L^1(\mathbb{R})$,

$$\iint_{\mathbb{R}^2} f(\xi - \eta) \psi(\xi) \bar{\psi}(\eta) d\xi d\eta \geq 0.$$

Taking the choice $\psi(\xi) = e^{ix\xi} e^{-\epsilon\xi^2/2}$ we infer that

$$0 \leq \iint_{\mathbb{R}^2} f(\xi - \eta) e^{ix(\xi-\eta)} e^{-\epsilon\xi^2/2} e^{-\epsilon\eta^2/2} \, d\xi \, d\eta$$

$$= \iint_{\mathbb{R}^2} f(u) e^{ixu} e^{-\epsilon(u+\eta)^2/2} e^{-\epsilon\eta^2/2} \, du \, d\eta$$

$$= \sqrt{\frac{\pi}{\epsilon}} \int_{\mathbb{R}} f(u) e^{ixu} e^{-\epsilon u^2/4} \, du$$

$$= \text{const.} \times m_{\epsilon/4}(x),$$

which proves that $m_\epsilon(x) \geq 0$. We claim that $m_\epsilon \in L^1(\mathbb{R})$ and that the Fourier transform of the nonnegative function m_ϵ is precisely $e^{-\epsilon u^2} f(\xi)$. To see this, we use Fubini to establish the following identity where $\delta > 0$

(5.4.3)
$$\int_{\mathbb{R}} m_\epsilon(x) e^{i\xi x} e^{-\delta x^2/2} \, dx = \frac{1}{2\pi} \int_{\mathbb{R}} \int_{\mathbb{R}} e^{ix(\xi-u)} e^{-\delta x^2/2} e^{-\epsilon u^2} f(u) \, dx \, du$$

$$= \int_{\mathbb{R}} \frac{e^{-(\xi-u)^2/2\delta}}{\sqrt{2\pi\delta}} f(u) e^{-\epsilon u^2} \, du.$$

Taking $\xi = 0$ and using Fatou's lemma and the continuity at $u = 0$ shows that

$$\int_{\mathbb{R}} m_\epsilon(x) \, dx \leq \liminf_{\delta \to 0} \int_{\mathbb{R}} m_\epsilon(x) e^{-\delta x^2/2} \, dx = f(0) < \infty.$$

Hence $m_\epsilon \in L^1(\mathbb{R})$ and $\|m_\epsilon\|_{L^1(\mathbb{R})} \leq f(0) < \infty$. Now we use (5.4.3) and the dominated convergence theorem to compute the Fourier transform with $\epsilon > 0$ fixed:

$$\hat{m}_\epsilon(\xi) = \int_{\mathbb{R}} m_\epsilon(x) e^{i\xi x} \, dx$$

$$= \lim_{\delta \to 0} \int_{\mathbb{R}} m_\epsilon(x) e^{i\xi x} e^{-\delta x^2/2} \, dx$$

$$= \lim_{\delta \to 0} \int_{\mathbb{R}} \frac{e^{-(\xi-u)^2/2\delta}}{\sqrt{2\pi\delta}} f(u) e^{-\epsilon u^2} \, du$$

$$= f(\xi) e^{-\epsilon \xi^2}$$

where we have used the continuity of f in the last line. Hence the product $f(\xi) e^{-\epsilon \xi^2}$ is the Fourier transform of the function m_ϵ. Taking $\epsilon \to 0$ and using Theorem 5.4.7, we see that f is the Fourier transform of a nonnegative finite measure, which was to be proved. ∎

Exercise 5.4.9. Show that the hypothesis of continuity in Bochner's theorem can be weakened to the hypothesis that $\xi = 0$ is a Lebesgue point for f, in the sense that $\lim_{a \to 0} a^{-1} \int_{-a}^{a} |f(0) - f(\xi)| \, d\xi = 0$.

Exercise 5.4.10. Show that the above argument can be generalized to $\mathbb{R}^d$, to obtain a characterization of the Fourier transform of a finite nonnegative measure as a positive definite function on $\mathbb{R}^d$.

5.5 CONVOLUTION SEMIGROUPS

Fourier analysis is particularly well suited to deal with one-parameter families of probability distributions that are closed under convolution.

Example 5.5.1. *Let P_t be the probability distribution on $\mathbb{R}^1$ with density $e^{-x^2/2t}/\sqrt{2\pi t}$. Then by direct computation, $P_t * P_s$ has the density $e^{-x^2/2(t+s)}/\sqrt{2\pi(t+s)}$, so that $P_t * P_s = P_{t+s}$.*

Definition 5.5.2. *A convolution semigroup of probability measures is a family $(P_t)_{t>0}$ with the properties*

(5.5.1) $$P_s * P_t = P_{s+t},$$

(5.5.2) $$P_t \to \delta_0 \quad \text{when} \quad t \to 0,$$

in the sense of weak convergence.

The Fourier transform of the convolution semigroup is defined by

(5.5.3) $$f_t(\xi) = \hat{P}_t(\xi) = \int_{\mathbb{R}} e^{i\xi x} P_t(dx).$$

This has the following properties.

Proposition 5.5.3.
- $\xi \to f_t(\xi)$ is continuous and bounded by 1.
- $f_t(\xi) \to 1$ uniformly on compact ξ-sets when $t \to 0$.
- $t \to f_t(\xi)$ is continuous for $t > 0$.
- $t \to f_t(\xi)$ is differentiable for $t > 0$ and $\psi(\xi) := \lim_{t \to 0} f'_t(\xi)$ exists.
- $f_t(\xi)$ satisfies the identity $f_t(\xi) = 1 + \psi(\xi) \int_0^t f_s(\xi)\, ds$, and $f_t(\xi) = e^{t\psi(\xi)}$.

Proof. The first property follows immediately from the definition (5.5.3). To prove the second, we have for any $\delta > 0$

$$|f_t(\xi) - 1| = \left| \int_{\mathbb{R}} \left[e^{i\xi x} - 1 \right] P_t(dx) \right| \leq \int_{|x| < \delta} |\xi x| P_t(dx) + 2 P_t(\{x : |x| > \delta\})$$

where the second term tends to zero when $t \to 0$. On any compact interval $|\xi| \leq M$, the first term is less than $M\delta$. Therefore

$$\limsup_{t \to 0} \sup_{|\xi| \leq M} |f_t(\xi) - 1| \leq M\delta.$$

But $\delta > 0$ is arbitrary, hence the limsup is zero, proving the uniform convergence. From (5.5.1) and (5.5.2) we have

(5.5.4) $$f_{t+s}(\xi) = f_t(\xi) f_s(\xi), \quad \lim_{t \to 0} f_t(\xi) = 1.$$

Letting $s \to 0$, we see that f_t is right continuous at each $t > 0$. But for small $s > 0$ we can write

$$f_{t-s}(\xi) = \frac{f_t(\xi)}{f_s(\xi)}$$

so that we can take $s \to 0$ and conclude that $t \to f_t$ is left continuous. In particular $t \to f_t$ is bounded and measurable, so that we may integrate (5.5.4) on an interval $[0, \delta]$ to obtain

(5.5.5) $$\int_t^{t+\delta} f_s(\xi)\, ds = f_t(\xi) \int_0^\delta f_s(\xi)\, ds$$

where $\delta > 0$ is chosen so that $|f_s(\xi) - 1| < 1/2$ for $0 < s < \delta$.

The formula (5.5.5) displays f_t as a differentiable function of t with a continuous derivative. Taking the derivative for $t > 0$ we obtain

$$f_{t+\delta}(\xi) - f_t(\xi) = f'_t(\xi) \int_0^\delta f_s(\xi)\, ds.$$

Hence there exists

$$\psi(\xi) = \lim_{t \to 0} f'_t(\xi) = \frac{f_\delta(\xi) - 1}{\int_0^\delta f_s(\xi)\, ds},$$

which displays ψ as a continuous function with $\psi(0) = 0$. Computing the derivative from (5.5.4), we obtain the differential equation $f'_t(\xi) = \psi(\xi) f_t(\xi)$. The unique solution satisfying $f_0(\xi) = 1$ is

$$f_t(\xi) = e^{t\psi(\xi)}. \qquad \blacksquare$$

The next goal is to prepare the proof of the following theorem of Lévy and Khintchine.

Theorem 5.5.4. *Suppose that $(P_t)_{t>0}$ is a convolution semigroup of probability measures on $\mathbb{R}$. Then there exists a unique Borel measure M and real numbers μ, σ, so that the Fourier transform has the representation*

$$\boxed{f_t(\xi) = \int_\mathbb{R} e^{i\xi x} P_t(dx) = e^{t\psi(\xi)}}$$

where

$$\boxed{\psi(\xi) = i\mu\xi - \frac{\sigma^2 \xi^2}{2} + \int_\mathbb{R} \left[e^{i\xi x} - 1 - i\xi \sin x \right] M(dx)}$$

and

$$\int_\mathbb{R} \frac{x^2}{1+x^2} M(dx) < \infty.$$

To prepare the proof, we define $F_n = P_{1/n}$ and $G_n = nx^2/(1+x^2) F_n$.

Lemma 5.5.5. *The total mass of G_n is uniformly bounded, specifically*

$$\sup_n \int_\mathbb{R} \frac{nx^2}{1+x^2} F_n(dx) < \infty.$$

Proof. We begin with the basic relation $1 - f_t(\xi) = \int_\mathbb{R}[1 - e^{ix\xi}]P_t(dx)$ that is integrated on the interval $[-1, 1]$ to yield

(5.5.6) $$\frac{1}{2}\int_{-1}^{1}(1 - f_t(\xi))\,d\xi = \int_\mathbb{R}\left(1 - \frac{\sin x}{x}\right)P_t(dx).$$

When we divide by t, the integrand on the left side tends boundedly to $\psi(\xi)$. On the right side, we use the inequality

$$1 - \frac{\sin x}{x} \geq \frac{1}{7}\frac{x^2}{1+x^2}.$$

Dividing both sides by $t = 1/n$ we obtain

$$\limsup_n \int_\mathbb{R} \frac{nx^2}{1+x^2}F_n(dx) \leq \frac{7}{2}\int_{-1}^{1}\psi(\xi)\,d\xi. \quad \blacksquare$$

Lemma 5.5.6. $\bar{F}(A) := \sup_n nF_n(|x| > A) < \infty$ for each $A > 0$ and $\bar{F}(A)$ tends to zero when $A \to \infty$.

Proof. We first prove this for the symmetrized distribution $F_n^\#(dx) = F_n(dx) * F_n(-dx)$, whose characteristic function is

$$f_{1/n}^\#(\xi) = e^{-(2/n)\psi(\xi)}.$$

We apply the technique of (5.5.6) with the interval $[-1, 1]$ replaced by the interval $[-2/A, 2/A]$ and use the inequality $1 - e^{-y} \leq y$ to write

$$F_n^\#(|\xi| > A) \leq 2A \int_{|\xi|<1/A}\left(1 - f_{1/n}^\#(\xi)\right)d\xi$$

$$\leq \frac{4A}{n}\int_{|\xi|<2/A}|\text{Re }\psi(\xi)|\,d\xi.$$

Therefore

$$\sup_n nF_n^\#(|x| > A) \leq 4A\int_{|\xi|<2/A}|\text{Re }\psi(\xi)|\,d\xi.$$

But ψ is continuous at $\xi = 0$, hence the right side of the final inequality tends to zero when $A \to \infty$.

To handle the general (nonsymmetric) case, note the following "symmetrization inequality" for any two independent random variables X, Y:

$$P[|X - Y| \geq A] \geq P[|X| \geq 2A]P[|Y| \leq A]$$

(draw the picture). Applying this to X, Y distributed as F_n, we obtain

$$F_n(|x| > 2A) \leq \frac{F_n^\#(|x| > A)}{F_n(|x| \leq A)}.$$

But the denominator tends to 1 when $n \to \infty$, from the basic hypothesis that $P_t \to \delta_0$, so that

$$\limsup_n nF_n(|x| > 2A) \leq \limsup_n nF_n^\#(|x| > A) \leq 4A\int_{|\xi|<2/A}|\text{Re }\psi(\xi)|\,d\xi,$$

which tends to zero when $A \to \infty$. $\quad \blacksquare$

Having done this, we now turn to

Lemma 5.5.7. *The measures G_n are tight: in detail*

$$\limsup_{A \to \infty} \sup_n \int_{|x|>A} \frac{nx^2}{1+x^2} F_n(dx) = 0.$$

Proof. We write

$$\int_{|x|>A} \frac{nx^2}{1+x^2} F_n(dx) \leq n F_n(|x|>A)$$

and we have shown above that the supremum over n tends to zero when $A \to \infty$. ∎

Proof of Theorem 5.5.4. Now we prove the representation theorem. From the definition

$$\psi(\xi) = \lim_{t \to 0} \frac{f_t(\xi) - 1}{t}$$

$$= \lim_n \int_{\mathbb{R}} (e^{ix\xi} - 1) n F_n(dx).$$

For any n we can write

(5.5.7) $$\int_{\mathbb{R}} (e^{ix\xi} - 1) n F_n(dx) = \int_{\mathbb{R}} \frac{e^{ix\xi} - 1 - i\xi \sin x}{x^2} (1 + x^2) G_n(dx) + i\mu_n \xi,$$

where $\mu_n = \int_{\mathbb{R}} \sin x F_n(dx)$. The integrand in (5.5.7) is defined by continuity at $x = 0$. Note that $G_n(\{0\}) = 0$ for all n. From Lemma 5.5.5, we may take a subsequence for which the total masses $G_n(\mathbb{R})$ converge to a limit. Since the measures G_n are tight from Lemma 5.5.7, we may take a further subsequence that converges weakly to a limit measure G. All of the limits below will be taken through this new subsequence. The integral term on the right side of (5.5.7) converges to

$$\int_{\mathbb{R}} \frac{e^{ix\xi} - 1 - i\xi \sin x}{x^2} (1 + x^2) G(dx)$$

while the left side converges to $\psi(\xi)$. Letting $\xi = 1$ and taking the imaginary part shows that μ_n converges to $\operatorname{Im} \psi(1)$. We have obtained the Lévy-Khintchine representation in Feller's form:

(5.5.8) $$\psi(\xi) = \int_{\mathbb{R}} \frac{e^{ix\xi} - 1 - i\xi \sin x}{x^2} (1 + x^2) G(dx) + i\mu\xi.$$

It remains to discuss the uniqueness of the pair (μ, G). The number μ is uniquely determined by $\mu = \operatorname{Im} \psi(1)$. To identify G, we form the convolution of ψ with the kernel $\frac{1}{2} e^{-|\xi|}$ whose Fourier transform is $1/(1+x^2)$. This results in the identity

$$-\psi(\xi) + \frac{1}{2} \int_{\mathbb{R}} e^{-|\xi - \eta|} \psi(\eta) \, d\eta = \int_{\mathbb{R}} e^{ix\xi} G(dx),$$

which shows that ψ uniquely determines the characteristic function of the finite measure G. But the uniqueness theorem for characteristic functions shows that G is thereby uniquely determined from its characteristic function, which completes the proof. ∎

5.6 THE BERRY-ESSÉEN THEOREM

Fourier analysis is extremely effective in obtaining sharp estimates on the *rate of convergence* in the central limit theorem. The following discussion is presented for the case of convolution powers of a single probability distribution, but it can be easily extended to the case of different distributions, under suitable hypotheses.

Let m be a probability measure on $\mathbb{R}$ with

(5.6.1) $$\int_{\mathbb{R}} x \, dm = 0, \quad \int_{\mathbb{R}} x^2 \, dm = 1, \quad \int_{\mathbb{R}} |x|^3 \, dm = m_3 < \infty.$$

From the central limit theorem, we know that the normalized convolution of the distribution function F converges to the normal distribution

(5.6.2) $$\lim_n (F * F * \cdots * F)(x\sqrt{n}) = \Phi(x) = \int_{-\infty}^{x} \frac{e^{-u^2/2}}{\sqrt{2\pi}} \, du.$$

The Berry-Esséen theorem gives the rate of convergence.

Theorem 5.6.1. *We have uniformly for $-\infty < x < \infty$*

(5.6.3) $$|(F * F * \cdots * F)(x\sqrt{n}) - \Phi(x)| \leq \frac{Cm_3}{\sqrt{n}}$$

where C is a universal constant.

The idea of the proof is to regularize the given distribution by convolution with a smooth density. Then we can apply the inversion formula directly to the regularized convolution, whose Fourier transform is the product of the given Fourier transform with the smoothing factor. For the smoothing density we choose the Fejér kernel

(5.6.4) $$K_T(x) = \frac{1 - \cos Tx}{\pi T x^2}, \quad \text{with} \quad \hat{K}_T(\xi) = \left(1 - \frac{|\xi|}{T}\right) 1_{[0,T]}(|\xi|).$$

Lemma 5.6.2. *Let G be a probability measure on $\mathbb{R}$ and set*

(5.6.5) $$\Delta(x) := G(x) - \Phi(x), \quad \eta = \sup_x |\Delta(x)|, \quad \eta_T = \sup_x |(\Delta * K_T)(x)|.$$

Then

$$\eta \leq 2\eta_T + \frac{24m}{\pi T}, \quad \text{where } m = \sup_x \Phi'(x) = \frac{1}{\sqrt{2\pi}}.$$

Proof. Since $\Delta(x)$ vanishes at $\pm\infty$, the supremum occurs at some point x_0 where we have $\Delta(x_0 \pm 0) = \pm\eta$. If $\Delta(x_0 \pm 0) = +\eta$, we propose to estimate the convolution

$$(\Delta * K_T)(x_0 + \delta) = \left(\int_{-\delta}^{\delta} + \int_{|x|\geq\delta}\right) \Delta(x_0 + \delta - x) K_T(x) \, dx$$

where δ is to be chosen. If $-\delta \leq x \leq \delta$, then

$$\Delta(x_0 + \delta - x) = G(x_0 + \delta - x) - \Phi(x_0 + \delta - x)$$
$$\geq G(x_0 \pm 0) - [\Phi(x_0) + m(\delta - x)]$$
$$= \eta + m(x - \delta).$$

We use the estimate on the Fejér kernel

$$\int_{|x|\geq\delta} K_T(x)\,dx \leq \int_{|x|\geq\delta} \frac{2}{\pi T x^2}\,dx = \frac{4}{\pi T\delta}$$

and the fact that $\int_{-\delta}^{\delta} x K_T(x)\,dx = 0$; thus

$$\int_{-\delta}^{\delta} \Delta(x_0 + \delta - x) K_T(x)\,dx \geq (\eta - m\delta)\left(1 - \frac{4}{\pi T\delta}\right).$$

On the interval $|x| \geq \delta$ we use the bound $\Delta(x) \geq -\eta$ to write

$$\int_{|x|\geq\delta} \Delta(x) K_T(x)\,dx \geq -\eta \int_{|x|\geq\delta} K_T(x)\,dx \geq -\eta \frac{4}{\pi T\delta}.$$

Adding the two, we obtain

$$(\Delta * K_T)(x_0 + \delta) = \left(\int_{-\delta}^{\delta} + \int_{|x|\geq\delta}\right) \Delta(x_0 + \delta - x) K_T(x)\,dx$$

$$\geq (\eta - m\delta)\left(1 - \frac{4}{\pi T\delta}\right) - \eta \frac{4}{\pi T\delta}.$$

The proof is completed by choosing $\delta = \eta/2m$ to obtain $(\Delta * K_T)(x_0) \geq \eta/2 - 12m/\pi T$, which immediately gives the required result. If $\Delta(x_0 \pm 0) = -\eta$, then we apply the above argument to $-\Delta(x)$. ∎

Proof of the Theorem. Let $G(x) = F_n(x) = (F * \cdots * F)(x\sqrt{n})$, whose Fourier transform is $\hat{F}(\xi/\sqrt{n})^n$. The Fourier transform of $F_n * K_T$ is obtained by taking the product with $(1 - |\xi|/T)1_{[0,T]}(|\xi|)$, in particular the product is absolutely integrable on $\mathbb{R}$. Hence we can apply the inversion formula to write

$$(F_n * K_T)([a, x]) = \frac{1}{2\pi} \int_{-T}^{T} \frac{e^{-i\xi x} - e^{-i\xi a}}{-i\xi} \left(1 - \frac{|\xi|}{T}\right) \hat{F}\left(\frac{\xi}{\sqrt{n}}\right)^n d\xi.$$

Applying the inversion formula to $\Phi * K_T$ and subtracting, we have

$$(F_n * K_T - \Phi * K_T)([a, x]) = \frac{1}{2\pi} \int_{-T}^{T} \frac{e^{-i\xi x} - e^{-i\xi a}}{-i\xi} \left(1 - \frac{|\xi|}{T}\right) \left(\hat{F}\left(\frac{\xi}{\sqrt{n}}\right)^n - e^{-\xi^2/2}\right) d\xi.$$

For each fixed $T > 0$, $n < \infty$ the integrand is an integrable function, so that we can apply the Riemann-Lebesgue lemma to let $a \to -\infty$ and obtain

(5.6.6)

$$(F_n * K_T - \Phi * K_T)((-\infty, x]) = \frac{1}{2\pi} \int_{-T}^{T} \frac{e^{-i\xi x}}{-i\xi} \left(1 - \frac{|\xi|}{T}\right) \left(\hat{F}\left(\frac{\xi}{\sqrt{n}}\right)^n - e^{-\xi^2/2}\right) d\xi.$$

It remains to estimate the integrand and apply Lemma 5.6.2. From the definition of the Fourier transform,

$$\left|\hat{F}(\xi) - 1 + \frac{\xi^2}{2}\right| = \left|\int_{\mathbb{R}} \left(e^{i\xi x} - 1 - i\xi x + \frac{\xi^2 x^2}{2}\right) F(dx)\right|$$

$$\leq \int_{\mathbb{R}} \frac{|\xi^3 x^3|}{6} F(dx)$$

$$= m_3 \frac{|\xi|^3}{6}.$$

We apply the same estimates to $e^{-\xi^2/2}$ and use Hölder's inequality to obtain

$$|e^{-\xi^2/2} - 1 + \xi^2/2| \leq \frac{|\xi|^3}{6} \int_{\mathbb{R}} |x|^3 e^{-x^2/2} \frac{dx}{\sqrt{2\pi}}$$

$$= \frac{2|\xi|^3}{3\sqrt{2\pi}}$$

$$\leq \frac{2m_3|\xi|^3}{3\sqrt{2\pi}}$$

$$\leq m_3 \frac{|\xi|^3}{2}.$$

Thus

$$\left|\hat{F}(\xi) - e^{-\xi^2/2}\right| \leq m_3|\xi|^3.$$

Furthermore

$$\hat{F}\left(\frac{\xi}{\sqrt{n}}\right) \leq 1 - \frac{\xi^2}{2n} + \frac{m_3|\xi|^3}{6n^{\frac{3}{2}}}$$

$$\leq 1 - \frac{\xi^2}{2n} + \frac{\xi^2}{6n} \qquad |\xi| \leq \frac{\sqrt{n}}{m_3}$$

$$= 1 - \frac{\xi^2}{3n} \qquad |\xi| \leq \frac{\sqrt{n}}{m_3}$$

$$\leq e^{-\xi^2/3n} \qquad |\xi| \leq \frac{\sqrt{n}}{m_3}.$$

We now estimate the integrand in (5.6.6) by the telescoping sum

$$\left|\hat{F}\left(\frac{\xi}{\sqrt{n}}\right)^n - e^{-\xi^2/2}\right| = \left|\left(\hat{F}\left(\frac{\xi}{\sqrt{n}}\right) - e^{-\xi^2/2n}\right) \sum_{j=1}^n \hat{F}\left(\frac{\xi}{\sqrt{n}}\right)^{j-1} \exp\left[-\frac{\xi^2}{2n}(n-j)\right]\right|$$

$$\leq \frac{m_3|\xi|^3}{n^{3/2}} ne^{-\xi^2/3} \qquad |\xi| \leq \frac{\sqrt{n}}{m_3}$$

$$= \frac{m_3|\xi|^3}{n^{1/2}} e^{-\xi^2/3} \qquad |\xi| \leq \frac{\sqrt{n}}{m_3}.$$

Applying Lemma 5.6.2 with $T = \sqrt{n}/4m_3$, we have

$$2\pi|F_n(x\sqrt{n}) - \Phi(x)| \leq 2 \int_{-T}^{T} \left(1 - \frac{|\xi|}{T}\right) \frac{m_3|\xi|^2}{2\sqrt{n}} e^{-\xi^2/3} d\xi + \frac{24}{T\pi^{3/2}}$$

$$\leq \frac{m_3}{\sqrt{n}} \int_{\mathbb{R}} |\xi|^2 e^{-\xi^2/3} d\xi + \frac{12m_3}{\sqrt{n}\pi^{3/2}}$$

which is of the required form with $2\pi C = (3/2)^{3/2}/\sqrt{2\pi} + 12/\pi^{3/2}$. ∎

Remark. We can obtain a lower bound on the best possible constant by recalling the DeMoivre-Laplace limit theorem from Chapter 1: In this case the explicit asymptotic analysis shows that $F_n(0+0) - F_n(0-0) \sim 1/\sqrt{n\pi} = \sqrt{2/\pi}1/\sqrt{2n}$. Since $m_3 = 1$, we see that the best constant in (5.6.3) satisfies $C \geq \sqrt{2/\pi} \sim .79$.

5.6.1 Extension to Different Distributions

The Berry-Esséen theorem extends naturally to a sequence of probability measures (F_k) that satisfy

(5.6.7) $$\int_{\mathbb{R}} x \, dF_k = 0, \qquad \int_{\mathbb{R}} x^2 \, dF_k = a_k^2 < \infty, \qquad \int_{\mathbb{R}} |x|^3 \, dF_k = m_k < \infty$$

and the additional conditions that

(5.6.8) $$A_k^2 := a_1^2 + \cdots + a_k^2 \to \infty$$
$$\frac{a_k}{A_k} \to 0$$
$$m_k \le \lambda a_k^2$$

for some $\lambda > 0$. Under these hypotheses we claim that

(5.6.9) $$|(F_1 * \cdots * F_n)(xA_n) - \Phi(x)| \le \frac{C}{A_n}$$

for a suitable constant C.

To prove (5.6.9), we apply Lemma (5.6.2) with $G = F_1 * \cdots * F_n$. The Fourier transform of the left side of (5.6.9) is $\hat{F}_1(\xi/A_n) \cdots \hat{F}_1(\xi/A_n) - e^{-\xi^2/2}$. Now if $\hat{\Phi}(\xi) = e^{-\xi^2/2}$ is the Fourier transform of the standard normal distribution, we can write $\hat{\Phi}(\xi) = \hat{\Phi}_1(\xi/A_n) \cdots \hat{\Phi}_n(\xi/A_n)$ where $\hat{\Phi}_j(\xi) = e^{-\xi^2 a_j^2/2}$ is the Fourier transform of a normal distribution with mean zero and variance a_j^2, so that we can write

$$\hat{F}_1(\xi/A_n) \cdots \hat{F}_n(\xi/A_n) - \hat{\Phi}_1(\xi/A_n) \cdots \hat{\Phi}_n(\xi/A_n)$$
$$= \sum_{j=1}^n \left(\hat{F}_j(\xi/A_n) - \hat{\Phi}_1(\xi/A_n) \right) \prod_{i<j} \hat{\Phi}_i(\xi/A_n) \prod_{i>j} \hat{F}_i(\xi/A_n).$$

Arguing as before, we have for sufficiently large n,

$$|\hat{F}_j(\xi/A_n)| \le \exp[-a_j^2 \xi^2 / 3A_n^2], \qquad |\hat{\Phi}_j(\xi/A_n)| \le \exp[-a_j^2 \xi^2 / 3A_n^2],$$

so that

$$|\hat{F}_1(\xi/A_n) \cdots \hat{F}_n(\xi/A_n) - e^{-\xi^2/2}| \le e^{-\xi^2/3} \sum_{j=1}^n |\hat{F}_j(\xi/A_n) - \hat{\Phi}_j(\xi/A_n)|.$$

Now

$$\left| \hat{F}_j(\xi/A_n) - 1 + \xi^2/2A_n^2 \right| = \left| \int_{\mathbb{R}} \left(e^{i\xi x} - 1 - i\frac{\xi x}{A_n} + \frac{\xi^2 x^2}{2A_n^2} \right) F_j(dx) \right| \le \frac{|\xi|^3 m_j}{6A_n^3}$$

$$\left| \hat{\Phi}_j(\xi/A_n) - 1 + \xi^2/2A_n^2 \right| \le \frac{4|\xi|^3 a_j^3}{\sqrt{2\pi} A_n^3} \le \frac{2|\xi|^3 m_j}{A_n^3}$$

so that

$$|\hat{F}_j(\xi/A_n) - \exp[-\xi^2 a_j^2/2A_n^2]| \leq \frac{3|\xi|^3 m_j}{A_n^3} \leq \frac{3|\xi|^3 \lambda a_j^2}{A_n^3},$$

$$\sum_{j=1}^{n} |\hat{F}_j(\xi/A_n) - \exp[-\xi^2 a_j^2/2A_n^2]| \leq \frac{3|\xi|^3 \lambda}{A_n}.$$

Applying Lemma (5.6.2) we have

$$\sup_{x \in \mathbb{R}} |G(x) - \Phi(x)| \leq \frac{\lambda}{3A_n} \int_{-T}^{T} \left(1 - \frac{|\xi|}{T}\right) |\xi|^2 e^{-\xi^2/3} \, d\xi + \frac{C}{T}.$$

Choosing $T = A_n$ balances the two error terms and leads to the desired result.

5.7 THE LAW OF THE ITERATED LOGARITHM

The central limit theorem, which gives the limiting distribution of normalized sums of independent random variables, has an almost-everywhere counterpart, as follows:

$$(5.7.1) \qquad \limsup_{k \to \infty} \frac{S_k(t)}{\sqrt{2A_k^2 \log \log A_k}} = +1 \qquad \text{a.e. } t.$$

The proof of (5.7.1) depends on careful bounds for the distribution of $S_k(t)/A_k$ and a simple estimate for the distribution of the maximum of $S_1(t), \ldots, S_k(t)$. We will prove the following result:

Theorem 5.7.1. *Suppose that* $\{X_k(t)\}_{k \in \mathbb{Z}^+}$ *are independent functions on* $\mathbb{T} = [0, 1]$ *with* $\int_{\mathbb{T}} X_k(t) \, dt = 0$, $\int_{\mathbb{T}} X_k(t)^2 \, dt = a_k^2$, $\int_{\mathbb{T}} |X_k(t)|^3 \, dt < \infty$ *and that the conditions (5.6.8) are satisfied. Suppose further that for each* $x \in \mathbb{R}$ *and* $k \in \mathbb{Z}^+$

$$(5.7.2) \qquad |\{t : X_k(t) < -x\}| = |\{t : X_k(t) > x\}|.$$

Then (5.7.1) holds.

The distribution of the maximum is estimated as follows.

Lemma 5.7.2. *Let* $S_k^*(t) = \max\{S_1(t), \ldots, S_k(t)\}$ *where (5.7.2) is satisfied. Then*

$$(5.7.3) \qquad \left|\{t : S_k^*(t) > x\}\right| \leq 2 \left|\{t : S_k(t) > x\}\right|.$$

Proof. Let $A_j = \{t : S_1(t) \leq x, \ldots, S_{j-1}(t) \leq x, S_j(t) > x\}$. Then the event $\{t : S_k^*(t) > x\}$ is written as the disjoint union $\cup_{j=1}^{k} A_j$. From the symmetry hypothesis (5.7.2) we have for any $n_1 \leq n_2$, $|\{t : \sum_{j=n_1}^{n_2} X_j(t) \geq 0\}| \geq \frac{1}{2}$. Now

$$S_j(t) > x, X_{j+1}(t) + \cdots + X_k(t) \geq 0 \implies S_k(t) > x$$

hence $A_j \cap \{X_{j+1} + \cdots + X_k \geq 0\} \subset A_j \cap \{S_k > x\}.$

Then

$$|\{t : S_k(t) > x\}| \geq \sum_{j=1}^{k} |A_j \cap \{t : S_k(t) > x\}|$$

$$\geq \sum_{j=1}^{k} |A_j \cap \{t : X_{j+1}(t) + \cdots + X_k(t) \geq 0\}|$$

$$= \sum_{j=1}^{k} |A_j| |\{t : X_{j+1}(t) + \cdots + X_k(t) \geq 0\}|$$

$$\geq \frac{1}{2} \sum_{j=1}^{k} |A_j|$$

$$= \frac{1}{2} |\{t : S_k^*(t) > x\}|. \qquad \blacksquare$$

Lemma 5.7.3. *If* $x_k \to \infty$ *so that* $x_k^2/2 - \log A_k \to -\infty$, *then*

$$\left|\left\{t : \frac{S_k(t)}{A_k} > x\right\}\right| = \exp\left[-x_k^2(1 + o(1))\right].$$

Proof. From (5.6.9) we have

$$\left|\left|\left\{t : \frac{S_k(t)}{A_k} > x\right\}\right| - (1 - \Phi(x))\right| \leq \frac{C}{A_k}$$

where C is a constant. When $x \to \infty$ we have $1 - \Phi(x) = \exp[-x^2/2(1 + o(1))]$. The hypothesis on x_k is equivalent to $1/A_k = o(e^{-x_k^2/2})$ and therefore the error term can be absorbed into the Gaussian term when this is satisfied. $\blacksquare$

We also need the first and second Borel-Cantelli lemmas, as follows.

Lemma 5.7.4. *Suppose that* (B_k) *are measurable sets with* $\sum_{k=1}^{\infty} |B_k| < \infty$. *Then*

$$\left|\left\{t : \sum_{k=1}^{\infty} 1_{B_k}(t) < \infty\right\}\right| = 1.$$

Proof. The assertion is that, almost surely, only a finite number of the events B_k occur. The proof comes from applying the monotone convergence theorem to write

$$\int_{\mathbb{T}} \sum_{k=1}^{\infty} 1_{B_k}(t) \, dt = \sum_{k=1}^{\infty} |B_k| < \infty.$$

Hence $\sum_{k=1}^{\infty} 1_{B_k}(t) < \infty$ for almost all t. $\blacksquare$

Lemma 5.7.5. *Suppose that* (B_k) *are measurable sets with* $\sum_{k=1}^{\infty} |B_k| = \infty$ *and that* (B_k) *are mutually independent:* $|\cap_{i=1}^{k} B_i^{\pm}| = \Pi_{i=1}^{k} |B_i^{\pm}|$ *for any choice of* $\pm$, *where* $B^+ = B$ *and* $B^- = B^c$. *Then*

$$\left|\left\{t : \sum_{k=1}^{\infty} 1_{B_k}(t) = +\infty\right\}\right| = 1.$$

Proof. The assertion is that, almost surely, infinitely many of the events occur. To prove this, use the monotone convergence theorem with $a = \log 2$ to write

$$\int_{\mathbb{T}} \exp\left(-a \sum_{k=1}^{\infty} 1_{B_k}(t)\right) dt = \int_{\mathbb{T}} \prod_{i=1}^{\infty} 2^{-1_{B_k}(t)} dt$$

$$= \prod_{i=1}^{\infty} \left(1 - \frac{1}{2}|B_k|\right)$$

$$\leq \prod_{i=1}^{\infty} \exp\left(-\frac{1}{2}|B_k|\right)$$

$$= \exp\left(-a \sum_{k=1}^{\infty} |B_k|\right)$$

$$= 0.$$

Thus $\exp(-a \sum_{k=1}^{\infty} 1_{B_k}(t)) = 0$ for a.e. t, hence $|\{t : \sum_{k=1}^{\infty} 1_{B_k}(t) < \infty\}| = 0$. ∎

Proof of (5.7.1). Given $\theta > 1$, define a sequence of integers (n_k) by the recipe that

$$n_k = \max\{n : A_n \leq \theta^k\}.$$

Thus $A_{n_k} \leq \theta^k < A_{n_{k+1}}$ and it follows that $A_{n_k} \sim \theta^k$ and $A_{n_{k+1}}/A_{n_k} \to \theta$ when $k \to \infty$. Now from Lemma 5.7.3 we have for any $\delta > 0$,

$$|\{t : S_{n_k}(t) > (1+\delta)A_{n_k}\sqrt{2\log\log A_{n_k}}\}| = \exp[-(1+\delta)^2 \log\log A_{n_k}(1+o(1))]$$

$$= \left(\frac{1}{\log A_{n_k}}\right)^{(1+\delta)^2(1+o(1))}$$

$$\sim \left(\frac{1}{\log A_{n_k}}\right)^{(1+\delta)^2(1+o(1))}.$$

Since $\delta > 0$, this is the general term of a convergent series, so that by the first Borel-Cantelli Lemma 5.7.4, for a.e. t only a finite number of these events occur. Thus for $k > k(t)$ we have

(5.7.4) $$S_{n_k}(t) \leq (1+\delta)A_{n_k}\sqrt{2\log\log A_{n_k}}.$$

Now by Lemma 5.7.2, given $\delta > 0$, choose $\theta > 1$ so that $\delta^2/(\theta-1) > 1$. Then the last series of terms converges and the first Borel-Cantelli Lemma 5.7.4 shows that for a.e. t we have for $k > k(t)$

(5.7.5) $$\max_{n_k \leq j < n_{k+1}} (S_j(t) - S_{n_k}(t)) \leq (1+\delta)A_{n_k}\sqrt{2\log\log A_{n_k}}.$$

Adding (5.7.4) and (5.7.5) we have for $k > k(t)$ and $n_k \leq j < n_{k+1}$

$$S_j(t) = (S_j(t) - S_{n_k}(t)) + S_{n_k}(t)$$

$$\leq \delta A_{n_k}\sqrt{2\log\log A_{n_k}} + (1+\delta)A_{n_k}\sqrt{2\log\log A_{n_k}}$$

$$\leq (1+2\delta)A_j\sqrt{2\log\log A_j}.$$

Thus $\limsup_j S_j/A_j\sqrt{2\log\log A_j} \leq (1+2\delta)$. But $\delta > 0$ was arbitrary therefore the limsup in equation (5.7.1) is less than or equal to 1. Applying the same reasoning to the sequence $-S_j$ shows that the liminf is greater than or equal to -1.

To prove the lower bound, we consider the independent events
$$B_k := \{t : S_{n_k}(t) - S_{n_{k-1}}(t) \geq (1-\delta)A_{n_k}\sqrt{2\log\log A_{n_k}}\}.$$
Then
$$|B_k| = \exp\left(-(1-\delta)^2 \frac{n_k}{n_k - n_{k-1}} \log\log A_{n_k}(1+o(1))\right)$$
$$= \left(\frac{1}{k\log\theta}\right)^{(1-\delta)^2\theta^k/(\theta^k-\theta^{k-1})(1+o(1))}$$
(5.7.6)
$$= \left(\frac{1}{k\log\theta}\right)^{(1-\delta)^2/(1-\theta^{-1})(1+o(1))}.$$

Given $\delta > 0$, we choose $\theta > 1$ so large that $1 - \theta^{-1} > (1-\delta)^2$. Then the last term in (5.7.6) is the general term of a *divergent* series, to which we can apply the second Borel-Cantelli Lemma 5.7.5 to conclude that for infinitely many indices $k \to \infty$

(5.7.7) $$S_{n_k}(t) - S_{n_{k-1}}(t) \geq (1-\delta)A_{n_k}\sqrt{2\log\log A_{n_k}}.$$

But since we have shown in the first part that the liminf is greater than -1, we have for *all* sufficiently large k,

(5.7.8) $$S_{n_{k-1}}(t) \geq -2A_{n_{k-1}}\sqrt{2\log\log A_{n_k}}.$$

Adding (5.7.7) and (5.7.8) and dividing by the right side, we have
$$\liminf_k \frac{S_{n_k}(t)}{A_{n_k}\sqrt{2\log\log A_{n_k}}} \geq (1-\delta) - \frac{2}{\sqrt{\theta}}.$$

Now we rechoose θ so that $1 - \delta - 2/\sqrt{\theta} > 1 - 2\delta$. This proves that for any $\delta > 0$ there is a subsequence $j \to \infty$ so that $S_j(t)/A_j\sqrt{2\log\log A_j} \geq (1-2\delta)$. Hence the limsup of this ratio is greater than or equal to 1, which was to be proved. ∎

CHAPTER 6

INTRODUCTION TO WAVELETS

6.1 MOTIVATION AND HEURISTICS

Classical Fourier analysis may be viewed as the problem of reconstructing a function f from dilations of a fixed sinusoidal function $x \to e^{2\pi i x}$ by writing $f(x) = \int_{\mathbb{R}} e^{2\pi i \xi x} \hat{f}(\xi)\, d\xi$. The Fourier transform $\hat{f}(\xi)$ may be thought of as the amount of the sinusoidal oscillation $e^{2\pi i \xi x}$ present in the function f. The Fourier representation is instrumental in analyzing translation-invariant operators such as convolution operators and linear differential operators with constant coefficients, where we can write

$$\int_{\mathbb{R}} f(x-y) K(y)\, dy = \int_{\mathbb{R}} \hat{K}(\xi) e^{2\pi i \xi x} \hat{f}(\xi)\, d\xi,$$

$$p\left(\frac{d}{dx}\right) f(x) = \int_{\mathbb{R}} p(2\pi i \xi) e^{2\pi i \xi x} \hat{f}(\xi)\, d\xi.$$

However classical Fourier analysis suffers from the defect of nonlocality: The behavior of a function in an open set, no matter how small, influences the global behavior of the Fourier transform. We have also remarked on the simultaneous nonlocalizability in connection with the uncertainty principle.

The theory of wavelets is concerned with the representation of a function in terms of a two-parameter family of dilates and translates of a fixed function that, in general, is not sinusoidal, for example:

$$f(x) = \int_{\mathbb{R}^2} |a|^{-\frac{1}{2}} \psi\left(\frac{x-b}{a}\right) W_\psi f(a,b)\, da\, db$$

where $W_\psi f$ is a suitably defined transform of f.

Alternatively one may envision a series expansion

$$f(x) = \sum_{j,k} c_{j,k} 2^{j/2} \psi(2^j x - k)$$

where we sum over the dilates in geometric progression. The factors of $|a|^{-1/2}$ and $2^{j/2}$ are inserted to preserve the L^2-norm of the basic wavelet ψ.

In this chapter we will describe the properties of wavelets in one dimension, making full use of the tools of Fourier analysis.

6.1.1 Heuristic Treatment of the Wavelet Transform

The wavelet transform of f with respect to ψ is defined by the integral

$$W_\psi f(a,b) = \int_\mathbb{R} f(y) \bar{\psi}\left(\frac{y-b}{a}\right) \frac{dy}{\sqrt{|a|}}.$$

It is straightforward to compute this transform and the inverse transform on the Fourier exponentials $f(x) = e^{2\pi i \xi x}$; from the definition of the Fourier transform, we have

$$W_\psi f(a,b) = \int_\mathbb{R} e^{2\pi i \xi y} \bar{\psi}\left(\frac{y-b}{a}\right) \frac{dy}{\sqrt{|a|}}$$

$$= \sqrt{|a|} \int_\mathbb{R} e^{2\pi i \xi (b+az)} \bar{\psi}(z)\, dz$$

$$= \sqrt{|a|} e^{2\pi i \xi b} \bar{\hat{\psi}}(a\xi).$$

Now we form the adjoint operator

$$W_\psi^* W_\psi f(x) = \int_\mathbb{R} (W_\psi f)(a,b) \psi\left(\frac{x-b}{a}\right) \frac{db}{\sqrt{|a|}}$$

$$= \sqrt{|a|}\,\bar{\hat{\psi}}(a\xi) \int_\mathbb{R} e^{2\pi i \xi b} \psi\left(\frac{x-b}{a}\right) \frac{db}{\sqrt{|a|}}$$

$$= \sqrt{|a|}\,\bar{\hat{\psi}}(a\xi)\sqrt{|a|} \int_\mathbb{R} e^{2\pi i \xi (x-az)} \psi(z)\, dz$$

$$= |a| |\hat{\psi}(a\xi)|^2 e^{2\pi i \xi x}$$

$$\int_\mathbb{R} W_\psi^* W_\psi f(x) \frac{da}{a^2} = e^{2\pi i \xi x} \int_\mathbb{R} \frac{|\hat{\psi}(a\xi)|^2}{|a|}\, da.$$

The final integral is independent of ξ, which is seen by making the substitution $v = a\xi$, from which we obtain the inversion formula

$$f(x) = e^{2\pi i \xi x} = \frac{\int_\mathbb{R} (W_\psi^* W_\psi f/a^2)\, da}{\int_\mathbb{R} |\hat{\psi}(v)|^2/|v|\, dv}.$$

This leads us to impose the normalization $\int_\mathbb{R} |\hat{\psi}(v)|^2/|v|\, dv = 1$, in order to obtain the wavelet representation

$$\boxed{f = \int_\mathbb{R} W_\psi^* W_\psi f \frac{da}{a^2}}$$

valid when $f(x) = e^{2\pi i \xi x}$. It now remains to investigate this inversion procedure for arbitrary $f \in L^2(\mathbb{R})$.

6.2 WAVELET TRANSFORM

Let $\psi \in L^2(\mathbb{R})$. The dilated-translated function is defined by

$$\psi_{a,b}(x) = |a|^{-1/2} \psi\left(\frac{x-b}{a}\right), \qquad 0 \neq a \in \mathbb{R}, b \in \mathbb{R. \tag{6.2.1}}$$

This function is obtained from ψ by first dilating by the factor a and then translating by b. Clearly $\|\psi_{a,b}\|_2 = \|\psi\|_2$.

Definition 6.2.1. $\psi \in L^2(\mathbb{R})$ is a continuum wavelet if

$$\langle \psi, \psi \rangle_w := \int_{\mathbb{R}} |\hat{\psi}(\xi)|^2 \frac{d\xi}{|\xi|} < \infty. \tag{6.2.2}$$

The wavelet transform of $f \in L^2(\mathbb{R})$ by ψ is defined by

$$W_\psi f(a, b) = \int_{\mathbb{R}} \bar{\psi}_{a,b}(x) f(x)\, dx. \tag{6.2.3}$$

From the Cauchy-Schwarz inequality, we see that $W_\psi f$ is a bounded function with $|W_\psi f(a, b)| \leq \|\psi\|_2 \|f\|_2$. The intuitive meaning of $W_\psi f(a, b)$ is the amount of the dilated-translated waveform $\psi_{a,b}$ that is present in the function f.

Remark. If, in addition, $\psi \in L^1(\mathbb{R})$, then the integrability condition (6.2.2) implies that $\int_{\mathbb{R}} \psi(x)\, dx = 0$. Indeed, $\hat{\psi}$ is continuous at $\xi = 0$ with $\hat{\psi}(0) = \int_{\mathbb{R}} \psi$. If this is nonzero, then the integral (6.2.2) is divergent.

The following is a form of Parseval's theorem for the wavelet transform:

Proposition 6.2.2. Suppose that ψ is a continuum wavelet with $\langle \psi, \psi \rangle_w = 1$. Then for any $f, g \in L^2(\mathbb{R})$, we have

$$\int_{\mathbb{R}} f(x) \bar{g}(x)\, dx = \int_{\mathbb{R}} \int_{\mathbb{R}} W_\psi f(a, b) \bar{W}_\psi g(a, b) \frac{da\, db}{a^2}. \tag{6.2.4}$$

Proof. Let $\tilde{\psi}(x) = \bar{\psi}(-x)$. Then $W_\psi f(a, b)$ is the convolution of f with $\tilde{\psi}_{a,0}$, whose Fourier transform is $\sqrt{|a|}\hat{\tilde{\psi}}(a\xi)$. Hence the Fourier transform of $W_\psi f$ is $\hat{f}(\xi)\sqrt{|a|}\hat{\tilde{\psi}}(a\xi)$, and similarly for $W_\psi g$. Therefore from Parseval's theorem for the Fourier transform we have

$$\int_{\mathbb{R}} W_\psi f(a, b) \bar{W}_\psi g(a, b)\, db = \int_{\mathbb{R}} \hat{f}(\xi) \bar{\hat{g}}(\xi) |a| |\hat{\psi}(a\xi)|^2\, d\xi. \tag{6.2.5}$$

We integrate both sides with respect to $da/|a|^2$, apply the Fubini theorem to the right side, and use the definition of $\langle \psi, \psi \rangle_w$ to remove this constant factor. The remaining integral is transformed by another application of Parseval's theorem in the form $\int_{\mathbb{R}} f\bar{g} = \int_{\mathbb{R}} \hat{f}\bar{\hat{g}}$, which completes the proof. ∎

This proposition can be interpreted as the statement that $f \to W_\psi f$ is an isometry from $L^2(\mathbb{R}; dx)$ to $L^2(\mathbb{R}^2; da\, db/|a|^2)$, where the inner product is defined as

$$((F, G)) := \int_{\mathbb{R}^2} F(a, b) \bar{G}(a, b) \frac{da\, db}{a^2}.$$

Theorem 6.2.3. *Suppose that ψ is a continuum wavelet with $\langle \psi, \psi \rangle_w = 1$. Then for any $f \in L^2(\mathbb{R})$, we have the L^2 inversion formula*

(6.2.6)
$$f(x) = \int_{\mathbb{R}^2} W_\psi f(a, b) \psi_{a,b}(x) \frac{da\, db}{a^2}$$
$$= \lim_{\epsilon \to 0, A, B \to \infty} \int_{\epsilon < |a| < A, |b| < B} W_\psi f(a, b) \psi_{a,b}(x) \frac{da\, db}{a^2}.$$

Proof. Writing $S(\epsilon, A, B)f$ for the integral in (6.2.6), we note that this integral is absolutely convergent for each $0 < A < B$ and $\epsilon > 0$, since each factor is in $L^2(\mathbb{R}^2; da\, db/a^2)$. To prove the required convergence, we first note that

$$\|f - S(\epsilon, A, B)f\|_2 = \sup_{\|g\|_2 = 1} |(f - S(\epsilon, A, B)f, g)|.$$

Applying Fubini's theorem, we see that

$$(S(\epsilon, A, B)f, g) = \int_{\mathbb{R}} \bar{g}(x) \left(\int_{\epsilon < |a| < A, |b| < B} W_\psi f(a, b) \psi_{a,b}(x) \frac{da\, db}{a^2} \right) dx$$
$$= \int_{\epsilon < |a| < A, |b| < B} W_\psi f(a, b) \bar{W}_\psi g(a, b) \frac{da\, db}{a^2}$$

so that by (6.2.4) and Cauchy-Schwarz,

$$|(f - S(\epsilon, A, B)f, g)| = \left| \int_{\{\epsilon < |a| < A, |b| < B\}^c} W_\psi f(a, b) \bar{W}_\psi g(a, b) \frac{da\, db}{a^2} \right|$$
$$\leq \left(\int_{\{\epsilon < |a| < A, |b| < B\}^c} |W_\psi f(a, b)|^2 \frac{da\, db}{a^2} \right)^{1/2} \left(\int_{\mathbb{R}^2} |W_\psi g|^2 \frac{da\, db}{a^2} \right)^{1/2}$$
$$= \left(\int_{\{\epsilon < |a| < A, |b| < B\}^c} |W_\psi f(a, b)|^2 \frac{da\, db}{a^2} \right)^{1/2} \|g\|_2.$$

When $\epsilon \to 0$ and $A, B \to \infty$ the region of integration decreases to the empty set, hence the integral tends to zero by the dominated convergence theorem. This completes the proof that $\|S(\epsilon, A, B)f - f\|_2 \to 0$. ∎

We now give some examples of continuum wavelets.

Example 6.2.4. *The standard Haar function is defined by $\psi(x) = +1$ for $0 \leq x < 1/2$, $\psi(x) = -1$ for $1/2 \leq x < 1$ and $\psi(x) = 0$ otherwise.*

The Fourier transform is computed as

$$\hat{\psi}(x) = \int_0^{1/2} e^{-2\pi i \xi x} \, dx - \int_{1/2}^1 e^{-2\pi i \xi x} \, dx$$

$$= \frac{e^{-i\pi\xi} - 1}{-2\pi i \xi} - \frac{e^{-2\pi i \xi} - e^{-i\pi\xi}}{-2\pi i \xi}$$

$$= \frac{1 - 2e^{-i\pi\xi} + e^{-2\pi i \xi}}{2\pi i \xi}$$

$$= \frac{(1 - e^{-i\pi\xi})^2}{2\pi i \xi}.$$

Clearly the integrability condition (6.2.2) is satisfied, so that ψ is a continuum wavelet.

In the next example, we specify the continuum wavelet in terms of its Fourier transform.

Example 6.2.5. Let $\hat{\psi}(\xi) = 1_{[\frac{1}{4}, \frac{1}{2}]}(|\xi|)$.

Clearly the integrability condition (6.2.2) is satisfied. The continuum wavelet is

$$\psi(x) = \int_{1/4}^{1/2} e^{2\pi i x \xi} \, d\xi + \int_{-1/2}^{-1/4} e^{2\pi i x \xi} \, d\xi$$

$$= \frac{e^{i\pi x} - e^{i\pi x/2}}{2\pi i x} + \frac{e^{-i\pi x/2} - e^{-i\pi x}}{2\pi i x}$$

$$= \frac{\sin \pi x}{\pi x} - \frac{\sin \pi x/2}{\pi x}.$$

However this continuum wavelet is not integrable, since $\int_{\mathbb{R}} |\psi(x)| dx = +\infty$.

The following two examples include the normalization $\langle \psi, \psi \rangle_w = 1$.

Example 6.2.6. *A Gaussian wavelet is defined by* $\psi(x) = Cxe^{-\pi x^2}$.

The Fourier transform is computed as $\hat{\psi}(\xi) = -iC\xi e^{-\pi \xi^2}$. Clearly the integrability condition (6.2.2) is satisfied. The normalization is computed from the integral $\int_{\mathbb{R}} |\hat{\psi}(\xi)|^2 / |\xi| \, d\xi = 2C^2 \int_0^\infty \xi e^{-2\pi \xi^2} \, d\xi = C^2/2\pi$, thus $C = \sqrt{2\pi}$.

Example 6.2.7. *The Mexican hat wavelet is defined through its Fourier transform by* $\hat{\psi}(\xi) = C\xi^2 e^{-\pi \xi^2}$.

The continuum wavelet ψ can be computed directly as the derivative of the previous example. Thus $\psi(x) = C(1/2\pi - x^2)e^{-\pi x^2}$. The normalization is obtained by noting that $\int_{\mathbb{R}} |\hat{\psi}(\xi)|^2 / |\xi| \, d\xi = 2C^2 \int_0^\infty \xi^3 e^{-2\pi \xi^2} \, d\xi = C^2/4\pi^2$, thus $C = 2\pi$.

Figures 6.1.1 and 6.1.2 illustrate Examples 6.2.6 and 6.2.7.

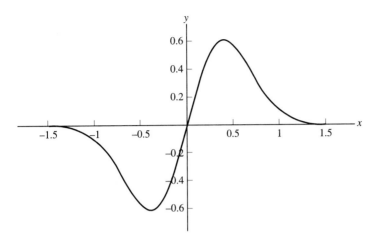

FIGURE 6.1.1
Gaussian wavelet.

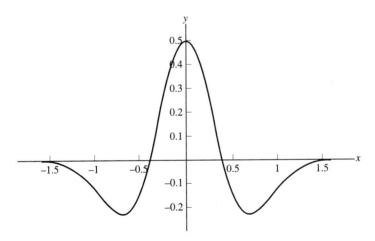

FIGURE 6.1.2
Mexican hat wavelet.

It is instructive to examine the inverse wavelet transform in the particular case of the Gaussian example, where $\psi(x) = \sqrt{2\pi}\,xe^{-\pi x^2}$. Define the partial inverse transform by

(6.2.7) $$S_\epsilon f(x) = \int_\mathbb{R} \int_{|a|>\epsilon} W_\psi f(a,b) \psi_{a,b}(x) \frac{da\,db}{a^2}.$$

This can be computed through the Fourier transform by writing

$$\widehat{S_\epsilon f}(\xi) = \int_{\mathbb{R}} S_\epsilon f(x) e^{-2\pi i \xi x}\, dx$$

$$= \left(\int_{|a|>\epsilon} \frac{da}{a^2} |a| |\hat{\psi}(a\xi)|^2\, da \right) \hat{f}(\xi)$$

$$= 2\pi \xi^2 \left(\int_{|a|>\epsilon} |a| e^{-2\pi^2 a^2 \xi^2}\, da \right) \hat{f}(\xi)$$

$$= e^{-2\pi^2 \epsilon^2 \xi^2} \hat{f}(\xi).$$

But this is the Fourier transform of the convolution $f * g_\epsilon$, where g_ϵ is the Gaussian density $g_\epsilon(x) = e^{-\pi x^2/2\epsilon^2}/\sqrt{2\epsilon^2}$, which is an approximate identity in the sense of Chapter 2, from which we conclude that for any homogeneous Banach space B, we have $S_\epsilon f \to f$ in norm. In particular, if f is bounded and uniformly continuous we have $\|S_\epsilon f - f\|_\infty \to 0$ when $\epsilon \to 0$. If $f \in L^p(\mathbb{R})$, $1 \le p < \infty$, then $\|S_\epsilon f - f\|_p \to 0$ when $\epsilon \to 0$ and $S_\epsilon f(x) \to f(x)$ for almost every $x \in \mathbb{R}$. This follows from the results on Gaussian summability in Chapter 2. These desirable properties are not shared by the partial inversion of the Fourier transform, for example.

Exercise 6.2.8. *For the Mexican hat wavelet, define the partial inversion by (6.2.7) and explicitly compute $S_\epsilon f$ as the convolution with an integrable function, in particular verify that $\|S_\epsilon f - f\|_p \to 0$ in case of $f \in L^p(\mathbb{R})$, $1 \le p < \infty$ or $f \in B_{uc}(\mathbb{R})$ in case $p = \infty$.*

Exercise 6.2.9. *Formulate the wavelet transform in n dimensions, beginning with $\psi \in L^2(\mathbb{R}^n)$ satisfying $\int_{\mathbb{R}^n} |\hat{\psi}(\xi)|^2 d\xi/|\xi|^n < \infty$ and defining $\psi_{a,b}(x) = \psi((x-b)/a)/|a|^{n/2}$ for $b \in \mathbb{R}^n$ and $0 \neq a \in \mathbb{R}$.*

6.2.0.1 Wavelet characterization of smoothness

We can use the wavelet transform to characterize the smoothness of $f \in L^2(\mathbb{R})$ as measured by the Sobolev norm

$$\|f\|_{2,s}^2 = \int_{\mathbb{R}} |\xi|^{2s} |\hat{f}(\xi)|^2\, d\xi.$$

Exercise 6.2.10. *If $\|f\|_{2,s} < \infty$, prove that f has k continuous derivatives, where $k < s - \frac{1}{2}$.*

Hint: Apply Cauchy-Schwarz to the Fourier integral representation of $f^{(k)}$.

The next proposition applies to continuum wavelets that possess a certain number of vanishing moments: $\int_{\mathbb{R}} x^k \psi(x)\, dx = 0$. The result states that for this class of wavelet expansions, the Sobolev norm is equivalent to a weighted L^2 norm of the wavelet transform.

Proposition 6.2.11. *Suppose that ψ is a continuum wavelet with $\langle \psi, \psi \rangle_w = 1$ and*

$$C_{\psi,s} := \int_{\mathbb{R}} \frac{|\hat{\psi}(\xi)|^2}{|\xi|^{1+2s}} d\xi < \infty$$

for some $s > 0$. Then

(6.2.8) $$\int_{\mathbb{R}} \int_{\mathbb{R}} |W_\psi f(a,b)|^2 \frac{da\, db}{|a|^{2+2s}} = C_{\psi,s} \|f\|_{2,s}^2.$$

Proof. Returning to (6.2.5) with $f = g$, we divide both sides by $|a|^{2+2s}$ and integrate with respect to $a \in \mathbb{R}$. Thus

$$\int_{\mathbb{R}} \int_{\mathbb{R}} |W_\psi f(a,b)|^2 \frac{da\, db}{|a|^{2+2s}} = \int_{\mathbb{R}} \int_{\mathbb{R}} |\hat{f}(\xi)|^2 \frac{|\hat{\psi}(a\xi)|^2}{|a|^{1+2s}} da\, d\xi$$

$$= \int_{\mathbb{R}} |\hat{f}(\xi)|^2 \left(\int_{\mathbb{R}} \frac{|\hat{\psi}(v)|^2}{|v|^{1+2s}} dv \right) |\xi|^{2s} d\xi$$

$$= C_{\psi,s} \int_{\mathbb{R}} |\xi|^{2s} |\hat{f}(\xi)|^2 d\xi,$$

which completes the proof. ∎

6.3 HAAR WAVELET EXPANSION

In this section we develop the properties of the Haar wavelet expansion, which is the oldest and most basic example of an orthonormal wavelet (to be defined in the next section).

The Haar series expansion can be naturally motivated by the search for an orthogonal series representation of Lebesgue's differentiation theorem for a locally integrable function $f \in L^1_{\text{loc}}$:

$$f(x) = \lim_{b \to x, a \to x} \frac{1}{b-a} \int_a^b f(y) \, dy \quad \text{a.e.} \quad x \in \mathbb{R}.$$

We will systematically describe this connection in the following subsections.

6.3.1 Haar Functions and Haar Series

We begin with the basic Haar function

$$\psi(x) = 1 \quad \text{if } 0 \leq x < \tfrac{1}{2}, \qquad \psi(x) = -1 \quad \text{if } \tfrac{1}{2} \leq x < 1,$$

and

$$\psi(x) = 0 \quad \text{otherwise.}$$

Clearly $\int_0^1 \psi(x)\, dx = 0$ and $\int_0^1 \psi(x)^2 \, dx = 1$.

A doubly indexed family of Haar functions is defined by writing

(6.3.1) $$\boxed{\psi_{jk}(x) = 2^{j/2} \psi(2^j x - k), \quad j, k = 0, \pm 1, \pm 2, \ldots.}$$

We will prove that $\{\psi_{jk}\}_{j,k \in \mathbb{Z}^2}$ form an orthonormal basis of $L^2(\mathbb{R})$.

Proof. The first task is to prove orthonormality, namely

$$\int_{\mathbb{R}} \psi_{jk}(x)\psi_{j'k'}(x)\,dx = \begin{cases} 0 & \text{if } (j,k) \neq (j',k') \\ 1 & \text{if } (j,k) = (j',k'). \end{cases}$$

A direct proof of orthogonality if $j = j'$ is seen from the fact that if $k \neq k'$,

$$\int_{\mathbb{R}} \psi(2^j x - k)\psi(2^j x - k')\,dx = 2^{-j} \int_{\mathbb{R}} \psi(y)\psi(y + k - k')\,dy = 0$$

since the integrand is identically zero. Otherwise we can assume that $j < j'$, and write

$$\int_{\mathbb{R}} \psi(2^j x - k)\psi(2^{j'} x - k')\,dx = \int_{\mathbb{R}} \psi(y)\psi(2^{j'-j}y + 2^{j'-j}k - k')\,dy$$

$$= \int_0^{1/2} \psi(2^{j'-j}y + k'')\,dy - \int_{1/2}^1 \psi(2^{j'-j}y + k'')\,dy.$$

But both of these integrals are zero, since $\int_0^1 \psi = 0$. Finally the normalization in case $(j,k) = (j',k')$ is established by computing

$$\int_{\mathbb{R}} \psi_{jk}(x)^2\,dx = 2^j \int_{\mathbb{R}} \psi(2^j x - k)^2\,dx = \int_{\mathbb{R}} \psi(y)^2\,dy = 1. \qquad\blacksquare$$

The Fourier/Haar coefficients of $f \in L^2(\mathbb{R})$ are defined by

(6.3.2) $$c_{jk} = C_{jk}(f) := \int_{\mathbb{R}} f(x)\psi_{jk}(x)\,dx,$$

leading to the Haar series

(6.3.3) $$f(x) \sim \sum_{j,k} c_{jk} \psi_{jk}(x).$$

From the orthonormality of $\{\psi_{jk}\}$ and Bessel's inequality, we conclude that $\sum_{jk} |c_{jk}|^2 < \infty$. The completeness of $L^2(\mathbb{R})$ further assures that the series (6.3.3) converges in $L^2(\mathbb{R})$. It now remains to identify the sum of the series with the given function $f \in L^2(\mathbb{R})$.

6.3.2 Haar Sums and Dyadic Projections

In order to identify the sum of the Haar series, we introduce the dyadic projection operator P_n as follows. Consider the dyadic partition $\mathcal{F}_n$, consisting of the sets $I_{kn} := ((k-1)/2^n, k/2^n]$, where $n = 0, \pm 1, \pm 2, \ldots$ and $k = 0, \pm 1, \pm 2, \ldots$. P_n is the projection of $f \in L^2(\mathbb{R})$ onto the space $L^2(\mathbb{R}, \mathcal{F}_n, dx)$, which consists of L^2 functions that are constant on each of the intervals I_{kn}. In detail, we have

(6.3.4) $$P_n f(x) = 2^n \int_{I_{kn}} f(y)\,dy \qquad \text{if } x \in I_{kn}.$$

Formula (6.3.4) can be written explicitly in terms of the *scaling function*

(6.3.5) $\qquad \phi(x) = 1 \text{ if } 0 \leq x \leq 1 \qquad$ and $\phi(x) = 0 \quad$ otherwise

(6.3.6) $$P_n f(x) = 2^n \sum_{k \in \mathbb{Z}} \left(\int_{\mathbb{R}} \phi(2^n y - k) f(y)\,dy \right) \phi(2^n x - k).$$

Indeed, the function $x \to \phi(2^n x - k) = 0$ unless $x \in (k/2^n, (k+1)/2^n]$. To be more succinct, we have

(6.3.7) $$P_n f(x) = \int_{\mathbb{R}} K_n(x, y) f(y)\, dy$$

where

(6.3.8) $$K_n(x, y) = 2^n \sum_{k \in \mathbb{Z}} \phi(2^n x - k) \phi(2^n y - k)$$

$$= \begin{cases} 2^n & \text{if } x, y \in I_{kn} \text{ for some } k \\ 0 & \text{otherwise.} \end{cases}$$

The projection operators are (i) increasing and (ii) converge to the identity:

$$\cdots \leq P_{-1} \leq P_0 \leq P_1 \leq \cdots \leq P_n \leq P_{n+1} \to I \quad (n \to \infty)$$

in the sense that (i) $P_n f = f$ implies $P_{n+1} f = f$ and (ii) $\lim_{n \to \infty} P_n f = f$, a.e. and in $L^2(\mathbb{R})$ (to be proved in the next subsection).

We now seek a representation of the operator $P_{n+1} - P_n$, which is the projection onto the orthogonal complement $L^2(\mathcal{F}_{n+1}) \ominus L^2(\mathcal{F}_n)$. To do this, note that any square of the form $((k-1)/2^n, k/2^n] \times ((k-1)/2^n, k/2^n]$ (where $K_n = 2^n$) is decomposed into four smaller squares; on these smaller squares we have $K_{n+1}(x, y) = 2^{n+1}$ on each of the smaller squares $((k-1)/2^n, (k-1/2)/2^n] \times ((k-1)/2^n, (k-1/2)/2^n]$ and $((k-1/2)/2^n, k/2^n] \times ((k-1/2)/2^n, k/2^n]$ whereas $K_{n+1}(x, y) = 0$ on the smaller squares $((k-1)/2^n, (k-1/2)/2^n] \times ((k-1/2)/2^n, k/2^n]$ and $((k-1/2)/2^n, k/2^n] \times ((k-1)/2^n, (k-1/2)/2^n]$. Hence

$$L_n(x, y) := K_{n+1}(x, y) - K_n(x, y)$$
$$= 2^{n+1} - 2^n \quad \text{on } \left(\frac{k-1}{2^n}, \frac{k-1/2}{2^n}\right] \times \left(\frac{k-1}{2^n}, \frac{k-1/2}{2^n}\right]$$
$$= 2^{n+1} - 2^n \quad \text{on } \left(\frac{k-1/2}{2^n}, \frac{k}{2^n}\right] \times \left(\frac{k-1/2}{2^n}, \frac{k}{2^n}\right]$$
$$= 0 - 2^n \quad \text{on } \left(\frac{k-1}{2^n}, \frac{k-1/2}{2^n}\right] \times \left(\frac{k-1/2}{2^n}, \frac{k}{2^n}\right]$$
$$= 0 - 2^n \quad \text{on } \left(\frac{k-1/2}{2^n}, \frac{k}{2^n}\right] \times \left(\frac{k-1}{2^n}, \frac{k-1/2}{2^n}\right].$$

This can be conveniently represented in terms of the Haar function by writing

$$L_n(x, y) = \sum_{k \in \mathbb{Z}} 2^n \psi(2^n x - k) \psi(2^n y - k).$$

Thus

(6.3.9) $$L_n(x, y) = \sum_{k \in \mathbb{Z}} \psi_{nk}(x) \psi_{nk}(y).$$

For a fixed value of n, the functions $\{\psi_{nk}\}_{k \in \mathbb{Z}}$ form an orthonormal basis of the space $L^2(\mathcal{F}_{n+1}) \ominus L^2(\mathcal{F}_n)$. For each $x \in \mathbb{R}$, the series (6.3.9) contains exactly one nonzero term, hence convergence is trivial.

This provides the desired representation of $P_{n+1} - P_n$ in terms of orthogonal functions, namely

(6.3.10)
$$P_{n+1}f - P_n f = \sum_{k \in \mathbb{Z}} \psi_{nk}(x) \left(\int_{\mathbb{R}} f(y) \psi_{nk}(y) \, dy \right).$$

Hence we can write the original projection operator in the form

$$P_{n+1}f = P_0 f + \sum_{j=0}^{n} (P_{j+1}f - P_j f)$$

$$= P_0 f + \sum_{j=0}^{n} \sum_{k \in \mathbb{Z}} \left(\int_{\mathbb{R}} f(y) \psi_{jk}(y) \, dy \right) \psi_{jk}(x)$$

$$= P_0 f + \sum_{j=0}^{n} \left(\sum_{k \in \mathbb{Z}} \psi_{jk} \otimes \psi_{jk} \right) f,$$

and thus the one-sided Haar series representation

(6.3.11)
$$f = P_0 f + \sum_{j=0}^{\infty} \sum_{k \in \mathbb{Z}} \left(\int_{\mathbb{R}} f(y) \psi_{jk}(y) \, dy \right) \psi_{jk}(x).$$

In the following sections, we will abstract this to a more general setting, noting that the subspaces $V_n := L^2(\mathbb{R}, \mathcal{F}_n, dx)$ have the following properties:

(i) $\bigcup_{n=0}^{\infty} V_n$ is dense in $L^2(\mathbb{R})$
(ii) $f \in V_n$ if and only if $f(2^{-n} \cdot) \in V_0$
(iii) $\{\phi(x - k)\}_{k \in \mathbb{Z}}$ is an orthonormal basis of V_0.

This can also be extended to a bilateral family of subspaces by considering V_n for $n < 0$, namely larger and larger dyadic intervals. As above, the orthogonal projection on the subspace $L^2(\mathbb{R}, \mathcal{F}_{n+1}, dx) \ominus L^2(\mathbb{R}, \mathcal{F}_n, dx)$ continues to be represented by the formula (6.3.10), and the above nesting properties of the subspaces can be modified to

(i') $\bigcup_{n=-\infty}^{\infty} V_n$ is dense in $L^2(\mathbb{R})$, $\bigcap_{n=-\infty}^{\infty} V_n = \{0\}$.

The latter property is evident from the fact that if $f \in V_n$ for all $n < 0$, then f is constant on each interval $[0, 2^{|n|})$, hence must be identically constant for $x > 0$; but $f \in L^2(\mathbb{R})$ means that the constant must be zero.

Before passing on to more general wavelet expansions, we note that the Haar function ψ and the scaling function ϕ are related by the identity

(6.3.12)
$$\psi(x) = \phi(2x) - \phi(2x - 1),$$

whereas ϕ satisfies the identity

(6.3.13) $$\phi(x) = \phi(2x) + \phi(2x-1).$$

Much of the challenge of constructing more general wavelets will reduce to the suitable generalization of these simple relations.

6.3.3 Completeness of the Haar Functions

To prove the validity of the two-sided Haar representation (6.3.3), we go back to (6.3.10) and write

(6.3.14) $$P_{n+1}f(x) - P_{-m}f(x) = \sum_{j=-m}^{n} \sum_{k \in \mathbb{Z}} c_{jk} \psi_{jk}(x).$$

It remains to prove that $P_{-m}f \to 0$ and $P_{n+1}f \to f$ when $m, n \to \infty$. First we prove that the operators P_n have uniformly bounded operator norms.

Lemma 6.3.1. *For any $f \in L^2(\mathbb{R})$ and $n \in \mathbb{Z}$, we have $\|P_n f\|_2 \leq \|f\|_2$.*

Proof. From the definition of $P_n f$, we apply Cauchy-Schwarz:

$$x \in I_{kn} \implies |P_n f(x)|^2 \leq 2^n \int_{I_{kn}} |f(y)|^2 \, dy$$

$$\int_{I_{kn}} |P_n f(x)|^2 \, dx \leq \int_{I_{kn}} |f(x)|^2 \, dx$$

$$\int_{\mathbb{R}} |P_n f(x)|^2 \, dx \leq \int_{\mathbb{R}} |f(x)|^2 \, dx. \quad \blacksquare$$

We use the notation $C_0(\mathbb{R})$ to denote continuous functions vanishing at infinity and $C_{00}(\mathbb{R})$ to denote continuous functions of compact support.

Lemma 6.3.2.

(i) *If $g \in C_0(\mathbb{R})$, we have $\|P_{-m}g\|_\infty \to 0$ when $m \to \infty$.*
(ii) *If $f \in L^2(\mathbb{R})$, we have $\|P_{-m}f\|_2 \to 0$ when $m \to \infty$.*

Proof. If $g \in C_{00}(\mathbb{R})$ has support in $[-K, K]$, we can write

$$0 \leq x \leq 2^m \implies |P_{-m}g(x)| = 2^{-m} \int_0^K |g| \to 0,$$

and similarly for $-2^m \leq x \leq 0$. Hence $\|P_{-m}g\|_\infty \to 0$. But these functions are dense in $C_0(\mathbb{R})$; given $f \in C_0(\mathbb{R})$ and $\epsilon > 0$, there exists $g \in C_{00}(\mathbb{R})$, $h \in C_0(\mathbb{R})$ so that $f = g + h$ with $\|h\|_\infty < \epsilon$. Then $\limsup_{m \to \infty} \|P_{-m}f\|_\infty \leq \limsup_{m \to \infty} \|P_{-m}h\|_\infty < \epsilon$, which proves the required convergence.

If $f \in L^2(\mathbb{R})$, for any $\epsilon > 0$, $f = g + h$, where g is continuous and has support in $[-K, K]$ for some $K > 0$, and $\|h\|_2 < \epsilon$. Then for $2^m > K$, we have

$$-2^m \leq x \leq 2^m \implies |P_{-m}g(x)| = 2^{-m} \int_{-K}^{K} |g| \leq 2^{-m}\sqrt{2K}\|g\|_2$$

$$\|P_{-m}g\|_2 \leq \sqrt{4K} 2^{-m/2} \|g\|_2$$

$$\|P_{-m}f\|_2 \leq \|P_{-m}g\|_2 + \|P_{-m}h\|_2$$

$$\leq \|P_{-m}g\|_2 + \epsilon$$

$$\limsup_{m \to \infty} \|P_{-m}f\|_2 \leq \epsilon$$

where we have used Lemma 6.3.1 in the last line, in the form $\|P_{-m}h\|_2 \leq \|h\|_2$. Since this holds for every $\epsilon > 0$, we conclude that $P_{-m}f \to 0$ when $m \to \infty$. ∎

The above proof contains the following general principle: If a sequence of bounded linear operators has uniformly bounded operator norms and converges to a bounded operator on a dense subset of a Banach space, then it converges on the entire space.

To prove that $P_n f \to f$ when $n \to \infty$, we first prove that this holds on the dense set of continuous functions with compact support.

Lemma 6.3.3. *If $f \in C_{00}(\mathbb{R})$, then $P_n f \to f$ uniformly and in $L^2(\mathbb{R})$, when $n \to \infty$.*

Proof. Let f be supported in $[-K, K]$, where we may suppose that $K \geq 1$. Given $\epsilon > 0$, from the uniform continuity of f, there exists $\delta > 0$ so that $|f(y) - f(x)| < \epsilon/K$ whenever $|x - y| < \delta$. If $2^{-n} < \delta$, we have $|P_n f(x) - f(x)| \leq \epsilon/\sqrt{2K} \leq \epsilon$ for all x, which proves the uniform convergence. Integrating over the support of f, we have

$$\int_{\mathbb{R}} |P_n f(x) - f(x)|^2 dx \leq \int_{-K}^{K} \epsilon^2/2K \leq \epsilon^2,$$

which proves that $\|P_n f - f\|_2 \to 0$ when $n \to \infty$. ∎

We have thus proved the following theorem.

Theorem 6.3.4. *The normalized Haar functions $\{\psi_{jk}\}_{j,k \in \mathbb{Z}}$ form an orthonormal basis of $L^2(\mathbb{R})$, in particular we have the L^2 convergent expansion*

(6.3.15)
$$f = \sum_{j,k=-\infty}^{\infty} \psi_{jk}(x) \left(\int_{\mathbb{R}} f(y) \psi_{jk}(y) \, dy \right).$$

6.3.3.1 Haar series in C_0 and L_p spaces

The Haar series is well defined for any locally integrable function, hence it makes sense to study the convergence in other spaces of functions. We have treated the L^2 convergence of the Haar series by relating the partial sum to the fundamental theorem of calculus. These ideas can also be used to discuss the uniform convergence of the Haar series in spaces of continuous functions, as well as the norm convergence in $L^p(\mathbb{R})$, $1 \leq p < \infty$.

We first treat the convergence in the space $C_0(\mathbb{R})$, consisting of continuous functions with $\lim_{|x|\to\infty} f(x) = 0$. This Banach space contains as a dense subspace the set of continuous functions with compact support, on which we have proved Lemma 6.3.3. It remains to prove that the operators P_n are uniformly bounded. We prove a more general estimate on the (larger) space of bounded continuous functions.

Lemma 6.3.5. *For any $f \in B_c(\mathbb{R})$, we have $|P_n f(x)| \leq \|f\|_\infty$.*

Proof.
$$x \in I_{kn} \Longrightarrow |P_n f(x)| \leq 2^n \int_{I_{kn}} |f(y)|\, dy \leq \|f\|_\infty. \qquad \blacksquare$$

From Lemma 6.3.2, we have for any continuous function g with compact support, $\|P_{-m}g\| \to 0$ when $m \to \infty$. Since these are dense in $C_0(\mathbb{R})$, from the uniform boundedness of $\|P_n\|$ we obtain $\|P_{-m}f\|_\infty \to 0$ when $m \to \infty$. Meanwhile, Lemmas 6.3.3 and 6.3.5 show that $P_n f \to f$ in the supremum norm when $n \to \infty$. This leads to the following general proposition on uniform convergence.

Proposition 6.3.6. *If $f \in C_0(\mathbb{R})$, then the Haar series (6.3.15) converges uniformly on the entire real line.*

If f is merely bounded and uniformly continuous, we cannot expect a uniformly convergent expansion on the entire real line, as shown by the following.

Exercise 6.3.7. *Let $f(x) \equiv 1$. Prove that the Haar series expansion (6.3.15) is identically zero, especially not convergent to f.*

Exercise 6.3.8. *Suppose that $f \in B_{uc}(\mathbb{R})$, the space of bounded and uniformly continuous functions. Prove directly that the one-sided Haar series (6.3.11) converges uniformly to f.*

To treat convergence in $L^p(\mathbb{R})$, we first prove uniform boundedness.

Lemma 6.3.9. *Let $f \in L^p(\mathbb{R})$, $1 \leq p < \infty$. Then $\|P_n f\|_p \leq \|f\|_p$ for all $n \in \mathbb{Z}$.*

Proof. Set $p' = p/(p-1)$ if $p > 1$. Then Hölder's inequality gives

$$x \in I_{kn} \Longrightarrow |P_n f(x)| \leq 2^n \left(\int_{I_{kn}} |f(y)|^p\, dy\right)^{1/p} 2^{-n/p'}$$

$$|P_n f(x)|^p \leq 2^{np} \left(\int_{I_{kn}} |f(y)|^p\, dy\right) 2^{-np/p'}$$

$$\int_{I_{kn}} |P_n f(x)|^p\, dx \leq 2^{-n} 2^{np} 2^{-np/p'} \int_{I_{kn}} |f(y)|^p\, dy$$

$$= \int_{I_{kn}} |f(y)|^p\, dy.$$

Summing on $k \in \mathbb{Z}$ gives the result. This proof also applies in case $p = 1$, by setting $1/\infty = 0$ whenever p' appears. $\blacksquare$

To prove L^p convergence, we must check that $P_n f \to f$ and $P_{-m} f \to 0$.

Lemma 6.3.10. *Let $1 \le p < \infty$. Then $\|P_n f - f\|_p \to 0$ when $n \to \infty$.*

Proof. The space of continuous functions with compact support is dense in $L^p(\mathbb{R})$. From Lemma 6.3.3 we have uniform convergence on this space. In particular if f is supported in $[-K, K]$, then for $n > N(\epsilon)$

$$\int_{\mathbb{R}} |P_n f(x) - f(x)|^p dx \le 2K \epsilon^p,$$

which shows that $\|P_n f - f\|_p < \epsilon(2K)^{1/p}$. ∎

It remains to consider $P_{-m} f$, $m \to \infty$. This puts a new restriction on p.

Lemma 6.3.11. *Let $1 < p < \infty$. Then $\|P_{-m} f\|_p \to 0$ when $m \to \infty$.*

Proof. It suffices to check this for g continuous with compact support in $[-K, K]$. If $2^m > K$, then $P_{-m} g$ is constant on $(-2^m, 0)$, $(0, 2^m)$ and zero elsewhere, so that from Hölder's equality,

$$0 \le x \le 2^m \implies |P_{-m} g(x)| = 2^{-m} |\int_0^K g(y) \, dy|$$

$$|P_{-m} g(x)|^p = 2^{-mp} \int_0^K |g(y)|^p \, dy \, (2K)^{p/p'}$$

$$\int_0^\infty |P_{-m} g(x)|^p \, dx = 2^m 2^{-mp} |\int_{-K}^K |g(y)|^p \, dy| \, (2K)^{p/p'},$$

which tends to zero when $m \to \infty$. The contribution from the negative axis is estimated in the same fashion. ∎

Hence we conclude the following.

Proposition 6.3.12. *Let $1 < p < \infty$. For any $f \in L^p(\mathbb{R})$, the Haar series (6.3.15) converges in the norm of $L^p(\mathbb{R})$.*

In the case $p = 1$ there is a simple example to show that this proposition is sharp.

Exercise 6.3.13. *Let $f = 1_{[0,1]}$. Prove that the Haar series (6.3.15) is not convergent in $L^1(\mathbb{R})$.*

However this anomaly is not present for the one-sided Haar series.

Exercise 6.3.14. *Let $1 \le p < \infty$ and let $f \in L^p(\mathbb{R})$. Prove that the one-sided Haar series (6.3.11) converges in the norm of $L^p(\mathbb{R})$.*

6.3.3.2 Pointwise convergence of Haar series

Since the projection operator P_n agrees with the average over dyadic intervals, it follows from Lebesgue's theorem that $P_n f(t) \to f(t)$ for almost every $t \in \mathbb{R}$. In particular if f is continuous at t, then we have $\lim_{n \to \infty} P_n f(t) = f(t)$. If f has a jump discontinuity at a dyadic rational t, then we note that $K_n(t, y) = 0$ for $y < t$ and sufficiently large n. Then we can write

$$P_n f(t) - f(t+0) = \int_t^\infty [f(y) - f(t+0)] K_n(t, y) \, dy \to 0, \qquad n \to \infty$$

to conclude that

$$P_n f(t) \to f(t+0) \qquad t = k/2^N, \quad n \to \infty.$$

One can also confirm the absence of a possible Gibbs phenomenon for Haar series. Indeed, the kernel $K_n(x, y) \geq 0$ with $\int_{\mathbb{R}} K_n(x, y) \, dy = 1$. Therefore if $f \in L^\infty(\mathbb{R})$

$$-\|f\|_\infty \leq P_n f(x) \leq \|f\|_\infty,$$

which implies that for any sequence $x_n \to x$, we must have

$$-\|f\|_\infty \leq \liminf_n P_n f(x_n) \leq \limsup_n P_n f(x_n) \leq \|f\|_\infty.$$

Exercise 6.3.15. *Suppose that $f(t) = 1$ for $0 \leq t < 1/3$ and that $f(t) = 0$ for $1/3 \leq t < 1$. Show that we have $\liminf_n P_n f(1/3) < \limsup_n P_n f(1/3)$, so that the Haar series diverges at $t = 1/3$.*

6.3.4 *Construction of Standard Brownian Motion

The Haar wavelet expansion can be used to make an effective construction of the *standard Brownian motion process*. By definition, this is an indexed family of real-valued functions $X_t(\omega)$ where $0 \leq t \leq 1$ and $\omega \in \Omega$, where $(\Omega, \mathcal{F}, P)$ is a measure space of total measure 1. In this context, the functions $\omega \to X_t(\omega)$ are called *random variables*. They are assumed to have the following properties:

1. For each $0 \leq s < t \leq 1$, $X_t - X_s$ has a normal distribution with mean zero and variance $t - s$: in detail

$$P[\omega : X_t(\omega) - X_s(\omega) < y] = \frac{1}{\sqrt{2\pi(t-s)}} \int_{-\infty}^y e^{-u^2/2(t-s)} \, du.$$

2. For any subdivision $0 = t_0 < t_1 < \cdots < t_N \leq 1$, the random variables $X_{t_1} - X_{t_0}, \ldots, X_{t_N} - X_{t_{N-1}}$ are independent.
3. For a.e. ω, the function $t \to X_t(\omega)$ is continuous, with $X_0 = 0$.

From properties (1), (2), it follows that the random variables $X_{t_1}, \ldots, X_{t_n}$ have a joint normal distribution with mean values zero and covariance matrix defined by $E[X_{t_i} X_{t_j}] = \min(t_i, t_j)$.

Exercise 6.3.16. *Prove this.*

The Haar functions ψ_{jk} are not continuous, so we would not expect to be able to construct the Brownian motion as a Haar series. But the functions $t \to \int_0^t \psi_{jk}(s) \, ds$ are continuous and can be used to construct the Brownian motion. In order to prove the distributional properties (1) and (2), we will first consider a general orthonormal basis of the space $L^2[0, 1]$.

The Brownian motion will be constructed as the infinite series

(6.3.16) $$X_t(\omega) = \sum_{n=0}^{\infty} Z_n(\omega) \int_0^t \phi_n(s) \, ds.$$

Here (ϕ_n) is an orthonormal basis of the space $L^2[0, 1]$ and (Z_n) is a sequence of independent standard normal random variables; in detail

$$P[Z_n < y] = \frac{1}{\sqrt{2\pi}} \int_{-\infty}^y e^{-u^2/2} \, du \qquad y \in \mathbb{R}, \quad n = 0, 1, 2, \ldots.$$

Lemma 6.3.17. *Suppose that (ϕ_n) is an orthonormal basis of $L^2[0, 1]$ and (Z_n) is a sequence of independent standard normal random variables. Then (6.3.16) converges in $L^2(\Omega)$ to a limit $X_t(\omega)$, which satisfies properties (1) and (2).*

Proof. Let the inner product in the space $L^2[0, 1]$ be denoted by $\langle , \rangle$ and let $1_{[s,t]}$ be the indicator function of the interval $[s, t]$. With these notations, we can write $\int_s^t \phi_n(u) \, du = \langle 1_{[s,t]}, \phi_n \rangle$ so that we can compute the variance of the sum (6.3.16) as

$$\int_\Omega (X_t - X_s)^2 \, dP = \sum_{n=0}^{\infty} \left(\int_s^t \phi_n(u) \, du \right)^2$$

$$= \sum_{n=0}^{\infty} \langle 1_{[s,t]}, \phi_n \rangle^2$$

$$= \| 1_{[s,t]} \|^2$$

$$= t - s$$

where we have used Parseval's identity for the orthonormal basis (ϕ_n). This proves that the series (6.3.16) converges in $L^2(\Omega)$; the partial sums of the series are normally distributed with mean zero, so that the limit is also a normally distributed random with mean zero and the asserted variance, proving (1). To prove (2), we note that the partial sums of the series define a Gaussian distribution on $\mathbb{R}^N$ so that the independence can be inferred from the covariance function by showing that the increments are orthogonal in pairs. Now if $s < t \leq u < v$, we have

$$\int_\Omega (X_v - X_u)(X_t - X_s) \, dP = \sum_{n=0}^{\infty} \int_s^t \phi_n(w) \, dw \int_u^v \phi_n(w) \, dw$$

$$= \sum_{n=0}^{\infty} \langle 1_{[s,t]}, \phi_n \rangle \langle 1_{[u,v]}, \phi_n \rangle$$

$$= \langle 1_{[s,t]}, 1_{[u,v]} \rangle$$

$$= 0$$

where we have used the bilinear version of Parseval's identity and the disjointness of the interval $[s, t]$ from the interval $[u, v]$. This proves the pairwise orthogonality. Since the vector is multivariate normal, the independence is thereby proved. ■

6.3.5 *Haar Function Representation of Brownian Motion

The normalized Haar functions $\psi_{jk}(t)$ together with the constant function provide a convenient orthonormal basis of the Hilbert space $L^2(0, 1)$. We find it convenient to relabel them as follows:

$$\psi_0(t) = 1, \qquad 0 \le t \le 1.$$

If $n \ge 1$, then we can write $n = 2^j + k$ for $j = 0, 1, 2, \ldots$ and $k = 0, 1, \ldots, 2^j - 1$ and we set

$$\psi_n(t) = \psi_{jk}(t) = 2^{j/2}\psi(2^j t - k).$$

From the one-sided Haar series representation (6.3.11), we see that $\{\psi_n(t)\}_{n=0,1,2,\ldots}$ is an orthonormal basis of $L^2(0, 1)$.

To display the Haar series representation of Brownian motion, we introduce a sequence of independent standard normal random variables Z_n, $n \ge 0$ with

(6.3.17) $$P[Z_n \le x] = \frac{1}{\sqrt{2\pi}} \int_{-\infty}^{x} e^{-u^2/2}\, du.$$

These may be defined on a probability space, denoted $(\Omega, \mathcal{F}, P)$. The Brownian motion is sought in the form

(6.3.18) $$X_t(\omega) = Z_0(\omega) \int_0^t \psi_0(s)\,ds + \sum_{j=0}^{\infty} \left(\sum_{k=0}^{2^j - 1} Z_{2^j + k}(\omega) \int_0^t \psi_{2^j + k}(s)\, ds \right).$$

It is immediate from the orthonormal basis properties of ψ_n that for each $t \in [0, 1]$, the series (6.3.18) converges in $L^2(\Omega, \mathcal{F}, P)$. From this it is immediate from the proofs of (1), (2) above that X_t has the required distributional properties of Brownian motion.

6.3.6 *Proof of Continuity

We will now prove property (3) of Brownian motion, by showing that the series (6.3.18) converges uniformly for almost all $\omega \in \Omega$.

Lemma 6.3.18. *There exists* $M = M(\omega) < \infty$ *so that*

$$P\left[\omega : \sup_n \frac{|Z_n(\omega)|}{\sqrt{\log n}} \le M(\omega)\right] = 1.$$

Proof. From the normal distribution (6.3.17), we have the bound

$$P[|Z_n| > x] \le e^{-x^2/2}, \qquad n = 0, 1, 2, \ldots, x > 0.$$

Setting $x = 2\sqrt{\log n}$, we have the bound $P[|Z_n| \ge 2\sqrt{\log n}] \le n^{-2}$, which is the general term of a convergent series. Therefore by the first Borel-Cantelli lemma there exists $n_0(\omega) < \infty$ a.e. so that $n \ge n_0(\omega)$ implies $|Z_n(\omega)| \le 2\sqrt{\log n}$. Hence we can set $M(\omega) = \max\{2, (|Z_n(\omega)|)/\sqrt{\log n}, n \le n_0(\omega)\}$. ∎

Lemma 6.3.19.

$$\left|\sum_{k=0}^{2^j-1} Z_{2^j+k}(\omega) \int_0^t \psi_{2^j+k}(s)\,ds\right| \leq M_1(\omega)\sqrt{j}2^{-j/2}, \quad M_1(\omega) := 2M(\omega)\sqrt{2\log 2}.$$

Proof. For fixed j, the functions $S_{jk}(t) := \int_0^t \psi_{jk}(s)\,ds$ are polygonal functions supported on disjoint intervals of length 2^{-j} with $0 \leq S_{jk}(t) \leq 2^{-j} \times 2^{j/2}$. Hence $\sum_{0 \leq k \leq 2^j-1} S_{jk}(t) \leq 2^{-j/2}$ and $\sum_{0 \leq k \leq 2^j-1} |S_{jk}(t) - S_{jk}(s)| \leq |t-s|2^{j/2}$. Thus

$$\left|\sum_{k=0}^{2^j-1} Z_{2^j+k}(\omega) \int_0^t \psi_{2^j+k}(s)\,ds\right| \leq \max_{0 \leq k \leq 2^j-1} |Z_{2^j+k}(\omega)| \sum_{0 \leq k \leq 2^j-1} S_{jk}(t)$$

$$\leq \sqrt{j}M_1(\omega)2^{-j/2}$$

as required. ∎

From this it follows that the jth dyadic block of the series (6.3.18) is bounded by a constant multiple of $\sqrt{j}2^{-j/2}$, the general term of a convergent numerical series. Therefore by the Weierstrass M test, this series of dyadic blocks converges uniformly to a continuous function, denoted $t \to X_t(\omega)$. This proves property (3), hence we have proved the existence of the Brownian motion process. ∎

6.3.7 *Lévy's Modulus of Continuity

The method used to prove continuity can be easily extended to obtain a modulus of continuity, first established by Paul Lévy (1948). This is encapsulated in the following theorem.

Theorem 6.3.20. *There exists $M_1 = M_1(\omega)$ so that if $|t-s| \leq \delta < \frac{1}{2}$, then*

$$|X_t(\omega) - X_s(\omega)| \leq M_1(\omega)\sqrt{\delta \log\left(\frac{1}{\delta}\right)}.$$

Furthermore there exist intervals (s_n, t_n) with $t_n - s_n \to 0$ such that $[X_{t_n} - X_{s_n}]/\sqrt{(t_n - s_n)\log(1/(t_n - s_n))}$ is bounded below by a positive constant.

Proof. We write the increment of (6.3.18) in two parts:

$$X_t(\omega) - X_s(\omega) = Z_0(\omega)(t-s) + \left(\sum_{j=0}^{L} + \sum_{j=L+1}^{\infty}\right) \sum_{k=0}^{2^j-1} Z_{2^j+k}(\omega) \int_s^t \psi_{2^j+k}(u)\,du$$

where L will be chosen in terms of δ. Now

$$\left|\sum_{j=0}^{L}\sum_{k=0}^{2^j-1} Z_{2^j+k}(\omega) \int_s^t \psi_{2^j+k}(u)\,du\right| \leq |s-t|M_1(\omega)\sum_{j=0}^{L}\sqrt{j}2^{j/2},$$

$$\left|\sum_{j=L+1}^{\infty}\sum_{k=0}^{2^j-1} Z_{2^j+k}(\omega) \int_s^t \psi_{2^j+k}(u)\,du\right| \leq M_1(\omega)\sum_{j=L+1}^{\infty}\sqrt{j}2^{-j/2}.$$

The two sums are estimated by the elementary inequalities

(6.3.19) $$\sum_{j=0}^{L} \sqrt{j} 2^{j/2} \le C_1 \sqrt{L} 2^{L/2}, \qquad \sum_{j=L+1}^{\infty} \sqrt{j} 2^{-j/2} \le C_2 \sqrt{L} 2^{-L/2}.$$

so that if $|t - s| \le \delta$, then

$$|X_t - X_s| \le |Z_0(\omega)|\delta + M_2(\omega)(\delta\sqrt{L}2^{L/2} + \sqrt{L}2^{-L/2}).$$

The final two terms are balanced by taking $\delta 2^L \sim 1$, or specifically, $L = [\log_2(1/\delta)]$, with the result

$$|X_t(\omega) - X_s(\omega)| \le Y_{00}(\omega)\delta + 2M_2(\omega)\sqrt{\delta \log_2(1/\delta)},$$

which completes the proof of the upper bound, since $\delta \le \sqrt{\delta}$ for $0 \le \delta \le 1$.

To prove the second statement, we consider the independent events

$$A_k^n = \left\{\omega : X_{k2^{-n}} - X_{(k-1)2^{-n}} \le c\sqrt{\frac{n}{2^n}}\right\}, \qquad 1 \le k \le 2^n,$$

where c is to be chosen. We use independence, the tail of the normal distribution, and the elementary inequality $1 - x \le e^{-x}$ to write

$$P\left[\bigcap_{k=1}^{2^n} A_k^n\right] = \prod_{k=1}^{2^n}\left(1 - \int_{c\sqrt{n}}^{\infty} e^{-u^2/2}\frac{du}{\sqrt{2\pi}}\right)$$

$$\le \prod_{k=1}^{2^n}\left(1 - e^{-nc^2/2(1+o(1))}\right)$$

$$\le \exp(-2^n e^{-nc^2/2}(1 + o(1)))$$

$$= \exp(-(2e^{-c^2/2})^n (1 + o(1))).$$

It suffices to choose $0 < c < \sqrt{2\log 2}$ so that $2e^{-c^2/2} > 1$, and we have the general term of a convergent series, and by the first Borel-Cantelli lemma the series $\sum_{n=1}^{\infty} 1_{\bigcap_{k=1}^{2^n} A_k^n} < \infty$ for almost all ω. Therefore for n sufficiently large $\bigcap_{k=1}^{2^n} A_k^n$ fails to occur, in particular for some k, $X_{k/2^n} - X_{(k-1)/2^n} \ge c\sqrt{n/2^n}$, which proves that the Lévy modulus is a sharp lower bound also. ∎

Exercise 6.3.21. *Prove that the elementary estimates (6.3.19) hold with the constant $C_1 = C_2 = 1/(1 - 2^{-1/2})$.*

Hint: Compare a sum with an integral, which can be estimated by partial integration.

6.4 MULTIRESOLUTION ANALYSIS

In this section we return to the construction of general wavelets. The main features of the Haar wavelet expansion can be abstracted as follows.

Definition 6.4.1. *An orthonormal wavelet is a function $\Psi \in L^2(\mathbb{R})$ such that the doubly indexed set $\{2^{j/2}\Psi(2^j t - k)\}_{j,k \in \mathbb{Z}}$ is an orthonormal basis of $L^2(\mathbb{R})$.*

We have already seen that the Haar function provides an example of an orthonormal wavelet. To develop a systematic method for producing orthonormal wavelets, we introduce another notion, which generalizes the Haar construction.

Definition 6.4.2. *A multiresolution analysis (MRA) is an increasing sequence of subspaces $\{V_n\} \subset L^2(\mathbb{R})$ defined for $n \in \mathbb{Z}$ with*

$$\cdots \subset V_{-1} \subset V_0 \subset V_1 \subset \cdots$$

together with a function $\Phi \in L^2(\mathbb{R})$ such that

(i) $\cup_{n=-\infty}^{\infty} V_n$ is dense in $L^2(\mathbb{R})$, $\quad \cap_{n=-\infty}^{\infty} V_n = \{0\}$
(ii) $f \in V_n$ if and only if $f(2^{-n} \cdot) \in V_0$
(iii) $\{\Phi(x - k)\}_{k \in \mathbb{Z}}$ is an orthonormal basis of V_0.

Φ *is called the scaling function of the MRA.*

Clearly V_0 is uniquely defined by Φ through (iii), and V_n is further uniquely determined through (ii). However we do not require that Φ be unique; a given family $\{V_n\}$ may have several different possible choices of Φ.

The job of the theory is to show that there exist other nontrivial examples of multiresolution analyses, to construct the corresponding orthonormal wavelet bases and to discuss their properties.

Example 6.4.3. *Let V_n be the set of $f \in L^2(\mathbb{R})$, which are constant on the dyadic intervals $I_{kn} = [(k-1)/2^n, k/2^n)$.*

Clearly all of the properties are satisfied, with the Haar scaling function $\Phi(x) = 1_{[0,1)}(x)$.

Example 6.4.4. *Let V_n be the set of $f \in L^2(\mathbb{R})$, which are continuous and linear on each dyadic interval I_{kn}.*

It is straightforward to see that properties (i) and (ii) of Definition 6.4.2 are satisfied. The choice of a scaling function is less obvious and will be obtained in this section. This example is related to piecewise linear spline approximation.

In order to develop scaling functions for more general MRA systems, we first develop the necessary properties of orthonormal systems and Riesz systems.

6.4.1 Orthonormal Systems and Riesz Systems

Let H be a Hilbert space with inner product $(\,,\,)$. A set of vectors (x_n) is an *orthonormal system*, by definition, if $(x_n, x_m) = \delta_{mn}$.

Lemma 6.4.5. *The set (x_n) is orthonormal if and only if for every finite set of complex numbers (a_n), we have*

(6.4.1) $$\left\|\sum_n a_n x_n\right\|^2 = \sum_n |a_n|^2.$$

Proof. If (x_n) is orthonormal, then the left side of (6.4.1) is the finite sum

$$\sum_{m,n} a_m \bar{a}_n (x_m, x_n) = \sum_n a_n \bar{a}_n = \sum_n |a_n|^2.$$

Conversely, if (6.4.1) holds, first we choose $a_n = \delta_{nN}$ to obtain $(x_N, x_N) = 1$. Then choosing $a_n = \delta_{nM} - \delta_{nN}$ with $M \neq M$ gives $2 = \|x_N - x_M\|^2 = 2 - (x_N, x_M) - (x_M, x_N)$ hence $0 = (x_N, x_M) + (x_M, x_N)$. Replacing x_M by ix_M we obtain $0 = (x_N, x_M) - (x_M, x_N)$, from which the result follows. ∎

This leads us to formulate a more general concept.

Definition 6.4.6. *Let H be a Hilbert space. A set of vectors (x_n) is, by definition, a Riesz system, if there exist constants $0 < c \leq C < \infty$ such that for any finite set of complex numbers (a_n)*

(6.4.2) $$c \sum_n |a_n|^2 \leq \left\|\sum_n a_n x_n\right\|^2 \leq C \sum_n |a_n|^2.$$

Clearly any orthonormal system is a Riesz system, where $c = C = 1$. If (x_n) is a Riesz system, then the vectors (x_n) are linearly independent: $\sum_n a_n x_n = 0$, implies that $a_n = 0$ for all n.

Example 6.4.7. *Let $H = L^2(\mathbb{R})$ and $x_n(t) = \Lambda(t-n)$ where Λ is the tent function*

$$\Lambda(t) = (1 - |t|) 1_{[-1,1]}(t).$$

To verify the Riesz property, we note that the linear combination $A(t) := \sum_n a_n \Lambda(t-n)$ is piecewise linear with $A(n) = a_n$ for all n. Hence

$$\int_{\mathbb{R}} |A(t)|^2 dt = \sum_{n \in \mathbb{Z}} \int_n^{n+1} |(n+1-t)a_n + (t-n)a_{n+1}|^2 dt$$

$$= \frac{1}{3} \sum_{n \in \mathbb{Z}} (|a_n|^2 + |a_{n+1}|^2 + \operatorname{Re} a_n \bar{a}_{n+1}).$$

We use the Cauchy inequality $|2ab| \leq |a|^2 + |b|^2$ to obtain the upper bound

$$\int_{\mathbb{R}} |A(t)|^2 dt \leq \frac{1}{2} \sum_{n \in \mathbb{Z}} (|a_n|^2 + |a_{n+1}|^2) = \sum_{n \in \mathbb{Z}} |a_n|^2$$

and the lower bound

$$\int_{\mathbb{R}} |A(t)|^2 dt \geq \frac{1}{6} \sum_{n \in \mathbb{Z}} (|a_n|^2 + |a_{n+1}|^2) = \frac{1}{3} \sum_{n \in \mathbb{Z}} |a_n|^2.$$

Therefore (6.4.2) is satisfied with $c = \frac{1}{3}, C = 1$.

The next proposition characterizes Riesz systems and orthonormal systems in terms of the Fourier transform.

Proposition 6.4.8. *Let $\Phi \in L^2(\mathbb{R})$ and $0 < c \leq C < \infty$. The following two conditions are equivalent.*

(i) The periodized square of the Fourier transform satisfies the double inequality

(6.4.3)
$$c \leq \sum_{l \in \mathbb{Z}} |\hat{\Phi}(\xi + l)|^2 \leq C \quad \text{a.e. } \xi \in \mathbb{R}$$

(ii) $\{\Phi(t - m)\}_{m \in \mathbb{Z}}$ is a Riesz system with constants (c, C).

Recalling that orthonormality is characterized by $c = C = 1$, we obtain the following useful characterization.

Corollary 6.4.9. *The translates $\{\Phi(t - m)\}_{m \in \mathbb{Z}}$ of $\Phi \in L^2(\mathbb{R})$ are orthonormal if and only if $\sum_{l \in \mathbb{Z}} |\hat{\Phi}(\xi + l)|^2 = 1$ for almost every $\xi \in \mathbb{R}$.*

The sum in (6.4.3) is well defined a.e., since we may compute the integral over the unit interval as

$$\int_0^1 \sum_{l \in \mathbb{Z}} |\hat{\Phi}(\xi + l)|^2 \, d\xi = \sum_{l \in \mathbb{Z}} \int_l^{l+1} |\hat{\Phi}(\xi)|^2 \, d\xi = \int_{\mathbb{R}} |\hat{\Phi}(\xi)|^2 \, d\xi < \infty$$

hence the integrand is finite a.e. and defines a 1-periodic function.

Before giving the proof of Proposition 6.4.8, we give some examples of the computational power of these relations.

Example 6.4.10. *Haar scaling function:* Let $\Phi(t) = 1_{[0,1)}(t)$.

Clearly the translates $\Phi(t - m)$ are orthonormal, hence $c = C = 1$. The Fourier transform is computed explicitly as $\hat{\Phi}(\xi) = e^{-i\pi\xi} \sin(\pi\xi)/\pi\xi$, so that (ii) gives for a.e. ξ

$$1 = \sum_{l \in \mathbb{Z}} \frac{\sin^2 \pi \xi}{\pi^2 (\xi + l)^2},$$

which is equivalent to the partial fraction expansion of the function $\csc^2 \pi \xi$. Noting that the series on the right converges uniformly on each finite interval, we infer that the series defines a continuous function, hence the equality holds for *every* ξ.

Example 6.4.11. *Shannon scaling function:* Let $\Phi(t) = \sin(\pi t)/\pi t$ for $t \neq 0$, with $\Phi(0) = 1$.

The Fourier transform is $\hat{\Phi}(\xi) = 1_{[-\frac{1}{2}, \frac{1}{2}]}(\xi)$, thus we have $\sum_{l \in \mathbb{Z}} |\hat{\Phi}(\xi + l)|^2 = 1$ a.e., since for $\xi \notin \mathbb{Z}$, all terms in the series are zero except for one which $= 1$. Hence $\{\Phi(t - n)\}_{n \in \mathbb{Z}}$ is an orthonormal system in $L^2(\mathbb{R})$. This is the orthonormal system that occurs in the Shannon sampling formula, studied in Chapter 4 and to be redone in the context of wavelets.

Proof of Proposition 6.4.8. We first establish an identity for the Riesz sum in terms of the Fourier transform. The Fourier transform of $\Phi(t-n)$ is $e^{-2\pi i n\xi}\hat{\Phi}(\xi)$, so that the Fourier transform of $\sum_{n\in\mathbb{Z}} a_n \Phi(t-n)$ is $A(\xi)\hat{\Phi}(\xi)$ where A is the 1-periodic function

$$A(\xi) = \sum_{n\in\mathbb{Z}} a_n e^{-2\pi i n\xi}.$$

From Parseval's identity for Fourier series, we have $\int_0^1 |A(\xi)|^2 \, d\xi = \sum_{n\in\mathbb{Z}} |a_n|^2$. From Parseval's identity for Fourier transforms, it follows that

$$(6.4.4) \qquad \left\| \sum_{n\in\mathbb{Z}} a_n \Phi(t-n) \right\|^2 = \int_{\mathbb{R}} |A(\xi)|^2 |\hat{\Phi}(\xi)|^2 \, d\xi$$

$$= \sum_{l\in\mathbb{Z}} \int_l^{l+1} |A(\xi)|^2 |\hat{\Phi}(\xi)|^2 \, d\xi$$

$$= \sum_{l\in\mathbb{Z}} \int_0^1 |A(\xi)|^2 |\hat{\Phi}(\xi+l)|^2 \, d\xi$$

$$= \int_0^1 |A(\xi)|^2 \left(\sum_{l\in\mathbb{Z}} |\hat{\Phi}(\xi+l)|^2 \right) d\xi.$$

To prove that (i) implies (ii) in Proposition 6.4.8, we simply note that the integrand in parentheses in (6.4.4) is bounded above by C and thus the integral is bounded above by $C \int_0^1 |A(\xi)|^2 = C \sum_n |a_n|^2$, similarly for the lower bound, which proves (ii). To prove that (ii) implies (i), we use the above transformations to rewrite the Riesz condition (6.4.2) in the form

$$(6.4.5) \qquad c \leq \frac{\int_0^1 |A(\xi)|^2 \left(\sum_{l\in\mathbb{Z}} |\hat{\Phi}(\xi+l)|^2 \right) d\xi}{\int_0^1 |A(\xi)|^2 \, d\xi} \leq C.$$

This holds for every trigonometric polynomial $A(\xi)$. Taking a sequence A_N that converges boundedly to the indicator function of the interval $[a, b] \subset (0, 1)$ (the partial sums of the Fourier series of $1_{[a,b]}$ will suffice for this purpose), we obtain

$$c \leq \frac{1}{b-a} \int_a^b \left(\sum_{l\in\mathbb{Z}} |\hat{\Phi}(\xi+l)|^2 \right) d\xi \leq C.$$

This holds for every interval (a, b). Taking a sequence with $(a, b) \to \{x\}$ and applying Lebesgue's differentiation theorem, we obtain (i). ∎

Example 6.4.12. *Returning to the tent function $\Lambda(t)$, we have*

$$\left(\sum_{l\in\mathbb{Z}} |\hat{\Lambda}(\xi+l)|^2 \right) = \sum_{l\in\mathbb{Z}} \left(\frac{\sin \pi\xi}{\pi(\xi+l)} \right)^4.$$

This sum can be evaluated by repeated differentiation of the series from Example 6.4.10, namely

$$\pi^2 \csc^2 \pi\xi = \sum_{l \in \mathbb{Z}} \frac{1}{(\xi + l)^2}$$

to obtain the identity

$$\sum_{l \in \mathbb{Z}} \left(\frac{\sin \pi\xi}{\pi(\xi + l)} \right)^4 = \frac{1}{3}(1 + 2\cos^2 \pi\xi),$$

which again reaffirms that $c = 1/3$, $C = 1$ for this Riesz system.

Exercise 6.4.13. *Check the details of this computation.*

We can use Corollary 6.4.9 to estimate the support of the Fourier transform of a scaling function.

Corollary 6.4.14. *Suppose that $\Phi \in L^2(\mathbb{R})$ and that $\{\Phi(t-k)\}_{k \in \mathbb{Z}}$ is an orthonormal set. Then $|\mathrm{supp}\,\hat{\Phi}| \geq 1$, with equality if and only if $|\hat{\Phi}| = 1_K$ for some measurable set K with $|K| = 1$.*

Proof. From Corollary 6.4.9 we have $\sum_{l \in \mathbb{Z}} |\hat{\Phi}(\xi + l)|^2 = 1$ a.e., hence $|\hat{\Phi}(\xi)| \leq 1$ a.e. From Parseval's identity

$$|\mathrm{supp}\,\hat{\Phi}| = \int_{\mathrm{supp}\,\hat{\Phi}} d\xi \geq \int_{\mathrm{supp}\,\hat{\Phi}} |\hat{\Phi}(\xi)|^2 \, d\xi = 1,$$

which proves that $|\mathrm{supp}\,\hat{\Phi}| \geq 1$. If equality holds, then the middle terms give

$$0 = \int_{\mathrm{supp}\,\hat{\Phi}} (1 - |\hat{\Phi}(\xi)|^2) \, d\xi.$$

But the integrand is nonnegative a.e., hence $1 - |\hat{\Phi}(\xi)|^2 = 0$ a.e. on the support of $\hat{\Phi}$, which means that $|\hat{\Phi}| = 1_K$ a.e., where $|K| = |\mathrm{supp}\,\hat{\Phi}| = 1$. ∎

Proposition 6.4.8 allows us to obtain the following orthogonalization procedure to generate scaling functions from a Riesz sequence.

Proposition 6.4.15. *Let $\Phi \in L^2(\mathbb{R})$ be such that $\{\Phi(t - m)\}_{m \in \mathbb{Z}}$ is a Riesz sequence. Then there exist complex numbers b_n with $\sum_{n \in \mathbb{Z}} |b_n|^2 < \infty$ such that $\{\Phi_1(t - m)\}_{m \in \mathbb{Z}}$ is an orthonormal sequence, where $\Phi_1(t) := \sum_{n \in \mathbb{Z}} b_n \Phi(t - n)$. Furthermore, the span of $\{\Phi_1(t - n)\}_{n \in \mathbb{Z}}$ equals the span of $\{\Phi(t - n)\}_{n \in \mathbb{Z}}$.*

Proof. From Proposition 6.4.8, it suffices to find b_n such that $\sum_{l \in \mathbb{Z}} |\hat{\Phi}_1(\xi + l)|^2 = 1$ a.e. From the definition of Φ_1, we have

$$\hat{\Phi}_1(\xi) = \sum_{n\in\mathbb{Z}} b_n e^{-2\pi in\xi} \hat{\Phi}(\xi)$$

$$:= B(\xi)\hat{\Phi}(\xi)$$

$$\sum_{l\in\mathbb{Z}} |\hat{\Phi}_1(\xi+l)|^2 = \sum_{l\in\mathbb{Z}} |B(\xi+l)|^2 |\hat{\Phi}(\xi+l)|^2$$

$$= |B(\xi)|^2 \sum_{l\in\mathbb{Z}} |\hat{\Phi}(\xi+l)|^2.$$

Therefore we must choose the constants b_n so that

$$|B(\xi)|^2 := \left|\sum_{n\in\mathbb{Z}} b_n e^{-2\pi in\xi}\right|^2 = \frac{1}{\sum_{l\in\mathbb{Z}} |\hat{\Phi}(\xi+l)|^2}.$$

Clearly there are many possible solutions. The simplest one is to take the positive square root, leading to

(6.4.6)
$$\hat{\Phi}_1(\xi) = \frac{\hat{\Phi}(\xi)}{\sqrt{\sum_{l\in\mathbb{Z}} |\hat{\Phi}(\xi+l)|^2}}.$$

This is clearly the Fourier transform of an L^2 function, since the denominator is bounded above and below by the Riesz condition. To prove the last statement, we need to study the equation

(6.4.7)
$$\sum_{n\in\mathbb{Z}} a_n \Phi(t-n) = \sum_{n\in\mathbb{Z}} c_n \Phi_1(t-n)$$

and to show that, given $(a_n) \in l^2(\mathbb{Z})$, we can solve for $(c_n) \in l^2(\mathbb{Z})$ and conversely. In terms of Fourier transforms, this is written

$$\left(\sum_{n\in\mathbb{Z}} a_n e^{-2\pi in\xi}\right) \hat{\Phi}(\xi) = \left(\sum_{n\in\mathbb{Z}} c_n e^{-2\pi in\xi}\right) \hat{\Phi}_1(\xi).$$

Recalling the relation between Φ_1 and Φ is a special case, with $a_n = b_n$, $c_n = \delta_{n0}$. Making this substitution we see that (6.4.7) is implied by the identity

(6.4.8)
$$A(\xi) := \sum_{n\in\mathbb{Z}} a_n e^{-2\pi in\xi} = \left(\sum_{n\in\mathbb{Z}} b_n e^{-2\pi in\xi}\right)\left(\sum_{n\in\mathbb{Z}} c_n e^{-2\pi in\xi}\right).$$

$$:= B(\xi)C(\xi).$$

But we have shown above that $C^{-1} \leq |\sum_{n\in\mathbb{Z}} b_n e^{-2\pi in\xi}| \leq c^{-1}$ from the Riesz property. Hence, given $(a_n) \in l^2(\mathbb{Z})$, we may solve (6.4.8) uniquely by taking c_n as the Fourier coefficients of the 1-periodic function $A(\xi)/B(\xi)$. Conversely, given c_n, one simply refers to (6.4.8) and chooses a_n as the Fourier coefficients of the right side. ∎

The above proof shows that the set of functions described by the left side of (6.4.7) when $(a_n) \in l^2(\mathbb{Z})$ is identical to the set of functions described by the right side when $(c_n) \in l^2(\mathbb{Z})$. But the set on the right is a closed subspace of $L^2(\mathbb{R})$, from the orthonormality of $\{\Phi_1(t-n)\}_{n\in\mathbb{Z}}$. Hence the set on the left is also a closed subspace, which is the closed linear span of $\{\Phi(t-n)\}_{n\in\mathbb{Z}}$.

Example 6.4.16. *In the case of the tent function* $\Phi = \Lambda$, *the orthogonalized Fourier transform is obtained as*

$$\hat{\Phi}_1(\xi) = \frac{(\sin \pi\xi/\pi\xi)^2}{\sqrt{(1 + 2\cos^2 \pi\xi)/3}}.$$

This corresponds to the MRA of Example 6.4.4, where the functions are continuous and piecewise linear on each dyadic interval. We write $\Phi_1(t) = \sum_n b_n \Lambda(t - n)$, which is clearly a piecewise linear continuous function with $\Phi_1(n) = b_n$. The coefficients are obtained from the Fourier expansion

(6.4.9) $$\frac{1}{\sqrt{(1 + 2\cos^2 \pi\xi)/3}} = \sum_{n \in \mathbb{Z}} b_n e^{2\pi i n \xi}.$$

Since the left side is a real analytic function, the Fourier coefficients have an exponential decay.

Exercise 6.4.17. *Show that there exist constants* $K > 0, \beta > 0$ *so that* $|b_n| \leq K e^{-\beta|n|}$ *and obtain an estimate for* β.

6.4.2 Scaling Equations and Structure Constants

The axioms describing an MRA system are not completely independent of one another, as we will show. First we note a simple consequence of properties (ii) and (iii) from Definition 6.4.2.

Proposition 6.4.18. *For each* $j \in \mathbb{Z}$, $\{2^{j/2}\Phi(2^j t - k)\}_{k \in \mathbb{Z}}$ *is an orthonormal basis of* V_j.

Proof. From property (ii), V_j and V_0 are isomorphic by virtue of the map $x \to 2^{-j}x$. The indicated functions are clearly orthonormal. Now we pull back to V_0 and use property (iii). ∎

In order to proceed further, we discuss the consequences of the inclusion $V_0 \subset V_1$. Since V_1 is spanned by translates of $\{\Phi(2t - n)\}_{n \in \mathbb{Z}}$, we have the L^2 convergent sum

(6.4.10) $$\Phi(t) = \sum_{n \in \mathbb{Z}} a_n \Phi(2t - n)$$

where the *structure constants* satisfy $\sum_{n \in \mathbb{Z}} |a_n|^2 < \infty$. Relation (6.4.10) is called the *scaling equation* and will be instrumental in the sequel.

Example 6.4.19. *If* $\Phi(t) = 1_{[0,1)}(t)$, *then clearly* $\Phi(t) = \Phi(2t) + \Phi(2t - 1)$ *is the scaling equation, with structure constants* $a_0 = 1, a_1 = 1$ *and* $a_n = 0$ *otherwise.*

Exercise 6.4.20. *Suppose that* $\Phi \in L^1(\mathbb{R}) \cap L^2(\mathbb{R})$ *satisfies the equation* $\Phi(t) = \Phi(2t) + \Phi(2t - 1)$. *Prove that for some* c, $\Phi(t) = c 1_{[0,1)}(t)$ *a.e.*

Hint: Take the Fourier transform of both sides and iterate to solve for $\hat{\Phi}$.

The next exercise shows that a similar two-scale difference equation can have radically different solution behavior.

Exercise 6.4.21. Suppose that $\Phi \in L^1(\mathbb{R})$ satisfies the functional equation $2\Phi(t) = 3\Phi(3t) + 3\Phi(3t - 1)$. Prove that $\Phi(t) \equiv 0$.

Hint: Show that $\hat{\Phi} \neq 0$ must agree with the Fourier transform of the Cantor measure.

The next example shows that the existence of a scaling equation does not follow from the orthogonality properties.

Example 6.4.22. Let $\Phi(t) = 1_{[-\frac{1}{2}, \frac{1}{2}]}(t)$. We suppose that (6.4.10) holds for some $(a_n) \in l^2(\mathbb{Z})$ and obtain a contradiction.

To see this, we have from the orthogonality of $\{\Phi(2t - n)\}_{n \in \mathbb{Z}}$

$$a_n = 2 \int_{\mathbb{R}} \Phi(t) \Phi(2t - n)\, dt.$$

But this integral is nonzero unless $n = 0, \pm 1$, in which case we obtain $a_0 = 1$, $a_{\pm 1} = \frac{1}{2}$. But this leads to a contradiction on the interval $\frac{1}{2} \leq t \leq \frac{3}{4}$, where the left side of (6.4.10) is zero but the right side is nonzero.

We record some properties of the structure constants.

Proposition 6.4.23. *The structure constants obey the following properties:*

(6.4.11) $\qquad a_k = 2 \int_{\mathbb{R}} \Phi(t) \bar{\Phi}(2t - k)\, dt, \qquad k \in \mathbb{Z}$

(6.4.12) $\qquad \sum_{k \in \mathbb{Z}} |a_k|^2 = 2$

(6.4.13) $\qquad \sum_{k' \in \mathbb{Z}} a_{k'} \bar{a}_{2k+k'} = 2\delta_{k0} \qquad$ (Kronecker delta).

If also $\Phi \in L^1(\mathbb{R})$, $\int_{\mathbb{R}} \Phi \neq 0$ and (6.4.10) converges in $L^1(\mathbb{R})$, then

(6.4.14) $\qquad \sum_{k \in \mathbb{Z}} a_k = 2.$

Proof. Since $a_k/\sqrt{2}$ are the Fourier coefficients of $\Phi \in V_1$ with respect to the orthonormal basis $\sqrt{2}\Phi(2t - k)$, we have $a_k/\sqrt{2} = \int_{\mathbb{R}} \Phi(t)\sqrt{2}\Phi(2t - k)\, dt$, as required. Parseval's theorem gives $\sum_{k \in \mathbb{Z}} |a_k|^2/2 = \|\Phi\|^2 = 1$. To prove (6.4.13), we begin with (iii) of definition 6.4.2, in the form

$$\int_{\mathbb{R}} \Phi(t - k) \bar{\Phi}(t)\, dt = \delta_{0k}.$$

Substitute (6.4.10) and use Parseval's identity and orthogonality to write

$$\delta_{0k} = \sum_{k',k''} a_{k'} \bar{a}_{k''} \int_{\mathbb{R}} \Phi(2t - 2k - k') \bar{\Phi}(2t - k'') \, dt$$

$$= \frac{1}{2} \sum_{2k+k'=k''} a_{k'} \bar{a}_{k''},$$

which is the same as (6.4.13). In particular, taking $k = 0$ we retrieve a new proof of (6.4.12).

If, in addition, we have $\Phi \in L^1$ with $\int_{\mathbb{R}} \Phi \neq 0$, then we integrate (6.4.10) term-by-term to obtain

$$\int_{\mathbb{R}} \Phi(t) \, dt = \sum_{k \in \mathbb{Z}} a_k \int_{\mathbb{R}} \Phi(2t - k) \, dt$$

$$= \frac{1}{2} \sum_{k \in \mathbb{Z}} a_k \int_{\mathbb{R}} \Phi(t) \, dt,$$

which we divide by $\int_{\mathbb{R}} \Phi$ to obtain (6.4.14). ∎

It is often useful to work with (6.4.10) in the Fourier domain. The Fourier transform of $\Phi(2t - n)$ is easily obtained:

$$\int_{\mathbb{R}} \Phi(2t - n) e^{-2\pi i t \xi} \, dt = \frac{1}{2} \int_{\mathbb{R}} \Phi(u) \exp\left[-2\pi i \xi \left(\frac{n + u}{2}\right)\right] du$$

$$= \frac{1}{2} e^{-i n \pi \xi} \hat{\Phi}\left(\frac{\xi}{2}\right)$$

so that (6.4.10) is written

(6.4.15) $$\boxed{\hat{\Phi}(\xi) = m_0\left(\frac{\xi}{2}\right) \hat{\Phi}\left(\frac{\xi}{2}\right)}$$

where the *scaling filter* is defined by

(6.4.16) $$\boxed{m_0(\xi) := \frac{1}{2} \sum_{n \in \mathbb{Z}} a_n e^{-2\pi i n \xi}.}$$

The existence of a scaling equation can be formulated in the Fourier domain as follows where $L^2(\mathbb{R}/\mathbb{Z})$ denotes 1-periodic functions that are square-integrable on any period.

Proposition 6.4.24. $\Phi \in L^2(\mathbb{R})$ *satisfies a scaling equation* (6.4.10) *with* $(a_n) \in l^2(\mathbb{Z})$ *if and only if there exists* $m_0 \in L^2(\mathbb{R}/\mathbb{Z})$ *so that* (6.4.15) *holds, in which case* (6.4.16) *holds. In particular* $a_n = 2 \int_{-1/2}^{1/2} m_0(\xi) e^{2\pi i n \xi} \, d\xi$.

Proof. If Φ satisfies (6.4.10), then we can take the Fourier transform of both sides to obtain (6.4.15). Conversely, if (6.4.15) holds with $m_0 \in L^2(\mathbb{R}/\mathbb{Z})$, we define $a_n = 2 \int_{-1/2}^{1/2} m_0(\xi) e^{2\pi i n \xi} \, d\xi$, so that (6.4.16) holds. The Plancherel theorem ensures that the map $\Phi \to \hat{\Phi}$ is bijective. Since the right side of (6.4.15) is the Fourier transform of $\sum_n a_n \Phi(2t - n)$, it follows that $\Phi(t) = \sum_n a_n \Phi(2t - n)$, which was to be proved. ∎

Example 6.4.25. *The Shannon scaling function is*

$$\Phi(t) = \frac{\sin(\pi t)}{\pi t} = \int_{-\frac{1}{2}}^{\frac{1}{2}} e^{2\pi i t \xi} \, d\xi.$$

The scaling equation can be obtained at the level of Fourier transforms by solving (6.4.15) as follows:

$$1_{[-\frac{1}{2},\frac{1}{2}]}(\xi) = m_0\left(\frac{\xi}{2}\right) 1_{[-\frac{1}{2},\frac{1}{2}]}\left(\frac{\xi}{2}\right).$$

Therefore we need to choose the coefficients so that $m_0(\xi) = 1$ for $|\xi| < \frac{1}{4}$ and $m_0(\xi) = 0$ for $\frac{1}{4} \leq |\xi| \leq \frac{1}{2}$. The structure constants are obtained from (6.4.10) as the Fourier coefficients

$$a_n = 2 \int_{\frac{1}{4} \leq |\xi| \leq \frac{1}{2}} e^{-2\pi i n \xi} \, d\xi.$$

Thus $a_0 = 1$ and $a_n = (-2/n\pi)\sin(n\pi/2)$.

Exercise 6.4.26. *Consider the spline function* $\Lambda(t) = (1 - |t|)1_{[-1,1]}(t)$ *with* $\hat{\Phi}(\xi) = (\sin \pi \xi / \pi \xi)^2$. *Show that* $\hat{\Lambda}$ *satisfies a scaling equation (6.4.10) and exhibit the scaling filter* $m_0(\xi)$. *Use this to infer that its orthogonalization, defined by* $\hat{\Phi}(\xi) = \hat{\Lambda}(\xi)/\sqrt{\sum_{l \in \mathbb{Z}} |\hat{\Lambda}(\xi + l)|^2}$, *also satisfies a scaling equation (6.4.10).*

6.4.3 From Scaling Function to MRA

We now prove an important theorem, showing the existence of MRA systems under useful hypotheses.

Theorem 6.4.27. *Suppose that* $\Phi \in L^2(\mathbb{R})$ *is such that*

(i) *The translates* $\{\Phi(t - m)\}_{m \in \mathbb{Z}}$ *are orthonormal.*
(ii) $\Phi(t) = \sum_{n \in \mathbb{Z}} a_n \Phi(2t - n)$, *an L^2-convergent sum, with* $\sum_{n \in \mathbb{Z}} |a_n|^2 < \infty$.
(iii) *The Fourier transform* $\hat{\Phi}(\xi)$ *is continuous at* $\xi = 0$ *with* $|\hat{\Phi}(0)| = 1$.

Define $V_j = \text{span } \{\Phi(2^j t - k)\}_{k \in \mathbb{Z}}$. *Then* $\{V_j\}$ *defines an MRA.*

Proof. The scaling equation (ii) implies that $V_j \subset V_{j+1}$. Now let $\Phi_{jk}(t) = 2^{j/2}\Phi(2^j t - k)$ and let P_j be the orthogonal projection on the space V_j. In detail

$$P_j f = \sum_{k \in \mathbb{Z}} (f, \Phi_{jk}) \Phi_{jk}.$$

These projection operators satisfy the bounds $\|P_j f\| \leq \|f\|$. To prove the MRA property, it suffices to show that $\lim_{j \to \infty} P_j f = f$ and $\lim_{j \to -\infty} P_j f = 0$ for all $f \in L^2(\mathbb{R})$. This is done in two separate lemmas.

Lemma 6.4.28. *For any* $f \in L^2(\mathbb{R})$, $\lim_{j \to -\infty} P_j f = 0$.

Proof. Since $\|P_j\| = 1$, it suffices to prove the result on a dense set, e.g., L^2 functions with compact support. If f has support in $[-R, R]$, then

$$\begin{aligned}
\|P_j f\|^2 &= \sum_{k \in \mathbb{Z}} |(P_j f, \Phi_{jk})|^2 \\
&= \sum_{k \in \mathbb{Z}} |(f, \Phi_{jk})|^2 \\
&\leq \sum_{k \in \mathbb{Z}} \left(\int_{-R}^{R} |f|^2 \right) \left(\int_{-R}^{R} |\Phi_{jk}|^2 \right) \\
&= \|f\|^2 \sum_{k \in \mathbb{Z}} 2^j \left(\int_{-R}^{R} |\Phi(2^j s - k)|^2 \, ds \right) \\
&= \|f\|^2 \sum_{k \in \mathbb{Z}} \int_{-k - 2^j R}^{-k + 2^j R} |\Phi(u)|^2 \, du.
\end{aligned}$$

If $2^j R < \frac{1}{2}$, then these integrals are over disjoint intervals whose union is written $U_j = \bigcup_{k \in \mathbb{Z}} (-k - 2^j R, -k + 2^j R)$, with $\cap_j U_j = \mathbb{Z}$, which has Lebesgue measure zero. Therefore

$$\|P_j f\|^2 \leq \|f\|^2 \int_{U_j} |\Phi(u)|^2 du \to 0, \quad j \to -\infty$$

by Lebesgue's dominated convergence theorem. ∎

To proceed further, we now turn to the Fourier domain and prove a useful identity.

Lemma 6.4.29. *Let $f \in L^2(\mathbb{R})$ with a Fourier transform $\hat{f}$ that is bounded and supported in $[-R, R]$ for some $R > 0$. Then for $2^{j-1} > R$ we have*

(6.4.17) $$\|P_j f\|^2 = \int_{-R}^{R} |\hat{f}(\xi)|^2 |\hat{\Phi}(2^{-j}\xi)|^2 \, d\xi.$$

Proof. We use Parseval's identity to write

$$\begin{aligned}
\|P_j f\|^2 &= \sum_{k \in \mathbb{Z}} |(P_j f, \Phi_{jk})|^2 \\
&= \sum_{k \in \mathbb{Z}} |(f, \Phi_{jk})|^2 \\
&= \sum_{k \in \mathbb{Z}} \left| \int_{-R}^{R} \hat{f}(\xi) \overline{\hat{\Phi}_{jk}(\xi)} \, d\xi \right|^2 \\
&= \sum_{k \in \mathbb{Z}} \left| \int_{-R}^{R} \hat{f}(\xi) e^{-2\pi i k \xi 2^{-j}} 2^{-j/2} \overline{\hat{\Phi}(2^{-j}\xi)} \, d\xi \right|^2
\end{aligned}$$

where we have used the fact that the Fourier transform of the function Φ_{jk} is explicitly written $2^{-j/2} e^{-2\pi i k \xi 2^{-j}} \hat{\Phi}(2^{-j}\xi)$. Now if $2^{j-1} > R$, the last integral is equal to the integral on the interval $[-2^{j-1}, 2^{j-1}]$, where the functions $\{2^{-j/2} e^{2\pi i k \xi 2^{-j}}\}_{k \in \mathbb{Z}}$ form an orthonormal basis. Moreover, $\xi \to \hat{f}(\xi) \hat{\Phi}(2^{-j}\xi) \in L^2(-2^{j-1}, 2^{j-1})$, so that by Parseval's theorem for

Fourier series, we have for $2^{j-1} > R$

$$\|P_j f\|^2 = \int_{-2^{j-1}}^{2^{j-1}} |\hat{f}(\xi)|^2 |\hat{\Phi}(\xi)|^2 \, d\xi$$
$$= \int_{-R}^{R} |\hat{f}(\xi)|^2 |\hat{\Phi}(2^{-j}\xi)|^2 \, d\xi,$$

which completes the proof. ∎

Corollary 6.4.30. *Suppose that the scaling function satisfies the additional condition that $\hat{\Phi}(\xi)$ is continuous at $\xi = 0$ with $|\hat{\Phi}(0)| = 1$. Then for any $f \in L^2(\mathbb{R})$, $\|P_j f - f\| \to 0$ when $j \to \infty$.*

Proof. Since P_j is a contraction, it suffices to prove this on the dense set of f whose Fourier transforms have compact support and are bounded. Furthermore, from the projection property, we have $\|P_j f\|^2 = \|f\|^2 - \|f - P_j f\|^2$, so we must show that $\|P_j f\| \to \|f\|$. Using the hypothesis, we see that $|\hat{\Phi}(2^{-j}\xi)|$ converges uniformly to 1 on compact sets, so that (6.4.17) gives for $2^{j-1} > R$

$$\|P_j f\|^2 = \int_{-R}^{R} |\hat{f}(\xi)\hat{\Phi}(2^{-j}\xi)|^2 \, d\xi$$
$$\to \int_{-R}^{R} |\hat{f}(\xi)|^2 \, d\xi$$
$$= \|f\|^2,$$

which completes the proof. ∎

Combining the above lemmas and corollaries completes the proof of the theorem. The following exercise provides a one-parameter generalization of the Shannon wavelet.

Exercise 6.4.31. *Let $K = [a-1, a]$ where $0 < a < 1$ and set $\hat{\Phi} = 1_K$. Prove that Φ is the scaling function of an MRA.*

Hint: First check that Φ has orthonormal translates. To find the scaling relation, it suffices to find $m_0 \in L^2(\mathbb{R}/\mathbb{Z})$, by solving $\hat{\Phi}(2\xi) = m_0(\xi)\hat{\Phi}(\xi)$.

Example 6.4.32. *One should not infer from the continuity at $\xi = 0$ that $\hat{\Phi}$ is continuous elsewhere, much less that $\Phi \in L^1(\mathbb{R})$. Consider the Shannon scaling function, where $\hat{\Phi} = 1_{[-\frac{1}{2}, \frac{1}{2}]}$, which is continuous at $\xi = 0$ but is discontinuous at $\xi = \pm\frac{1}{2}$.*

6.4.3.1 Additional remarks

We note some additional relations between the above notions.

- If we have an MRA, the condition $|\hat{\Phi}(0)| = 1$ actually follows from the apparently weaker condition that $\hat{\Phi}(\xi)$ is continuous at $\xi = 0$. To see this, apply Lemma 6.4.29 to a function f whose Fourier transform is bounded with compact support. Since we assume an MRA, it follows that $P_j f \to f$ when $j \to \infty$. Taking limits in (6.4.17), we have for $2^j > R$,

$$\|f\|^2 = |\hat{\Phi}(0)|^2 \int_{-R}^{R} |\hat{f}(\xi)|^2 \, d\xi = |\hat{\Phi}(0)|^2 \|f\|^2$$

by Parseval's identity. Hence $|\hat{\Phi}(0)| = 1$, as promised.

- Theorem 6.4.27 remains true if one weakens the condition $|\hat{\Phi}(0)| = 1$ to $\hat{\Phi}(0) \neq 0$.

To see this, suppose that f is orthogonal to $\cup_{j \in \mathbb{Z}} V_j$. Then $P_j f = 0$ for all $j \in \mathbb{Z}$. Given $\epsilon > 0$, there exists g whose Fourier transform is bounded and supported in $[-R, R]$ for some $R > 0$ and so that $\|f - g\| < \epsilon$. Hence $\|P_j g\| = \|P_j(g - f)\| < \epsilon$. Applying Lemma 6.4.29 yields the estimate

$$\epsilon^2 \geq \|P_j g\|^2 = \int_{-R}^{R} |\hat{g}(\xi)|^2 |\hat{\Phi}(2^{-j}\xi)|^2$$

$$\to |\hat{\Phi}(0)|^2 \|g\|^2$$

$$\geq |\hat{\Phi}(0)|^2 (\|f\| - \epsilon)^2.$$

This holds for every $\epsilon > 0$, which is a contradiction if ϵ is sufficiently small.

- The continuity condition (iii) in Theorem 6.4.27 can be weakened to

$$\lim_{j \to \infty} |\hat{\Phi}(2^{-j}\xi)| = 1 \quad \text{a.e.} \quad \xi \in \mathbb{R},$$

and this condition is also necessary.

Indeed, the sufficiency is apparent from application of Lemma 6.4.29 to f, whose Fourier transform is bounded with compact support. To see the necessity, we anticipate a result from the next section, that $|m_0(\xi)| \leq 1$ a.e. From this it follows that

$$|\hat{\Phi}(\xi)| \leq \left|\hat{\Phi}\left(\frac{\xi}{2}\right)\right| \leq \cdots \leq \left|\hat{\Phi}\left(\frac{\xi}{2^j}\right)\right|.$$

In addition $1 = \sum_{l \in \mathbb{Z}} |\hat{\Phi}(\xi + l)|^2 \geq |\hat{\Phi}(\xi)|^2$, so that we have the existence of the limit $g(\xi) = \lim_{j \to \infty} |\hat{\Phi}(2^{-j}\xi)|$ and $g(\xi) \leq 1$. Applying Lemma 6.4.29 to f with $\hat{f} = 1_{[-1,1]}$, we see that

$$2 = \lim_{j \to \infty} \int_{-1}^{1} |\hat{\Phi}(\xi)|^2 d\xi = \int_{-1}^{1} g(\xi) \, d\xi,$$

where we have applied the Lebesgue dominated convergence theorem, thanks to the bound $|g(\xi)| \leq 1$. Hence $\int_{-1}^{1} (1 - |g(\xi)|) \, d\xi = 0$ where the integrand is nonnegative, hence $g(\xi) = 1$ a.e.

- If the scaling function $\Phi \in L^1(\mathbb{R}) \cap L^2(\mathbb{R})$, then the conditions of Theorem 6.4.27 are necessary as well as sufficient.

    To see this, note that $\Phi \in L^1(\mathbb{R})$ implies that $\hat{\Phi}$ is continuous, especially at $\xi = 0$. If Φ generates an MRA, then we can apply Lemma 6.4.29 to $f \neq 0$ whose

Fourier transform is bounded with compact support. Taking $j \to \infty$ in (6.4.17), we obtain $\|f\| = |\hat{\Phi}(0)| \|f\|$, hence $|\hat{\Phi}(0)| = 1$, as promised.

- If the scaling function $\Phi \in L^1(\mathbb{R}) \cap L^2(\mathbb{R})$, then $\hat{\Phi}(l) = 0$ for $0 \neq l \in \mathbb{Z}$ and $\sum_{k \in \mathbb{Z}} \Phi(t-k) = \hat{\Phi}(0)$ a.e.

  Indeed, from the orthonormality relation, we have a.e. $1 \geq |\hat{\Phi}(\xi)|^2 + |\hat{\Phi}(\xi + l)|^2$. But $\hat{\Phi}$ is continuous, hence we can take $\xi \to 0$ avoiding the exceptional set to obtain $|\hat{\Phi}(l)| \leq 0$, which was to be proved. From this it also follows from the Poisson summation formula that the periodized scaling function $\sum_{k \in \mathbb{Z}} \Phi(t-k) = \hat{\Phi}(0)$ a.e., since its Fourier coefficients are all zero except for one term.

- If the scaling function has compact support with $\int_\mathbb{R} \Phi(x)\,dx = 1$, then the Fourier transform argument of Lemma 6.4.29 can be avoided. This is formulated as follows.

Proposition 6.4.33. *Suppose that Φ is the scaling function for a compact MRA with $\int_\mathbb{R} \Phi(x)\,dx = 1$. Then $\cup_{j \in \mathbb{Z}} V_j$ is dense in $L^2(\mathbb{R})$.*

Proof. The orthogonal projection P_j onto V_j is given by

$$P_j f = 2^j \sum_{\gamma \in \mathbb{Z}} \Phi(2^j x - \gamma) \left(\int_\mathbb{R} f(y) \Phi(2^j y - \gamma)\,dy \right)$$

and satisfies

(6.4.18) $$\int_\mathbb{R} (f - P_j f) P_j f = 0, \quad \|f\|^2 = \|f - P_j f\|^2 + \|P_j f\|^2.$$

Therefore, to show that $P_j f \to f$ in L^2, it suffices to prove that

(6.4.19) $$\|P_j f\| \to \|f\|.$$

Since the operators P_j have norm 1, it suffices to prove (6.4.19) on a dense set, e.g., linear combinations of $f = I_A$, where $A = [a,b]$. If supp $\Phi \in [-M, M]$

$$\|P_j 1_A\|^2 = 2^j \sum_{\gamma \in \mathbb{Z}} \left(\int_A \Phi(2^j y - \gamma)\,dy \right)^2$$

$$= 2^{-j} \sum_{\gamma \in \mathbb{Z}} \left(\int_{2^j A} \Phi(x - \gamma)\,dx \right)^2$$

$$= 2^{-j} \left(\sum_{\gamma \in [2^j a + M, 2^j b - M]} + \sum_{\gamma \notin [2^j a - M, 2^j b + M]} + \sum_{\text{dist}(\gamma, 2^j A) \leq M} \right) \left(\int_{2^j A} \Phi(x - \gamma)\,dx \right)^2.$$

In the first term the integral is one, while in the second term the integral is zero. In the third term the integral is less than $\|\Phi\|_1$ and the number of such terms is less than $4M$. Thus

$$2^{-j} \sum_{\gamma \in \mathbb{Z}} \left(\int_{2^j A} \Phi(x - \gamma)\,dx \right)^2 = 2^{-j} \sum_{2^j a + M \leq \gamma \leq 2^j b - M} 1 + O(2^{-j})$$

$$= (b - a) + O(2^{-j}).$$

Hence $\|P_j 1_A\|_2^2 \to b - a = \|1_A\|_2^2$, which completes the proof. ∎

6.4.4 Meyer Wavelets

The previous examples of scaling functions include the Haar wavelet—where the scaling function Φ has compact support but is not smooth. At the other extreme we have the Shannon wavelet, where the scaling function is infinitely differentiable but has slow decay at infinity. Theorem 6.4.27 allows one to construct a large class of new scaling functions, including the Meyer wavelets, where the scaling function is infinitely differentiable and rapidly decreasing (Schwartz class $\mathcal{S}$). This general class also includes the Shannon scaling function from Example 6.4.25 as an extreme case.

We begin with a function $\Theta(\xi)$ defined on the interval $0 \leq \xi \leq 1$ satisfying the following properties:

(6.4.20) $$0 \leq \Theta(\xi) \leq 1,$$

(6.4.21) $$\Theta(\xi) + \Theta(1-\xi) = 1,$$

(6.4.22) $$\xi \to \Theta(\xi) \text{ is monotone decreasing,}$$

(6.4.23) $$\Theta(\xi) = 1 \quad 0 \leq \xi \leq \tfrac{1}{3}.$$

The symmetry condition (6.4.21) implies that $\Theta(\xi) = 0$ for $\tfrac{2}{3} \leq \xi \leq 1$, $\Theta(\tfrac{1}{2}) = \tfrac{1}{2}$, while the monotonicity condition (6.4.22) shows that $\Theta(\xi) \geq \tfrac{1}{2}$ for $0 \leq \xi \leq \tfrac{1}{2}$. We extend Θ to the real line by setting $\Theta(\xi) = \Theta(-\xi)$ for $-1 \leq \xi \leq 0$ and setting $\Theta(\xi) = 0$ for $|\xi| > 1$. The resulting function is even on the entire axis and satisfies $0 \leq \Theta(\xi) \leq 1$. Now we define

(6.4.24) $$\Phi(t) = \int_{-1}^{1} \sqrt{\Theta(\xi)} e^{2\pi i t \xi} \, d\xi, \quad t \in \mathbb{R}.$$

Proposition 6.4.34. *$\Phi(t)$ is the scaling function of an MRA system satisfying the conditions of Theorem 6.4.27 and is of class C^∞ with all derivatives bounded: $|\Phi^{(l)}(t)| \leq C_l$. If, in addition, $\xi \to \sqrt{\Theta(\xi)}$ is of class C^k, then $|t^k \Phi^{(l)}(t)| \leq C_{kl}$ for all real t and $l \in \mathbb{Z}^+$. In particular if $\xi \to \sqrt{\Theta(\xi)}$ is of class C^∞, then $\Phi \in \mathcal{S}$.*

Proof. We first check the conditions of Theorem 6.4.27. Since Θ is supported in an interval of length 2, the sum $\sum_{l \in \mathbb{Z}} |\hat{\Phi}(\xi + l)|^2 = \sum_{l \in \mathbb{Z}} \Theta(\xi + l)$ consists of at most two nonzero terms, of the form

$$\Theta(\xi) + \Theta(\xi + 1) = \Theta(-\xi) + \Theta(1 + \xi) = 1,$$

which proves the orthonormality of $\{\Phi(t-m)\}_{m \in \mathbb{Z}}$. To prove the scaling equation, define

$$m_0(\xi) = \sqrt{\Theta(2\xi)} \quad |\xi| \leq \tfrac{1}{2}$$

and extend m_0 to the real line as a 1-periodic function: $m_0(\xi + 1) = m_0(\xi)$, especially $m_0 \in L^2(\mathbb{R}/\mathbb{Z})$. In addition $\Theta(\xi) = 1$ whenever $\Theta(2\xi) \neq 0$, so that the equation $\hat{\Phi}(2\xi) = m_0(\xi)\hat{\Phi}(\xi)$ holds for $|\xi| \leq \tfrac{1}{3}$. In addition $m_0(\xi)$ is zero for $\tfrac{1}{3} \leq \xi \leq \tfrac{1}{2}$, so that we have the scaling equation $\hat{\Phi}(2\xi) = m_0(\xi)\hat{\Phi}(\xi)$ for all $\xi \in \mathbb{R}$, since both $\hat{\Phi}(\xi)$ and $\hat{\Phi}(2\xi)$ are zero when $|\xi| > \tfrac{2}{3}$, while m_0 is zero when $\tfrac{1}{3} < |\xi| < \tfrac{2}{3}$. Finally, the condition (6.4.23) guarantees that $\hat{\Phi}$ is continuous at $\xi = 0$ with $|\hat{\Phi}(0)| = 1$. Hence by Theorem 6.4.27 there exists an MRA corresponding to Φ. Since $\hat{\Phi}$ has compact support, Φ is of class C^∞ and we have the bounds $|\Phi^{(l)}(t)| \leq \int_{-1}^{1} |2\pi \xi|^l |\hat{\Phi}(\xi)| \, d\xi < \infty$. If in addition we have k continuous derivatives, we can write

$$(-2\pi it)^k \Phi^{(l)}(t) = \int_{-1}^{1} e^{2\pi it\xi} \left(\frac{d}{d\xi}\right)^k [(2\pi i\xi)^l \hat{\Phi}(\xi)] \, d\xi,$$

which completes the proof. ∎

Example 6.4.35. *If* $\Theta(\xi) = 1$ *for* $-\frac{1}{2} < \xi < \frac{1}{2}$, *then* $\Theta(\xi) = 0$ *for* $|\xi| > \frac{1}{2}$ *with* $\Theta(\pm\frac{1}{2}) = \frac{1}{2}$. *This gives the Shannon scaling function* $\Phi(t) = (\sin \pi t)/\pi t$.

6.4.5 From Scaling Function to Orthonormal Wavelet

Once we have a scaling function satisfying the hypotheses of Theorem 6.4.27, it is relatively straightforward to construct a corresponding orthonormal wavelet, namely a function $\Psi \in L^2(\mathbb{R})$ so that $\{2^{j/2}\Psi(2^j t - k)\}_{k \in \mathbb{Z}}$ is an orthonormal basis of the orthogonal complement $V_{j+1} \ominus V_j$. This general construction will specialize to yield the Haar function in case $\Phi = 1_{[0,1)}$.

To describe the defining equations on Ψ, it suffices to take $j = 0$. Since $\Psi \in V_1$, which is spanned by $\{\Phi(2t - n)\}_{n \in \mathbb{Z}}$, there exists an L^2-convergent expansion

(6.4.25) $$\Psi(t) = \sum_{n \in \mathbb{Z}} b_n \Phi(2t - n).$$

But Ψ must be orthogonal to V_0, namely

(6.4.26) $$\int_{\mathbb{R}} \Phi(t - k)\bar{\Psi}(t) \, dt = 0 \quad \forall k \in \mathbb{Z}.$$

These are translated into the Fourier domain as follows:

(6.4.27) $$\sum_{l \in \mathbb{Z}} |\hat{\Psi}(l + \xi)|^2 = 1,$$

(6.4.28) $$\hat{\Psi}(\xi) = m_1\left(\frac{\xi}{2}\right) \hat{\Phi}\left(\frac{\xi}{2}\right),$$

(6.4.29) $$\int_{\mathbb{R}} \hat{\Phi}(\xi)\bar{\hat{\Psi}}(\xi) e^{-2\pi ik\xi} \, d\xi = 0 \quad \forall k \in \mathbb{Z},$$

(6.4.30) $$m_1(\xi) := \frac{1}{2} \sum_{n \in \mathbb{Z}} b_n e^{-2\pi in\xi}.$$

The 1-periodic function m_1 is called the *wavelet filter*. It allows us to pass directly from the scaling function to the wavelet via (6.4.25).

It remains to periodize these relations. At the same time we formulate the periodized version of (6.4.15) from the previous discussion. The idea is that the scaling equations allow us to rewrite the orthogonality relations as identities on the circle $\mathbb{R}/\mathbb{Z}$ in terms of the scaling filter and wavelet filter.

Proposition 6.4.36. *Suppose that* Φ *is a scaling function of an MRA with scaling filter* m_0 *defined by (6.4.16). Then* $m_0(\xi)$ *satisfies (6.4.31) a.e.*

(i) If Ψ is an orthonormal wavelet with respect to Φ, then the 1-periodic functions $m_0(\xi), m_1(\xi)$ satisfy the relations (6.4.32)–(6.4.33), a.e.

(6.4.31) $$|m_0(\xi)|^2 + \left|m_0\left(\xi + \tfrac{1}{2}\right)\right|^2 = 1,$$

(6.4.32) $$|m_1(\xi)|^2 + \left|m_1\left(\xi + \tfrac{1}{2}\right)\right|^2 = 1,$$

(6.4.33) $$m_0(\xi)\bar{m}_1(\xi) + m_0\left(\xi + \tfrac{1}{2}\right)\bar{m}_1\left(\xi + \tfrac{1}{2}\right) = 0.$$

(ii) Conversely, given $m_1 \in L^2(\mathbb{R}/\mathbb{Z})$ satisfying (6.4.32) and (6.4.33), if we define Ψ by (6.4.28), then $\{\Psi(2^j t - k)\}_{k \in \mathbb{Z}}$ is an orthonormal system in $V_1 \ominus V_0$.

The equations (6.4.31) and (6.4.32) suggest the term *quadrature mirror filter* for the functions $m_0(\xi), m_1(\xi)$, since the point $\xi + \tfrac{1}{2}$ is the mirror image of ξ in the circle of unit circumference, with respect to which the quadratic functional equations are satisfied.

Proof. We apply Corollary 6.4.9 to formula (6.4.15) summing separately over the odd and even $l \in \mathbb{Z}$.

$$1 = \sum_{l \in \mathbb{Z}} |\hat{\Phi}(\xi + l)|^2$$

$$= \sum_{k \in \mathbb{Z}} |\hat{\Phi}(\xi + 2k)|^2 + \sum_{k \in \mathbb{Z}} |\hat{\Phi}(\xi + 2k + 1)|^2$$

$$= \sum_{k \in \mathbb{Z}} \left|m_0\left(\frac{\xi + 2k}{2}\right)\right|^2 \left|\hat{\Phi}\left(\frac{\xi + 2k}{2}\right)\right|^2 + \sum_{k \in \mathbb{Z}} \left|m_0\left(\frac{\xi + 2k + 1}{2}\right)\right|^2 \left|\hat{\Phi}\left(\frac{\xi + 2k + 1}{2}\right)\right|^2$$

$$= \left|m_0\left(\frac{\xi}{2}\right)\right|^2 \sum_{k \in \mathbb{Z}} \left|\hat{\Phi}\left(\frac{\xi}{2} + k\right)\right|^2 + \left|m_0\left(\frac{\xi}{2} + \frac{1}{2}\right)\right|^2 \sum_{k \in \mathbb{Z}} \left|\hat{\Phi}\left(\frac{\xi}{2} + \frac{1}{2} + k\right)\right|^2$$

$$= \left|m_0\left(\frac{\xi}{2}\right)\right|^2 + \left|m_0\left(\frac{\xi}{2} + \frac{1}{2}\right)\right|^2$$

where we have applied Corollary 6.4.9 twice in the last line, thus proving (6.4.31). To prove (6.4.32), we replace Φ by Ψ on the left and m_0 by the 1-periodic function m_1 on the right. Applying (6.4.27) and (6.4.28), we see that with these replacements, all of the above computations apply and we obtain (6.4.32). To prove (6.4.33), we periodize (6.4.29) by writing

$$0 = \sum_{l \in \mathbb{Z}} \int_l^{l+1} \hat{\Phi}(\xi)\bar{\hat{\Psi}}(\xi) e^{-2\pi i n \xi} d\xi$$

$$= \sum_{l \in \mathbb{Z}} \int_0^1 \hat{\Phi}(\xi + l)\bar{\hat{\Psi}}(\xi + l) e^{-2\pi i n \xi} d\xi$$

$$= \int_0^1 \left(\sum_{l \in \mathbb{Z}} \hat{\Phi}(\xi + l)\bar{\hat{\Psi}}(\xi + l)\right) e^{-2\pi i n \xi} d\xi.$$

The Fourier coefficients of the indicated 1-periodic function are all zero, hence

(6.4.34) $$\sum_{l \in \mathbb{Z}} \hat{\Phi}(\xi + l)\bar{\hat{\Psi}}(\xi + l) = 0 \quad \text{a.e. } \xi.$$

Now we can apply the same transformations that were used to prove (6.4.31) above: in detail

$$0 = \sum_{l \in \mathbb{Z}} \hat{\Phi}(\xi + l) \bar{\hat{\Psi}}(\xi + l)$$

$$= \sum_{k \in \mathbb{Z}} \hat{\Phi}(\xi + 2k) \bar{\hat{\Psi}}(\xi + 2k) + \sum_{k \in \mathbb{Z}} \hat{\Phi}(\xi + 2k + 1) \bar{\hat{\Psi}}(\xi + 2k + 1)$$

$$= \sum_{k \in \mathbb{Z}} m_0\left(\frac{\xi}{2} + k\right) \bar{m}_1\left(\frac{\xi}{2} + k\right) \left|\hat{\Phi}\left(\frac{\xi}{2} + k\right)\right|^2$$

$$+ \sum_{k \in \mathbb{Z}} m_0\left(\frac{\xi}{2} + k + \frac{1}{2}\right) \bar{m}_1\left(\frac{\xi}{2} + k + \frac{1}{2}\right) \left|\hat{\Phi}\left(\frac{\xi}{2} + k + \frac{1}{2}\right)\right|^2$$

$$= m_0\left(\frac{\xi}{2}\right) \bar{m}_1\left(\frac{\xi}{2}\right) \sum_{k \in \mathbb{Z}} \left|\hat{\Phi}\left(\frac{\xi}{2} + k\right)\right|^2$$

$$+ m_0\left(\frac{\xi}{2} + \frac{1}{2}\right) \bar{m}_1\left(\frac{\xi}{2} + \frac{1}{2}\right) \sum_{k \in \mathbb{Z}} \left|\hat{\Phi}\left(\frac{\xi}{2} + \frac{1}{2} + k\right)\right|^2$$

$$= m_0\left(\frac{\xi}{2}\right) \bar{m}_1\left(\frac{\xi}{2}\right) + m_0\left(\frac{\xi}{2} + \frac{1}{2}\right) \bar{m}_1\left(\frac{\xi}{2} + \frac{1}{2}\right),$$

which completes the direct proof. Conversely, if (6.4.32) and (6.4.33) hold, then we can compute $\sum_{l \in \mathbb{Z}} |\hat{\Phi}(\xi + l)|^2$ following the proof of (6.4.32) above, using the orthonormality of $\Phi(t - m)$. Similarly, the proof of (6.4.33) demonstrates that $\sum_{l \in \mathbb{Z}} \hat{\Phi}(\xi + l) \bar{\hat{\Psi}}(\xi + l) = 0$. ∎

This lemma may be paraphrased by the statement that the matrix

$$M = \begin{pmatrix} m_0(\xi) & m_0\left(\xi + \frac{1}{2}\right) \\ m_1(\xi) & m_1\left(\xi + \frac{1}{2}\right) \end{pmatrix}$$

is unitary.

Example 6.4.37. *Haar wavelets*

We illustrate the above formulation in the case of the Haar wavelet, where

$$\Phi(t) = 1, \quad 0 \le t < 1$$

$$\Psi(t) = \begin{cases} 1 & 0 \le t < \frac{1}{2} \\ -1 & \frac{1}{2} \le t < 1 \end{cases}.$$

The structure constants are obtained from the scaling relations

$$\Phi(t) = \Phi(2t) + \Phi(2t - 1), \qquad \Psi(t) = \Phi(2t) - \Phi(2t - 1).$$

Hence $a_0 = 1$, $a_1 = 1$, $b_0 = 1$, $b_1 = -1$ and otherwise $a_j = b_j = 0$. The scaling filter and wavelet filters are given by (6.4.16) and (6.4.30):

$$m_0(\xi) = \tfrac{1}{2}[1 + e^{-2\pi i \xi}], \qquad m_1(\xi) = \tfrac{1}{2}[1 - e^{-2\pi i \xi}].$$

The Fourier transform of the scaling function and wavelet are

$$\hat{\Phi}(\xi) = \int_0^1 e^{-2\pi it\xi}\, dt = \frac{1 - e^{-2\pi i\xi}}{2\pi i\xi} = e^{-i\pi\xi}\frac{\sin\pi\xi}{\pi\xi},$$

$$\hat{\Psi}(\xi) = \left(\int_0^{\frac{1}{2}} - \int_{\frac{1}{2}}^1\right) e^{-2\pi it\xi}\, dt = \frac{1 - 2e^{-i\pi\xi} + e^{-2\pi i\xi}}{2\pi i\xi} = \frac{(1 - e^{-i\pi\xi})^2}{2\pi i\xi}$$

$$= ie^{-i\pi\xi}\frac{\sin^2(\pi\xi/2)}{(\pi\xi/2)}.$$

Equivalently

$$m_0(\xi) = \frac{\hat{\Phi}(2\xi)}{\hat{\Phi}(\xi)} = e^{-i\pi\xi}\cos(\pi\xi),$$

$$m_1(\xi) = \frac{\hat{\Psi}(2\xi)}{\hat{\Phi}(\xi)} = ie^{-i\pi\xi}\sin\pi\xi.$$

Returning to the theory, we now solve for the wavelet Ψ by means of the function m_1. We give the first row of the matrix M and must find the second row. The orthogonality condition (6.4.33) requires that we have

$$\left(m_1(\xi), m_1\left(\xi + \tfrac{1}{2}\right)\right) = \alpha(\xi)\left(\bar{m}_0\left(\xi + \tfrac{1}{2}\right), -\bar{m}_0(\xi)\right)$$

for some 1-periodic complex-valued function $\alpha(\xi)$. The normalizations (6.4.31) and (6.4.32) further require that $|\alpha(\xi)| = 1$. Finally, the substitution $\xi \to \xi + \tfrac{1}{2}$ shows that α must satisfy the half period condition $\alpha(\xi + \tfrac{1}{2}) = -\alpha(\xi)$. Thus we find the general solution

(6.4.35) $\quad m_1(\xi) = \bar{m}_0\left(\xi + \tfrac{1}{2}\right)\alpha(\xi), \quad$ where $\alpha\left(\xi + \tfrac{1}{2}\right) = -\alpha(\xi), \; |\alpha(\xi)| \equiv 1.$

It is immediately verified that this choice of m_1 satisfies the conditions of Proposition 6.4.36. Therefore the wavelet Ψ can be obtained through its Fourier transform as

(6.4.36) $\quad \hat{\Psi}(\xi) = m_1\left(\frac{\xi}{2}\right)\hat{\Phi}\left(\frac{\xi}{2}\right) = \bar{m}_0\left(\frac{\xi}{2} + \frac{1}{2}\right)\alpha\left(\frac{\xi}{2}\right)\hat{\Phi}\left(\frac{\xi}{2}\right).$

Clearly we have infinitely many choices for α. A unique choice is dictated by the normalization that the Haar scaling function $\Phi = 1_{[0,1)}$ give the standard Haar function ψ. Thus we choose $\alpha(\xi) = -e^{-2\pi i\xi}$, which satisfies all of the conditions. Computing in detail, we have

$$m_1(\xi) = -e^{-2\pi i\xi}\bar{m}_0\left(\xi + \frac{1}{2}\right)$$

$$= -\frac{1}{2}e^{-2\pi i\xi}\sum_{n\in\mathbb{Z}}\bar{a}_n e^{2\pi in(\xi+1/2)}$$

$$= \frac{1}{2}\sum_{n\in\mathbb{Z}}\bar{a}_n(-1)^{n+1} e^{2\pi i(n-1)\xi}$$

$$= \frac{1}{2}\sum_{m\in\mathbb{Z}}\bar{a}_{1-m}(-1)^m e^{-2\pi im\xi}.$$

Referring to (6.4.25), we have the explicit representation of $\Psi(t)$ in terms of $\{\Phi(2t-n)\}_{n\in\mathbb{Z}}$, namely

(6.4.37) $$\Psi(t) = \sum_{n\in\mathbb{Z}} (-1)^n \bar{a}_{1-n} \Phi(2t-n).$$

This explicit formula displays the orthonormal wavelet in terms of the scaling function and the structure constants. In case of the Haar scaling function ($a_0 = 1, a_1 = 1$) we obtain the standard Haar function $\Psi(t) = \Phi(2t) - \Phi(2t-1)$. In the Fourier domain (6.4.37) is written

(6.4.38) $$\hat{\Psi}(\xi) = -e^{-\pi i \xi} \bar{m}_0\left(\frac{\xi}{2} + \frac{1}{2}\right) \hat{\Phi}\left(\frac{\xi}{2}\right).$$

Finally, we prove that the spaces $V_{j+1} \ominus V_j$ are spanned by the set consisting of $\{\Psi(2^j t - k)\}_{k\in\mathbb{Z}}$. By scaling, it suffices to prove this in case $j = 0$.

Proposition 6.4.38. *Any $f \in V_1 \ominus V_0$ can be represented by its Fourier transform as*

$$\hat{f}(\xi) = \bar{m}_0\left(\xi + \frac{1}{2}\right) \nu(\xi) \hat{\Phi}\left(\frac{\xi}{2}\right)$$

where $\nu \in L^2(\mathbb{R}/\mathbb{Z})$ satisfies $\nu(\xi + \frac{1}{2}) = -\nu(\xi)$. In particular, we can write

(6.4.39) $$f(t) = \sum_{n\in\mathbb{Z}} c_n \Psi(t-n)$$

where $\sum_{n\in\mathbb{Z}} |c_n|^2 < \infty$.

Proof. Since $f \in V_1$, we have $f(t) = \sum_{n\in\mathbb{Z}} a_n \Phi(2t-n)$ for some $\alpha \in l_2(\mathbb{Z})$. The orthogonality to V_0 further requires $\int_\mathbb{R} f(t) \overline{\Phi(t-k)} dt = 0$ for all $k \in \mathbb{Z}$. In terms of Fourier transforms, we have

$$\hat{f}(\xi) = \frac{1}{2} \sum_{n\in\mathbb{Z}} \alpha_n e^{-in\pi\xi} \hat{\Phi}\left(\frac{\xi}{2}\right) := C\left(\frac{\xi}{2}\right) \hat{\Phi}\left(\frac{\xi}{2}\right),$$

$$0 = \int_\mathbb{R} \hat{f}(\xi) \overline{\hat{\Phi}(\xi)} e^{-2\pi i k \xi} d\xi \quad \forall k \in \mathbb{Z}.$$

Now we can apply the same computations as those following (6.4.34) to conclude that

$$m_0(\xi) \bar{C}(\xi) + m_0\left(\xi + \frac{1}{2}\right) \bar{C}\left(\xi + \frac{1}{2}\right) = 0.$$

But we have already seen from the proof of Proposition 6.4.36 that the general solution of this equation is obtained as (6.4.35). Finally, this can be written as a multiple of $\hat{\Phi}(\xi/2)$ defined in (6.4.38), expressed as

$$\hat{f}(\xi) = \bar{m}_0\left(\xi + \frac{1}{2}\right) \nu(\xi) \hat{\Phi}\left(\frac{\xi}{2}\right) = -\alpha(\xi) e^{2\pi i \xi} \hat{\Phi}(\xi),$$

which translates into the t domain as (6.4.39). ∎

In conclusion, we note some other basic properties of the scaling filter m_0.

Proposition 6.4.39. *If m_0 is the scaling filter for an MRA, then $|m_0(\xi)| \le 1$ and $m_0(0) = 1$, $m_0\left(\frac{1}{2}\right) = 0$.*

Proof. This follows from (6.4.31) and then setting $\xi = 0$ in the scaling equation in the Fourier domain (6.4.15). ∎

6.4.5.1 Direct proof that $V_1 \ominus V_0$ is spanned by $\{\Psi(t-k)\}_{k \in \mathbb{Z}}$

Proposition 6.4.40. *Any $f \in V_1$ is equal to its projection on the set of functions spanned by $\{\Phi(t-k)\}_{k \in \mathbb{Z}}$, $\{\Psi(t-k)\}_{k \in \mathbb{Z}}$.*

Proof. Since V_1 is spanned by $\{\Phi(2t-l)\}_{l \in \mathbb{Z}}$, it suffices to prove that for each $l \in \mathbb{Z}$,

$$\Phi(2t-l) = \sum_{k \in \mathbb{Z}} c_k \Phi(t-k) + \sum_{k \in \mathbb{Z}} d_k \Psi(t-k)$$

where c_k, d_k are the generalized Fourier coefficients of $\Phi(2t-l)$ with respect to the orthogonal system $\{\Phi(t-k)\}_{k \in \mathbb{Z}}$, $\{\Psi(t-k)\}_{k \in \mathbb{Z}}$. Computing directly, we have

$$c_k = \int_{\mathbb{R}} \Phi(2t-l) \bar{\Phi}(t-k)\, dt$$

$$= \frac{1}{2} \int_{\mathbb{R}} \hat{\Phi}\left(\frac{\xi}{2}\right) e^{-i\pi \xi l} \bar{\hat{\Phi}}(\xi) e^{-2\pi i k \xi}\, d\xi$$

$$= \frac{1}{2} \int_{\mathbb{R}} \hat{\Phi}\left(\frac{\xi}{2}\right) e^{-i\pi \xi l} \bar{\hat{\Phi}}\left(\frac{\xi}{2}\right) \bar{m}_0\left(\frac{\xi}{2}\right) e^{-2\pi i k \xi}\, d\xi$$

$$= \int_{\mathbb{R}} |\hat{\Phi}(\xi)|^2 \bar{m}_0(\xi) e^{-2\pi i \xi (2k+l)}\, d\xi$$

$$= \int_0^1 \bar{m}_0(\xi) e^{-2\pi i \xi (2k+l)}\, d\xi,$$

where we have changed $\xi \to 2\xi$ in the next-to-last line and used periodization in the last line. Similarly

$$d_k = \int_0^1 \bar{m}_1(\xi) e^{-2\pi i \xi (2k+l)}\, d\xi$$

$$= -\int_0^1 m_0\left(\xi + \frac{1}{2}\right) e^{2\pi i \xi} e^{-2\pi i \xi (2k+l)}\, d\xi$$

$$= (-1)^{1-l-2k} \int_0^1 m_0(\eta) e^{-2\pi i \eta (2k+l-1)}\, d\eta.$$

Thus c_k is the $2k+l$th Fourier coefficient of $\bar{m}_0$ and d_k is $\pm$ the $2k+l-1$th Fourier coefficient of m_0. Hence by Parseval,

$$\sum_{k \in \mathbb{Z}} (|c_k|^2 + |d_k|^2) = \int_0^1 |m_0(\eta)|^2\, d\eta = \frac{1}{2}.$$

On the other hand,

$$\int_{\mathbb{R}} |\Phi(2t-l)|^2\, dt = \frac{1}{2} \int_{\mathbb{R}} |\Phi(y)|^2\, dy = \frac{1}{2},$$

which completes the proof. ∎

6.4.5.2 Null integrability of wavelets without scaling functions

Any wavelet may be expected to have integral zero. Indeed, (6.4.38) shows that $\hat{\Psi}(0) = m_0\left(\frac{1}{2}\right)\hat{\Phi}(0) = 0$, by Proposition 6.4.39. In particular, if $\Psi \in L^1(\mathbb{R})$, then we must have $\int_\mathbb{R} \Psi = \hat{\Psi}(0) = 0$. Yet this may not be strictly true if the scaling function does not belong to the space $L^1(\mathbb{R})$. For example, the Shannon wavelet has $\Phi(t) = (\sin \pi t)/\pi t$, which fails to be integrable, from which it follows that the associated wavelet defined by (6.4.38) also fails to be integrable.

The following proposition generalizes the property of null integrability in two directions: (i) to wavelets that do not necessarily belong to an MRA and (ii) to higher dimensions. The proof assumes only orthonormality. We use the notation S_j for the cube of lattice points k defined by the inequalities $-2^{j-1} \le k_i < 2^{j-1}$ for $i = 1, \ldots, d$. Thus $\mathrm{card}(S_j) = 2^{jd}$.

Proposition 6.4.41. *Suppose that $\Psi \in L^1(\mathbb{R}^d) \cap L^2(\mathbb{R}^d)$ has the property that $\{\Psi_{jk}\} := \{2^{jd/2}\Psi(2^j t - k)\}_{1 \le j \le N, k \in \mathbb{Z}^d}$ is an orthonormal set and that $Q \supset [-1,1]^d$ is a cube centered at 0 with $\int_{Q^c}|\Psi| < \frac{1}{2}|\int_{\mathbb{R}^d}\Psi| > 0$. Then $\sqrt{N} \le 2\sqrt{|2Q|}/|\int_{\mathbb{R}^d}\Psi|$. In particular, if $\{\Psi_{jk}\}_{j \in \mathbb{Z}^+, k \in \mathbb{Z}^d}$ is an orthonormal set, then $\int_{\mathbb{R}^d}\Psi = 0$.*

Proof. Let $\int_{\mathbb{R}^d}\Psi = Re^{i\theta}$ with $R > 0$. Replacing Ψ by $e^{-i\theta}\Psi$, we preserve the orthonormality of Ψ_{jk} while achieving $\int_{\mathbb{R}^d}\Psi = R > 0$. Now let $Q \supset [-1,1]^d$ be a cube centered at 0 so that $|\int_{Q^c}\Psi| < R/2$. Then for any set A with $Q \subset A$, we have $|\int_{A^c}\Psi| < R/2$ and $|\mathrm{Re}\int_A \Psi - \mathrm{Re}\int_{\mathbb{R}^d}\Psi| \le R/2$, so that $\mathrm{Re}\int_A \Psi \ge R - R/2 = R/2$. Now let

$$(6.4.40) \qquad \Upsilon_j := \sum_{k \in S_j}\Psi(2^j x - k) \implies \|\Upsilon_j\|_2^2 = \sum_{k \in S_j} 2^{-jd} = 1.$$

On the other hand,

$$\int_{2Q}\Upsilon_j = \sum_{k \in S_j}\int_{2Q}\Psi(2^j x - k)\, dx$$

$$= \sum_{k \in S_j} 2^{-jd}\int_{2^{j+1}Q - k}\Psi(u)\, du.$$

But if $Q = [-M, M]^d$ and $x \in Q$ and $j \ge 0$, then $|(x_i + k_i)/2^{j+1}| \le M$, so that $Q \subset 2^{j+1}Q - k$ for $k \in S_j$ and $x \in Q$. This means that

$$(6.4.41) \qquad \mathrm{Re}\int_{2Q}\Upsilon_j \ge \frac{R}{2}.$$

On the other hand, the orthonormality of Ψ_{jk} implies that $\{\Upsilon_j\}$ is an orthonormal sequence, so that for any $N \ge 1$ we have

$$(6.4.42) \qquad \left\|\frac{\sum_{j=1}^N \Upsilon_j}{\sqrt{N}}\right\|_2 = 1.$$

On the other hand, (6.4.41) shows that

$$(6.4.43) \qquad \mathrm{Re}\int_{2Q}\frac{\sum_{j=1}^N \Upsilon_j}{\sqrt{N}} \ge \frac{R}{2}\sqrt{N}.$$

We combine these and apply the Cauchy-Schwarz inequality in the last line:

$$\frac{R}{2}\sqrt{N} \leq \operatorname{Re} \int_{2Q} \frac{\sum_{j=1}^{N} \Upsilon_j}{\sqrt{N}}$$

$$\leq \left| \int_{\mathbb{R}^d} 1_{2Q} \frac{\sum_{j=1}^{N} \Upsilon_j}{\sqrt{N}} \right|$$

$$\leq \sqrt{|2Q|}$$

from which the first statement follows. To prove the second, we take $\sqrt{N} > 2\sqrt{|2Q|}/R$ to obtain a contradiction, hence $R = 0$. ∎

6.5 WAVELETS WITH COMPACT SUPPORT

In this section we develop the tools to construct MRA wavelets whose scaling functions are differentiable and vanish outside of a finite interval. This will be done by a passage from the scaling filter to the scaling function.

Clearly, a necessary condition for a compactly supported scaling function is that the scaling filter be a trigonometric polynomial. This is also sufficient, formalized as follows.

Proposition 6.5.1. *Let Φ be the scaling function of an MRA satisfying the hypotheses of Theorem 6.4.27.*

(i) *If $\Phi(t) = 0$ for $|t| \geq M$, then the structure constants satisfy $a_n = 0$ for $|n| \geq 3M$. In particular the scaling filter is a trigonometric polynomial:*

(6.5.1) $$m_0(\xi) = \frac{1}{2} \sum_{|k| < 3M} a_k e^{-2\pi i k \xi}.$$

(ii) *Conversely, suppose that the scaling filter is a trigonometric polynomial (6.5.1). Then the scaling function can be obtained from the infinite product*

(6.5.2) $$\hat{\Phi}(\xi) = \prod_{j=1}^{\infty} m_0(\xi/2^j),$$

and Φ has compact support.

Proof. From orthogonality, we have $a_n = 2 \int_{-M}^{M} \Phi(t)\bar{\Phi}(2t - n)\, dt$. If $|n| \geq 3M$ then the support of $\Phi(2t - n)$ is disjoint from $[-M, M]$ and the integrand is identically zero, hence $a_n = 0$ for $|n| \geq 3M$. Conversely, suppose that m_0 is a trigonometric polynomial. Since $m_0(0) = 1$, the infinite product (6.5.2) converges and we have for any N

$$\hat{\Phi}(\xi) = \left(\prod_{j=1}^{N} m_0\left(\frac{\xi}{2^j}\right) \right) \hat{\Phi}(\xi 2^{-N})$$

where the last factor tends to 1 when $N \to \infty$, hence (6.5.2) holds.

From the construction, $\hat{\Phi}(\xi) = \lim_N \Pi_N(\xi)$, where $\Pi_N(\xi) := \Pi_{j=1}^N m_0(\xi/2^j)$. Now $\Pi_N(\xi)$ is a finite linear combination of terms of the form

$$e^{-2\pi i \xi \sum_{j=1}^N 2^{-j} r_j}$$

where $|r_j| \leq M$. This is the Fourier transform of a linear combination of δ-measures concentrated at the points $x = \sum_{j=1}^N 2^{-j} r_j \in [-M, M]$. Hence the scaling function Φ is the weak limit of measures concentrated on $[-M, M]$, in particular of compact support. ∎

6.5.1 From Scaling Filter to Scaling Function

Having obtained some conditions in terms of the Fourier transform of the scaling function and its scaling filter, we now attempt to go in the other direction. Beginning with the scaling filter $m_0(\xi)$ we attempt to construct the scaling function. The next theorem does not assume that m_0 is a trigonometric polynomial.

Theorem 6.5.2. *Suppose that $m_0(\xi)$ is a 1-periodic function on the line that satisfies the following conditions:*

(6.5.3) $\qquad m_0(0) = 1, \; |m_0(\xi)| \geq c > 0 \; \text{for} \; |\xi| \leq \dfrac{1}{4},$

(6.5.4) $\qquad |m_0(\xi)|^2 + \left| m_0\left(\xi + \dfrac{1}{2}\right) \right|^2 \equiv 1,$

(6.5.5) $\qquad |1 - m_0(\xi)| \leq \dfrac{C}{\log^2(1/|\xi|)} \qquad |\xi| \leq \dfrac{1}{2}.$

Then the infinite product $\prod_1^\infty m_0(\xi/2^k)$ converges and defines an L^2 function $\hat{\Phi}$ for which $\{\Phi(t-k)\}_{k \in \mathbb{Z}}$ is orthonormal and Φ is the scaling function of an MRA.

We remark that although condition (6.5.3) can be weakened, it cannot be dispensed with entirely, as shown by the next example.

Example 6.5.3. *Let $m_0(\xi) = \frac{1}{2}(1 + e^{-6\pi i \xi})$.*

Computing the infinite product explicitly, we find that for $\xi \neq 0$, $\hat{\Phi}(\xi) = (1 - e^{-6\pi i \xi})/6\pi i \xi$, which is the Fourier transform of $\Phi = \frac{1}{3} 1_{[0,3]}$, whose translates do not form an orthonormal sequence.

Exercise 6.5.4. *(a) Check this calculation. (b) Check that $\|\Phi\|_2 < 1$. (c) Show that $\{\Phi(t-k)\}_{k \in \mathbb{Z}}$ does not form a Riesz sequence.*

We will see below that the lower bound in (6.5.3) can be replaced by a weaker condition, which is in fact necessary.

Proof of the Theorem. We will break the proof into four distinct steps, which will reveal the points where the different assumptions are invoked. The first step is a general statement about L^2 functions. If $\Phi \in L^2(\mathbb{R})$, let

$$\mathcal{M} := \{\Phi \in L^2(\mathbb{R}) : \|\Phi\|_2 = 1 \quad \text{and} \quad \int_{\mathbb{R}} \Phi(t)\bar{\Phi}(t-k)\,dt = 0, \ \forall k \in \mathbb{Z} \setminus \{0\}\}.$$

Step 1. $\mathcal{M}$ is a closed subset of $L^2(\mathbb{R})$.

To see this, we let $\Phi_j \in \mathcal{M}$ with $\Phi_j \to \Phi \in L^2(\mathbb{R})$. Using the notation $L_k\Phi(t) = \Phi(t-k)$, a norm-preserving operator, we have

$$(\Phi, L_k\Phi) - (\Phi_j, L_k\Phi_j) = (\Phi - \Phi_j, L_k\Phi) + (\Phi_j, L_k\Phi - L_k\Phi_j)$$
$$|(\Phi, L_k\Phi) - (\Phi_j, L_k\Phi_j)| \leq \|\Phi - \Phi_j\|_2 \|L_k\Phi\|_2 + \|\Phi_j\|_2 \|L_k\Phi - L_k\Phi_j\|_2$$
$$= \|\Phi - \Phi_j\|_2 \|\Phi\|_2 + \|\Phi_j\|_2 \|\Phi - \Phi_j\|_2$$
$$\to 0.$$

Taking $k = 0$ shows that $\|\Phi\|_2 = 1$. Taking $k \neq 0$ proves that $(\Phi, L_k\Phi) = 0$, i.e., $\Phi \in \mathcal{M}$.

Step 2. We define an inductive process beginning with the Shannon wavelet. Let $\Phi_0(t) = (\sin \pi t)/\pi t$, with $\hat{\Phi}_0(\xi) = 1_{[-\frac{1}{2},\frac{1}{2}]}$, and define for $j \geq 1$.

$$\hat{\Phi}_j(\xi) = \hat{\Phi}_0\left(\frac{\xi}{2^j}\right) m_0\left(\frac{\xi}{2}\right) \cdots m_0\left(\frac{\xi}{2^j}\right).$$

We claim that $\Phi_j \in \mathcal{M}$ for all $j \geq 0$.

To see this, we will use mathematical induction to prove that $\{\Phi_j(t-k)\}_{k\in\mathbb{Z}}$ is orthonormal. For $j = 0$ we have $\sum_{l\in\mathbb{Z}} |\hat{\Phi}_0(\xi + l)|^2 = 1$, so that $\Phi_0(t-k)_{k\in\mathbb{Z}}$ is an orthonormal set, i.e., $\Phi_0 \in \mathcal{M}$.

Assuming that $\Phi_{j-1} \in \mathcal{M}$ we write $\hat{\Phi}_j(\xi) = m_0(\xi/2)\hat{\Phi}_{j-1}(\xi/2)$ and compute

$$\sum_{l\in\mathbb{Z}} |\hat{\Phi}_j(\xi + l)|^2 = \left(\sum_{l \text{ even}} + \sum_{l \text{ odd}}\right) \left|m_0\left(\frac{\xi+l}{2}\right) \hat{\Phi}_{j-1}\left(\frac{\xi+l}{2}\right)\right|^2$$
$$= \left|m_0\left(\frac{\xi}{2}\right)\right|^2 \sum_{l \text{ even}} \left|\hat{\Phi}_{j-1}\left(\frac{\xi+l}{2}\right)\right|^2$$
$$+ \left|m_0\left(\frac{\xi}{2} + \frac{1}{2}\right)\right|^2 \sum_{l \text{ odd}} \left|\hat{\Phi}_{j-1}\left(\frac{\xi+l}{2}\right)\right|^2.$$

The first sum can be written $\sum_{l\in\mathbb{Z}} |\hat{\Phi}_{j-1}(\xi/2 + l)|^2$, which equals 1 by the induction hypothesis. The second sum can be written $\sum_{l\in\mathbb{Z}} |\hat{\Phi}_{j-1}((\xi + 1)/2 + l)|^2$, which also equals 1 by the induction hypothesis. Hence (6.5.4) shows that we have orthonormality for all j, in particular $\int_{\mathbb{R}} |\hat{\Phi}_j|^2 = 1$, completing Step 2.

Step 3. The infinite product (6.5.2) converges uniformly on compact sets to $\tilde{\Phi} \in L^2(\mathbb{R})$ with $\|\tilde{\Phi}\|_2 \leq 1$ and $\lim_{\xi \to 0} \tilde{\Phi}(\xi) = 1$.

It suffices to prove uniform convergence on the interval $|\xi| \leq M$ with $M > 1$. This follows from the estimates

$$\left|m_0\left(\frac{\xi}{2^j}\right) - 1\right| \leq \frac{C}{\log^2(M2^{-j})}, \qquad |\xi| \leq M, 2^{j-1} \geq M$$

$$= \frac{C}{(\log M - j\log 2)^2}$$

$$\leq \frac{C}{(j\log 2)^2},$$

which is the general term of a convergent numerical series. To prove that $\tilde{\Phi} \in L^2(\mathbb{R})$, we use Fatou's lemma:

$$\|\tilde{\Phi}\|_2 = \|\lim_j \hat{\Phi}_j\|_2 \leq \liminf_j \|\hat{\Phi}_j\|_2 = 1,$$

which completes Step 3.

We define Φ as the L^2 function such that $\hat{\Phi} = \tilde{\Phi}$. It remains to prove that $\Phi \in \mathcal{M}$.

Step 4. $\|\Phi_j - \Phi\|_2 \to 0$ when $j \to \infty$.

To see this we will first prove that for some $C_1 > 0$, $|\hat{\Phi}_j(\xi)| \leq C_1 \hat{\Phi}(\xi)$ for all $\xi \in \mathbb{R}$ and all sufficiently large j. From Step 3, the infinite product is uniformly convergent on $[-\frac{1}{2}, \frac{1}{2}]$, hence for some M we have $\Pi_{j>M} |m_0(\xi 2^{-j})| > \frac{1}{2}$ for all $|\xi| \leq \frac{1}{2}$. Hence

$$|\hat{\Phi}(\xi)| \geq \frac{1}{2} \prod_{j=1}^{M} |m_0(\xi 2^{-j})| \geq \frac{1}{2} c^M, \qquad |\xi| \leq \frac{1}{2}$$

where $c = \inf_{|\xi| \leq 1/4} |m_0(\xi)|$. Hence $\inf_{|\xi| \leq 1/2} |\hat{\Phi}(\xi)| \geq C > 0$. From the definition of Φ_j in Step 2, we see that on the interval $|\xi| > 2^{j-1}$, $\hat{\Phi}_j$ is zero, whereas on $|\xi| \leq 2^{j-1}$ we have $\hat{\Phi}(\xi) = \hat{\Phi}_j(\xi)\hat{\Phi}(\xi/2^j)$. We can solve this to write for all j,

(6.5.6) $$\left|\hat{\Phi}_j(\xi)\right| = \left|\frac{\hat{\Phi}(\xi)}{\hat{\Phi}(\xi/2^j)}\right| \leq C^{-1}|\hat{\Phi}(\xi)|1_{[-2^{j-1}, 2^{j-1}]}(\xi) \leq C^{-1}|\hat{\Phi}(\xi)|$$

where $C = \inf_{[-1/2, 1/2]} |\hat{\Phi}(\xi)|$, so we have proved the stated domination. Thus $|\hat{\Phi}_j(\xi) - \hat{\Phi}(\xi)|$ tends to zero and is bounded by an L^2 function, which shows that $\|\hat{\Phi}_j - \hat{\Phi}\|_2 \to 0$, by the dominated convergence theorem. Parseval's identity shows further that $\|\Phi_j - \Phi\|_2 \to 0$ which completes Step 4.

Combining these results, we see that $\Phi \in \mathcal{M}$, namely the integer translates of Φ form an orthonormal sequence. The uniform convergence from Step 2 shows that $\hat{\Phi}$ is continuous at $\xi = 0$ with $\hat{\Phi}(0) = 1$, while the scaling identity follows from the definition of $\hat{\Phi}$:

$$\hat{\Phi}(\xi) = \prod_{j=1}^{\infty} m_0\left(\frac{\xi}{2^j}\right) = m_0\left(\frac{\xi}{2}\right) \hat{\Phi}\left(\frac{\xi}{2}\right).$$

Applying Theorem 6.4.27 completes the proof. ∎

Exercise 6.5.5. *Suppose that the 1-periodic function m_0 satisfies (6.5.4) and (6.5.5).*

Define an operator P on nonnegative measurable functions by

$$Pf(\xi) = \left|m_0\left(\frac{\xi}{2}\right)\right|^2 f\left(\frac{\xi}{2}\right) + \left|m_0\left(\frac{\xi}{2} + \frac{1}{2}\right)\right|^2 f\left(\frac{\xi}{2} + \frac{1}{2}\right).$$

(i) Check that $P1 = 1$.
(ii) If g satisfies the equation $g(\xi) = m_0(\xi/2)g(\xi/2)$, define $e(\xi) = \sum_{l \in \mathbb{Z}} |g(\xi + l)|^2$ and prove that $Pe = e$.
(iii) Define $\hat{\Phi}(\xi) := \prod_{j=1}^{\infty} m_0(\xi/2^j)$. Assuming that the unique solution of $Pf = f$ is $f \equiv \text{const}$, prove that $\{\Phi(t - k)\}_{k \in \mathbb{Z}}$ is an orthogonal sequence.

Hint: For part (ii), copy Step 2 of the above proof.

6.5.2 Explicit Construction of Compact Wavelets

In order to construct wavelets with compact support, it remains to exhibit trigonometric polynomials with the properties (6.5.3) and (6.5.4), since (6.5.5) is automatically satisfied for any trigonometric polynomial. We begin with a trigonometric polynomial $C(\xi)$, which satisfies the following conditions

(6.5.7)
$$C(\xi) \geq 0, C(\xi) > 0 \text{ for } |\xi| \leq \frac{1}{4}, \quad C(0) = 1, \quad C(\xi) + C\left(\xi + \frac{1}{2}\right) = 1.$$

For example $C(\xi) = \cos^2 \pi \xi$ satisfies these conditions.

Exercise 6.5.6. Show that any trigonometric polynomial that satisfies (6.5.7) is of the form

(6.5.8) $$C(\xi) = \frac{1}{2} + \sum_{k=1}^{N} (a_k \cos(2k+1)2\pi\xi + b_k \sin(2k+1)2\pi\xi)$$

for some N and suitable values of the real constants a_k, b_k.

Once we have found a solution of (6.5.7), we need to find the scaling filter m_0 as a suitable *square root*. This is accomplished by applying the following lemma on factorization of trigonometric polynomials.

Lemma 6.5.7. *Fejér and Riesz:* Suppose that $t(x) = \sum_{-n}^{n} c_k e^{ikx}$ is a nonnegative trigonometric polynomial. Then there exists a trigonometric polynomial $q(x)$ so that $t(x) = q(x)\bar{q}(x)$ for all $x \in \mathbb{R}$.

Proof. We first prove the lemma in case $t(x) > 0$ for all x. Since $t(x)$ is real, the coefficients must be Hermitian-symmetric: $c_{-k} = \bar{c}_k$ for $-n \leq k \leq n$. Define a polynomial in the complex plane:

$$P(z) := c_{-n} + \cdots + c_n z^{2n} = \bar{c}_n + \cdots + \bar{c}_{-n} z^{2n}.$$

Clearly we have

$$t(x) = e^{-inx} P(e^{ix}), \quad z^{2n} P\left(\frac{1}{z}\right) = \bar{P}(\bar{z}).$$

This implies that, whenever α is a zero of P in the exterior of the unit circle, then $1/\bar{\alpha}$ is a zero in the interior of the unit circle, of the same multiplicity. Furthermore, there are no zeros on the unit circle, since $t(x) > 0$. Factoring out the zeros and simplifying, we have

$$P(z) = Cz^m \prod_k (z - \alpha_k)\left(z - \frac{1}{\bar{\alpha}_k}\right)$$

$$t(x) = e^{-inx} P(e^{ix}) = Ce^{ipx} \prod_k \left(\frac{(e^{ix} - \alpha_k)(e^{-ix} - \bar{\alpha}_k)}{\bar{\alpha}_k}\right)$$

for suitable integers m, p. Since $t(x) > 0$, we must have $p = 0$ and $C/\prod_k \bar{\alpha}_k > 0$. This leads us to choose

$$q(x) = c\prod_k (e^{ix} - \alpha_k), \qquad c := \sqrt{\frac{C}{\prod_k \bar{\alpha}_k}},$$

which completes the proof in case $t(x) > 0$. In the general case when $t(x) \geq 0$, we apply the above construction to the trigonometric polynomial $\epsilon + t(x)$, to obtain a trigonometric polynomial $q_\epsilon(x)$, which depends continuously on ϵ. In particular, we can take $q_0(x) = \lim_\epsilon q_\epsilon(x)$ to obtain the required trigonometric polynomial. ∎

We present several concrete formulas for generating the required trigonometric polynomials

6.5.2.1 Daubechies recipe

This begins with the identity $\cos^2 \pi\xi + \sin^2 \pi\xi = 1$, which is raised to an odd power:

$$1 = (\cos^2 \pi\xi + \sin^2 \pi\xi)^{2N+1} = \sum_{k=0}^{2N+1} \binom{2N+1}{k} \sin^{2k}(\pi\xi) \cos^{4N+2-2k}(\pi\xi).$$

This sum has an even number of terms. We take the first half of the terms and define

(6.5.9) $$C(\xi) := \sum_{k=0}^{N} \binom{2N+1}{k} \sin^{2k}(\pi\xi) \cos^{4N+2-2k}(\pi\xi).$$

When we replace ξ by $\xi + \frac{1}{2}$ the sines turn into cosines and the cosines turn into sines, yielding the remaining terms in the binomial expansion; thus $C(\xi) + C(\xi + \frac{1}{2}) = 1$, as required. Clearly $C(0) = 1$ and $C(\xi) > 0$ for all ξ, so that the conditions (6.5.7) are satisfied.

It is also useful to write this in the factored form

(6.5.10) $$C(\xi) = (\cos^2 \pi\xi)^{N+1} \sum_{k=0}^{N} \binom{2N+1}{k} \sin^{2k}(\pi\xi) \cos^{2N-2k}(\pi\xi)$$

$$= (\cos^2 \pi\xi)^{N+1}[(\cos^2 \pi\xi)^N + \cdots + \binom{2N+1}{N}(\sin^2 \pi\xi)^N].$$

The second factor can be rewritten as a polynomial in $\sin^2 \pi\xi$ with positive coefficients, which will be shown below. Therefore the maximum of the second factor is attained when $\sin^2 \pi\xi = 1$, $\cos^2 \pi\xi = 0$, and we obtain the useful bound

$$C(\xi) \leq (\cos^2 \pi\xi)^{N+1} \binom{2N+1}{N}.$$

The last factor is the middle term of the binomial expansion of $(1 + 1)^{2N+1}$ and can be conveniently expressed by Laplace's method as asymptotic to $2^{2N+1}/\sqrt{N\pi}$.

Exercise 6.5.8. *Show that (6.5.9) can be written as a trigonometric sum in the form*

$$C(\xi) = \frac{1}{2} + \sum_{k=1}^{m} a_k \cos(2k+1)2\pi\xi$$

for suitable values of the constants a_k.

Exercise 6.5.9. *Use Laplace's method for integrals to show that*

$$\binom{2N+1}{N} \sim 2^{2N+1}/\sqrt{N\pi} \qquad N \to \infty.$$

Hint: Begin with the integral representation

$$\frac{1}{2^{2N+1}}\binom{2N+1}{N} = \frac{1}{2\pi}\int_{-\pi}^{\pi} e^{-i\theta} \cos^{2N+1}\theta \, d\theta.$$

We still need to show that the Daubechies recipe can represented as a polynomial with positive coefficients. In fact

Lemma 6.5.10. *Let $P_N(y) = \sum_{k=0}^{N} \binom{2N+1}{k} y^k (1-y)^{N-k}$ for $0 \le y \le 1$. Then $P_N(y) = \sum_{k=0}^{N} \binom{N+k}{k} y^k$. In particular $P_N(y) \le P_N(1) = \binom{2N+1}{N}$.*

Proof. From the binomial theorem, we have

$$1 = (y + (1-y))^{2N+1}$$

$$= \sum_{k=0}^{2N+1} \binom{2N+1}{k} y^k (1-y)^{2N+1-k}$$

$$= (1-y)^{N+1} \sum_{k=0}^{N} \binom{2N+1}{k} y^k (1-y)^{N-k}$$

$$+ y^{N+1} \sum_{k=N+1}^{2N+1} \binom{2N+1}{k} y^{k-N-1} (1-y)^{2N+1-k}$$

$$= (1-y)^{N+1} P_N(y) + y^{N+1} P_N(1-y).$$

In particular $(1-y)^{N+1} P_N(y) = 1 + O(y^{N+1})$, $y \to 0$. Now let

$$Q_N(y) = \sum_{k=0}^{N} \binom{N+k}{k} y^k = (1-y)^{-(N+1)} + O(y^{N+1}), \qquad y \to 0,$$

which is the first $N+1$ terms of power series expansion of $(1-y)^{-(N+1)}$. Thus

$$(1-y)^{N+1} Q_N(y) = 1 + O(y^{N+1}), \qquad y \to 0$$

hence the difference polynomial satisfies

$$(1-y)^{N+1}[P_N(y) - Q_N(y)] = O(y^{N+1}), \qquad y \to 0.$$

INTRODUCTION TO WAVELETS   333

In particular, the first N derivatives of the polynomial $P_N - Q_N$ evaluated at $y = 0$ are all zero, hence the polynomial is identically zero, and we have proved that $P_N(y) - Q_N(y) \equiv 0$, as required. Since the coefficients of Q_N are positive, it follows immediately that for $0 \le y \le 1$, $P_N(y) \le P_N(1) = \binom{2N+1}{N}$. ∎

Exercise 6.5.11. *Show that we have the following values of $P_N(y)$.*

$P_1(y) = 1 + 2y$, $\qquad\qquad P_1(1) = 3 = \binom{3}{1}$

$P_2(y) = 1 + 3y + 6y^2$, $\qquad\qquad P_2(1) = 10 = \binom{5}{2}$

$P_3(y) = 1 + 4y + 10y^2 + 20y^3$, $\qquad\qquad P_3(1) = 35 = \binom{7}{3}$

$P_4(y) = 1 + 5y + 15y^2 + 35y^3 + 70y^4$, $\qquad\qquad P_4(1) = 126 = \binom{9}{4}$

$P_5(y) = 1 + 6y + 21y^2 + 56y^3 + 126y^4 + 252y^5$, $\qquad\qquad P_5(1) = 462 = \binom{11}{5}$

$P_6(y) = 1 + 7y + 28y^2 + 84y^3 + 210y^4 + 462y^5 + 924y^6$, $\quad P_6(1) = 1716 = \binom{13}{6}$

In the above Daubechies recipe, we used an odd exponent in the binomial theorem so that we could conveniently group the terms into two separate groups. The following exercise shows that one can make a parallel computation for even exponents also.

Exercise 6.5.12. *For $N \ge 2$, define*

$$R_N(y) = \sum_{k=0}^{N-1} \binom{2N}{k} y^k (1-y)^{N-k} + \frac{1}{2}\binom{2N}{N} y^N (1-y)^N.$$

Show that $1 = (1-y)^N R_N(y) + y^N R_N(1-y)$ and deduce that R_N is a polynomial of degree $N - 1$, specifically, $R_N(y) = P_{N-1}(y)$, defined above, for $N \ge 2$.

6.5.2.2 Hernández-Weiss recipe

This begins with the function

(6.5.11) $$g_m(\xi) = 1 - \frac{\int_0^\xi \sin^{2m+1}(2\pi t)\, dt}{\int_0^{\frac{1}{2}} \sin^{2m+1}(2\pi t)\, dt}$$

defined for $-\frac{1}{2} \le \xi \le \frac{1}{2}$ and extended periodically. Clearly $0 \le g_m(\xi) \le 1$ with $g_m(0) = 1$, $g_m\left(\frac{1}{2}\right) = 0$ and $g_m(\xi) > 0$ for $0 \le |\xi| < \frac{1}{2}$. A direct computation shows further that

$g_m(\xi) + g_m\left(\xi + \frac{1}{2}\right)$

$= \dfrac{2\int_0^{\frac{1}{2}} \sin^{2m+1}(2\pi t)\, dt - \int_0^\xi \sin^{2m+1}(2\pi t)\, dt - \int_0^{\xi+\frac{1}{2}} \sin^{2m+1}(2\pi t)\, dt}{\int_0^{\frac{1}{2}} \sin^{2m+1}(2\pi t)\, dt}$

$= \dfrac{\int_0^{\frac{1}{2}} \sin^{2m+1}(2\pi t)\, dt - \int_0^\xi \sin^{2m+1}(2\pi t)\, dt - \int_{\frac{1}{2}}^{\xi+\frac{1}{2}} \sin^{2m+1}(2\pi t)\, dt}{\int_0^{\frac{1}{2}} \sin^{2m+1}(2\pi t)\, dt}$

$= 1$

since the integrand is odd about $t = \frac{1}{2}$. Thus the conditions (6.5.7) are satisfied.

Exercise 6.5.13. *Show that (6.5.11) can be written as a trigonometric sum in the form*

$$g_m(\xi) = \frac{1}{2} + \sum_{k=1}^{m} a_k \cos(2k+1)2\pi\xi$$

for suitable values of the constants a_k.

6.5.3 Smoothness of Wavelets

We now investigate the smoothness properties of the compact wavelets, which are constructed in terms of the scaling function Φ. The estimates obtained below, although not optimal, do show that one can obtain compactly supported wavelets of any degree of smoothness, beginning with a suitable trigonometric polynomial $m_0(\xi)$.

We begin with the function $C(\xi) = |m_0(\xi)|^2$, which satisfies the properties (6.5.7). Following the Daubechies example from (6.5.10) we can write

$$C(\xi) = (\cos^2 \pi \xi)^{N+1} P_N(\sin^2 \pi \xi),$$

where P_N is a polynomial of degree N that satisfies a bound $|P_N(y)| \leq K_N$ for some constant K_N.

The scaling function satisfies the relation

$$|\hat{\Phi}(\xi)|^2 = \prod_{l=1}^{\infty} C\left(\frac{\xi}{2^l}\right) \leq \prod_{l=1}^{j+1} C\left(\frac{\xi}{2^l}\right) \qquad (j \geq 1).$$

A pointwise estimate is obtained by using the identity $\sin\theta = 2\sin(\theta/2)\cos(\theta/2)$ to write

$$|\hat{\Phi}(\xi)|^2 \leq \prod_{l=1}^{j+1} C\left(\frac{\xi}{2^l}\right)$$

$$\leq \left(\cos\frac{\pi\xi}{2} \cdots \cos\frac{\pi\xi}{2^{j+1}}\right)^{2(N+1)} K_N^{j+1}$$

$$= \left(\frac{\sin \pi\xi}{2^{j+1}\sin\left(\frac{\pi\xi}{2^{j+1}}\right)}\right)^{2(N+1)} K_N^{j+1}.$$

Now restrict ξ to lie in the dyadic shell defined by the inequalities $2^{j-1} \leq |\xi| \leq 2^j$. With this restriction, we have $\pi\xi/2^{j+1} \leq \pi/2$ and we can use the inequality $\sin\theta \geq 2\theta/\pi$ to underestimate the denominator. At the same time we use the upper bound $|\sin\theta| \leq 1$ in the numerator. Combining these with the bound for K_N we have

$$|\hat{\Phi}(\xi)| \leq \left(\frac{1}{2^j}\right)^{(N+1)} K_N^{(j+1)/2}, \qquad 2^{j-1} \leq |\xi| \leq 2^j$$

$$\leq \left(\frac{2^{-j/2}}{(N\pi)^{j/4}}\right)\left(\frac{2^{N+1/2}}{(N\pi)^{1/4}}\right).$$

Therefore

$$\int_{2^{j-1} \leq |\xi| \leq 2^j} |\hat{\Phi}(\xi)| \, d\xi \leq \left(\frac{2^{j/2}}{(N\pi)^{j/4}}\right) \left(\frac{2^{N+1/2}}{(N\pi)^{1/4}}\right)$$

$$= \left(\frac{2^{1/2}}{(N\pi)^{1/4}}\right)^j \left(\frac{2^{N+1/2}}{(N\pi)^{1/4}}\right)$$

$$= C2^{-j\alpha}, \qquad \alpha := \frac{\log_2 N\pi - 2}{4}.$$

To complete the proof of smoothness, it remains to prove the following proposition from classical harmonic analysis, related to the Littlewood-Paley method.

Proposition 6.5.14. *Suppose that $f \in L^2(\mathbb{R})$ and that for some $\alpha > 0$, $C > 0$, we have the system of inequalities*

(6.5.12) $$\int_{2^{j-1} \leq |\xi| \leq 2^j} |\hat{f}(\xi)| \, d\xi \leq C2^{-\alpha j}, \qquad j = 0, 1, 2, \ldots$$

Then

(i) If $0 < \alpha < 1$, then f satisfies a Hölder condition with exponent α.

(ii) If $K < \alpha < K + 1$ for some $K \in \mathbb{Z}^+$, then f has K continuous derivatives and $f^{(K)}$ satisfies a Hölder condition with exponent $\alpha - K$.

Proof. The hypothesis (6.5.12) shows that $\hat{f} \in L^1(\mathbb{R})$, since

$$\int_{\mathbb{R}} |\hat{f}(\xi)| \, d\xi \leq \int_{|\xi| \leq 1} |\hat{f}(\xi)| \, d\xi + C \sum_{j \geq 1} 2^{-\alpha j} < \infty.$$

In particular f is a.e. equal to a continuous function. To prove the Hölder continuity, first suppose that $0 < \alpha < 1$. Define the index $j = j(h)$ so that $(1/2)^{j+1} \leq |h| < 1/2^j$. From (6.5.12) we see that $\hat{f} \in L^1(\mathbb{R})$ so that we can write the absolutely convergent integral

$$f(x+h) - f(x) = \int_{\mathbb{R}} e^{2\pi i x \xi} \left(e^{2\pi i h \xi} - 1\right) \hat{f}(\xi) \, d\xi$$

$$= I + II + III$$

where

$$I = \int_{|\xi| \leq \frac{1}{2}} e^{2\pi i x \xi} \left(e^{2\pi i h \xi} - 1\right) \hat{f}(\xi) \, d\xi$$

$$II = \int_{\frac{1}{2} \leq |\xi| \leq 2^j} e^{2\pi i x \xi} \left(e^{2\pi i h \xi} - 1\right) \hat{f}(\xi) \, d\xi$$

$$III = \int_{|\xi| \geq 2^j} e^{2\pi i x \xi} \left(e^{2\pi i h \xi} - 1\right) \hat{f}(\xi) \, d\xi.$$

It suffices to prove the Hölder condition when $|h| \leq 1$. Making use of the inequality $|e^{i\theta} - 1| \leq \theta$, we have

$$|I| \leq \pi|h| \int_{|\xi| \leq \frac{1}{2}} |\hat{f}(\xi)| \, d\xi \leq \pi|h|^\alpha \|\hat{f}\|_1$$

$$|II| \leq 2\pi|h| \sum_{k=0}^{j} \int_{2^{k-1} \leq |\xi| \leq 2^k} |\xi| |\hat{f}(\xi)| \, d\xi$$

$$\leq C|h| \sum_{k=0}^{j} 2^k \, 2^{-\alpha k}$$

$$= C|h| \sum_{k=0}^{j} 2^{k(1-\alpha)}$$

$$= C(\alpha)|h| \left(2^{(j+1)(1-\alpha)} - 1\right)$$

$$\leq C(\alpha)|h|^\alpha$$

where we have used the definition of $j = j(h)$ in the last line. Finally, we have

$$|III| \leq 2 \sum_{k=j+1}^{\infty} \int_{2^{k-1} \leq |\xi| \leq 2^k} |\hat{f}(\xi)| \, d\xi$$

$$\leq 2C \sum_{k=j+1}^{\infty} 2^{-k\alpha}$$

$$= 2C(\alpha) 2^{-j\alpha}$$

$$\leq 2C(\alpha)|h|^\alpha,$$

which completes the proof in case $0 < \alpha < 1$. Now if $\alpha > 1$ we see immediately that $\int_\mathbb{R} |\xi| |\hat{f}(\xi)| \, d\xi < \infty$ so that we have the absolutely convergent integral

$$f'(x) = \int_\mathbb{R} 2\pi i \xi e^{2\pi i \xi} \hat{f}(\xi) \, d\xi$$

from which we can repeat the above reasoning to show that f' satisfies the appropriate Hölder condition. The higher derivatives are handled in the same manner. ∎

We can summarize the above results in the following form.

Theorem 6.5.15. *For any preassigned integer s, there exists a compactly supported scaling function Φ with continuous derivatives up to order s, so that the corresponding MRA generates a wavelet Ψ of compact support with continuous derivatives of order s.*

6.5.3.1 A negative result

It is interesting to note the impossibility of constructing wavelets that are simultaneously infinitely differentiable and of compact support. The following discussion includes the case of MRA wavelets, but also applies to wavelets which may be constructed by other means.

Proposition 6.5.16. *Suppose that $\psi \in L^2(\mathbb{R})$ has the property that $\{2^{j/2}\psi(2^j x - k)\}_{j,k \in \mathbb{Z}}$ is an orthogonal set.*

(i) *If $\psi \in L^1 \cap L^\infty \cap C$, then $\int_\mathbb{R} \psi = 0$.*
(ii) *If for some $m \in \mathbb{Z}^+$, $\delta > 0$, we have $\psi, \psi' \ldots, \psi^{(m)} \in L^1 \cap L^\infty \cap C$ with $\psi^{(m)}(x) = O((\frac{1}{|x|})^{m+1+\delta})$, $|x| \to \infty$, then $0 = \int_\mathbb{R} \psi = \int_\mathbb{R} x\psi = \cdots = \int_\mathbb{R} x^m \psi$.*

Proof. Letting $a = k/2^j \in Q_2$ be an arbitrary dyadic rational number, we have from the definition of orthogonality

$$0 = \int_\mathbb{R} \psi(x)\bar{\psi}(2^j x - k)\, dx$$

$$= \int_\mathbb{R} \psi(x)\bar{\psi}(2^j(x - a))\, dx$$

$$0 = \int_\mathbb{R} \psi(a + y2^{-j})\bar{\psi}(y)\, dy$$

where we have made a change of variable in the integral. Letting $j \to \infty$ with $k = a2^j$ (a fixed), we use the dominated convergence theorem to conclude that $0 = \psi(a) \int_\mathbb{R} \bar{\psi}(y)\, dy$. If $\int_\mathbb{R} \bar{\psi} \neq 0$, then $\psi(a) = 0$, $\forall a \in Q_2$ hence $\psi \equiv 0$, a contradiction, proving part (i).

To prove part (ii), let

$$\theta_1(x) = \int_{-\infty}^{x} \psi(y)\, dy,\ \theta_2(x) = \int_{-\infty}^{x} \theta_1(y)\, dy, \ldots, \theta_m(x) = \int_{-\infty}^{x} \theta_{m-1}(y)\, dy.$$

Clearly $\theta_1(x) = O(1/|x|)^{m+\delta}$, $x \to -\infty$. But since we have proved in (i) that $\int_\mathbb{R} \psi = 0$, we can write $\theta_1(x) = -\int_x^\infty \psi(y)\, dy$, showing also that $\theta_1(x) = O(1/|x|)^{m+\delta}$, $x \to +\infty$. Hence we can integrate by parts and write

$$\int_\mathbb{R} x\psi(x)\, dx = \int_\mathbb{R} x\theta_1'(x)\, dx = -\int_\mathbb{R} \theta_1(x)\, dx.$$

Now we use orthogonality and partial integration to write

$$0 = \int_\mathbb{R} \psi(a + y2^{-j})\bar{\psi}(y)\, dy$$

$$= \int_\mathbb{R} \psi(a + y2^{-j})\bar{\theta}_1'(y)\, dy$$

$$= -\int_\mathbb{R} \psi'(a + y2^{-j})\bar{\theta}_1(y)\, dy.$$

Letting $j \to \infty$ with $a \in Q_2$ fixed, we have $0 = \psi'(a)\int_\mathbb{R} \bar{\theta}_1(y)\, dy = -\psi'(a)\int_\mathbb{R} x\psi(x)\, dx$. If $\int_\mathbb{R} x\psi(x)\, dx \neq 0$, we conclude that $\psi'(a) \equiv 0$, $\forall a \in Q_2$, which proves that ψ is a linear function, a contradiction. Proceeding inductively, suppose that we have shown that

$$0 = \int_\mathbb{R} \theta_1(y)\, dy = \cdots = \int_\mathbb{R} \theta_{m-1}(y)\, dy.$$

From the hypothesis on ψ, we obtain the bounds

$$\theta_1(y) = O\left(\frac{1}{y}\right)^{m+\delta}, \ldots, \theta_m(y) = O\left(\frac{1}{y}\right)^{1+\delta}, \quad |y| \to \infty.$$

Partial integration shows that

$$\int_{\mathbb{R}} x^m \psi(x)\, dx = -m \int_{\mathbb{R}} x^{m-1} \theta_1(x)\, dx = \cdots = (-1)^m m! \int_{\mathbb{R}} \theta_m(x)\, dx.$$

From the orthogonality relation we have for any $a \in Q_2, j \in \mathbb{Z}$

$$0 = \int_{\mathbb{R}} \psi(a + y2^{-j}) \bar{\psi}(y)\, dy = (-1)^m \int_{\mathbb{R}} \psi^{(m)}(a + y2^{-j}) \bar{\theta}_m(y)\, dy.$$

Letting $j \to \infty$ with a fixed and using the hypothesis on $\psi^{(m)}$ proves that

$$0 = \psi^{(m)}(a) \int_{\mathbb{R}} \bar{\theta}_m(y)\, dy = \frac{\psi^{(m)}(a)}{m!} \int_{\mathbb{R}} y^m \bar{\psi}(y)\, dy,$$

from which it follows that if $\int_{\mathbb{R}} y^m \psi \neq 0$, then ψ is a polynomial, a contradiction. ∎

Corollary 6.5.17. *If $\psi \in C_{00}^\infty$ and $\{2^{j/2} \psi(2^j x - k)\}_{j,k \in \mathbb{Z}}$ is an orthogonal set, then $\psi(x) = 0$.*

Proof. From the previous proposition, we have $\int_{\mathbb{R}} x^m \psi(x)\, dx = 0$ for all $m \in \mathbb{Z}^+$. Since ψ has compact support, we can find a sequence of polynomials $p_n(x)$ that converge uniformly to $\bar{\psi}(x)$ on the bounded interval of support. From this it follows that $\int_{\mathbb{R}} |\psi(x)|^2 dx = 0$, which shows that $\psi \equiv 0$, as required. ∎

We can obtain the same conclusion if we have a suitable decay rate.

Corollary 6.5.18. *Suppose that $\psi \in S$ and that for each $A > 0$*

$$\lim_{M \to \infty} \frac{A^M}{M!} \int_{\mathbb{R}} |x|^M |\psi(x)|\, dx = 0.$$

If $\{2^{j/2} \psi(2^j x - k)\}_{j,k \in \mathbb{Z}}$ is an orthogonal set, then $\psi(x) = 0$.

Proof. From the hypotheses we can apply the previous proposition to conclude that $\int_{\mathbb{R}} x^n \psi(x)\, dx = 0$ for all n. Now we can estimate $\hat{\psi}$ by applying Taylor's theorem with remainder to the complex exponential function. Thus

$$\left| e^{-2\pi i \xi x} - \sum_{n=0}^{N-1} \frac{(-2\pi i \xi x)^n}{n!} \right| \leq \frac{(2\pi \xi x)^N}{N!}.$$

Multiplying by ψ and integrating, we obtain

$$|\hat{\psi}(\xi)| \leq \frac{|2\pi \xi|^N}{N!} \int_{\mathbb{R}} |x^N| |\psi(x)|\, dx.$$

Applying the hypothesis shows that the right side tends to zero, hence $\hat{\psi}(\xi) \equiv 0$, so that the result follows from the uniqueness of the Fourier transform. ∎

6.5.4 Cohen's Extension of Theorem 6.5.1

We have seen by an example that the positivity condition on the scaling filter in (6.5.3) cannot be totally omitted. Nevertheless it can be relaxed to a condition that is both necessary and sufficient for the existence of a scaling function of an MRA system.

To motivate Cohen's condition, note that the set $K = [-\frac{1}{2}, \frac{1}{2}]$ has the following properties:

(6.5.13) $$\bigcup_{n=1}^{\infty} 2^n K = \mathbb{R}$$

(6.5.14) $$\sum_{l \in \mathbb{Z}} 1_K(\xi + l) = 1 \quad \text{a.e.}$$

Furthermore if (6.5.3) holds, then

(6.5.15) $$m_0\left(\frac{\xi}{2^j}\right) \neq 0, \quad j \geq 1, \xi \in K.$$

Condition (6.5.14) was instrumental in proving that the sequence of approximants Φ_j satisfies the orthonormality condition for each j, while (6.5.13) was necessary to identify $\lim_j \hat{\Phi}_j$ with the infinite product $\hat{\Phi}$ on the entire real line. Condition (6.5.14) states that K is *congruent to* $[-\frac{1}{2}, \frac{1}{2}]$, modulo 1. More precisely, for a.e. $\xi \in \mathbb{R}$, there is precisely one $l \in \mathbb{Z}$ so that $\xi + l \in K$. As a simple example, consider $K = [0, \frac{1}{2}] \cup [\frac{3}{2}, 2]$. This can be transformed into $[-\frac{1}{2}, \frac{1}{2}]$ by translating the second interval by 2 to the left. More general examples can be obtained by further cutting and pasting.

Now suppose that m_0 is one-periodic and satisfies (6.5.4) and (6.5.5) of Theorem 6.5.2. Then the infinite product (6.5.2) converges uniformly on compact sets. Further, suppose that $K \subset \mathbb{R}$ is a compact set that satisfies the three conditions (6.5.13)–(6.5.15). Defining Ψ so that $\hat{\Phi}_0 = 1_K$, we see that Φ_0 has orthonormal translates, since $\sum_{l \in \mathbb{Z}} |\hat{\Phi}_0(\xi + l)|^2 = \sum_{l \in \mathbb{Z}} 1_K(\xi + l) = 1$ a.e. Hence we can begin the inductive process of Theorem 6.5.2 and define $\hat{\Phi}_j(\xi) = \hat{\Phi}_0(\xi 2^{-j}) m_0(\xi/2) \cdots m_0(\xi/2^j)$, which also satisfies the orthonormality condition. The limit function $\hat{\Phi} \in L^2(\mathbb{R})$ by Fatou's lemma. If $\xi 2^{-j} \notin K$ then $\hat{\Phi}_j(\xi) = 0$. Otherwise we may solve for $\hat{\Phi}_j(\xi) = \hat{\Phi}(\xi)/\hat{\Phi}(\xi 2^{-j})$ and copy the argument in the proof of Theorem 6.5.2 to conclude that we have an estimate on the real line: $|\hat{\Phi}_j(\xi)| \leq C |\hat{\Phi}(\xi)|$, from which we can apply dominated convergence to deduce that $\|\hat{\Phi}\|_2 = 1$, and that Φ has orthonormal translates. Thus we have proved half of the following theorem.

Theorem 6.5.19. *Suppose that m_0 is 1-periodic and satisfies (6.5.4) and (6.5.5). Suppose further that there exists a compact set $K \subset \mathbb{R}$ such that (6.5.13)–(6.5.15) hold. Then Φ, defined by the infinite product (6.5.2), defines the scaling function of an MRA.*

Conversely, suppose that $\Phi \in L^1(\mathbb{R}) \cap L^2(\mathbb{R})$ is the scaling function of an MRA for which $\sum_{l \in \mathbb{Z}} |\hat{\Phi}(\xi + l)|^2 = 1, \forall \xi \in \mathbb{R}$, and that the scaling filter m_0 satisfies (6.5.5). Then there exists a set $K \subset \mathbb{R}$, which is a finite union of closed intervals such that (6.5.13)–(6.5.15) hold.

Proof. The direct assertion has been proved in the above discussion. To prove the converse, we begin with the orthonormality condition of Φ:

$$\sum_{l \in \mathbb{Z}} |\hat{\Phi}(\xi + l)|^2 = 1, \quad \xi \in \mathbb{R}.$$

In particular, for each $\xi \in [-\frac{1}{2}, \frac{1}{2}]$, there exists $k = k(\xi) \in \mathbb{Z}$ so that $|\hat{\Phi}(\xi + k)| > 0$. But $\Phi \in L^1(\mathbb{R})$ implies that $\hat{\Phi}$ is continuous. Hence there exists $\delta = \delta_\xi > 0$ so that $|\hat{\Phi}(\xi' + k)| \geq C(\xi) > 0$ for $|\xi' - \xi| < \delta$. Since $|\hat{\Phi}(0)| = 1$, we can take $k(0) = 0$. From the Heine-Borel theorem, we can extract a finite system of these open intervals which covers $[-\frac{1}{2}, \frac{1}{2}]$. Now we form a finite union $\cup_{j=0}^N V_j$ that covers $[-\frac{1}{2}, \frac{1}{2}]$, beginning with $V_0 = [-\delta_0, \delta_0]$; replacing V_j by $\bar{V}_j \ominus (V_0 \cup \cdots \cup V_{j-1}) \cap [-\frac{1}{2}, \frac{1}{2}]$, we may assume that int V_j are disjoint. Now define $K := \cup_{j=0}^N (V_j + k(\xi_j))$. Since $0 \in V_0 \subset K$, we see that $\cup_{n-1}^\infty 2^n K = \mathbb{R}$, as required. Since $\{V_j\}$ is a finite partition of $[-\frac{1}{2}, \frac{1}{2}]$, we have $\sum_{j=0}^N 1_{V_j}(\xi) = 1$ a.e. on $[-\frac{1}{2}, \frac{1}{2}]$. Then we compute the 1-periodic function

$$\sum_{l \in \mathbb{Z}} 1_K(\xi + l) = \sum_{l \in \mathbb{Z}} \left(\sum_{j=0}^N 1_{V_j + k(\xi_j)}(\xi + l) \right)$$

$$= \sum_{j=0}^N \left(\sum_{l \in \mathbb{Z}} 1_{V_j + k(\xi_j)}(\xi + l) \right)$$

$$= \sum_{j=0}^N 1_{V_j}(\xi)$$

$$= 1 \quad \text{a.e. } \xi \in [-\tfrac{1}{2}, \tfrac{1}{2}]$$

where we have used the fact that $\xi + l \in V_j + k(\xi_j)$ precisely once, when $l = k(\xi_j)$ (otherwise we would contradict $V_j \subset [-\frac{1}{2}, \frac{1}{2}]$). But this sum is 1-periodic, so that it is 1 almost everywhere on $\mathbb{R}$. Finally to prove (6.5.15) we note that from the above construction $|\hat{\Phi}(\xi)| \geq C_j$ when $\xi \in V_j$, hence for all $\xi \in K$, we have $|\hat{\Phi}(\xi + \xi_j)| \geq C = \min\{C_0, \ldots, C_N\}$. Referring to (6.5.2), we see that for any $J \in \mathbb{Z}^+, \xi \in K$, $0 < C \leq |\hat{\Phi}(\xi)| \leq \Pi_{j=1}^J m_0(\xi 2^{-j})$, so that we must have $m_0(\xi 2^{-j}) \neq 0$ for all $\xi \in K$ and $j \geq 1$, completing the proof. ∎

Remark. The sum $\sum_{l \in \mathbb{Z}} |\hat{\Phi}(\xi + l)|^2 \equiv 1$ under mild additional conditions, especially whenever this series is uniformly convergent on compact sets. This will hold if, for example, we have an estimate of the form $\hat{\Phi}(\xi) = O(|\xi|^{-\beta})$, $|\xi| \to \infty$, with $\beta > \frac{1}{2}$. This will be satisfied whenever $\Phi' \in L^1(\mathbb{R})$. Cohen's theorem yields an important corollary that gives an optimal estimate for the interval on which m_0 must remain positive.

Corollary 6.5.20. *Suppose that m_0 satisfies (6.5.4) and (6.5.5).*

(i) *If $|m_0(\xi)| \geq c > 0$ for $|\xi| \leq \frac{1}{6}$, then the infinite product (6.5.2) generates an MRA.*

(ii) *There exists m_0 for which $|m_0(\xi)| > 0$ for $|\xi| < \frac{1}{6}$ for which (6.5.2) does not generate an MRA.*

Proof. It suffices to apply Cohen's theorem to the set $K = [-\frac{1}{3}, \frac{2}{3}]$. Clearly (6.5.13) and (6.5.15) are satisfied so that the direct part of the theorem applies. To see that this is optimal, consider the example $m_0(\xi) = (1 + e^{-6\pi i \xi})/2$. From Exercise 6.5.4 we see that the infinite product (6.5.2) is the Fourier transform of $1_{[0,3]}/3$, which does not have orthonormal translates. ∎

6.6 CONVERGENCE PROPERTIES OF WAVELET EXPANSIONS

If the scaling function satisfies additional regularity properties, we can expect that the series expansion $f(t) = \sum_{jk} c_{jk} \psi_{jk}(t)$ will converge uniformly when f is continuous or in $L^p(\mathbb{R})$ when $f \in L^p(\mathbb{R})$, $1 < p < \infty$. Indeed, this has already been demonstrated in detail for the Haar series expansion in Section 6.3. In the present section we will describe a class of scaling functions for which one has these extended convergence properties, as well as the properties of a.e. convergence.

Going beyond the qualitative fact of convergence, one can also discuss the *speed of convergence*, when f has additional regularity. Indeed, in the case of Fourier series in Chapter 1, we saw that the L^2-Hölder continuity of f is equivalent to a rate of convergence result in the L^2 norm. The corresponding results for continuous functions in the supremum norm are attributed to Jackson and Bernstein in the case of trigonometric series. We will see below that corresponding results apply to a class of MRA systems associated with a suitably regular scaling function Φ.

6.6.1 Wavelet Series in L^p Spaces

We begin with a scaling function that satisfies the estimate

$$(6.6.1) \qquad |\Phi(t)| \leq K(2|t|),$$

where $K : [0, \infty) \to \mathbb{R}$ is a monotone decreasing integrable function. In particular, $\Phi \in L^1(\mathbb{R})$ and $|\int_{\mathbb{R}} \Phi(t)\, dt| = 1$, by the remarks following the proof of Theorem 6.4.27. The projection operator P_j is represented by the series

$$(6.6.2) \qquad P_j f(t) = \sum_{k \in \mathbb{Z}} \Phi_{jk}(t) \left(\int_{\mathbb{R}} f(s) \bar{\Phi}_{jk}(s)\, ds \right)$$

$$= 2^j \sum_{k \in \mathbb{Z}} \Phi(2^j t - k) \int_{\mathbb{R}} f(s) \bar{\Phi}(2^j s - k)\, ds$$

$$= 2^j \int_{\mathbb{R}} \Phi(2^j t, 2^j s) f(s)\, ds$$

where the *wavelet kernel*

$$(6.6.3) \qquad \boxed{\Phi(t, s) := \sum_{k \in \mathbb{Z}} \Phi(t - k) \bar{\Phi}(s - k).}$$

Proposition 6.6.1. *The wavelet kernel $\Phi(t, s)$ enjoys the following properties:*

$$(6.6.4) \qquad \Phi \in L^1_{\mathrm{loc}}(\mathbb{R}^2),$$

$$(6.6.5) \qquad \Phi(t, s) = \bar{\Phi}(s, t),$$

$$(6.6.6) \qquad \int_{\mathbb{R}} |\Phi(t, s)|\, ds \leq C < \infty \quad \text{and} \quad \int_{\mathbb{R}} \Phi(t, s)\, ds = 1,$$

$$(6.6.7) \qquad |\Phi(t, s)| \leq CK(|t - s|).$$

Proof. From (6.6.3), we have

$$|\Phi(t,s)| \leq \sum_{k,l} |\Phi(t-k)||\Phi(s-l)| < \infty,$$

where the right side is the product of two periodic, locally integrable functions, which proves (6.6.4). From the definition (6.6.3), (6.6.5) is immediate. Now

$$\int_{\mathbb{R}} |\Phi(t,s)|\,ds \leq \sum_{k \in \mathbb{Z}} |\Phi(t-k)| \int_{\mathbb{R}} |\Phi(s-k)|\,ds = \|\Phi\|_1 \sum_{k \in \mathbb{Z}} |\Phi(t-k)| \leq C$$

and

$$\int_{\mathbb{R}} \Phi(t,s)\,ds = \left(\int_{\mathbb{R}} \bar{\Phi}\right) \sum_{k \in \mathbb{Z}} \Phi(t-k).$$

But $\hat{\Phi}(l) = 0$ for $l \in \mathbb{Z}$, so that the Poisson summation formula shows that the last sum has the constant value $\hat{\Phi}(0) = \int_{\mathbb{R}} \Phi$, which proves (6.6.6). To prove (6.6.7) we write

$$\Phi(t,s) \leq \left(\sum_{l:|l-t| \geq |s-t|/2} + \sum_{l:|l-s| \geq |s-t|/2}\right) K(2|t-l|)K(2|s-l|)$$

$$\leq K(|s-t|) \sum_{l \in \mathbb{Z}} K(2|s-l|) + K(|s-t|) \sum_{l \in \mathbb{Z}} K(2|t-l|)$$

$$\leq CK(|t-s|),$$

which completes the proof. ∎

The main purpose of this section is to prove that $P_j f \to f$ when $j \to \infty$ and $P_j f \to 0$ when $j \to -\infty$. We can reduce this to the study of P_0 by introducing the *dilation operator* J_r, defined by

(6.6.8) $$J_r f(t) = f(2^r t), \qquad r \in \mathbb{Z}.$$

Proposition 6.6.2. *We have the following properties:*

(i) *For any $r \in \mathbb{Z}$, we have the commutation relation $P_j J_r = J_r P_{j-r}$ and the norm relation $\|J_j f\|_p = 2^{-j/p} \|f\|_p$, $1 \leq p \leq \infty$.*

(ii) *There exists a constant C_p such that for each $j \in \mathbb{Z}$ and for each $f \in L^p(\mathbb{R})$, $1 \leq p \leq \infty$,*

$$\|P_j f\| \leq C_p \|f\|_p.$$

Proof. From (6.6.2), we make the substitution $2^j s = u$ to obtain

$$P_j f(t) = \int_{\mathbb{R}} \Phi(2^j t, u) f(u 2^{-j})\,du$$

$$= P_0(J_{-j}f)(2^j t)$$

$$= J_j P_0 J_{-j} f(t),$$

which is equivalent to the stated commutation relation. The norm relation follows from the identity

$$\|J_j f\|_p^p = \int_{\mathbb{R}} |f(2^j t)|^p \, dt = 2^{-j} \int_{\mathbb{R}} |f(s)|^p \, ds.$$

To prove (ii), we first do the case $j = 0$. Then

$$|P_0 f(t)| = \left| \int_{\mathbb{R}} \Phi(t,s) f(s) \, ds \right| \leq C \int_{\mathbb{R}} K(|t-s|) |f(s)| \, ds \leq C \|K\|_1 \|f\|_\infty$$

$$\|P_0 f\|_p \leq C \|K \star f\|_p \leq C \|K\|_1 \|f\|_p,$$

where we have used Young's inequality for convolutions. To treat the case $j \neq 0$, we use part (i) to write

$$\|P_j f\|_p = \|J_j (P_0 J_{-j}) f\|_p$$
$$= 2^{-j/p} \|P_0 J_{-j} f\|_p$$
$$\leq C \|K\|_1 2^{-j/p} \|J_{-j} f\|_p$$
$$= C \|K\|_1 \|f\|_p. \qquad \blacksquare$$

These bounds allow us to formulate and prove a general theorem on the convergence of the small-scale projection operators.

Theorem 6.6.3. *Suppose that Φ is the scaling function of an MRA and satisfies the bound (6.6.1).*

(i) *If $f \in B_{uc}(\mathbb{R})$, then $\|P_j f - f\|_\infty \to 0$ when $j \to \infty$.*
(ii) *If $f \in L^p(\mathbb{R})$ for $1 \leq p < \infty$, then $\|P_j f - f\|_p \to 0$ when $j \to \infty$.*

Proof. First we note that $P_j 1 = 1$, which follows from $\int_{\mathbb{R}} \Phi(t,s) \, ds = 1$. This allows one to write $f(t) - P_j f(t) = 2^j \int_{\mathbb{R}} \Phi(2^j t, 2^j s)[f(t) - f(s)] \, ds$. Since f is uniformly continuous, given $\epsilon > 0$, let $\delta > 0$ be chosen so that $|f(t) - f(s)| < \epsilon/2C$ for $|s-t| < \delta$. We write

$$f(t) - P_j f(t) = 2^j \left(\int_{|s-t| \leq \delta} + \int_{|s-t| > \delta} \right) [f(t) - f(s)] \Phi(2^j t, 2^j s) \, ds.$$

We apply the bound $\int_{\mathbb{R}} |\Phi(t,s)| \, ds \leq C$ in the first integral to conclude that this term is less than $\epsilon/2$, for all j. To estimate the second integral, we use the boundedness to obtain the upper bound

$$2\|f\|_\infty \int_{|s-t|>\delta} 2^j |\Phi(2^j t, 2^j s)| \, ds$$
$$\leq C \|f\|_\infty \int_{|s-t|>\delta} 2^j K(2^j |s-t|) \, ds$$
$$= C \|f\|_\infty \int_{|u-2^j t|>2^j \delta} K(|u - 2^j t|) \, du$$
$$= C \|f\|_\infty \int_{|v|>2^j \delta} K(v) \, dv,$$

which tends to zero when $j \to \infty$, by the dominated convergence theorem. This is a uniform bound, independent of $t \in \mathbb{R}$, from which we obtain the asserted uniform convergence.

To prove the L^p convergence, we first discuss the case $p = 1$. From the uniform boundedness $\|P_j f\|_1 \le C\|f\|_1$, it suffices to prove the L^1 convergence on the dense set of continuous functions with compact support in $[-R, R]$. For such an f, we have

$$\|f - P_j f\|_1 = \int_{|t| \le 2R} |f(t) - P_j f(t)|\, dt + \int_{|t| > 2R} |P_j f(t)|\, dt.$$

The first integral tends to zero by virtue of the uniform convergence already proved. To estimate the second integral, we write

$$\int_{|t|>2R} |P_j f(t)|\, dt \le \|f\|_\infty \int_{|t|>2R} \left(\int_{|s|<R} 2^j |\Phi(2^j t, 2^j s)|\, ds \right) dt$$

$$\le C \int_{|s|<R} \int_{|t|>2R} 2^j K(2^j |s-t|)\, dt\, ds$$

$$\le C \int_{|s|<R} \int_{|t-s|>R} 2^j K(2^j |s-t|)\, dt\, ds$$

$$\le C \int_{|s|<R} \int_{|u|>R} 2^j K(2^j |u|)\, du\, ds$$

$$\le 2CR \int_{|v|>2^j \delta} K(v)\, dv \to 0 \quad j \to \infty,$$

which completes the proof of L^1 convergence.

To treat the case $1 < p < \infty$, it again suffices to deal with continuous functions with compact support. In this case we have the bounds

$$|f(t) - P_j f(t)|^p \le \|f - P_j f\|_\infty^{p-1} |f(t) - P_j f(t)|$$

$$\int_\mathbb{R} |f(t) - P_j f(t)|^p \le \|f - P_j f\|_\infty^{p-1} \int_\mathbb{R} |f(t) - P_j f(t)|\, dt$$

$$= \|f - P_j f\|_\infty^{p-1} \|f - P_j f\|_1,$$

which tends to zero, by the convergence in case $p = 1$, already proved. This proves the theorem. ∎

Remark. An alternative approach to the proof of Theorem 6.6.3 is to utilize theorems on approximate identities, which were proved in Chapter 2. We begin with the bounds

$$|P_j f(t) - f(t)| \le C 2^j \int_\mathbb{R} |f(t) - f(s)| K(2^j |t-s|)|\, ds,$$

which shows that the left side is majorized by $K_j \star f - f$ where K_j is a family of kernels defined by $K_j(s) = 2^j K(2^j s)$, to which we can apply Theorem 2.2.21 from Chapter 2.

It is interesting to compare the relative simplicity of the above proofs of L^p convergence with the difficulty in proving L^p convergence of trigonometric series. For trigonometric series we have two main obstacles: (i) convergence does not hold in case $p = 1$ and (ii) it is nontrivial to prove uniform L^p boundedness of the operators P_j. In the case of MRA wavelets, the basic hypothesis (6.6.1) on the decay of the scaling function easily provides the necessary uniform bounds, as we have seen.

6.6.1.1 Large scale analysis

To complete the analysis of L^p convergence of general wavelet series, it remains to prove that $P_j f \to 0$ when $j \to -\infty$. As in the case of Haar series, we expect only that this will take place for $L^p(\mathbb{R})$, $1 < p < \infty$ and in the space $C_0(\mathbb{R})$.

Proposition 6.6.4. *(i) If $f \in C_0(\mathbb{R})$, then $\|P_j f\|_\infty \to 0$ when $j \to -\infty$. (ii) If $f \in L^p(\mathbb{R})$, $1 < p < \infty$, then $\|P_j f\|_p \to 0$ when $j \to -\infty$.*

Proof. We begin with $f \in C_{00}(\mathbb{R})$. If $f(t) = 0$ for $|t| > R$, we can write

$$(6.6.9) \qquad P_{-m} f(t) = 2^{-m} C \int_{-R}^{R} f(s) \Phi(2^{-m} t, 2^{-m} s) \, ds$$

$$\leq 2^{-m} C \int_{-R}^{R} |f(s)| K(2^{-m}|t - s|) \, ds$$

$$\leq 2^{-m} C \|f\|_\infty 2 R K(0) \to 0, \qquad m \to \infty.$$

But $C_{00}(\mathbb{R})$ is dense in $C_0(\mathbb{R})$ where we have the estimate $\|P_j f\|_\infty \leq C \|f\|_\infty$.

To prove the L^p convergence, it suffices to take $f \in C_{00}(\mathbb{R})$. For $|t| \leq R$ the estimate (6.6.9) shows that $\int_{-R}^{R} |P_{-m} f(t)|^p \, dt \to 0$. For $t > R$ we make the substitution $v = 2^{-m}(t-s)$ to write

$$|P_{-m} f(t)| \leq \|f\|_\infty \int_{-2^{-m}(t+R)}^{2^{-m}(t+R)} |K(v)| \, dv$$

$$\leq \|f\|_\infty 2^{-m} R K(2^{-m}(t - R))$$

$$\int_R^\infty |P_{-m} f(t)|^p \, dt \leq \|f\|_\infty^p 2^{-mp} \int_R^\infty |K(2^{-m}(t - R))|^p \, dt$$

$$= \|f\|_\infty^p 2^{-mp} 2^m \int_0^\infty |K(s)|^p \, ds$$

$$= \|f\|_\infty^p 2^{m(1-p)} \|K\|_p^p \to 0$$

with a similar estimate for $t < -R$. ∎

In exact parallel with the case of Haar series, the large scale projection operators to do not behave well on $L^1(\mathbb{R})$.

Exercise 6.6.5. *Let $f \in L^1(\mathbb{R})$. Prove that $\int_\mathbb{R} P_j f(t) \to \int_\mathbb{R} f(t) \, dt$ when $j \to \infty$.*

This means that we must restrict the range of p when formulating a general L^p convergence theorem for wavelet series. Similarly, we must restrict to $C_0(\mathbb{R})$, since the identity $P_j 1 = 1$ shows that $P_j f \to 0$ is false in general when $f \in B_{uc}(\mathbb{R})$ for $j \to -\infty$.

Combining Proposition 6.6.4 with Theorem 6.6.3, gives a complete picture of the convergence of one-dimensional wavelet series in the spaces $C_0(\mathbb{R})$ and $L^p(\mathbb{R})$, $1 < p < \infty$. This can be restated as a separate theorem.

Theorem 6.6.6. *Suppose that the scaling function Φ satisfies (6.6.1).*

(i) *If $f \in C_0(\mathbb{R})$, then the sum $\sum_{j=-m}^{n} \sum_{k \in \mathbb{Z}} \Phi_{jk}(t) \int_\mathbb{R} f(s) \bar{\Phi}_{jk}(s) \, ds$ converges uniformly to f when $m, n \to \infty$.*

(ii) *If* $f \in L^p(\mathbb{R})$, $1 < p < \infty$, *then the sum* $\sum_{j=-m}^{n} \sum_{k \in \mathbb{Z}} \Phi_{jk}(t) \int_{\mathbb{R}} f(s) \bar{\Phi}_{jk}(s) \, ds$ *converges to* f *in* $L^p(\mathbb{R})$ *when* $m, n \to \infty$.

6.6.1.2 Almost-everywhere convergence

We can prove that the projection operators $P_j f$ converge almost everywhere when $j \to \infty$, by using the techniques appropriate to monotone kernels from Chapter 2.

Theorem 6.6.7. *Suppose that the scaling function* Φ *satisfies (6.6.1) and let* $f \in L^1(\mathbb{R})$. *Then* $P_j f(t) \to f(t)$ *for every* t *in the Lebesgue set of* f, *in particular almost everywhere.*

Recall that the Lebesgue set of f, Leb (f), is the set of t for which $\lim_{a \to 0} a^{-1} \int_{-a}^{a} |f(t) - f(t+u)| \, du = 0$.

Proof. From the previous computations, we write

$$|f(t) - P_j f(t)| \leq 2^j \int_{\mathbb{R}} |f(t) - f(s)| K(2^j |t-s|) \, ds$$

$$= \left(\int_0^\infty + \int_{-\infty}^0 \right) |f(t) - f(t + 2^{-j} v)| K(v) \, dv.$$

To treat the first integral, we define

$$G(u) := \int_0^u |f(t) - f(t+w)| \, dw.$$

If $t \in \text{Leb}(f)$, then $G(u)/u \to 0$ when $u \to 0$ and we have the bound $|G(u)| \leq C|u|$ for all u. Then we can set $\epsilon = 2^{-j}$ and obtain

(6.6.10)
$$\int_0^\infty K(v)|f(t) - f(t+\epsilon v)| \, dv = \int_0^\infty K(v) G'(\epsilon v) \, dv$$

$$= \int_0^\infty K(v) d\left(\frac{G(\epsilon v)}{\epsilon} \right)$$

$$= -\int_0^\infty v dK(v) \frac{G(\epsilon v)}{\epsilon v} \, dv.$$

The monotonicity of K implies that $vK(v) \to 0$ and a partial integration shows that

$$-\int_0^\infty v dK(v) \, dv = \int_0^\infty K(v) \, dv < \infty$$

so that we can apply the dominated convergence theorem to conclude that (6.6.10) tends to zero with ϵ. A similar analysis applied to the integral for $-\infty < v < 0$ proves that $P_j f(t) \to f(t)$ when $t \to \infty$. ∎

Corollary 6.6.8. *Suppose that* $f \in L^1(\mathbb{R})$ *and that* f *is continuous at* t. *Then* $P_j f(t) \to f(t)$ *when* $t \to \infty$.

Indeed, if f is a continuity point, then it is also a Lebesgue point of f.

Exercise 6.6.9. *Extend Theorem 6.6.3 to* $f \in L^p$, $1 < p < \infty$.

We also note that if $f \in L^1(\mathbb{R})$, then $P_j f(t) \to 0$ when $j \to -\infty$. This follows from the estimate

$$|P_j f(t)| \leq 2^j \int_{\mathbb{R}} K(2^j|s-t|)|f(s)|\, ds$$
$$\leq 2^j K(0) \|f\|_1$$
$$\to 0 \qquad j \to -\infty.$$

This observation can be combined with Theorem 6.6.3 to obtain a statement on the almost-everywhere convergence of the bilateral wavelet series.

Corollary 6.6.10. *Suppose that the scaling function Φ satisfies (6.6.1). Then for any $f \in L^1(\mathbb{R})$, the wavelet sum $\sum_{j=-m}^{n} c_{jk} \Psi_{jk}(t)$ converges for a.e. t, when $m, n \to \infty$.*

6.6.1.3 Convergence at a preassigned point

It is natural to expect that the wavelet series will converge to the normalized value of f in case f has a simple discontinuity. In the case of Fourier series one needs additional regularity conditions in order to ensure this convergence. In the case of MRA wavelets it is sufficient that the wavelet kernel satisfy the mild normalization condition

(6.6.11) $$\int_t^{\infty} \Phi(t,s)\, ds = \frac{1}{2} = \int_{-\infty}^t \Phi(t,s)\, ds.$$

Proposition 6.6.11. *Suppose that the scaling function Φ satisfies (6.6.1) and that the wavelet kernel satisfies (6.6.11). Suppose that $f \in L^p(\mathbb{R})$ for some $1 \leq p < \infty$ and that there exist the one-sided limits $f(t+0) = \lim_{s \downarrow t} f(s)$ and $f(t-0) = \lim_{s \uparrow t} f(s)$. Then*

$$\lim_{j \to \infty} P_j f(t) = \frac{1}{2}[f(t+0) + f(t-0)].$$

Proof. Using (6.6.11), we can write

$$\frac{1}{2}[f(t+0) + f(t-0)] - P_j f(t) = 2^j \int_t^{\infty} \Phi(2^j t, 2^j s)[f(t+0) - f(s)]\, ds$$
$$+ 2^j \int_{-\infty}^t \Phi(2^j t, 2^j s)[f(t+0) - f(s)]\, ds.$$

From the proof of Theorem 6.6.3 we see that both of these integrals tend to zero when $j \to \infty$. ∎

6.6.2 Jackson and Bernstein Approximation Theorems

In this subsection we formulate results that relate the speed of convergence of wavelet series to the smoothness of f. We focus attention on the rate of decay of $\|P_j f - f\|_p$.

In order to measure the smoothness of a function, we introduce the L^p *modulus of continuity*:

(6.6.12)
$$\omega_p(f;\delta) := \sup_{0<h<\delta} \|f(\cdot) - f(\cdot - h)\|_p.$$

This is certainly defined if $f \in L^p(\mathbb{R})$, but may also be defined more generally, e.g., if $f \equiv 1$. The elementary properties are detailed as follows:

Proposition 6.6.12. *The L^p modulus of continuity satisfies the following conditions:*

(i) $\delta \to \omega_p(f;\delta)$ is monotone increasing.
(ii) If $f \in L^p(\mathbb{R})$, then $\omega_p(f;\delta) \to 0$ when $\delta \to 0$.
(iii) $\omega_p(f;\delta_1 + \delta_2) \leq \omega_p(f;\delta_1) + \omega_p(f;\delta_2)$
(iv) $\omega_p(f_1 + f_2;\delta) \leq \omega_p(f_1;\delta) + \omega_p(f_2;\delta)$
(v) $\omega_p(J_a f;\delta) = 2^{-a/p} \omega_p(f;2^a \delta)$

Proof. Property (i) comes directly from the definition. Property (ii) is immediate for continuous functions with compact support, which are dense in $L^p(\mathbb{R})$, hence the general case. Properties (iii) and (iv) follow from the triangle inequality for the L^p norm. Property (v) is a direct computation when we recall the defining properties of $J_a f$ from (6.6.8). ∎

We define the space

(6.6.13) $\quad MC_p(\mathbb{R}) := \{f : \omega_p(f;\delta) < \infty \quad \text{for all} \quad \delta > 0\}$

Exercise 6.6.13. *If $f \in MC_p(\mathbb{R})$, prove that $|f|^p \in L^1_{\text{loc}}(\mathbb{R})$.*

In order to prove suitable approximation theorems, we need to consider a smaller class of scaling functions, defined by an estimate of the form

(6.6.14)
$$|\Phi(t)| \leq \frac{C}{(1+|t|)^B} \quad B > 2.$$

Lemma 6.6.14. *If Φ satisfies (6.6.14), then the kernel function $\Phi(t,s)$ satisfies the estimate*

(6.6.15)
$$|\Phi(t,s)| \leq \frac{C}{(1+|t-s|)^B}.$$

Proof. The proof is identical to the proof of (6.6.7) in Proposition 6.6.1. ∎

The direct approximation (Jackson's estimate) is the following statement.

Theorem 6.6.15. *Suppose that the scaling function satisfies (6.6.14). Then there exists $C = C_p$ such that for all $f \in MC_p(\mathbb{R})$*

(6.6.16)
$$\|f - P_j f\|_p \leq C \omega_p(f; 2^{-j}).$$

Corollary 6.6.16. *If $\omega_p(f; \delta) \leq C\delta^\alpha$ for some α with $0 < \alpha \leq 1$, then $\|P_j f - f\|_p \leq C 2^{-j\alpha}$.*

This is a direct counterpart of the corresponding result for trigonometric approximation, proved in Chapter 1 in case $p = 2$ or $p = \infty$.

Proof of the theorem. First we do the case $j = 0$. We pick $a > 0, b > 0$ so that $B = a + b$ and $ap > p + 1, bp' > 1$. Applying (6.6.15), Hölder's inequality, and the Fubini theorem, we obtain

$$\|f - P_0 f\|_p^p \leq \int_{\mathbb{R}} \left(\int_{\mathbb{R}} |f(t) - f(s)| |\Phi(t, s)| \, ds \right)^p dt$$

$$\leq C \int_{\mathbb{R}} \left(\int_{\mathbb{R}} \frac{|f(t) - f(t+u)|}{(1+|u|)^B} \, du \right)^p dt$$

$$\leq C \int_{\mathbb{R}} \left(\int_{\mathbb{R}} \frac{|f(t) - f(t+u)|^p}{(1+|u|)^{ap}} \, du \right) \left(\int_{\mathbb{R}} \frac{du}{(1+|u|)^{bp'}} du \right)^{p/p'} dt$$

$$\leq C \int_{\mathbb{R}} \frac{\omega_p(f; |u|)^p}{(1+|u|)^{ap}} \, du.$$

On the interval $|u| \leq 1$ we apply the monotonicity of ω_p to majorize this contribution to the integral by $C\omega_p(1)$. On the interval $|u| > 1$ we apply the subadditivity property (iii) to write $\omega_p(f; u) \leq 2|u|\omega_p(f; 1)$, hence this contribution to the integral is majorized by $C\omega_p(f; 1) \int_{|u|>1} |u|^p/(1+|u|)^{ap} du < \infty$, since $ap > p + 1$.

To prove the estimate for $j \geq 1$, we use the scaling properties of J_j.

$$P_j f - f = J_j (P_0 - I) J_{-j} f$$

$$\|P_j f - f\|_p = 2^{-j/p} \|(P_0 - I) J_{-j} f\|_p$$

$$\leq C 2^{j/p} \omega_p(J_{-j} f; 1)$$

$$= C\omega_p(f; 2^{-j}),$$

where we have used the scaling property (v) in the last line. The proof is complete. ∎

Remark. This proof can be rendered more transparent by replacing the Hölder inequality with Minkowski's integral estimate: beginning with

$$|P_0 f(t) - f(t)| \leq \int_{\mathbb{R}} |f(t+u) - f(t)| K(u) du,$$

we have

$$\|P_0 f - f\|_p \leq \int_{\mathbb{R}} \omega_p(f, u) K(u) \, du$$

$$\leq \omega_p(f; 1) \int_{|u| \leq 1} K(u) \, du + 2\omega_p(f; 1) \int_{|u| \geq 1} |u| K(u) \, du,$$

which provides the required bound.

This result can be reformulated in terms of an estimate of the *best approximation* by elements in the space V_j. Let $s_j(f) = \inf_{g \in V_j} \|f - g\|_p$, which is the distance from f to the subspace V_j. If $p = 2$, then $P_j f$ provides the best approximation, so that

$s_j(f) = \|f - P_j f\|_2$. In general, we have $\|f - P_j f\|_p \geq s_j(f)$. On the other hand, given $\epsilon > 0$, let $g \in V_j$ so that $\|g - f\|_p \leq s_j(f) + \epsilon$. Then we can write

$$f - P_j f = (f - g) + (g - P_j f)$$
$$= (f - g) + P_j(g - f)$$
$$\|f - P_j f\|_p \leq \|f - g\|_p + \|P_j\|_{p,p} \|g - f\|_p$$
$$\leq C\|g - f\|_p$$
$$\leq C(s_j(f) + \epsilon)$$

where we have used the uniform L^p boundedness of the projection operators P_j. But $\epsilon > 0$ was arbitrary so that we have the two-sided bound

$$s_j(f) \leq \|f - P_j f\|_p \leq C s_j(f)$$

leading to the following restatement of Theorem 6.6.15.

Corollary 6.6.17. *Suppose that the scaling function Φ satisfies the estimate (6.6.14). Then the distance from $f \in MC_p(\mathbb{R})$ to the subspace V_j satisfies the bound*

$$s_j(f) \leq C_p \omega_p(f; 2^{-j}).$$

We now formulate the Bernstein inequality, for which we impose a condition of smoothness, namely that the scaling function have a continuous derivative $\Phi'(t)$ which satisfies an estimate

(6.6.17) $$|\Phi'(t)| \leq C|\Phi(t)|$$

for some $C > 0$. From this it follows that the wavelet kernel is estimated by

$$|\Phi(t+h, s) - \Phi(t, s)| = \left|\sum_{l \in \mathbb{Z}} (\Phi(t+h-l) - \Phi(t-l))\bar{\Phi}(s-l)\right|$$
$$\leq C|h|K(|t-s|).$$

Theorem 6.6.18. *Suppose that the scaling function satisfies (6.6.1) and (6.6.17). Then there exists a constant C_p so that for any $f \in V_j \cap L^p$, we have*

(6.6.18) $$\boxed{\omega_p(f; \delta) \leq C_p \min(2^j \delta, 1)\|f\|_p.}$$

Proof. First we do the case $j = 0$. Then $f = P_0 f \in V_0$ and we can write

$$f(t+h) - f(t) = \int_{\mathbb{R}} [\Phi(t+h, s) - \Phi(t, s)] f(s)\, ds$$

$$|f(t+h) - f(t)| \leq |h| \int_{\mathbb{R}} K(|t-s|)|f(s)|ds = |h|K \star |f|$$

$$\left(\int_{\mathbb{R}} |f(t+h) - f(t)|^p dt\right)^{1/p} \leq |h|\|K\|_1 \|f\|_p,$$

which proves the result in case $|h| \leq 1$. If $|h| > 1$, then $\omega_p(f; h) \leq 2\|f\|_p$, which completes the proof in case $j = 0$.

If $j \neq 0$, then $f \in V_j$ iff $J_{-j}f \in V_0$ and we can use the scaling properties of ω_p to write for $f \in V_j$

$$\omega_p(f; \delta) = 2^{-j/p}\omega_p(J_{-j}f; 2^j\delta)$$
$$\leq C2^{-j/p} \min(2^j\delta, 1)\|J_{-j}f\|_p$$
$$= C \min(2^j\delta, 1)\|f\|_p. \blacksquare$$

The Bernstein inequality (6.6.18) can be combined with the Jackson inequality to obtain a characterization of the L^p Hölder continuity in terms of the speed of convergence to zero of $\|P_j f - f\|_p$. The method follows closely the corresponding proofs in Chapter 1.

Proposition 6.6.19. *Suppose that the scaling function satisfies (6.6.1) and (6.6.17). Then $f \in L^p(\mathbb{R})$ satisfies $\|P_j f - f\|_p \leq C2^{-j\alpha}$ for some $0 < \alpha < 1$, if and only if f satisfies the L^p Hölder condition $\omega_p(f; \delta) \leq C\delta^\alpha$.*

Proof. The direct statement that $\|P_j f - f\|_p \leq C2^{-j\alpha}$ is an immediate consequence of Jackson's estimate, as previously noted. To prove the converse, we define $Q_j f = P_{j+1} - P_j f$, hence $\|Q_j f\|_p \leq C2^{-j\alpha}$. Given $h > 0$, let $m = m(h)$ be defined by the inequalities $2^{m-1} \leq 1/h < 2^m$. Then we write

$$f(t) = P_0 f(t) + \sum_{j=0}^{m-1} Q_j f(t) + \sum_{j=m}^{\infty} Q_j f(t)$$

$$f(t+h) - f(t) = P_0 f(t+h) - P_0 f(t) + \sum_{j=0}^{m-1}[Q_j f(t+h) - Q_j f(t)]$$

$$+ \sum_{j=m}^{\infty}[Q_j f(t+h) - Q_j f(t)].$$

The first term is Lipschitz continuous, from (6.6.17). To estimate the first sum we have from (6.6.18)

$$\sum_{j=0}^{m-1} \|Q_j f(\cdot + h) - Q_j f\|_p \leq \sum_{j=0}^{m-1} 2^j h \|Q_j f\|_p$$

$$\leq Ch \sum_{j=0}^{m-1} 2^j 2^{-\alpha j}$$

$$\leq Ch 2^{m(1-\alpha)}$$

$$= Ch^\alpha.$$

To estimate the second sum we write

$$\sum_{j=m}^{\infty} \|Q_j f(t+h) - Q_j f(t)\|_p \leq 2 \sum_{j=m}^{\infty} \|Q_j f\|_p$$

$$\leq C \sum_{j=m}^{\infty} 2^{-j\alpha}$$

$$= C 2^{-m\alpha}$$

$$= Ch^{\alpha},$$

which completes the proof. ∎

The alert reader will note that the converse Proposition 6.6.19 has been stated only for $0 < \alpha < 1$, whereas the direct result following Jackson's inequality is valid for $0 < \alpha \leq 1$. Indeed, we see that the proof of the converse proposition breaks down if $\alpha = 1$.

Exercise 6.6.20. *Suppose that* $\|P_j f - f\|_p \leq C 2^{-j}$. *Prove that* $\omega_p(f; \delta) \leq C\delta \log(1/\delta)$ *for* $0 < h < \frac{1}{2}$.

6.7 WAVELETS IN SEVERAL VARIABLES

We conclude this introduction to wavelets with a glimpse of the multidimensional theory. The one major change in passing from one to several variables is the need for a *system of wavelets*, in contrast with the single wavelet Ψ, which suffices in the one-dimensional case. In order to motivate this, we consider two important generalizations of the one-dimensional Haar expansion.

6.7.1 Two Important Examples

Without leaving the one-dimensional setting, we can already see the need for several wavelets when we try to generalize the Haar expansion to describe an orthogonal expansion associated with a b-adic subdivision of the real line, where $b \in \{2, 3, \ldots\}$. Consider the projection operator

$$P_j f(x) := b^j \int_{k/b^j}^{(k+1)/b^j} f(y) \, dy, \qquad \frac{k}{b^j} \leq x < \frac{k+1}{b^j}$$

and the limit relation $\lim_{j \to \infty} P_j f(x) = f(x)$ a.e. where $f \in L^2(\mathbb{R})$. In order to transform this into a wavelet framework, we begin with the Haar scaling function $\Phi(t) = 1_{[0,1)}(t)$ in the context of the scaling equation

$$\Phi(t) = \Phi(bt) + \Phi(bt - 1) + \cdots + \Phi(bt - (b-1)).$$

Defining $V_j = P_j(L^2(\mathbb{R}))$, $W_j = V_{j+1} \ominus V_j$, we need to exhibit a basis of the space W_j, the orthogonal complement of V_j in V_{j+1}. It is clear that on any interval $[k/b^j, (k+1)/b^j)$ the space W_j has dimension $b-1$. Beginning with any basis, we can apply the Gram-Schmidt procedure to obtain an orthonormal basis. For example, in case $b = 3$, we define

$$\Psi^1(t) = \begin{cases} 1 & \text{if } 0 \leq t < \frac{1}{3} \\ -1 & \text{if } \frac{1}{3} \leq t < \frac{2}{3} \\ 0 & \text{if } \frac{2}{3} \leq t < 1 \end{cases}$$

$$\Psi^2(t) = \begin{cases} 1 & \text{if } 0 \leq t < \frac{2}{3} \\ -2 & \text{if } \frac{2}{3} \leq t < 1 \end{cases}.$$

It is clear that $\int_0^1 \Phi(t)\Psi^1(t)\,dt = 0 = \int_0^1 \Phi(t)\Psi^2(t)\,dt$ and that $\int_0^1 \Psi^1(t)\Psi^2(t)\,dt = 0$. Therefore we have an orthogonal basis of $V_1 \ominus V_0$ on $[0, 1]$. Normalizing these functions we obtain an orthonormal basis of $L^2(\mathbb{R})$ by $\{\Psi^1(3^j t - k)\}_{j \in \mathbb{Z}, k \in \mathbb{Z}}$, $\{\Psi^2(3^j t - k)\}_{j \in \mathbb{Z}, k \in \mathbb{Z}}$. This demonstrates the necessity of two different wavelets.

In general, we will need $b - 1$ different wavelets in order to generate $L^2(\mathbb{R})$. These can be obtained by applying the Gram-Schmidt orthogonalization to the Haar functions $\{H(bt - k)/2\}_{k=0}^{b-2}$. For example in case $b = 3$ we can take $\Psi^1(t) = H(3t/2)$, $\Psi^2(t) = 2H[(3t - 1)/2] + H[(3t)/2]$.

As a second example, we consider the problem of generalizing the Haar series to $L^2(\mathbb{R}^2)$. We begin with the projection operator

$$P_j f(x, y) := 2^{2j} \int_{k/2^j}^{(k+1)/2^j} \int_{l/2^j}^{(l+1)/2^j} f(s, t)\,ds\,dt, \quad (x, y) \in \left[\frac{k}{2^j}, \frac{k+1}{2^j}\right) \times \left[\frac{l}{2^j}, \frac{l+1}{2^j}\right)$$

and the limit relation $\lim_{j \to \infty} P_j f(x, y) = f(x, y)$, a.e. $(x, y) \in \mathbb{R}^2$. As above, we let $V_j = P_j(L^2(\mathbb{R}))$ and $W_j = V_{j+1} \ominus V_j$. The associated scaling function is $\Phi(t_1, t_2) = 1_{[0,1)}(t_1) 1_{[0,1)}(t_2)$ with scaling equation

$$\Phi(t_1, t_2) = \Phi(2t_1, 2t_2) + \Phi(2t_1 - 1, 2t_2) + \Phi(2t_1, 2t_2 - 1) + \Phi(2t_1 - 1, 2t_2 - 1).$$

Letting $H(t)$ be the standard Haar function, we consider the products

$$\Psi^{10}(t_1, t_2) = H(t_1) 1_{[0,1)}(t_2)$$

$$\Psi^{01}(t_1, t_2) = 1_{[0,1)}(t_1) H(t_2)$$

$$\Psi^{11}(t_1, t_2) = H(t_1) H(t_2).$$

It is clear that these functions are orthonormal and orthogonal to the scaling function $\Phi(t_1, t_2)$. We obtain an orthonormal basis of $L^2(\mathbb{R}^2)$ by considering the three sets of functions

$$\{\Psi^{10}(2^j t - k)\}_{j \in \mathbb{Z}, k \in \mathbb{Z}^2}, \{\Psi^{01}(2^j t - k)\}_{j \in \mathbb{Z}, k \in \mathbb{Z}^2}, \{\Psi^{11}(2^j t - k)\}_{j \in \mathbb{Z}, k \in \mathbb{Z}^2}.$$

This is the *two-dimensional Haar wavelet basis*, which is constructed in a canonical manner beginning with the one-dimensional Haar wavelet basis. If we perform the corresponding construction in $\mathbb{R}^d$, we would need a system of $2^d - 1$ basic products. This general construction is described in the following subsection.

6.7.1.1 Tensor product of wavelets

In order to generalize the previous example, we let $(\epsilon_1, \ldots, \epsilon_D)$ be a multiindex, where $\epsilon \in \{0, 1\}$. If $\{\Phi, \Psi\}$ is an MRA wavelet, the symbol Ψ^ϵ is interpreted as Φ for $\epsilon = 0$ and is interpreted as Ψ is $\epsilon = 1$.

Proposition 6.7.1. *Let $\{\Phi_\alpha, \Psi_\alpha\}_{1 \leq \alpha \leq d}$ be a set of one-dimensional MRA wavelet systems. Let V_j be the span of $\Phi_1(2^j t_1 - \gamma)\Phi_2(2^j t_2 - \gamma) \cdots \Phi_d(2^j t_d - \gamma)$ where $\gamma \in \mathbb{Z}^d$ and $j \in \mathbb{Z}$.*
Then $W_j := V_{j+1} \ominus V_j$ is spanned by

$$\{\Psi_1^{\epsilon_1}(2^j t_1 - \gamma) \cdots \Psi_d^{\epsilon_d}(2^j t_d - \gamma)\}$$

where $(\epsilon_1, \ldots, \epsilon_d)$ ranges over all $2^d - 1$ multiindices with $0 < \epsilon_1 + \cdots + \epsilon_d \leq d$.

6.7.2 General Formulation of MRA and Wavelets in $\mathbb{R}^d$

We now abstract these examples to formulate the MRA concept in $\mathbb{R}^d$.

Definition 6.7.2. *A $d \times d$ matrix A is called a dilation matrix if it has integer entries with eigenvalues larger than 1 in absolute value.*

In case $d = 1$, a dilation matrix is defined by an integer $b \in \{2, 3, \ldots\}$. The two-dimensional Haar system is associated to the dilation matrix

$$A = \begin{pmatrix} 2 & 0 \\ 0 & 2 \end{pmatrix}.$$

Definition 6.7.3. *A wavelet set with respect to a dilation matrix A is a set $\Psi^1, \ldots, \Psi^s \in L^2(\mathbb{R}^d)$ so that*

$$\{|\det A|^{j/2} \Psi^r(A^j x - \gamma)\}_{\gamma \in \mathbb{Z}^d, j \in \mathbb{Z}, 1 \leq r \leq s}$$

is an orthonormal basis of $L^2(\mathbb{R}^d)$.

Note that the factor of $|\det A|^{j/2}$ is inserted in order to preserve the norm of the functions Ψ^r.

Definition 6.7.4. *A d-dimensional MRA with respect to a dilation matrix A is an increasing sequence of subspaces $\{V_j\} \subset L^2(\mathbb{R})$ defined for $j \in \mathbb{Z}$ with*

$$\cdots \subset V_{-1} \subset V_0 \subset V_1 \subset \cdots$$

together with a scaling function $\Phi \in L^2(\mathbb{R})$ such that

(i) $\cup_{j=-\infty}^\infty V_j$ is dense in $L^2(\mathbb{R})$, $\cap_{j=-\infty}^\infty V_j = \{0\}$.
(ii) $f \in V_j$ if and only if $f(A^{-j} \cdot) \in V_0$.
(iii) $\{\Phi(x - \gamma)\}_{\gamma \in \mathbb{Z}^d}$ is an orthonormal basis of V_0.

In the previous section we have exhibited two examples of such MRA systems.

6.7.2.1 Notations for subgroups and cosets

If A is a dilation matrix, the image of $\mathbb{Z}^d$ under A is an additive subgroup of $\mathbb{Z}^d$, denoted $A(\mathbb{Z}^d)$. The quotient group $\mathbb{Z}^d/A(\mathbb{Z}^d)$ is obtained by applying the equivalence relation: $v \equiv w$ iff $v - w \in A(\mathbb{Z}^d)$ to the points of $\mathbb{Z}^d$ to obtain a collection of *cosets*. These cosets may be labelled by coset representatives $k_1, \ldots, k_m \in \mathbb{Z}^d$ with $k_1 = 0$. Thus we have the disjoint union

$$(6.7.1) \qquad \mathbb{Z}^d = \bigcup_{i=1}^{m}(k_i + A(\mathbb{Z}^d)) = \bigcup_{i=1}^{m}\bigcup_{\gamma \in \mathbb{Z}^d}(k_i + A(\gamma)).$$

Note that A maps $\mathbb{Z}^d$ to a proper subset of $\mathbb{Z}^d$, but A is invertible on $\mathbb{R}^d$, since $q := |\det A| \geq 1$. In particular, for any two sets $C_1, C_2 \subset \mathbb{R}^d$,

$$A(C_1 \cup C_2) = A(C_1) \cup A(C_2), \qquad A(C_1 \cap C_2) = A(C_1) \cap A(C_2),$$
$$A^{-1}(C_1 \cup C_2) = A^{-1}(C_1) \cup A^{-1}(C_2), \qquad A^{-1}(C_1 \cap C_2) = A^{-1}(C_1) \cap A^{-1}(C_2).$$

Proposition 6.7.5. *The number of elements in the quotient group $\mathbb{Z}^d/A(\mathbb{Z}^d)$ is equal to $|\det A|$.*

This will be proved using the following simple lemma on sets whose integer translates cover $\mathbb{R}^d$.

Lemma 6.7.6. *Suppose that $Q \subset \mathbb{R}^d$ has integer translates that cover $\mathbb{R}^d$: $\cup_{\gamma \in \mathbb{Z}^d}(Q + \gamma) = \mathbb{R}^d$. Then $|Q| \geq 1$ with equality iff these integer translates are a.e. disjoint:*

$$|Q| = 1 \quad \text{iff} \quad |Q \cap (Q + \gamma)| = 0, \quad \forall \gamma \in \mathbb{Z}^d, \quad \gamma \neq 0.$$

Proof. Let $Q_0 := [0, 1]^d$, $f(x) := \sum_{\gamma \in \mathbb{Z}^d} 1_Q(x - \gamma)$, an integer-valued periodic function on $\mathbb{R}^d$ with $f(x) \geq 1$, by hypothesis. Clearly

$$(6.7.2) \qquad |Q| = \int_{\mathbb{R}^d} 1_Q(x)\, dx = \sum_{\gamma \in \mathbb{Z}^d} \int_{Q_0 + \gamma} 1_Q(x)\, dx = \int_{Q_0} f(x)\, dx \geq 1.$$

If $|Q| = 1$, then (6.7.2) shows that $\int_{Q_0}(f(x) - 1)\, dx = 0$, hence $f(x) = 1$ a.e. $x \in \mathbb{R}^d$, hence for a.e. $x \in \mathbb{R}^d$ the series defining $f(x)$ contains only one nonzero term, in particular for any $\gamma \neq 0$ we have a.e. $1 = f(x) \geq 1_Q(x) + 1_Q(x - \gamma)$, which proves that $|Q \cap (Q + \gamma)| = 0$. Conversely, if $|Q \cap (Q + \gamma)| = 0$ for all $\gamma \neq 0$, then the series defining $f(x)$ contains a.e. at most one nonzero term, since $x - \gamma_1 \in Q$ and $x - \gamma_2 \in Q$ contradicts the hypothesis with $\gamma = \gamma_1 - \gamma_2$ and x replaced by $x - \gamma_1$. Thus $f(x) \leq 1$ a.e., which implies that $f(x) = 1$ a.e., hence (6.7.2) implies that $|Q| = 1$. ∎

Proof of Proposition 6.7.5. Using the disjoint union (6.7.1), we can write $\mathbb{R}^d$ as the a.e. disjoint union:

$$\mathbb{R}^d = \bigcup_{v \in \mathbb{Z}^d}(Q_0 + v) = \bigcup_{i=1}^{m}\bigcup_{\gamma \in \mathbb{Z}^d}(Q_0 + k_i + A(\gamma)).$$

Let Q be the a.e. disjoint union $Q := \cup_{i=1}^{m} A^{-1}(Q_0 + k_i)$, a set of measure $|Q| = m |\det(A^{-1})| = m/q$. Then for $\gamma \in \mathbb{Z}^d$ we can write

$$Q + \gamma = \bigcup_{i=1}^{m} A^{-1}(Q_0 + k_i + A(\gamma))$$

$$\bigcup_{\gamma \in \mathbb{Z}^d} (Q + \gamma) = \bigcup_{\gamma \in \mathbb{Z}^d} \bigcup_{i=1}^{m} A^{-1}(Q_0 + k_i + A(\gamma))$$

$$= A^{-1} \left(\bigcup_{\gamma \in \mathbb{Z}^d} \bigcup_{i=1}^{m} (Q_0 + k_i + A(\gamma)) \right)$$

$$= A^{-1}(\mathbb{R}^d)$$

$$= \mathbb{R}^d$$

so that the integer translates of Q cover $\mathbb{R}^d$. But $|Q \cap (Q + \gamma)| = 0$ for $\gamma \neq 0$, since

$$\left| A \left(Q \cap (Q + \gamma) \right) \right| = \left| A(Q) \cap (A(Q) + A(\gamma)) \right|$$

$$= \left| \bigcup_{i=1}^{m} (Q_0 + k_i) \cap \bigcup_{i=1}^{m} (Q_0 + k_i + A(\gamma)) \right|$$

$$= 0.$$

Therefore we can apply Lemma 6.7.6 to conclude that $1 = |Q| = m/q$, which was to be proved. ∎

6.7.2.2 Riesz systems and orthonormal systems in $\mathbb{R}^d$

It is straightforward to generalize the notion of Riesz system to $\mathbb{R}^d$.

Definition 6.7.7. *A set of functions $\{\Phi_k\} \in L^2(\mathbb{R}^d)$ is a Riesz system iff there exist constants $0 < c \leq C < \infty$ such that for any finite set of complex numbers (a_k)*

$$c \sum_k |a_k|^2 \leq \left\| \sum_k a_k \Phi_k \right\|_2^2 \leq C \sum_k |a_k|^2.$$

In particular we have an orthonormal sequence if and only if $c = C = 1$.

Proposition 6.7.8. *Let $F \in L^2(\mathbb{R}^d)$.*

- *$\{F(x - \gamma)\}_{\gamma \in \mathbb{Z}^d}$ is a Riesz system iff $c \leq \sum_{l \in \mathbb{Z}^d} |\hat{F}(\xi - l)|^2 \leq C$ a.e. $\xi \in \mathbb{R}^d$.*
- *The sequence is orthonormal iff $\sum_{l \in \mathbb{Z}^d} |\hat{F}(\xi - l)|^2 = 1$ a.e.*
- *If $G \in L^2(\mathbb{R}^d)$, then the set of translates $\{G(x - \gamma)\}_{\gamma \in \mathbb{R}^d}$ is orthogonal to F if and only if $\sum_{l \in \mathbb{Z}^d} \hat{G}(\xi - l) \hat{F}(\xi - l) = 0$ a.e. $\xi \in \mathbb{R}^d$.*

Proof. The Fourier transform of $\sum_k a_k F(x - k)$ is $\sum_k a_k e^{-2\pi i k \cdot \xi} \hat{F}(\xi) := \mathcal{A}(\xi)\hat{F}(\xi)$. Therefore from the Plancherel theorem and periodization, we have

$$\left\| \sum_k a_k F(x - k) \right\|^2 = \int_{\mathbb{R}^d} |\mathcal{A}(\xi)|^2 |\hat{F}(\xi)|^2 \, d\xi$$

$$= \sum_{l \in \mathbb{Z}^d} \int_{[0,1]^d} |\mathcal{A}(\xi)|^2 |\hat{F}(\xi + l)|^2 \, d\xi$$

$$= \int_{[0,1]^d} |\mathcal{A}(\xi)|^2 \left(\sum_{l \in \mathbb{Z}^d} |\hat{F}(\xi + l)|^2 \right) d\xi.$$

If $c \leq \sum_{l \in \mathbb{Z}^d} |\hat{F}(\xi + l)|^2 \leq C$, then Parseval's theorem for Fourier series yields

$$\sum_k \|a_k F(x - k)\|^2 \leq C \int_{[0,1]^d} |\mathcal{A}(\xi)|^2 \, d\xi = C \sum_{k \in \mathbb{Z}^d} |a_k|^2$$

and

$$\sum_k \|a_k F(x - k)\|^2 \geq c \int_{[0,1]^d} |\mathcal{A}(\xi)|^2 \, d\xi = c \sum_{k \in \mathbb{Z}^d} |a_k|^2,$$

which proves the Riesz property. On the other hand, if $\{F(x - k)\}_{k \in \mathbb{Z}^d}$ is a Riesz system, then for any trigonometric polynomial $\mathcal{A}$, we have

$$c \leq \frac{\int_{[0,1]^d} |\mathcal{A}(\xi)|^2 \left(\sum_{l \in \mathbb{Z}^d} |\hat{F}(\xi + l)|^2 \right) d\xi}{\int_{[0,1]^d} |\mathcal{A}(\xi)|^2 \, d\xi} \leq C.$$

First we take a sequence of trigonometric polynomials that converge boundedly to the indicator function $\Pi_{i=1}^d 1_{[a_i, b_i]}$, then we let $b_i \to a_i$ to conclude that for a.e. $\xi \in \mathbb{R}^d$, $c \leq \sum_{l \in \mathbb{Z}^d} |\hat{F}(\xi + l)|^2 \leq C$. The orthogonality statement is proved by writing

$$\left\| \sum_k a_k \bar{F}(x) G(x - k) \right\|^2 = \int_{\mathbb{R}^d} \mathcal{A}(\xi) \bar{\hat{F}}(\xi) \hat{G}(\xi - l) \, d\xi$$

$$= \sum_{l \in \mathbb{Z}^d} \int_{[0,1]^d} \mathcal{A}(\xi) \bar{\hat{F}}(\xi) \hat{G}(\xi - l) \, d\xi$$

$$= \int_{[0,1]^d} \mathcal{A}(\xi) \left(\sum_{l \in \mathbb{Z}^d} \bar{\hat{F}}(\xi) \hat{G}(\xi - l) \right) d\xi.$$

The inner sum is a.e. zero if and only if all inner products on the left are zero. ∎

6.7.2.3 *Scaling equation and structure constants*
Since V_1 is spanned by translates of $\{\Phi(Ax - \gamma)\}_{\gamma \in \mathbb{Z}^d}$, there exist constants a_γ so that we have the L^2 convergent sum

(6.7.3) $$\boxed{\Phi(x) = \sum_{\gamma \in \mathbb{Z}^d} a_\gamma \Phi(Ax - \gamma)}$$

where the *structure constants* satisfy $\sum_{\gamma \in \mathbb{Z}^d} |a_\gamma|^2 < \infty$. Relation (6.7.3) is the *scaling equation* for wavelets in $\mathbb{R}^d$ and plays the same role here as for $d = 1$. The scaling filter is defined by

(6.7.4)
$$m_0(\xi) := \frac{1}{|\det A|} \sum_{\gamma \in \mathbb{Z}^d} a_\gamma e^{-2\pi i \xi \cdot \gamma}.$$

The Fourier transformed equation, proved below, is written

(6.7.5)
$$\hat{\Phi}(\xi) = m_0((A^{-1})^\star \xi)\hat{\Phi}((A^{-1})^\star \xi).$$

Then we have the counterpart of Theorem 6.4.27.

Theorem 6.7.9. *Suppose that $\Phi \in L^2(\mathbb{R})$ is such that*

  (i) *The translates $\{\Phi(x - \gamma)\}_{\gamma \in \mathbb{Z}^d}$ are orthonormal.*
  (ii) *$\Phi(x) = \sum_{\gamma \in \mathbb{Z}^d} a_\gamma \Phi(Ax - \gamma)$, an L^2-convergent sum.*
  (iii) *The Fourier transform $\hat{\Phi}(\xi)$ is continuous at $\xi = 0$ with $|\hat{\Phi}(0)| = 1$. Define $V_j = \text{span}\{\Phi(A^j x - \gamma)\}_{\gamma \in \mathbb{Z}^d}$. Then $\{V_j\}$ defines an MRA system with respect to the dilation matrix A.*

The proof follows the one-dimensional development and is left to the following exercises. As before, the MRA property reduces to the behavior of the projection operators P_j, which satisfy the uniform boundedness property in L^2: $\|P_j\|_{2,2} = 1$.

Exercise 6.7.10. *For any $f \in L^2(\mathbb{R})$, $\lim_{j \to -\infty} P_j f = 0$.*

Hint: First prove this on the dense set of continuous function with compact support.

Exercise 6.7.11. *Let $f \in L^2(\mathbb{R}^d)$ with a Fourier transform $\hat{f}$ that is bounded and supported in $[-R, R]^d$ for some $R > 0$. Then for all j sufficiently large, we have*

(6.7.6)
$$\|P_j f\|^2 = \int_{[-R,R]^d} |\hat{f}(\xi)|^2 |\Phi(A^{-j}\xi)|^2 \, d\xi.$$

Hint: Imitate the proof of Lemma 6.4.29.

When we combine the continuity hypothesis (iii) with the identity in Exercise 6.7.11, we see that for a dense class of $f \in L^2(\mathbb{R}^d)$, $\lim_{j \to \infty} \|P_j f\|_2 = \|f\|_2$, which implies that $P_j f \to f$ for all $f \in L^2(\mathbb{R}^d)$.

6.7.2.4 Existence of the wavelet set

It now remains to construct the wavelet system $\{\Psi^1, \ldots, \Psi^s\}$ as previously announced. To do this, we first need to describe the orthogonal complement $W_1 = V_1 \ominus V_0$.

Lemma 6.7.12. *Suppose that Φ is the scaling function of an MRA with respect to the dilation matrix A. Let $f \in V_1$. Then the Fourier transform satisfies*

(6.7.7) $$\boxed{\hat{f}(A^*\xi) = m_f(\xi)\Phi(\xi)}$$

where $m_f \in L^2(\mathbb{R}^d/\mathbb{Z}^d)$ with

(6.7.8) $$\int_{[0,1]^d} |m_f(\xi)|^2 \, d\xi = \frac{\|f\|_2^2}{|\det A|} = \frac{\sum_{\gamma \in \mathbb{Z}^d} |a_\gamma|^2}{|\det A|^2}.$$

Proof. Any $f \in V_1$ has the L^2 convergent expansion

(6.7.9) $$f(x) = \sum_{\gamma \in \mathbb{Z}^d} a_\gamma \Phi(Ax - \gamma).$$

We must compute the Fourier transform of $\Phi(Ax - \gamma)$. We make the change of variable $y = Ax - \gamma$ to write

$$\int_{\mathbb{R}^d} \Phi(Ax - \gamma) e^{-2\pi i \xi \cdot x} \, dx = \frac{1}{|\det A|} \int_{\mathbb{R}^d} \Phi(y) e^{-2\pi i \xi \cdot A^{-1}(y+\gamma)} \, dy$$

$$= e^{-2\pi i \xi \cdot A^{-1}(\gamma)} \int_{\mathbb{R}^d} \Phi(y) e^{-2\pi i \xi \cdot A^{-1}(y)} \, d\xi$$

$$= e^{-2\pi i \xi \cdot A^{-1}(\gamma)} \hat{\Phi}((A^{-1})^*\xi).$$

We define

(6.7.10) $$m_f(\xi) = \frac{1}{|\det A|} \sum_{\gamma \in \mathbb{Z}^d} a_k e^{-2\pi i \gamma \cdot \xi}.$$

Taking the Fourier transform of (6.7.9), we have

(6.7.11) $$\hat{f}(\xi) = m_f((A^{-1})^*\xi) \hat{\Phi}((A^{-1})^*\xi).$$

Letting $\xi = A^*\eta$, we obtain (6.7.7) as written.

Now we compute the L^2 norm in two different ways. From (6.7.9) we have

$$\int_{\mathbb{R}^d} |f(x)|^2 \, dx = \sum_{\gamma \in \mathbb{Z}^d} |a_\gamma|^2 \int_{\mathbb{R}^d} |\Phi(Ax - \gamma)|^2 \, dx$$

$$= \frac{1}{|\det A|} \sum_{\gamma \in \mathbb{Z}^d} |a_\gamma|^2 = |\det A| \int_{[0,1]^d} |m_f(\xi)|^2 \, d\xi.$$

On the other hand, from (6.7.11), we have

$$\int_{\mathbb{R}^d} |\hat{f}(A^*\xi)|^2 \, d\xi = \int_{\mathbb{R}^d} |m_f(\xi)|^2 |\hat{\Phi}(\xi)|^2 \, d\xi$$

$$= \sum_{\gamma \in \mathbb{Z}^d} \int_{[0,1]^d} |m_f(\xi)|^2 |\hat{\Phi}(\xi + \gamma)|^2 \, d\xi$$

$$= \int_{[0,1]^d} |m_f(\xi)|^2 \, d\xi$$

where we have used the orthonormality of $\{\Phi(x - \gamma)\}$ in the last step. Combining these two calculations, we have

$$\int_{[0,1]^d} |m_f(\xi)|^2 \, d\xi = \int_{\mathbb{R}^d} |\hat{f}(A^*\xi)|^2 \, d\xi$$

$$= \frac{1}{|\det A|} \int_{\mathbb{R}^d} |f(x)|^2 \, dx$$

$$= \frac{\sum_{\gamma \in \mathbb{Z}^d} |a_\gamma|^2}{|\det A|^2}.$$

∎

The orthonormality properties of Φ and the wavelet basis are translated into unitary properties of the functions m_f, as follows. We rewrite the scaling equation (6.7.5) as

(6.7.12) $$\hat{\Phi}(A^*\xi) = m_0(\xi)\hat{\Phi}(\xi).$$

We denote by $\Gamma_1, \ldots, \Gamma_q$ a set of representatives for $A^*(\mathbb{Z}^d)$ and their antecedents by $k_1, \ldots, k_q$, defined by $A^*(k_i) = \Gamma_i$. The following identity is useful for dealing with orthogonality.

Lemma 6.7.13. *Let Φ be the scaling function of an MRA, with scaling filter m_0. Let $f, g \in V_1$ with $\hat{f}(A^*\xi) = m_f(\xi)\hat{\Phi}(\xi)$, $\hat{g}(A^*\xi) = m_g(\xi)\hat{\Phi}(\xi)$. Then*

(6.7.13) $$\sum_{l \in \mathbb{Z}^d} \hat{f}(A^*\xi + l)\overline{\hat{g}(A^*\xi + l)} = \sum_{i=1}^{q} m_f(\xi + k_i)\bar{m}_g(\xi + k_i).$$

Proof. We break up the sum (6.7.13) according to the cosets defined by $A^*(\mathbb{Z}^d)$. Each point of $\mathbb{Z}^d$ is uniquely represented as $l = A^*\gamma + \Gamma_i$, where $1 \leq i \leq q$ and $\gamma \in \mathbb{Z}^d$. Thus

$$\hat{f}(A^*\xi + A^*\gamma + \Gamma_i)\overline{\hat{g}(A^*\xi + A^*\gamma + \Gamma_i)} = m_f(\xi + \gamma + k_i)\bar{m}_g(\xi + \gamma + k_i)|\hat{\Phi}(\xi + \gamma + k_i)|^2$$

$$= m_f(\xi + k_i)\bar{m}_g(\xi + k_i)|\hat{\Phi}(\xi + \gamma + k_i)|^2$$

$$\sum_{l \in \mathbb{Z}^d} \hat{f}(A^*\xi + l)\overline{\hat{g}(A^*\xi + l)} = \sum_{i=1}^{q} \sum_{\gamma \in \mathbb{Z}^d} m_f(\xi + k_i)\bar{m}_g(\xi + k_i)|\hat{\Phi}(\xi + \gamma + k_i)|^2$$

$$= \sum_{i=1}^{q} m_f(\xi + k_i)\bar{m}_g(\xi + k_i) \sum_{\gamma \in \mathbb{Z}^d} |\hat{\Phi}(\xi + \gamma + k_i)|^2$$

$$= \sum_{i=1}^{q} m_f(\xi + k_i)\bar{m}_g(\xi + k_i)$$

where we have used the orthonormality of $\Phi(x - k)$ in the last step. ∎

This identity is used repeatedly in what follows.

Proposition 6.7.14. *Let Φ be the scaling function of an MRA with scaling filter m_0. Then the scaling filter satisfies the condition that for a.e. $\xi \in \mathbb{R}^d$, we have*

(6.7.14) $$\sum_{r=0}^{q-1} |m_0(\xi + k_r)|^2 = 1.$$

If $m_1, \ldots, m_{q-1}$ correspond to the wavelet basis $\{\Psi^1, \ldots, \Psi^{q-1}\}$, then the row vectors $(m_i(\xi + k_1), \ldots, m_i(\xi + k_q))$ are mutually orthogonal unit vectors in $\mathcal{C}^q$ for $i = 1, \ldots, q-1$ and are orthogonal to the vector $(m_0(\xi + k_1), \ldots, m_0(\xi + k_q))$. Conversely, if we have 1-periodic functions $m_1(\xi), \ldots, m_{q-1}(\xi)$ with the aforementioned orthogonality properties, then defining $f_1, \ldots, f_{q-1}$ by (6.7.7), we obtain a wavelet set.

Hence a d-dimensional wavelet is described by a $q \times q$ unitary matrix just as in the one-dimensional case, where $q = 2$.

Proof. We take $f = g = \Phi$ in (6.7.13), noting that the left side is a.e. equal to 1, by orthonormality, hence (6.7.14) follows. Now if we have a wavelet set $\Psi^1, \ldots, \Psi^{q-1}$, then we take $f = \Psi^i, g = \Phi$ to deduce the orthogonality of the first row with the other row vectors. Taking $f = \Psi^i, g = \Psi^j$ proves the mutual orthonormality of the remaining row vectors.
    Conversely, if we are given 1-periodic functions $m_1, \ldots, m_{q-1}$ with the aforementioned orthogonality properties, we apply (6.7.13) repeatedly to deduce that the functions $\{f_i(x - k)\}_{1 \leq i \leq q-1, k \in \mathbb{Z}^d}$ constitute an orthonormal family. ∎

Remark. It is noteworthy that the number of wavelets is equal to $q = |\det A|$, independent of any details of the scaling equation.

6.7.2.5 Proof that the wavelet set spans $V_1 \ominus V_0$

Proof. It remains to prove that any $f \in V_1 \ominus V_0$ is an L^2 convergent sum of linear combinations of integer translates of $\Psi^1, \ldots, \Psi^{q-1}$. Since V_1 is spanned by $\{\Phi(Ax - \gamma)\}_{\gamma \in \mathbb{Z}^d}$, it is sufficient to prove that functions of this form are equal to their projection on the integer translates of $\Phi, \Psi^1, \ldots, \Psi^{q-1}$. The following computation is carried out in the case $\gamma = 0$.

Using the notation $B = A^*$, the projection of $\Phi(Ax)$ is written

$$\sum_{k \in \mathbb{Z}^d} c_k \Phi(x - k) + \sum_{k \in \mathbb{Z}^d} d_k^1 \Psi^1(x - k) + \cdots + \sum_{k \in \mathbb{Z}^d} d_k^{q-1} \Psi^{q-1}(x - k).$$

Applying Parseval's identity, the scaling equation and periodization, we obtain

$$c_k = \int_{\mathbb{R}^d} \Phi(Ax) \bar{\Phi}(x - k)\, dx$$

$$= \frac{1}{q} \int_{\mathbb{R}^d} \hat{\Phi}(B^{-1}\xi) \bar{\hat{\Phi}}(\xi) e^{2\pi i k \cdot \xi}\, d\xi$$

$$= \frac{1}{q} \int_{\mathbb{R}^d} \bar{m}_0(B^{-1}\xi) |\hat{\Phi}(B^{-1}\xi)|^2 e^{2\pi i k \cdot \xi}\, d\xi$$

$$= \int_{\mathbb{R}^d} \bar{m}_0(\eta) |\hat{\Phi}(\eta)|^2 e^{2\pi i k \cdot B\eta}\, d\eta$$

$$= \int_{[0,1]^d} \bar{m}_0(\eta) e^{2\pi i Ak \cdot \eta}\, d\eta$$

$$= \frac{1}{q} \int_{B[0,1]^d} \bar{m}_0(B^{-1}\xi) e^{2\pi i \xi \cdot k}\, d\xi.$$

But $B[0, 1]^d$ is the disjoint union of q sets each of which is congruent to $[0, 1]^d$. Hence we can write

(6.7.15) $$c_k = \frac{1}{q} \int_{[0,1]^d} \left(\sum_{\alpha=0}^{q-1} \bar{m}_0(B^{-1}\xi + k_\alpha) \right) e^{2\pi i k \cdot \xi} d\xi$$

so that by Parseval's identity we have

(6.7.16) $$\sum_{k \in \mathbb{Z}^d} |c_k|^2 = \frac{1}{q^2} \int_{[0,1]^d} \left| \sum_{\alpha=0}^{q-1} \bar{m}_0(B^{-1}\xi + k_\alpha) \right|^2 d\xi.$$

We compute the other sums in the same fashion, replacing m_0 by m_β, $1 \leq \beta \leq q - 1$, thus obtaining

(6.7.17) $$\sum_{k \in \mathbb{Z}^d} |d_k^\beta|^2 = \frac{1}{q^2} \int_{[0,1]^d} \left| \sum_{\alpha=0}^{q-1} \bar{m}_\beta(B^{-1}\xi + k_\alpha) \right|^2 d\xi, \quad 1 \leq \beta \leq q - 1.$$

When we add the terms of (6.7.16) to those of (6.7.17) and apply the unitary properties of the matrix $m_\beta(\xi + k_\alpha)$, we find q terms, each equal to $1/q^2$, hence the sum is $1/q$. On the other hand, $\int_{\mathbb{R}^d} |\Phi(Ax)|^2 dx = 1/q$, and the proof is complete. ∎

Exercise 6.7.15. *Carry out the computaton for $\Phi(Ax - \gamma)$ with $\gamma \neq 0$.*

6.7.2.6 Cohen's theorem in $\mathbb{R}^d$

In order to construct a scaling function Φ from the scaling filter m_0, we formulate the extension of Cohen's Theorem 6.5.19 to the multidimensional case. As before, the main task is to prove that the infinite product $\prod_{j=1}^\infty m_0((A^*)^{-j}\xi)$ has orthonormal translates. We restrict attention to the case of trigonometric polynomials, which is equivalent to scaling functions of compact support.

Theorem 6.7.16. *Let A be a dilation matrix and suppose that m_0 is a trigonometric polynomial that satisfies $m_0(0) = 1$ and $\sum_{r=0}^{q-1} |m_0(\xi + k_r)|^2 = 1$. Suppose that there exists a set $K \subset \mathbb{R}^d$ with the properties that K contains a neighborhood of 0, $\sum_{\gamma \in \mathbb{Z}^d} 1_K(\xi + \gamma) = 1$ a.e. and $m_0((A^*)^{-j}\xi) \neq 0$ for $j = 1, 2, \ldots$. Define $\hat{\Phi}_0(\xi) = 1_K(\xi)$ and $\hat{\Phi}_j(\xi) = m_0((A^*)^{-1}\xi)\hat{\Phi}_{j-1}((A^*)^{-1}\xi)$. Then Φ_j converges in $L^2(\mathbb{R}^d)$ to Φ, which is the scaling function of an MRA.*

The proof follows exactly the steps 1 through 4 in the proof of Theorem 6.5.2, as modified in the proof of Cohen's theorem in one dimension. The details are left to the reader.

Exercise 6.7.17. *Prove that the scaling function Φ has compact support if and only if the scaling filter m_0 is a trigonometric polynomial.*

6.7.3 Examples of Wavelets in $\mathbb{R}^d$

We can construct an interesting class of d-dimensional wavelets by considering scaling functions that are the indicator functions of suitable measurable sets. If $\Phi = 1_Q$ is a scaling function, then orthonormality requires that (i) the integer translates of Q be a.e.

disjoint, and (ii) $|Q| = 1$. These imply that $f_Q(x) := \sum_{\gamma \in \mathbb{Z}^d} 1_Q(x + \gamma) = 1$ a.e., since (i) implies that $f_Q(x) \leq 1$ a.e. and (ii) implies that $1 = |Q| = \int_{[0,1]^d} f_Q(x)\, dx$. Hence $0 = \int_{[0,1]^d} (1 - f_Q(x))\, dx$, which proves that $f_Q(x) = 1$ a.e. Now the scaling relation is written

$$(6.7.18) \qquad 1_Q(x) = \sum_{\gamma \in \mathbb{Z}^d} a_\gamma 1_Q(Ax - \gamma).$$

Conversely, if we are given a set Q for which $\sum_{\gamma \in \mathbb{Z}^d} 1_Q(x+\gamma) = 1$ a.e. and which satisfies a scaling relation (6.7.18), then we obtain an MRA wavelet, since $1_Q \in L^1(\mathbb{R}^d)$ implies that Φ is continuous at $\xi = 0$ with $\hat{\Phi}(0) = 1$.

Exercise 6.7.18. *If $\Phi = 1_Q$ is a scaling function, prove that $a_\gamma = 0$ for all except a finite number, where $a_\gamma = 1$. Hence the set Q is identical to a finite sum of translates of an A-similar copy of itself.*

A large class of sets Q satisfying (6.7.18) are generated in the following manner: Given a dilation matrix A, let $S = \{\Gamma_1, \ldots, \Gamma_q\}$ be a set of representatives of the cosets $\Gamma_i + A(\mathbb{Z}^d)$, $1 \leq i \leq q$. The set Q is defined by

$$(6.7.19) \qquad Q = \left\{ x \in \mathbb{R}^d : x = \sum_{j=1}^{\infty} A^{-j}(s_j) \right\}$$

where $s_j \in S$. The series is clearly convergent, since we have the estimate $|A^{-j}x| \leq C\alpha^{-j}|x|$ for any dilation matrix, where $0 < \alpha < 1$ and $C > 0$. Furthermore Q satisfies a scaling equation, since for any $x \in Q$

$$x = A^{-1}(s_1) + A^{-1} \sum_{j=1}^{\infty} A^{-j}(s_{j+1}) = A^{-1}(s_1) + A^{-1}y$$

with $y \in Q$. This means that $x \in Q$ if and only if for some i, we have $Ax - \Gamma_i \in Q$ for some $i = 1, \ldots, q$. Thus $1_Q(x) = \sum_{i=1}^{q} 1_Q(Ax - \Gamma_i)$. The scaling relation (6.7.18) is satisfied with $a_\gamma = 1$ if $\gamma = \Gamma_i$ for some i and $a_\gamma = 0$ otherwise.

In order to prove the orthonormality, we appeal to Theorem 6.7.16. This requires that we study the properties of the scaling filter

$$(6.7.20) \qquad m_0(\xi) = \frac{1}{q} \sum_{i=1}^{q} e^{-2\pi i \xi \cdot k_i}.$$

If we can find a set K to satisfy the conditions of Theorem 6.7.16, then we can assert that 1_Q is the scaling function of an MRA wavelet. The details will depend on the choice of the representatives $k_1, \ldots, k_q$.

Example 6.7.19. *If*

$$A = \begin{pmatrix} 2 & 0 \\ 0 & 2 \end{pmatrix}$$

we choose $k_1 = (0, 0), k_2 = (1, 0), k_3 = (0, 1), k_4(1, 1)$ as a set of representatives.

Then
$$m_0(\xi) = m_0(\xi_1, \xi_2)$$
$$= \frac{1}{4}(1 + e^{-2\pi i\xi_1} + e^{-2\pi i\xi_2} + e^{-2\pi i(\xi_1+\xi_2)})$$
$$= \frac{1}{4}(1 + e^{-2\pi i\xi_1})(1 + e^{-2\pi i\xi_2}).$$

If we choose $K = [-\frac{1}{2}, \frac{1}{2}]^2$, then $m_0(\xi) \neq 0$ on K, so that the conditions of Theorem 6.7.2 are satisfied. The set Q is simply the set of binary expansions of pairs of real numbers in $[0, 1]$, hence $Q = [0, 1]^2$, so that we obtain the two-dimensional Haar wavelet, discussed previously.

Example 6.7.20. *With the same choice of A, let* $k_1 = (0, 0), k_2 = (1, 1), k_3 = (0, 1), k_4 = (1, 2)$.

In this case the scaling filter is
$$m_0(\xi) = \frac{1}{4}(1 + e^{-2\pi i(\xi_1+\xi_2)} + e^{-2\pi i\xi_2} + e^{-2\pi i(\xi_1+\xi_2)}).$$

Letting $K = [-\frac{1}{2}, \frac{1}{2}]^2$, we again infer that the $m_0(\xi) \neq 0$ on K. In this case the set Q is the rhombus obtained as the convex hull of the four points k_1, k_2, k_3, k_4.

REFERENCES

R. Banuelos and C. N. Moore. 1999. *Probabilistic Behavior of Harmonic Functions*. Basel: Birkhauser.

W. Beckner. 1975. Inequalities in Fourier analysis on $\mathbb{R}^n$. *Proceedings of the National Academy of Sciences* 72: 638–641.

M. Benedicks. 1985. On Fourier transforms of functions supported on sets of finite Lebesgue measure. *Journal of Mathematical Analysis and Applications* 106: 180–183.

D. Bernoulli. 1753. Reflexions et eclairements sur les nouvelles vibrations des cordes. *Mem de l'Academie des Sciences de Berlin* 9: 173ff.

S. N. Bernstein. 1912. On the best approximation of continuous functions by polynomials of a given degree. *Mem. Acad. Roy. Belgique* 4: 1–104.

P. Billingsley. 1999. *Convergence of Probability Measures*. New York: Wiley.

R. P. Boas. 1972. Summation formulas and band-limited signals. *Tohoku Journal of Mathematics* 24: 121–125.

S. Bochner. 1931. Ein konvergenzsatz für mehrvariablige Fouriersche Integrale. *Math Zeit* 34: 440–447.

S. Bochner. 1936. Summation of multiple Fourier series by spherical means. *Transactions of the American Mathematical Society* 40: 175–207.

L. Brandolini and L. Colzani. 1999. Localization and convergence of eigenfunction expansions. *Journal of Fourier Analysis and Applications* 5: 431–447.

A. Calderón and A. Zygmund. 1952. On the existence of certain singular integrals. *Acta Mathematica* 8: 85–139.

A. Calderón and A. Zygmund. 1956. On singular integrals. *American Journal of Mathematics* 18: 289–309.

L. Carleson. 1966. On convergence and growth of partial sums of Fourier series. *Acta Mathematica* 116: 135–157.

I. Daubechies. 1992. *Ten Lectures on Wavelets*. Vol. 61. CBS-NSF Regional Conference in Applied Mathematics, SIAM.

B. Davis. 1974. On the weak type (1, 1) inequality for conjugate functions. *Proceedings of the AMS* 44: 307–311.

M. Davis and S. Y. Chang. 1987. *Lectures on Bochner-Riesz Means*. LMS Lecture Notes No. 114. Cambridge: Cambridge University Press.

P. G. Dirichlet. 1829. Sur la convergence des séries trigonometriques qui servent a representer une fonction arbitraire entre des limites données. *J. Reine und Angewand Math.* 4: 157–169.
H. Dym and H. P. McKean. 1972. *Fourier Series and Integrals.* San Diego: Academic Press.
C. Fefferman. 1970. Inequalities for strongly singular convolution operators. *Acta Mathematica* 124: 9–36.
C. Fefferman. 1973. Pointwise convergence of Fourier series. *Annals of Mathematics* 98: 551–571.
G. B. Folland. 1995. *Introduction to Partial Differential Equations.* Princeton, NJ: Princeton University Press.
G. B. Folland and A. Sitaram. 1997. The uncertainty principle: A mathematical survey. *Journal of Fourier Analysis and Applications* 3: 207–238.
G. B. Folland. 1999. *Real Analysis.* New York: John Wiley.
J. B. Fourier. 1822. *Théorie Analytique de la Chaleur.* Paris: Dover reprint, 1955, New York.
L. Garding. 1997. *Some points of analysis and their history.* Providence, RI: American Mathematical Society.
F. Grunbaum. 1990. Trying to beat Heisenberg. In C. Sadosky (ed.), *Analysis and Partial Differential Equations.* New York: Marcel Dekker, pp 657–665.
G. H. Hardy. 1949. *Divergent Series.* Oxford: Clarendon Press.
E. Hernández and G. Weiss. 1996. *A First Course on Wavelets.* Boca Raton, FL: CRC Press.
R. Hunt. 1968. On the convergence of Fourier series. In *Proceedings of the Conference on Orthogonal Expansion and their Continuous Analogues.* Edwardsville, IL: Southern Illinois University Press, 234–255.
S. Igari. 1996. *Real Analysis—With an Introduction to Wavelet Theory.* Vol. 177. Providence, RI: AMS Translations of Mathematical Monographs.
C. Jordan. 1881. Sur la série de Fourier. *Comptes Rendus de l'Academie des Sciences, Paris* 92: 228–230.
M. Kac. 1939. Power series with large gaps. *American Journal of Mathematics* 61: 473–476.
J. P. Kahane and Y. Katznelson. 1966. Sur les ensembles de divergence des séries trigonométriques. *Studia Mathematica* 26: 305–306.
J. P. Kahane. 1968. *Some Random Series of Functions.* Lexington, MA: D.C. Heath.
J. P. Kahane. 1995. Le phénomène de Pinsky et la géométrie des surfaces. *Comptes Rendus de l'Academie des Sciences, Paris* 321: 1027–1029.
J. P. Kahane and P. G. Lemarié Rieusset. 1995. *Fourier Series and Wavelets.* Vol. 3. In *Studies in the Development of Modern Mathematics.* New York: Gordon Breach.
Y. Katznelson. 1976. *An Introduction to Harmonic Analysis.* New York: Dover Reprint.
D. G. Kendall. 1948. On the number of lattice points inside a random oval. *Quarterly Journal of Mathematics*, Oxford Series 19: 1–26.
A. Kolmogorov. 1935. Sur les fonctions harmoniques conjugées et les séries de Fourier. *Fundamenta Mathematica* 7: 23–28.
A. Kolmogorov. 1926. Une série de Fourier-Lebesgue divergente partout. *Comptes Rendus de l'Academie des Sciences de Paris* 183: 1327–1328.
T. Körner. 1989. *Fourier Analysis.* Cambridge: Cambridge University Press.
P. Lévy. 1948. *Processus stochastiques et mouvement brownien.* Paris: Gauthiers-Villars.
J. C. Maxwell. 1860. Illustrations of the dynamical theory of gases. *Philosophical Magazine and Journal of Science* 19: 19–32.
Y. Meyer. *Ondlettes.* Hermann; translated as *Wavelets and Operators.* Cambridge, UK: Cambridge University Press, 1990.
M. Pinsky. 1994. Pointwise Fourier inversion and related eigenfunction expansions. *Communications in Pure and Applied Mathematics* 47: 653–681.
M. Pinsky and M. Taylor. 1997. Pointwise Fourier inversion—a wave equation approach. *Journal of Fourier Analysis and Applications* 3: 647–703.

M. Pinsky, N. Stanton, and P. Trapa. 1993. Fourier series of radial functions in several variables. *Journal of Functional Analysis* 116: 111–132.

M. Riesz. 1927. Sur les fonctions conjugées. *Mathematische Zeitschrift* 27: 218–244.

W. Rudin. 1987. *Real and Complex Analysis*. New York: McGraw-Hill.

R. Salem and A. Zygmund. 1948. On lacunary trignometric series II. *Proceedings of the National Academy of Sciences* 33: 54–62.

K. T. Smith. 1956. A generalization of an inequality of Hardy and Littlewood. *Canadian Journal of Mathematics* 8: 157–170.

C. D. Sogge. 1993. *Fourier Integrals in Classical Analysis*. Cambridge, UK: Cambridge University Press.

E. M. Stein. 1976. Maximal functions: Spherical means. *Proceedings of the National Academy of Sciences USA* 73: 2174–2175.

E. M. Stein. 1970. *Singular Integrals and Differentiability Properties of Functions*. Princeton, NJ: Princeton University Press.

E. M. Stein. 1993. *Harmonic Analysis, Real-Variable Methods, Orthogonality and Oscillatory Integrals*. Princeton, NJ: Princeton University Press.

E. M. Stein and G. Weiss. 1971. *Introduction to Fourier Analysis on Euclidean Spaces*. Princeton, NJ: Princeton University Press.

R. Strichartz. 1994. Construction of orthonormal wavelets. In J. J. Benedetto and J. W. Frazer (eds.), *Wavelets, Mathematics and Applications*. Boca Raton, FL: CRC Press.

R. Strichartz. 2000. Gibbs' phenomenon and arclength. *Journal of Fourier Analysis and Applications* 6: 533–536.

D. W. Stroock. 1993. *Probability Theory*. Cambridge, UK: Cambridge University Press.

M. E. Taylor. 1996. *Partial Differential Equations*, 3-volume work. New York: Springer Verlag.

M. E. Taylor. 1999. *Schrödinger Equation and Gauss Sums*. Preprint by University of North Carolina, Chapel Hill.

E. C. Titchmarsh. 1986. *Introduction to the Theory of Fourier Integrals*, 3rd ed. New York: Chelsea.

P. Tomas. 1975. A restriction theorem for the Fourier transform. *Bulletin of the American Mathematical Society* 81: 477–478.

G. Watson. 1944. *A Treatise on the Theory of Bessel Functions*. Cambridge, UK: Cambridge University Press.

N. Wiener. 1933. *The Fourier Integral and Certain of Its Applications*. Cambridge, UK: Cambridge University Press.

P. Wojtaszczyk. 1997. *A Mathematical Introduction to Wavelets*. Cambridge, UK: Cambridge University Press, LMS Student Texts No. 37.

A. Zygmund. 1959. *Trigonometric Series*. 2nd ed., Cambridge, UK: Cambridge University Press.

A. Zygmund. 1974. On Fourier coefficients and transforms of functions of two variables. *Studia Mathematica* 50: 189–201.

NOTATIONS

$A \setminus B$ is the difference of the two sets, defined as $\{x \in A : x \notin B\}$

$A \triangle B$ is the symmetric difference of the two sets, defined as $(A \setminus B) \cup (B \setminus A)$

1_A is the indicator function of the set A, defined by $1_A(n) = 1$ if $n \in A$ and $1_A(n) = 0$ if $n \notin A$

"iff" means "if and only if"

$\mathbb{R} = \{x : -\infty < x < \infty\}$, the real number system

$\mathbb{R}^+ = \{x \in \mathbb{R} : 0 \leq x < \infty\}$, the nonnegative real numbers

$\mathbb{R}^n = \{(x_1, \ldots, x_n) : x_i \in \mathbb{R} \text{ for } 1 \leq i \leq n\}$, the n-dimensional Euclidean space

$\mathbb{Z} = \{0, \pm 1, \pm 2, \ldots\}$, the integers

$\mathbb{Z}^+ = \{0, 1, 2, \ldots\}$, the nonnegative integers

$\sum_{n \in \mathbb{Z}} a_n$ is the symmetric infinite sum, defined as $\lim_{N \to \infty} \sum_{n=-N}^{N} a_n$

$\mathbb{T} = \mathbb{R}/2\pi\mathbb{Z}$, the circle, identified with $(-\pi, \pi]$

$C(\mathbb{T})$ = the space of complex-valued continuous functions on $\mathbb{T}$, identified with continuous 2π-periodic functions on $\mathbb{R}$

$L^p(\mathbb{T})$ is the space of complex-valued measurable functions on $\mathbb{T}$ with $\|f\|_p := \left((2\pi)^{-1} \int_{\mathbb{T}} |f(\theta)|^p \, d\theta\right)^{1/p} < \infty$, where $1 \leq p < \infty$

$L^\infty(\mathbb{T})$ is the space of complex-valued measurable functions on $\mathbb{T}$ with $\|f\|_\infty := \operatorname{esssup}_{\theta \in \mathbb{T}} |f(\theta)| < \infty$

$L^1_{\text{loc}}(\mathbb{R})$ is the space of complex-valued measurable functions on $\mathbb{R}$ with $\int_{-M}^{M} |f(x)| \, dx < \infty$ for each $M > 0$

$\operatorname{Var}(f)$ is the total variation of the complex-valued measurable function f on $\mathbb{T}$, defined as the supremum of $\sum_i |f(x_{i+1}) - f(x_i)|$ taken over all finite partitions of $\mathbb{T}$

f_h is the translate of $f \in L^1(\mathbb{T})$, defined by $f_h(\theta) = f(\theta - h)$

$(f * g)(\theta) = (1/2\pi) \int_{\mathbb{T}} f(\phi) g(\theta - \phi) \, d\phi$, the convolution of $f, g \in L^1(\mathbb{T})$

369

$\langle f, g \rangle = \frac{1}{2\pi} \int_{\mathbb{T}} f(\theta) \bar{g}(\theta) \, d\theta$, the inner product of $f, g \in L^1(\mathbb{T})$.

$P_r(\theta) = \dfrac{1 - r^2}{1 + r^2 - 2r \cos \theta}$, the Poisson kernel, defined for $0 \leq r < 1, \theta \in \mathbb{T}$

$Q_r(\theta) = \dfrac{2r \sin \theta}{1 + r^2 - 2r \cos \theta}$, the conjugate Poisson kernel, defined for $0 \leq r < 1, \theta \in \mathbb{T}$

$\omega(f; h) = \sup_{\theta \in \mathbb{T}, |y| \leq h} |f(\theta + y) - f(\theta)|$ is the modulus of continuity of $f \in C(\mathbb{T})$

$\Omega_p(f; h) = \sup_{|y| \leq h} \left(\frac{1}{2\pi} \int_{\mathbb{T}} |f(\theta + y) - f(\theta)|^p \, d\theta \right)^{1/p}$, the L^p modulus of continuity of $f \in L^p(\mathbb{T})$

$\hat{f}(n) = \dfrac{1}{2\pi} \int_{\mathbb{T}} f(\theta) e^{-in\theta} \, d\theta$ is the nth Fourier coefficient of $f \in L^1(\mathbb{T})$.

$D_N(\theta) = \dfrac{\sin(N + \frac{1}{2})\theta}{\sin \frac{\theta}{2}}$ is the Dirichlet kernel, defined for $N \in \mathbb{Z}^+$ and $0 \neq \theta \in \mathbb{T}$

$L_n = \frac{1}{2\pi} \int_{\mathbb{T}} |D_n(\phi)| \, d\phi$ is the nth Lebesgue constant

$\text{Si}(x) = \frac{2}{\pi} \int_0^x \sin t / t \, dt$ is the Sine Integral function, defined for $x \in \mathbb{R}^+$

$S_N f(\theta) = \sum_{n=-N}^{N} \hat{f}(n) e^{in\theta}$ is the Nth symmetric partial sum of the Fourier series of $f \in L^1(\mathbb{T})$

$\sigma_N f(\theta) = 1/(N+1) \sum_{n=-N}^{N} (1 - |n|/(N+1)) \hat{f}(n) e^{in\theta}$ is the Nth Fejér mean of the Fourier series of $f \in L^1(\mathbb{T})$

$\tau_n(\theta) = 2\sigma_{2n-1}(\theta) - \sigma_{n-1}(\theta)$, the de la Vallée Poussin mean of order n

$P_r f(\theta) = \sum_{n \in \mathbb{Z}} r^{|n|} \hat{f}(n) e^{in\theta}$ is the Abel mean of the Fourier series of $f \in L^1(\mathbb{T})$

$[x]$ is the integer part of $x \in \mathbb{R}$

$(x) = x - [x]$ is the fractional part of $x \in \mathbb{R}$

$\text{card}(A)$ is the number of elements in the set A

$|A|$ is the Lebesgue measure of the set A

$\mathcal{P}_N$ is the space of trigonometric polynomials of degree N, functions of the form $f(\theta) = \sum_{k=-N}^{N} c_k e^{ik\theta}$

Λ_α is the set of Hölder continuous functions, with $\omega(f; h) \leq Ch^\alpha$ for some $C > 0$, $0 < \alpha \leq 1$ for all $h > 0$

Λ^* is the Zygmund class, $\{f \in C(\mathbb{T}) : |f(\theta + h) + f(\theta - h) - 2f(\theta)| \leq Ch, \forall h > 0, \theta \in \mathbb{T}\}$

$J_N f(x) = (1/2h_N) \int_0^{\pi/2} (\sin Nu / \sin u)^4 [f(x + 2u) + f(x - 2u)] \, du$, the Jackson mean of order 4 at level N, where $h_N = \int_0^{\pi/2} \left(\frac{\sin Nu}{\sin u} \right)^4 du$.

$f^{(r)}(x)$ is the rth derivative of the function f, where $r \in \mathbb{Z}^+$

$C^r(\mathbb{T}) = \{f \in C(\mathbb{T}) : f^{(r)} \in C(\mathbb{T})\}$

$C^{(r, \alpha)}(\mathbb{T}) = \{f \in C^r(\mathbb{T}) : f^{(r)} \in \Lambda_\alpha\}$

$(\Delta a)_n = a_n - a_{n-1}$, the first difference of the sequence $\{a_n\}$

$I_m(t) = \frac{1}{2\pi} \int_{-\pi}^{\pi} e^{-im\theta} e^{t \cos \theta} \, d\theta$, the mth modified Bessel function, defined for $m \in \mathbb{Z}^+$ and $t \in \mathbb{R}$

$f(t) = O(g(t))$, $t \to \infty$ means that there exists $M > 0$ such that $|f(t)| \leq Mg(t)$ for $t > M$

$f(t) = o(g(t))$, $t \to \infty$ means that $\lim_{t \to \infty} f(t)/g(t) = 0$

$\hat{f}(\xi) = (\mathcal{F}f)(\xi)$ is the Fourier transform of $f \in L^1(\mathbb{R}^n)$, defined by $\int_{\mathbb{R}^n} f(x) e^{-2\pi i \xi \cdot x} \, dx$

$(f * g)(x) = \int_{\mathbb{R}^n} f(y) g(x - y) \, dy$ is the convolution of $f, g \in L^1(\mathbb{R}^n)$

$\alpha = (\alpha_1, \ldots, \alpha_n)$ is a multiindex, where $\alpha_i \in \mathbb{Z}^+$

$|\alpha| = \alpha_1 + \cdots + \alpha_n$ is the norm of the multiindex

$D^\alpha f(x) = \dfrac{\partial^\alpha f}{\partial x_1^{\alpha_1} \ldots \partial x_n^{\alpha_n}}$ is the mixed partial derivative of f

$\|f\|_{k,m} = \sup_{x \in \mathbb{R}^n, |\alpha| \leq m} (1 + |x|)^k |D^\alpha f(x)|$

$\mathcal{S} = \{f : \|f\|_{k,m} < \infty, \forall k \in \mathbb{Z}^+, m \in \mathbb{Z}^+\}$

$H_t(x) = \dfrac{e^{-|x|^2/4t}}{(4\pi t)^{n/2}}$ is the heat kernel of $\mathbb{R}^n$

$k_T(x) = \dfrac{1 - \cos Tx}{\pi T x^2}$ is the Fejér kernel of $\mathbb{R}$

$P_y(x) = \dfrac{y}{\pi(x^2 + y^2)}$ is the Poisson kernel of $\mathbb{R}$

$B_{uc}(\mathbb{R}^n)$ is the space of complex-valued bounded and uniformly continuous functions on $\mathbb{R}^n$

$C_0(\mathbb{R}^n) = \{f \in B_{uc}(\mathbb{R}^n) : \lim_{|x| \to \infty} f(x) = 0\}$

$C_{00}(\mathbb{R}^n)$ is the set of continuous functions on $\mathbb{R}^n$ with compact support

B_k is the Banach space of complex-valued functions f on $\mathbb{R}$ with $\int_{\mathbb{R}} f(x)/(1 + |x|^k) \, dx < \infty$

$L^p(\mathbb{R}^n)$ is the space of complex-valued measurable functions on $\mathbb{R}^n$ with $\|f\|_p = \left(\int_{\mathbb{R}^n} |f(x)|^p \, dx\right)^{1/p} < \infty$

$\text{Leb}(f)$ is the Lebesgue set of $f \in L^1_{\text{loc}}(\mathbb{R}^n)$, defined as those $x \in \mathbb{R}^n$ for which $\lim_{r \to 0} r^{-n} \int_{|y-x| \leq r} |f(y) - f(x)| \, dy = 0$

$P(x, y)$ is the n-dimensional Poisson kernel, defined by $P(x, y) = \int_{\mathbb{R}^n} e^{2\pi i x \cdot \xi} e^{-2\pi y |\xi|} \, d\xi$

$D_M(x)$ is the Dirichlet kernel of $\mathbb{R}$, defined by $D_M(x) = \dfrac{\sin Mx}{\pi x}$ for $M > 0, 0 \neq x \in \mathbb{R}$

$S_{M,N} f(x)$ is the two-sided Fourier partial sum of $f \in L^1(\mathbb{R})$, defined by $\int_{-M}^{N} \hat{f}(\xi) e^{2\pi i x \xi} \, d\xi$

$S_N f(x) = S_{-N,N} f(x)$ is the symmetric Fourier partial sum of $f \in L^1(\mathbb{R})$

$F_c(\xi) = \int_0^\infty f(x) \cos(\pi \xi x / 2) \, dx$ is the Fourier cosine transform of $f \in L^1(\mathbb{R}^+)$

$F_s(\xi) = \int_0^\infty f(x) \sin(\pi \xi x / 2) \, dx$ is the Fourier sine transform of $f \in L^1(\mathbb{R}^+)$

$D_0(f) = \int_{\mathbb{R}} x^2 |f(x)|^2 \, dx / \int_{\mathbb{R}} |f(x)|^2 \, dx$ is the dispersion about zero of $f \in L^2(\mathbb{R})$

$H_k(x) = \left(\dfrac{d}{dt}\right)^k e^{(tx - t^2/2)}|_{t=0}$ is the kth Hermite polynomial

$R_j f(x) = \int_{\mathbb{R}^n} i \xi_j / |\xi| e^{2\pi i \xi \cdot x} \hat{f}(\xi) \, d\xi$ is the jth Riesz transform of $f \in L^2(\mathbb{R}^n)$

$S_R f(x)$ is the spherical partial sum of the Fourier integral of $f \in L^1(\mathbb{R}^n)$, defined by $\int_{|\xi| \leq R} \hat{f}(\xi) e^{2\pi i \xi \cdot x} \, d\xi$

$\bar{f}(x)$ is the periodization of $f \in L^1(\mathbb{R}^d)$, defined by $\bar{f}(x) = \sum_{n \in \mathbb{Z}^d} f(x-n)$

$K_{M,\alpha}$ is the Bochner-Riesz kernel, defined as the Fourier transform of $(1 - |\xi|^2/M^2)^\alpha 1_{[0,M]}(|\xi|)$.

$\psi_{a,b}(x)$ is the rescaled function defined by $|a|^{-1/2}\psi((x-b)/a)$.

$W_\psi f(a,b)$ is the wavelet transform of $f \in L^2(\mathbb{R})$, defined as the convolution of f with the rescaled function $\psi_{a,0}$

$\mathcal{F}_n$ is the dyadic partition of $\mathbb{R}$, consisting of the intervals $((k-1)/2^n, k/2^n]_{k \in \mathbb{Z}}$

V_n is the nth space in a multiresolution analysis

$P_n f$ is the orthogonal projection of f onto the space V_n

$\Phi(t)$ is the scaling function for a multiresolution analysis

$\Psi(t)$ is the orthonormal wavelet associated with the scaling function Φ

$m_0(\xi)$ is the scaling filter associated with the scaling function Φ

$m_1(\xi)$ is the wavelet filter associated to the wavelet Ψ

INDEX

A

Abel means, 45, 64, 261, 266
Abel sums, 49
Abel's lemma, 167
Abelian theorem, 54
Absolutely continuous functions, 21
Absolutely convergent trigonometric series, 3, 43
Almost-everywhere convergence, 20, 28, 49, 153, 346
Analytic function, 22, 172
Approximate identitites, 45, 97
Averaging, 118

B

Banach space, 75
Banuelos, R., 202
Beckner, W., 174, 175
Benedicks, M., 240
Bernoulli, D., 1
Bernstein's inequality, 122
Bernstein's theorem, 43, 70, 129, 350
Berry-Esséen theorem, 276
Bessel functions, 6, 81, 140, 154, 164
Bessel's inequality, 36
Boas, R. P., 228
Bochner, S., 139
Bochner's method of subordination, 106
Bochner's theorem on positive-definite functions, 220, 270
Bochner-Riesz summability, 152

Bounded linear operator, 75
Bounded variation, 27, 60
Borel-Cantelli lemmas, 281
Brandolini, L., 149
Brownian motion, 299

C

Calderón, A. P., 194, 196, 209
Calderón-Zygmund decomposition, 209
Calderón-Zygmund theorem, 211
Cantor measure, 20
Cantor-Lebesgue theorem, 87
Cantor's uniqueness theorem, 86
Carleson, L., 73, 118
Central limit theorem, 260
Cesàro average, 119
Cohen's theorem, 338, 262
Colzani, L., 149
Compactly supported wavelets, 326
Completeness of the Haar functions, 295
Completeness of the Hermite functions, 138
Conjugate function, 176
Conjugate Poisson kernel, 8, 179
Continuity theorem, 269
Continuum wavelet, 286
Contraction operator, 14
Convergence of wavelet expansions, 341
Convex function, 170
Convex sequence, 96
Convolution, 5, 14, 19, 92
Convolution semigroups, 272

D

Daubechies' formula, 331
de-Moivre Laplace theorem, 100
Dilation matrix, 354
Dini condition, 27, 89
Dini's theorem, 115
Dirac measure, 20
Directed set, 46
Dirichlet, P. G., 27
Dirichlet kernel, 15, 25, 112, 140, 223
Dirichlet-Jordan theorem, 27, 117
Distribution function, 181
Divergence of Fourier series, 73
Du Bois-Reymond, 73, 117
Dyadic projections, 292

E

Equidistribution mod 1, 57
Euler formula, 4
Exponential function, 6

F

Factorial function, 6
Fatou's theorem, 50, 214
Fefferman, C., 73
Fejér means, 45, 54, 62, 98, 120
Fejér-Riesz lemma, 330
Finite measures, 19, 119, 256, 273
Fourier, J. B., 2
Fourier coefficient, 4, 13, 19
Fourier cosine transform, 124
Fourier reciprocity formula, 4, 14, 93, 97, 231
Fourier sine transform, 125
Fourier transforms, 89, 102, 256, 323

G

Gap series, 262
Gaussian approximation, 80
Gaussian density function, 95, 257
Gaussian summability, 104, 232
Gaussian wavelet, 288
Gauss-Weierstrass kernel, 97
Generalized h-transform, 125
Gibbs-Wilbraham phenomenon, 31, 115
Grunbaum, A., 133

H

Haar function, 387, 306
Haar series, 291, 296
Haar wavelet, 291, 321
Hardy, G. H., 59
Hardy-Littlewood maximal function, 197
Hardy's Tauberian theorem, 59
Harmonic functions, 212
Hausdorff-Young inequality, 174
Heat flow, 2
Heat kernel, 97, 203, 223
Herglotz theorem, 215
Hermite polynomials, 134
Hernández-Weiss formula, 333
Higher-order approximation in $C(\mathbb{T})$, 66
Hilbert transform on $\mathbb{R}$, 184, 224
Hilbert transform on $\mathbb{T}$, 176, 224
Hölder condition, 13, 40, 63, 116
Hölder means, 53
Hölder's inequality, 170
Homogeneous Banach space, 47, 98
Hörmander condition, 211
Hunt, R., 73, 118
Huygens' principle, 148

I

Independent random variables, 261
Infinitely differentiable function, 21
Integration of Fourier series, 29
Isoperimetric inequality, 38

J

Jackson's theorem, 65, 348
Jordan, C., 27

K

Kac, M., 266
Kahane, J. P., 45, 73, 143
Katznelson, I., 73, 180
Kendall's theorem, 241
Kolmogorov, A. N., 73, 118, 179
Kolmogorov's inequality, 192

L

Landau's asymptotic formula, 243
Laplace asymptotic method, 6, 9, 80

Laplace's equation, 9, 106, 212
Large scale analysis, 345
Lattice points, 241
Laurent series, 5
Law of the iterated logarithm, 280
Lebesgue constants, 61, 75
Lebesgue differentiation theorem, 200
Lebesgue set, 124
Lévy modulus of continuity, 302
Lévy-Khintchine theorem, 273
Lipschitz continuity, 62
Lusin, N., 73

M

Marcinkiewicz, J., 169, 179
Marcinkiewicz interpolation theorem, 206
Maxwell's theorem, 109, 259
Mexican hat wavelet, 288
Meyer wavelets, 318
Modified Bessel function, 6, 81, 85
Modulus of continuity, 12, 20
Moore, C., 202
Multiple Fourier series, 230, 244
Multiple harmonic motion, 1
Multi-resolution analysis, 303

N

Negative binomial series, 52
Newtonian potential kernel, 109
Non-tangential convergence, 202
Null-integrability, 325

O

One-sided Fourier representation, 124
Orthogonality, 2, 137
Orthonormal wavelet, 319

P

Parseval's identity, 35
Periodic function, 13
Periodization, 223
Piecewise smooth function, 21, 145
Pinsky, M., 139, 145, 147
Plancherel's theorem, 128
Pointwise convergence criteria, 25, 115, 139, 233, 298, 347
Poisson kernel, 7, 15, 46, 90, 106, 189
Poisson summation formula on $\mathbb{R}^1$, 225

Poisson summation formula on $\mathbb{R}^d$, 238
Positive-definite function, 257

Q

Quotient groups, 355

R

Radial functions, 92, 157, 202, 245
Random walk, 252
Rapidly decreasing function, 96
Rates of convergence in $C(\mathbb{T})$, 61
Rates of convergence in $L^2(\mathbb{T})$, 39
Restriction theorem, 158
Riemann-Lebesgue lemma, 18, 94
Riemann localization, 31
Riemann means, 64
Riesz, M., 73
Riesz-Fischer theorem, 37
Riesz kernels, 195
Riesz potential kernel, 107
Riesz systems, 304, 356
Riesz-Thorin interpolation theorem, 169
Rotations, 194

S

Salem, R., 266
Scaling equation, 310, 357
Scaling filter, 312
Scaling functions, 304, 319
Schrödinger equation and Gauss sums, 247
Schwartz class, 96
Schwartz distribution, 146
Self-adjoint operator, 15
Shannon sampling theorem, 228
Shannon scaling function, 306
Simple harmonic motion, 1
Simultaneous non-localization, 240
Sine integral, 24, 114, 185
Singular integrals, 183, 194, 210
Skew-adjoint operator, 16, 176
Smith, K. T., 202
Sobolev norm, 43, 290
Spherical Fourier inversion, 139
Spherical maximal function, 203
Spherical mean value, 140
Spherical partial sum, 139, 233
Stationary phase method, 162
Stein, E. M., 157, 159, 175, 202, 204, 211
Stein's complex interpolation method, 159, 175

Stirling's formula, 6
Strichartz, R., 33
Stronger summability method, 52
Structure constants, 310, 357
Sublinear operator, 197, 207
Summability matrix, 51
Summation by parts, 11, 79
Symmetric partial sum, 4, 112
Systems of wavelets, 352
Smoothness of wavelets, 334
Subgroups, 354

T

Tail sum, 12
Tauberian theorem, 54
Taylor, M. E., 139, 147, 248
Tempered distribution, 108, 129
Three-lines theorem, 172
Tomas, P., 159
Translation error, 40

U

Uncertainty principle, 134
Uniform boundedness principle, 76, 84
Uniform continuity, 98
Uniform convergence, 4

Uniqueness of Fourier coefficients, 17
Upper half plane, 216

V

Vibrating string, 1
de la Vallée Poussin means, 68

W

Wave equation, 171
Wavelets, 284
Wavelet transform, 285
Weak convergence of measures, 268, 327
Weierstrass M-test, 4
Weiss, G., 157
Whittaker, E. T., 228
Wiener's theorem, 57, 110
Wiener covering lemma, 198

Y

Young's inequality for convolutions, 174

Z

Zygmund, A., 20, 44, 72, 209, 236, 262, 266
Zygmund's theorem, 44, 209